T0251326

GEOLOGIC AND MINE MODELLING USING TECHBASE AND LYNX

Geologic and Mine Modelling using Techbase and Lynx

MARTIN LLOYD SMITH
Mining Engineering, University of Idaho, Moscow, Idaho, USA

A. A. BALKEMA / ROTTERDAM / BROOKFIELD / 1999

This publication includes a CD-ROM with LYNX and TECHBASE® software to be used for educational purposes in conjunction with the examples of this text. Also included are readme files for software installation, complete user manuals for both products and the project databases used in the examples of this text. TECHBASE is a registered trademark of TECHBASE International.

LYNX Internet site: www.lynxgeo.com
TECHBASE® Internet site: www.techbase.com

The software mentioned in this book is now available for download on our Web site at: http://www.crcpress.com/e_products/downloads/default.asp

Published by
A.A. Balkema, P.O. Box 1675, 3000 BR Rotterdam, Netherlands
Fax: +31.10.4135947; E-mail: balkema@balkema.nl; Internet site: http://www.balkema.nl

A.A. Balkema Publishers, Old Post Road, Brookfield, VT 05036-9704, USA
Fax: 802.276.3837; E-mail: info@ashgate.com

ISBN 90 5410 691 3 hardbound edition
ISBN 90 5410 692 1 student paper edition

Printed in the Netherlands

Contents

Preface

I started this text as a series of detailed examples for class a composed of seniors and graduate students in mining engineering, geology, hydrology and geologic engineering. The text that grew out of this course reflects the educational interests of upper division engineering and geoscience students, i.e. it provides a basic understanding of how to apply complex software along with sufficient supporting theory; just enough theory to feel confident about the underlying algorithms and for users to be able to determine when they are treading on thin ice. I have tried to maintain this approach so that professionals in the field and self-motivated students can use this text and the accompanying CD to learn on their own. The result is a text that is highly example-driven, with detailed notes on navigating complex software procedures and modelling methodologies.

Probably the most unusual aspect of this text is the use of two competing software products: Techbase and Lynx. Why complicate an already complex topic by using application examples drawn from both products? Consider my own experience with using similar software: I started using mining software in 1980 while working for Morrison Knudsen when they were developing the Eagles system, which would become very popular among coal mining operations. Following my graduate work, which included a period of software development, I found myself developing a new curriculum centered around software usage. During this time, I used a number of competing packages including Eagles, Minex, Lynx, microLynx and Techbase and was exposed to numerous others. In the search for one package that could do it all, one thing became clear – none can. I settled on Techbase and Lynx largely because they complement each other nicely and illustrate a wide variety of problem solving methods. My objective has been to demonstrate the breadth and depth of mining and exploration software on topics ranging from geologic databases to mine production scheduling. Simply put, no one software product was sufficient for this text. For example, Techbase is geared towards exploration and database management but not toward detailed mine design, especially for underground mines. In contrast, Lynx is geared toward mine design, especially underground, but has limited database facilities.

It should be emphasized that this is not a book on how to use Techbase and Lynx. If that were the case, it would be best to have split the book in two. Instead, this is a book that uses high-end commercial software to demonstrate common procedures. The methodologies presented herein are not meant to be software specific, but general methodologies which are common across several related disciplines. Anyone completing a course of studies using this text should have no difficulty using competing packages; they all do pretty much the same thing in the same way and in the same order.

The material presented in this text is meant to be covered in two semesters. The first five chapters are of general interest to engineers and geoscientists as they cover topics from geologic databases to geostatistics. The remaining chapters are of more interest to mining students and professionals as they cover mining design to production scheduling. As a whole, the text is designed to support a senior capstone course centered around a project that includes geologic modelling, reserve analysis, mine design and production scheduling.

A number of people have made this book possible. Special thanks goes to my wife Carol Ann Smith for her editorial skills and support. Many of the algorithmic details were provided by Mark Stoakes of Lynx Geosystems and J.O. Williams of MineSoft. Special thanks go to Mike Norred of MineSoft and Pierre d'Hill of Lynx Geosystems who made it all possible by providing the software and user manuals for this work. Last but not least there are my parents, Ray and Bea Smith, to whom I can finally say, 'It's done'.

CHAPTER 1

Introduction

1.1 GEOLOGIC AND MINE MODELLING SOFTWARE

This text draws together disparate technical fields relating to computer databases, statistical analysis, geology, mining engineering and operations research within the context and limitations of two popular software packages. All of these technologies are used to work towards one primary objective, the examination of a mineralized body that might be defined as an ore deposit. The breadth of technical fields required for the evaluation of mineral projects is unique to this industry and has generated a class of computer packages which are almost as unique.

Mining software is used in a wide variety of applications: (1) databases to record geologic data coming from diverse sources such as surveys, drillholes, geochemical analysis, geologic structures, rock mass behavior, maintenance records, personnel files, and processing and production costs; (2) statistical analyses to manipulate and summarize the very large and complex data sets that are typical of exploration and mining operations; (3) a wide variety of estimation tools for creating spatially continuous models of geologic structures; (4) computer-aided design (CAD) facilities that are specialized for accurate and efficient modelling of geologic structures and mine openings; and (5) an array of specialized algorithms that serve as an aid for engineering design. No other combination of commercially available programs is capable of filling all of these requirements. Even on the basis of individual applications, mining, environmental and exploration professionals have specialized needs which are often not addressed by more widely used programs. For instance, in the case of drillhole derived data, there is the need to relate numerous different record types within a hole including collar location, downhole surveys of inclination, geology and various types of analysis from geochemical to geophysical. Statistical analysis is another example of the departure of geologic applications from the norm. While there are a number of superb statistical packages in the market, these are almost exclusively oriented towards data sets which are normal in their distribution and independent. Geologic data are frequently lognormally distributed and spatially dependent. Even in traditional areas of engineering design, CAD software must be specialized to work with irregular structures and calculate volumes of intersection. The goal of geologic and mine modelling software is to bring all of these technologies within one comprehensive system in which data can be passed from one aspect of analysis, such as volumetrics, to another related task such as estimating grades for the volumes. In summary, mining software addresses specialized needs within a wide variety of technologies and facilitates the passing of data between these technologies within a common database.

1.2 MODELLING AS A BASIS FOR DECISION MAKING AND DESIGN

The reader will note that the concept of 'modelling' is central to the material covered herein, and that the use of computers as the modelling tool is implicit. While modelling has become established as the basis for design in engineering, mining professionals, including engineers and geologists, are often occupied with the hands-on realities of projects and are less familiar with the role of modelling in their professions.

Modelling is the simplified representation of reality or of a proposed reality. Generally, simplification is achieved by reducing the representation to only those variables and features which are essential for the decision making or design process. In the case of CAD, the broader framework of graphical modelling has replaced drafting. Where engineering drawings are a means of modelling three dimension within two, CAD tools now make it possible to model directly in three dimensions, a development of immense significance to the modelling of geologic structures and mine openings. The evolution of CAD modelling can be characterized by increased interaction with the components of the model, i.e. of rendering a model which is closer to reality in its behavior within the context of the design problem. CAD models can now be made so that they are dynamic in that they react to external influences.

An example of dynamic CAD modelling in geotechnical engineering is the use of finite element analysis to evaluate the stability of slopes and mine openings. In the case of underground mine openings, such as a pillar, a two, or even three, dimensional model is generated representing the geometry of the opening. The surrounding rock mass is then descritized into an irregular mesh of elements in which each of these elements represents a specific volume of the rock mass as shown in Figure 1.1.

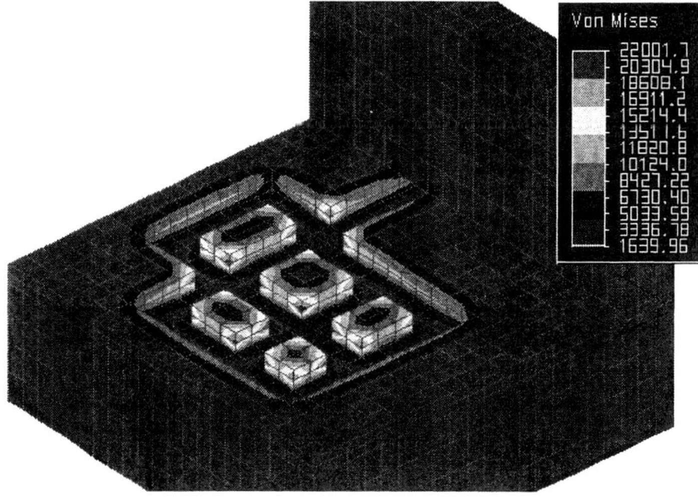

Figure 1.1. Finite element modelling of stresses about a 3D mine opening. (Figure in colour, see opposite page 54.)

Greater detail is provided by generating a finer mesh in areas of critical concern, such as the boundary of the mine opening. The elements are then assigned the strength characteristics that might exist in reality. An incumbent load is then placed on the model which is resolved into stresses for each element. In this case, the goal of modelling is to study the stress concentrations and displacement of the elements on the boundary of the mine opening.

This use of finite elements illustrates the importance of modelling. In general, modelling provides a means of evaluating designs. Furthermore, computer-based modelling makes it possible to evaluate designs without the time, expense or risk involved in using scaled physical models or pilot openings. In fact, many aspects of modelling cannot be physically represented and a computer model is the only option prior to the development of a prototype. Thus, the model provides the basis for experimentation in design, or to be more specific, in the re-design cycle. Returning to the example of finite element analysis of the stability of mine openings, the engineer can experiment with different designs to study their stability under a variety of conditions. In terms of the design, the shape of the opening can be altered and the benefits of adding artificial support can be included in the model if deemed necessary by the results of the analysis of the unsupported opening. Since the rock mass characteristics and stresses are uncertain, and since any model is only a simplification of reality, sensitivity analysis is a crucial element of the modelling/design process. Sensitivity analysis of a model is conducted by varying parameters which are uncertain over their probable range and studying the responsiveness, or sensitivity, of the model to changes in those parameters. In mining applications, the number of uncertain variables is commonly very large and are typically related to geologic conditions which are by their nature hidden from view and can only partially observed if they outcrop on the surface or are discovered by drilling or indirectly by geophysical means. Since the risk and investment are both so great in mining projects, modelling and sensitivity analysis is crucially important.

Modelling is used for all facets of evaluation and design in the mining industry, to name a few applications: statistical models of geochemical data to understand the relationships between mineral concentrations and geologic structures; geostatistical models of the spatial variability of geologic contacts, geotechnical characteristics and geochemical concentrations; two and three dimension graphical models of geologic structure and mine openings; and models of mining systems such as materials handling, geomechanical stability and ventilation. All of these models interact to some degree. This interaction, as well as the specialized nature of the models, has created a need for comprehensive mining software packages that integrate the modelling process.

While modelling has come relatively late to the mining industry, its influence on the evolution of the industry is likely to be even more profound than the revolutionary impact modelling has had on other aspects of engineering practice. In comparison to mining, the design process for most engineering systems is relatively simple and is fraught with far less uncertainty. This statement may initially be surprising to the reader since by comparison to other fields of engineering, mining engineering usually deals with broader issues of design. For instance, the sinking and construction of a shaft can be compared to the raising of a skyscraper. Certainly, the level of engineering detail is greater for a skyscraper, but the cost to investors is comparable while the profit that might be realized and the financial risk is greater in the case of the shaft. Additionally, the level of detail in the design of a shaft goes far beyond the shaft in itself since most mining systems are dependent on the siting, capacity, depth, longevity and design of the shaft. Because of the vast complexity of a mining project, mining engineering has had to limit itself to the broader issues, often leaving the details of design to other professionals. When the complexity of the overall mining system is combined with the uncertainty that is inherent in the mining environment, mining projects are often little more than a gentleman's lottery when compared to investing in a new chicken factory or an automobile plant. Two elements of the design process must be changed in order to reduce the risk involved in mining investment, the level of detail in the design and the impact of the uncertainty in the design parameters, specifically the geologic parameters. Both of these sources of risk can be reduced by incorporating modelling in the design process. Modelling can be used not only on the level of detailed design and sensitivity analysis as is rapidly becoming the norm in other engineering disciplines, but also to integrate the entire system from models of mineralization to materials handling systems. Thus, while computer modelling has facilitated design in other disciplines, it has the potential to fundamentally change the mining industry by rationalizing the design and decision making process.

1.3 OVERVIEW AND LIMITS OF GEOLOGIC AND MINE MODELLING

While it is the author's belief that modelling will revolutionize the mining industry, this remains to be seen. Currently, mining software is not sufficiently flexible, comprehensive or integrated to realize this prediction. In part, this is due to a lack of research and development, but research and development of mining software has been entirely driven by demand and competition in the private sector. Thus, the evolution of mining software will be no more rapid than is the acceptance of modelling in the industry. This text deals strictly within the limits of existing modelling technology, specifically, within the limits of two popular packages, Techbase and Lynx. These two packages cover the main modelling capabilities found in other comprehensive mining software systems and serve to demonstrate the overall strengths and weaknesses of geologic and mine modelling as it exists today.

The facilities included in most of these packages can be separated into the following categories:

1. A geologic database capable of storing and manipulating input data from surveys and drillholes as well as gridded estimates, triangulated surfaces, polygons and volume model components.

2. Statistical and geostatistical capabilities for data analysis and estimating grids or random point values.

3. Gridding, triangulation and interactive graphics for the modelling of geologic structures and mine openings.

4. Specialized programs to facilitate mine design.

5. Volumetric facilities for the calculation of geologic and mining component vol-

umes as well as volumes of intersection and the reporting of geologic and mining reserves.

6. Production scheduling facilities.

7. Facilities for the import and export of data to and from ASCII, spreadsheets, common graphical languages such as DXF and postscript and other valuable secondary analysis programs such as 3D data visualizers and popular statistical packages.

The quality of database facilities varies widely between the different software packages. All of the packages have the ability to enter and internally manipulate data derived from surveys and drillholes, but it is the accessibility, flexibility and variety of data types accommodated by the database that determines its value to the user. The database is the foundation of the entire package. The user should have the ability to directly view, access and manipulate data with few limits on the type of calculation (either algebraic or logical) to the point of being able to create new database structures. Unfortunately, the database facilities included in most mining software are very limited being specifically designed to perform a small number of predefined tasks. With many programs, importing, exporting and manipulating data can be very cumbersome. Often it is necessary to export data from the geologic database and manipulate it with other programs such as spreadsheets. Therefore, it is very important that the ability to import and export data be easy and flexible.

Not surprisingly, a weak database is indicative of a package with limited statistical analysis facilities. A limited suite of statistical programs is included with these packages, routines which are most commonly used in the preliminary analysis of geologic data. These usually include basic statistical analysis of distributions such as summary statistics, histograms and probability plots. While more advanced statistical capabilities are often included, such as multivariate analysis and variography, it is often best to export data for analysis in specialized statistical and geostatistical packages.

While it is not essential to integrate statistical analysis within a geologic and mine modelling package, it is necessary to have strong internal facilities for gridding and geologic interpretation. Grids and triangulation are used to model surfaces while geologic interpretation is used for more complex three-dimensional bodies that do not lend themselves to representation by one or more surfaces as do bedded deposits. All methods of geologic modelling generate a model which is smoother than reality. Gridding methods are based on weighted averages or on splines which result in surfaces which are even more regular than the data used to estimate the grid. Triangulation, while honoring the data, creates a faceted surface composed of planar patches. Interactive interpretation is based on nothing more than the geologist's interpretation of drillhole data projected onto a section. While interactive graphical interpretation makes possible a model that can be highly complex on the plane of the section, the shape of the volume model between sections will only be a linear interpolation, an unneccessary simplification. In any case, the distance between drillholes is commonly so great that a complex geologic interpretation is little more than a geofantasy. Here the limitation is not the modelling methods, but is the lack of a variety of data sources that can be used in interpretation. Currently, the inclusion of three-dimensional interpretations of geophysical measurements has the greatest potential for filling in the gaps left between drillholes.

A wide variety of modelling methods are available for mining engineering design, but routines for specific mining systems are generally not integrated into these mining software packages. Mining software for ventilation networks, simulation of materials handling systems, mine hydrology and drainage, finite element analysis and cost estimating have yet to be integrated into a comprehensive mine modelling system. Mining software packages focus on design in terms of generating graphical representations of pit limits, the layout of mining cuts and the layout and geometry of underground mine openings. Facilities which directly aid this process are included in the routines that will automate the generation of pit limits, pushbacks for haulroad design and the creation of underground development headings.

The need to generate geologic and mining reserves is the driving force behind the integration of mining packages. To generate geologic reserves, gridded estimates of grades must be intersected with geologic surface and volume models in order to determine both grade and tonnage of the deposit. In order to obtain mining reserves, the intersection procedure can be taken one step further by intersecting models of grade, ore volumes and mine openings. The only limits to this procedure are those involved in obtaining accurate grade estimates and geologic volume models, but these are areas of major concern whose limitations have more to do with education and the failure to integrate multiple data sources than the limitations imposed by the existing technology of modelling available in the software.

Long and short-term production scheduling consists of determining the order of extraction of ore and waste. The objective of long-term scheduling is to extract the highest valued ore as early as possible in the mine's life in order to maximize the project's cash flow while maintaining a reasonably constant production rate from year to year. The objectives of short-term production planning are to maintain a production of ore and waste which will meet the long term production goals, maximize the utilization of production equipment and provide as close to an ideal feed to the mill, or product to the customer, as possible. While production scheduling facilities in mining packages greatly facilitate this task by allowing volumes and grades from alternate mining cuts to be rapidly analyzed, the number of variables involved in ore blending to the mill in combination with the different possible locations, sizes and sequences of mining cuts makes short-term blending a time consuming and far from optimal procedure.

1.4 GEOLOGIC AND MINE MODELLING PROCEDURE

The progression of topics covered in this text follows the procedures that are used in the evaluation of a new mineral project, but the contents do not extend as far as a comprehensive feasibility study since the topics covered herein must remain within the bounds imposed by the software. Still, the logical flow of the chapters does follow the steps taken in all but the economic stages of a feasibility study.

Geologic database management – Chapter 2

The first stage in any project is to gather, compile, organize, validate and become familiar with the data. This process is especially important in the construction of geologic databases since the volume, variety, complexity and expense of collecting the data is so great. In terms of the sheer volume of data, geologic databases can be of enormous size. Digitized maps can consist of tens of thousands of point values, so many that data thinning routines must be used by mapping software so that graphics based on digitized data can be calculated and displayed within a reasonable period of time. Likewise, drillholes can be hundreds of meters in length with assays taken in the mineralized zones at intervals as short as a few centimeters. Since the entire length of the hole might be assayed in order to determine the location of mineralization, this can add up to thousands of data records per hole, and a project might include hundreds of holes. If the project progresses into a mining operation, analysis of the blasthole cuttings will also be included in the database in order to delineate diglines based on a cutoff grade. Blastholes will generate a huge data storage requirement.

The data pulled together in the database can come from many different sources. The most common data sources are from exploration drillholes and wells, digitized maps, geochemical soil surveys, geophysical measurements, maps of structural features and surveys of surface features and underground mine workings.

Drillhole data is the most heavily relied on source of information once a project reaches the stage that requires the estimation of geologic reserves, i.e. once other sources of information have indicated that a deposit of sufficiently great value to warrant major investment probably exists. Drillholes require a specialized data structure in which data records have a one-to-many relationship. Specifically, a hole will have one record for the location of the top, or collar, possibly several records to represent the change in inclination along the length of the hole and then many records down the hole to record changes in assay, lithology, alteration or other structural features (see Fig. 2.2). Except for the collar record, for which there will be only one per hole, these records will be recorded on depth intervals. The database software must be able to relate the position of the collar, the changes in inclination and the various depth intervals on which the hole data are recorded so that the position of each datum is known. Well data are similar to drillhole data except that they are usually used for hydrologic studies and might contain less precise information on lithology or soil characteristics.

Map data are commonly included in the database by digitizing older aerial or ground surveys or by using digital aerial surveys. A hardcopy topographic map is digitized by entering a series of *xy* coordinates along a contour line and then assigning a constant *z* to the points on the contour, but old map data can include much more than topography. Often, the project will predate the use of computers and existing mine workings, so geologic sections and surface facilities will have to be digitized and entered into the database. Once in the database, the digitized maps will have to be converted into a usable format: geologic sections will be converted into surfaces and volumes, topography will be converted into a surface, while other surface features, such as roads and facilities, can be converted to polygons or just left as map data in the form of points, traverse lines, etc. Sections showing existing mine workings can be treated in the same way as geology in that they can be digitized and converted into either volume models or polygons.

Geochemical and geophysical surveys are often ignored once a project moves past the exploration stage. This may be justified because of the difference in scale between exploration data and the level of detail that is need to define geologic reserves, but often valuable data is ignored due to a breakdown in communication. Drillhole data is by its very nature extremely sparse. Geochemical surveys can be mapped by gridding and used as a valuable guide in determining the extension of mineralization towards the surface. Geophysical data on the velocity of sound can be converted to *xyz* coordinates providing imprecise but dense data on densities and discontinuities that can be a valuable aid in filling in the blanks between drillholes during geologic interpretation and modelling.

Up to this point only the input data to the database has been considered. Usually, the memory requirements and complexity of the data that is generated from the output data is even greater than is required for the input data. There are a number of common data structures that are generated in geologic databases:

1. Tables of data which are random in their organization such that individual records represent a point in space with one or more associated variable values: examples of this include the drillhole and survey input data that was discussed above. Regular gridded (non-random) data can also be input data when they are imported from some other routine that was used to generate it, as is the case when geophysical particle velocities are converted into a grid.

2. Two-dimensional grids in which a continuous surface has been estimated using random sample values: these estimated surfaces are used to represent geologic structures such as lithologic contacts and faults. When the geologic structures of interest are bedded, this might be the only database structure that is used for calculating geologic reserves. This is commonly the case for industrial minerals.

3. 2D grids are not the only means of modelling a surface (or thickness); an alternative to using a grid is to use a triangle set. A grid is a matrix of values based on estimation or interpolation of actual data values. The disadvantage of a grid is that many estimation methods do not honor the original data. All grid estimation methods result in a surface which is smoothed. Triangle sets are defined by the input data in that the apex of each triangle's corner is a datum location. Thus triangles by definition honor the input data.

4. Three-dimensional grids are often referred to as block models. Unlike two-dimensional grids in which each record associates an offset, or *z* value, with a *xy* coordinate in 2D space, a block model associates values such as assay a volume with and *xyz* coordinate in 3D space. These data structures are more commonly associated with massive deposits whose geometry cannot be adequately described using a combination of 2D surfaces.

5. Other data structures are needed that are neither random, triangulated or gridded. These include entities such as boundaries, buildings and roads that can be represented as polygons, surveys that can be represented as polylines, point data such as outcrops and associated structural measurements of bedding strike and dip. In Techbase, polygon type data is contained in its own data structure. Lynx includes all of these in its map data structure which doesn't discriminate between input or internally generated data.

6. Volume components are part of the Lynx database and are used to represent geologic volumes, pit volumes, underground development headings and stopes. Vol-

ume components are defined by polygons on three parallel planes and the distance between those planes.

Statistical analysis – Chapter 3

Statistical techniques are an integral component of database development. They are used to evaluate sampling protocols, the need for additional data, sample spacing, identification of population domains and data validation during exploration. This fundamental statistical analysis also serves to familiarize the modeler with a large and complex data set providing a quantitative basis for how geologic modelling and estimation should proceed, specifically, what level of model detail is necessary, what types of estimation algorithms should be used and how they should be applied. It has been the author's experience that the engineers and geologists commonly involved in geologic modelling have little experience, and subsequently little faith in statistical methods, but lack of detailed statistical analysis prior to modelling is the surest route to failure in modelling and geostatistical estimation: the modeler who ignores using statistical methods to first familiarize himself with the data is proceeding blindly into a cycle of error and repetition, or worse, denial.

As noted earlier, the majority of mine modelling packages provide only the barest essentials in statistical routines. The advanced modeler must be prepared to work with general statistical packages. Still, most mining packages include routines sufficient for basic analysis.

Statistical analysis starts with examining the location and spread of the data. This is done by generating a scatter plot of the *xy* coordinates of drillhole collars and other surface data (see Fig. 3.1). Both areas of sparse and dense data concentration should be noted at this time.

Summary statistics and histograms of assays are used to identify variable distributions, check for outlier data and discriminate between different data populations (see Fig. 3.6). Highly skewed distributions, such as are common when dealing with trace elements, need to be identified since the normal summary statistics may lead to false estimates of the geologic reserves. Probability plots can be used to visually check on the nearness of a distribution to being either normal or lognormal (see Fig. 3.9). Outliers must be identified since they can have a very high influence on the statistical analysis. Outliers may be errors in assaying or data entry, or anomalous values pointing to areas requiring more detailed study. Areas of mineralization can be the result of more than one event. There may be more than one mineralized population which would appear in the histogram as multiple distributions. Histograms are generated for each lithology or alteration zone to identify significantly separate data populations requiring separate models.

Experimental variograms are generated to study the influence of spatial variability on variables that are going to be estimated into grids such as assays. The variograms are used as a statistical distance in determining optimal estimation weights for kriging procedures. Variograms are also invaluable as part of the statistical analysis during which they are used to evaluate the sampling distance, the stationarity of the variable, and as another means of establishing population domains (see Fig. 3.30).

Modelling geologic surfaces – Chapter 4

The structural information available for modelling lithologic contacts, faults, etc. is limited to relatively few drillhole intersections, outcrop measurements and mapped structural features from exploration drifts. From this sparse data set, continuous surfaces must be estimated before volumetric calculations can be made.

The first step in surface estimation is to generate a data set of the coordinates of the point of contact of the drillhole with the surface that will be modelled, i.e. drillhole compositing. In the process of compositing, multiple sample intervals for a drillhole that is uniform in one of its variables can be averaged to a single sample interval and the *xyz* coordinate for this composited sample can be calculated. This method can be used to determine the location of the top or bottom of a lithologic unit by compositing all assays within a continuous rock type and assigning the sample location to the top of the lithologic interval or to the bottom of the interval as shown in Figure 1.2.

As an alternative to compositing, drillhole intercepts with a lithologic unit, or grade cutoff, can be determined by projecting the location of the intercept normal to a plane to generate an offset to the plane. For example, the reference plane could be oriented to the strike, dip and plunge of a vein and given an origin that places it in the neighborhood of the vein. The offset for the hangingwall and the footwall can be

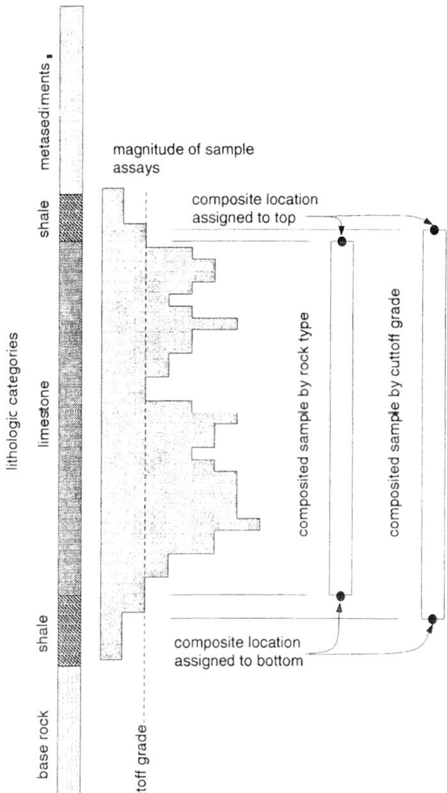

Figure 1.2. Methods of drillhole compositing to identify ore zone contact coordinates.

projected normal to the plane generating two offset values for each drillhole that penetrates the vein (see Fig. 4.11). The difference of these two offsets can be used as an approximation of the vein's thickness as long as the drillhole's inclination is not too far from being normal to the inclination of the vein at the point of contact. Additionally, the two offsets can be modelled as surfaces by either triangulation or gridding the offsets and the difference between the two surfaces can be used to estimate thickness and volume for the vein.

One of the technical difficulties in generating surfaces is choosing the estimation algorithm that is best suited to creating a realistic surface. There are a number of methods that can be used to generate a surface including triangulation, splines, and the weighted average methods referred to as inverse distance and kriging. Each algorithm has its advantages and disadvantages which make it a more or less suitable method depending on the nature of the samples and the geology of the surface. A large part of the challenge for the modeler is to intelligently apply the various estimation algorithms based on his understanding of the structure being modelled.

Not all geologic surfaces and volumes can be represented as a combination of surfaces. When the structure becomes too complex, a more interactive approach is needed. Consider the case of modelling a seam which has been heavily folded and faulted as shown in Figure 1.3. In this case both the method used for generating drillhole intercept data and surface estimation using that data will lead to problems. When generating the *xyz* coordinates for the intersection of the drillholes with the top and bottom of the seam, there might be more than one intercept for the top and bottom of the seam in a hole if folding or fault displacement is sufficiently great. Since all surface estimation algorithms assume a smooth transition between data points, seam overlap in the neighborhood of faults or folds will be modelled as a smooth change in elevation, not as a sharp discontinuity. In fact, for estimation methods based on moving averages, the smoothing may extend for a considerable distance

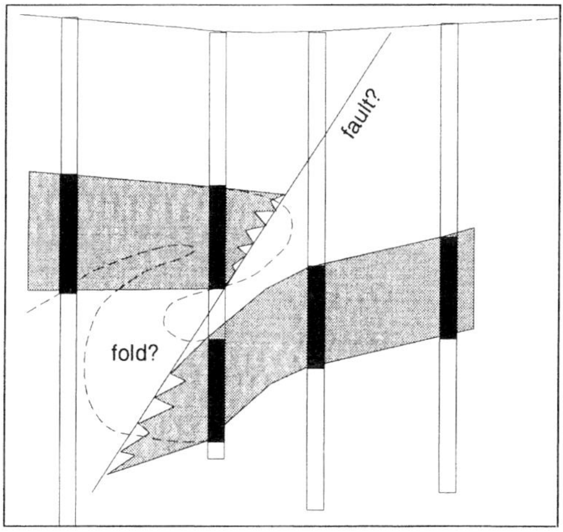

Figure 1.3. Example of a geologic structure too complex for surface modelling.

beyond the zone of faulting or folding, resulting in a model in which no discontinuity is apparent. In these cases, the modeler must be aware of the existence and location of discontinuities and directly intervene in the modelling process.

The most common approach to modelling complex structures is to use interactive graphics to either modify a surface that has been estimated using some estimation algorithm or to interactively generate the entire structure. An estimated surface can be modified by adding points showing a sharp discontinuity as a change in elevation of the surface on either side of the fold or fault and by removing the smoothing effect of the estimation algorithm by adjusting the estimated surface in the area of the discontinuity, but in any case a surface cannot be folded in order to create more than one elevation or thickness estimate at the same *xy* position in the plane of the surface. This requires three-dimensional volume modelling in which the structure is directly represented as a volume instead of a surface. This is accomplished by displaying structural data in cross-section. Typically, drillholes that lay within a certain distance of the section plane are projected onto the section with the structural data displayed along the trace of the drillhole. In the case of the coal deposit, the coal seams and interbedding would be displayed beside the trace. It is then the modeler's responsibility to generate a two-dimensional interpretation of the individual seams' appearance in the plane of the cross-section by digitizing a polygon in the section to represent the seam's geometry (see Fig. 4.30). Therefore, when the modeler observes the same seam appearing twice in the same hole, it can be interpreted as a fold or as a fault. The volume model is then created from the digitized polygons by linear interpolation between adjacent sections.

Estimation of grids, areas and volumes of intersection – Chapter 5
While Chapter 4 introduced the concept of using estimation algorithms for surface modelling, Chapter 5 delves into the details and application of these algorithms. In this context, the basic structure used to hold the model is either a two- or three-dimensional grid or matrix. An alternative approach is to represent a geologic structure directly from the data using triangles (2D) or prisms based on the triangles (3D), but this is a surface modelling method discussed in Chapter 4.

A geologic structure can be modelled using either a two- or three-dimensional data structure. Often either a two- or three-dimensional representation can be used. The most efficient data structure that is available for modelling is a grid. A grid is a matrix in which each element (data record) is associated with a locality having a fixed area or volume. In two dimensions a grid element (cell) corresponds to a specific area (see Fig. 4.10). Thus, in the case of modelling a coal seam, the cell would correspond to a rectangular area of the seam centered on a *xy* coordinate. All values estimated for that cell, such as elevation and thickness, would be assumed to be constant within the area of the cell. When the geologic structure becomes very massive or complex, it is necessary to use a three-dimensional grid. In three dimensions a grid element (block) corresponds to a specific volume centered on a *xyz* coordinate (see Fig. 5.14). All values for the block, such as assays, would be assumed to be constant within the volume of the block.

The various estimation algorithms are used to estimate values from the sample data into the grid. This is necessary since the sample data essentially consist of points in space which cannot be directly used for volumetric calculations. To demonstrate this, consider the problem of estimating the reserves available in a coal seam. If only the original data were used to estimate reserves, the seam thickness for each hole could be calculated and then the average thickness would be applied to the assumed extent of the seam to yield the estimated volume of coal. The obvious problem with this approach is that the location and the spatial density of the holes is being ignored: if many of the holes are located in an area of the seam that is thick (and therefore of greater interest for mining) while only a few holes are spread over a large tract of thin or discontinuous coal, then the estimate of reserves will be grossly exaggerated. Additionally, an estimate of the average thickness cannot be used for volumetric calculations associated with mine planning without repeating the averaging procedure for all the sample values that fall within the limits of a proposed mining cut. These problems are avoided if a continuous model of the seam's characteristics is used for volumetric calculations. In this case, the grid contains estimates for characteristics such as elevation, thickness and coal quality. The estimated cells can be limited to those that fall within the presumed limits of the seam. An estimation algorithm can be used that reflects the spatial distribution of the sample data around each cell so that the estimated values in each cell are accurate representations of the surrounding data (given that the quality and quantity of data is sufficient: garbage in, garbage out). Now the seam's reserves are based on the area of each cell that is within the presumed limit of the seam and the cell's estimated thickness. Likewise, volumetric calculations for a mining cut can be accomplished by locating all cells lying within the limit of a mining cut (see Fig. 5.43).

The same procedure is used for volumetric calculations in 3D. A 3D grid (block model) can be intersected with polygons, surfaces or volume models in the course of volumetric calculations (see Fig. 5.19). For example, a massive copper deposit might be represented as a block model with each block containing estimates of the copper content and volumes of ore and waste. The limits of the open pit used in mining the deposit can be represented as a 2D grid or surface as can the pre-mining surface topography. The block model can then be intersected with the post and pre-mining grids to determine mining reserves.

As mentioned earlier, a number of grid estimation routines are available to the modeler. These can be roughly divided into non-statistical and statistical procedures. Among the non-statistical estimation algorithms can be numbered polygonal areas of influence, triangulation, splines and inverse distance weighted averages. While non-statistically based estimation methods have fallen out of favor, they have their own advantages, not the least of which is simplicity and their traditional use dating back prior to the use of statistical estimation methods. Statistical methods include trend surfaces and kriging. Kriging algorithms are based on a geostatistical approach to estimation in which the estimation error can be calculated and used as an indication of estimation quality. In the geostatistical approach to grid estimation, the experimental variogram is represented by a model which can be used to determine the spatial variance (or alternatively the covariance) of a regionalized variable (in which the variable's value is a function of position) for the direction and distance of separation (see Fig. 5.30). In kriging, the model variogram is used as a measure of the statistical distance to determine the statistically optimal sample weights for estimating a grid value using a weighted average of the surrounding data.

For each estimation algorithm, there are a number of estimation parameters whose values can strongly influence the estimation process. For example, in weighted average estimation methods, such as inverse distance and kriging, these parameters include the search radius within which data points will be selected for estimation and the minimum and maximum number of samples that will be used to estimate the grid values. Changing any of these parameters can change the resulting estimate. So which combination of parameter values is optimal for estimation? Validation is an essential component of the estimation process that can be used to compare the quality of estimation resulting from using different estimation algorithms or estimation parameters. The validation procedure (Jackknifing) consists of generating point estimates at each of the datum locations using the same algorithm and parameter values that are being considered for use during grid estimation. Thus, there will be two values that can be compared at each datum location: the actual datum value and an estimate of that value from the surrounding data (the local datum value being excluded from the estimation). The difference between the estimate and actual datum value is the error of estimation or residual. Statistical analysis and maps of the residuals resulting from different estimation algorithms or parameter settings can be used to compared estimation quality. Trial and error can be used to select a reasonably good set of parameter settings.

Mine design – Chapters 6 and 7

Once the geometry of geologic structures has been modelled along with gridded estimates of their properties (block assay and cell thickness) a model of mine geometry can be intersected with the geologic models and grids to determine mining reserves and production schedules.

Engineering cost modelling is a key element of computer-aided mine design. In order to determine which blocks are to be mined, their net mining values must be calculated (remember that a block represents a discrete volume of ore and waste). The gross block value can be calculated from the estimates of ore tonnage, ore grades and net recovery. To determine the net mining value of each block, costing techniques must be used to estimate the mining and processing costs for each block and the resulting net block value. Unfortunately, cost modelling is not integrated into mine modelling software, although this could be done. As a result, cost engineering is not included in this text although several examples of the calculations used to assign net block values are given.

Open pit mine design is based on generating a pit that maximizes the net value of the project. Algorithms are used that select mining blocks that maximize the total value of the final pit subject to constraints on pit slopes. Several algorithms are available for pit optimization including the Moving Cone heuristic, graph theory and linear programming with each method having its own strengths and weaknesses depending on the geometry of the deposit, the complexity of the mining constraints and the goals of the modeler. None of these algorithms generate a final pit design which is necessarily acceptable in terms of operational practice. For instance, push-backs and lowering of slopes might be necessary to widen pit bottoms and include haul roads. Additionally, sensitivity analysis must be used to evaluate the sensitivity of the optimal pit to changes in key parameters such as pit slopes, metal commodity prices, recovery and costs. In any case, the optimal pit serves primarily as a guideline which in all probability will have to be extensively modified to produce an operationally feasible design.

The approach to underground mine design differs substantially from open pit design in that there is no optimization algorithm available that can be used to provide guidance in the ultimate layout, geometry and extent of the mine workings. While optimization procedures do exist for shaft siting, level interval and the like, they have not been incorporated into underground mining software and are surprisingly little used in practice. Therefore, computer-aided underground mine design is accomplished using specialized CAD tools and interactive graphics. Once the mine design is completed, the mine workings can be represented as polygons or volume components which can be intersected with the mining block model for volumetric calculations of tonnages of ore and waste in the development and production workings to obtain mining reserves. Mine development headings and stopes are represented by volume components. These volume components are intersected with the block model using numeric integration. In numeric integration, the modeler selects a 2D slice thickness. For each slice the area of intersection between the mine openings' volume components and the block model will be calculated. Since each slice has a thickness, tonnages of ore and waste can be calculated.

Different approaches can be used when estimating open pit mining reserves. The limits of the pit can be represented as either a series of polygons representing the limit of mining on each bench, as a 2D grid or as volume components. Again, each method of computing reserves has its own advantages and disadvantages.

Production scheduling – Chapter 8

Long and short term production scheduling is closely associated with the procedures used for mine design in Chapter 6, and to a lesser extent in Chapter 7 in that the same basic design methods are used. The main difference is that production goals over a series of time periods drive the design process instead of being a consequence of design as is the case in determining the maximum valued pit. Surprisingly, the process of determining a sequence of intermediate pits is far more complex that finding an ultimate pit.

Long-term production scheduling aims at determining mine design and production on the scale of years. Once the mine life, or equivalently the required production rate, is determined, the engineer must determine the limits of the mine workings for, say, each year. The production from these annual pits must meet a set of criteria which were not considered when the final mine design was determined. These criteria or goals are likely to include: mining higher valued ore earlier in order to maximize the time value of the operation, maintaining an even production of ore and waste from year to year, and meeting corporate or contractual requirements for annual mineral production. As will be seen, it is very difficult to meet multiple production goals and constraints over a series of periods.

Once a long-term production schedule has been established, the annual pit limits can be used as the basis for short-term production scheduling. Most of the goals and criteria for short-term schedules are the same as for long-term scheduling, especially the production of a tonnage of ore that will go towards meeting the year's production goal, but this ore must be mined without excessive waste haulage. Additionally, the levels of other ore characteristics might have to be held within limits set by contract or by the feed requirements of a mill. For example, in the case of a coal mine shipping on contract to a customer, the percent ash and sulfur of a coal shipment may not be allowed to exceed a given percentage or a penalty will be incurred. In the case of mill feed from a mine, the metallurgical recovery and mill efficiency will be a function of the feed characteristics: ore grades and contaminant levels exceeding an desired range will result in loss of production and increased costs.

Short term production schedules are the basis of operations from week to week in a mine. The planning engineer starts with the existing mine limit, possibly the long-term mine plan from the previous year, and defines new cuts in the benches, push-backs to accommodate stripping and extends haul roads in order to meet production goals. In practice, those ore blocks that are available to be mined are displayed along with the values that are pertinent to scheduling, such as the tonnage of ore and waste, the grade and any contaminant of special importance. The planning engineer then defines a cut that meets operational requirements, such as being accessible to a production shovel, by digitizing the cut on the display of mining blocks. The totals for tonnage of ore and waste and the average grades are calculated by intersecting the cut polygon with the block model on that bench and these results are displayed. The polygon can be modified and other cuts added to the volumetric calculation until a reasonable production plan has been found. Unfortunately, this is a trail-and-error procedure which can yield very unsatisfactory results when there are more than two or three variables involved in scheduling. Another problem that is inherent in this procedure is that the short term schedule is solved as a series of problems, but each subsequent production period is dependent on the results of the previous periods' production plans. As a result, it is possible for the planning engineer to paint himself into a corner and find that he is unable to meet production requirements towards the later part of the year without resorting to increased stripping of waste to expose more ore blocks that have the required grade and tonnage. Mine planning software allows the volumetric calculations to be accomplished rapidly and interactively to speed this trial-and-error methodology, allowing many alternatives to be examined, but the problem of determining an optimal production schedule over time by this approach is intractable due to the astronomically large number of solutions that need to be examined for a pit of even modest size. Possible solutions to this problem are suggested at the end of Chapter 8.

1.5 EXAMPLES AND CONVENTIONS USED IN TEXT

Following each of the topics discussed in the Chapters 2-8 are examples of their application using Techbase and/or Lynx. The examples themselves are meant to be instructive and are used to elaborate on the theory and practical applications of the methods being dicussed. They include instructions on how to navigate the menu system and input parameter values, but do not necessarily give specific values to use as input. Each example is followed by a corresponding Summary of procedure, which includes both a detailed description of navigating the menu and specific values to input when using the databases that have been supplied with this text. By this means, the reader is invited to study the examples in order to understand what is being done and why, and then to refer to the subsequent Summary of procedure in order to familiarize himself with the software or to use at a later date as a quick reference.

For the sake of clarity the following conventions have been used when referring to the use of the software:

To indicate the path to be taken in navigating the menus, bold face type and arrows are used. For example: **Techbase ⇒ Define ⇒ Database ⇒ Initialize** are the menu items that are selected in Techbase to get to the level 2 menu in the Define program for initializing a new database. Note that the menu systems are different in Techbase and Lynx and that in general bold face type refers either to an executable program or a menu item (a pushbutton in Lynx Version 4 or function key in Version 3).

At the program input level (level 3 menus in Techbase and entry forms in Lynx), the titles and descriptive labels used for parameter input are capitalized. For example, in Lynx's Background Hole Data Selection entry form (see Fig. 1.8) the fore and back thickness for drillhole projection on the section must be entered. In the text this is given as DISPLAY THICKNESS: FORE/BACK. When values are input either by toggling a menu item on/off, by selecting from a list or by keyboard input, the value that is input is indicated by enclosing it in <>. Thus, in the example given above for initializing a database, upon selecting **Initialize** (a menu item) from the level 2 menu of **Define** (an executable) at the level 3 menu the new database name would be entered as FILE: <whatever>. Similarly the fore and back plane thicknesses for Lynx's background drillhole display might be given as DISPLAY THICKNESS: FORE/BACK <20> <20>.

The titles of menus and data entry forms are given simply by capitalizing first letters as in Background Hole Data Selection.

1.6 AN INTRODUCTION TO TECHBASE

1.6.1 *Overview of Techbase*

Techbase is a package of programs encompassing applications from geologic database management to strip and open pit mine design. Techbase is available in DOS and Unix versions. The database is compatible between DOS and Unix. The core of Techbase is its superb database facilities which include a wide variety of data structures (tables) and variables (fields) that are flexible enough to accommodate virtually any type of input data. Not only are real, integer, date and text fields available, but complex calculated fields can also be defined using algebraic, boolean and logical operations based on RPN (Reverse Polish Notation). RPN is an efficient notation that many readers will be familiar with as the basis of HP calculators. ASCII data files in almost any logical format can be readily loaded into some type of Techbase tables. This is a tremendous advantage to anyone who must work with projects coming from a wide variety of sources. The reporting of data to ASCII files is flexible in report formatting and for summarizing data. Additional output file types include Techbase's own efficient transfer file type as well as export to dBase and a variety of other specialized file types. Graphic files (metafiles) can be converted to AutoCAD's DXF and a wide variety of drivers, such as postscript, are available for importing metafiles into drawing packages.

While Techbase's statistical routines cannot compete with a dedicated statistical package, its facilities are far superior to what is available in the majority of competing programs and have the additional advantage of being integrated with its powerful database.

The graphics facilities are based on gridded contour maps, posting of any of its data structures in plan view, digitization, cross-sectional posting of any data structure, a metafile editor, perspective views of 3D data, and a versatile polygon editor.

A very wide variety of modelling algorithms are available which are implemented to provide an experienced user with a great deal of flexibility. These estimation routines span the range of algorithms that are commonly used including: polygons of influence, polygonal gridding, triangulation and gridding with triangle sets, trend surfaces, inverse distance, minimum quadratic curvature (a spline method), and several types of kriging.

The mining facilities are somewhat basic but do include a Moving Cone and Lerches-Grossman ultimate pit generator, a facility for reporting mining reserves and a short-term production scheduling program. The open pit facilities are paralleled by a strip mine cut generator, mining and production reporting programs for layered deposits.

Additional facilities are available for slope stability analysis and ground water flow modelling.

1.6.2 *Limitations of Techbase*

Techbase does have its limitations and cannot be effectively used for all types of applications, but the same can be said of its competitors. One major weakness of the database as of Version 2.52 is its inability to work with highly inclined surfaces or volume models. Techbase's 2D grids are defined in plan view. As a result it is difficult to efficiently represent highly inclined faults, veins and seams. A volume modelling program is available as secondary software, but is not integrated into Techbase and seems to be of limited value. To a certain degree, 3D volume models can be approximated using polygons. While Version 2.52 does allow for the definition and storage of 3D polygons in the database, the supporting facilities do not recognize 3D polygons. This deficiency will probably have been rectified by the time of the publication of this edition.

Techbase's mine design facilities are limited. While the basic surface mining programs are there, they are not oriented towards engineering productivity for an operating mine. There are no specialized routines for common tasks such as haul road design and the generation of underground ramps. There are no underground mine design programs, but Techbase's 2D polygon facilities can be effectively used for the underground mining of thin layered deposits such as coal. Since most industrial mineral and coal mines, as well as many room and pillar metal mines, already use a CAD package for mine design, the use of Techbase in conjunction with a CAD program is an attractive option for these types of operations, especially if there is considerable CAD expertise among the engineering staff.

1.6.3 *Navigating Techbase*

To start Techbase enter <techbase> from the command line. This will bring up the main Techbase menu as shown in Figure 1.4. Items in the master menu can be selected either by clicking on them with the mouse, tabbing to the menu item (when selected it will be displayed in reverse video) and hitting enter, or hitting the one capitalized letter that is always used in a menu item's name. For instance, selection of **Techbase** will display a list of executable programs used in database management (Fig. 1.5). Selection of **eXit** will always result in going up one level in the menu hierarchy. Selecting any other item on this list will execute that program replacing the main Techbase menu with the menu specific to that program. For example, selection

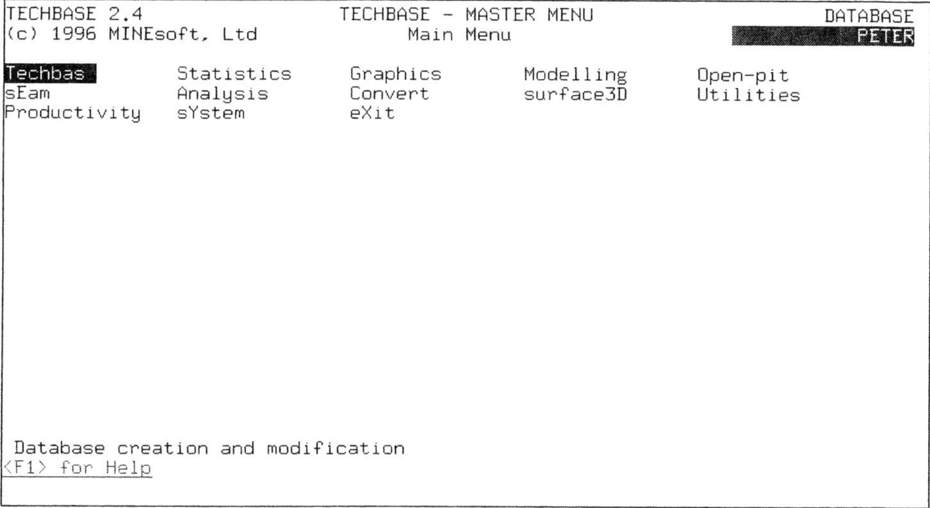

Figure 1.4. Techbase master menu.

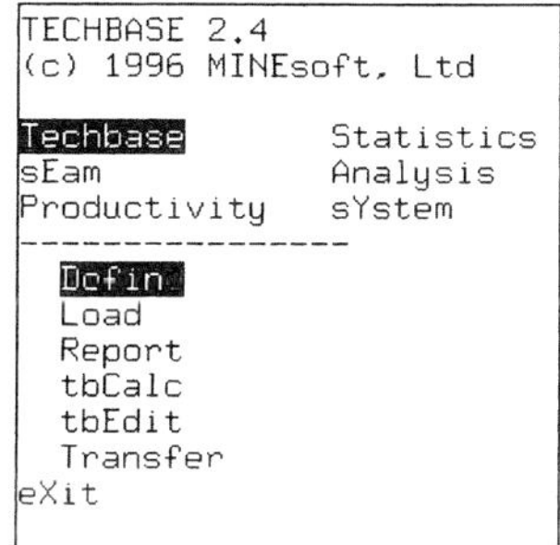

Figure 1.5. List of executables available under Techbase in the main Techbase menu.

of **Define** will bring up the Database Definition main menu (level 1), further selection of **Options** from that list will include the Options menu (level 2) in the window directly below the Database Definition main menu. The final level of the menu tree (level 3) in this program can be reached by selecting an item, such as **Output name**, from the Options listing (see Fig. 1.6). In Figure 1.6 the currently opened database is

```
TECHBASE 2.4            DEFINE - database definition                     DATABASE
(c) 1996 MINEsoft, Ltd           Main Menu                                  PETER

Options                 Database            Tables              Fields
Records                 Print               eXit

          ------------------------------------------------------------
                                 Option Menu

List                    Output name         Timestamp           View file
Runlog                  Save values YES     Case sensitive YES  Message output YES
Units ENGLISH           eXit
          ------------------------------------------------------------

Filename: define.out

  Append? NO


<F1> for Help           <CR> to Go          ^X to Cancel        ^T to Test
```

Figure 1.6. Menu display for changing output file name in the define program.

PETER. The active menu in the level 1 menu is **Options** and in level 2 **Output name** has been selected. The level 3 menu is usually used for input of parameter values. In this example the FILE NAME has been entered as <define.out> which is the default naming convention. Other items in the level 2 menu have their current settings displayed to their right. These items (in this menu **Save values**, **Case sensitive**, **Message output** and **Units**) are toggled between their available settings simply by selecting them. The available hot keys for this menu are listed along the bottom of the screen. Use F1 to get help on the items in the current menu. The level of detail that will be given depends on the menu level. Only a fraction of Techbase's capabilities are covered in this text, so the help key can be an invaluable aid to those interested in proceeding beyond the examples given herein. Inclusion of CR on the hot key listing is to simply indicate that the return key must be used to implement the parameter settings and continue. Hitting CR at level 3 will generally move the menu level up by one. To escape from the level 3 menu without inputting the parameter values use Ctrl X (^X) to cancel. Ctrl T (^T) can be used to test for the existence of the file being created so as to avoid overwriting it. Use of a hot key often generates a message from Techbase that will be displayed in the upper left corner along with instructions to 'Type space to continue'. Be sure to watch for these messages.

Even though Version 2.52 of Techbase is not completely implemented in a windows motif, the use and flow of its menu system is very similar and easy to use. As a general rule, the flow through the menu system is from top to bottom and left to right.

1.7 AN INTRODUCTION TO LYNX

1.7.1 *Overview of Lynx*

Lynx is highly oriented towards data visualization, geologic modelling and underground mine design with significant facilities for strip and open pit mining. Lynx is only available on Unix workstations. As a consequence, only those readers with access to a workstation will be able to repeat the examples given in this text.

Lynx's project database includes structures for drillholes, maps, and internally generated data such as triangle sets, grids and volume components. Drillhole and map data can be readily imported and exported to and from Lynx as ASCII files, but the file format is fairly strict requiring the use of predefined variable names. The map data structure is highly flexible. Map files are defined by their origin and orientation. Any data that could be drawn on a map can be included in a map file both for display and as data that can be used in modelling. This includes surface and underground surveys, contours, roads, pits, surface structures and even projected drillhole data. A large variety of data types are available whose attributes can be modified to change their appearance in display. For instance, Lynx is not limited to importing a digitized contour as only a series of *xy* coordinates of constant *z* value: instead, the *xyz* values can be imported as belonging to a specific contour line having a certain line type, width and color. In this way, an original source map's appearance can be closely honored. The ability to place a map at any position and orientation has powerful applications. Not only can a map be a plan view display of topographic contours, it can also be used to hold inclined section data. For instance, a map can be used to hold offset data from a vein's hanging and footwall drillhole intersepts. This offset data can then be triangulated and used for surface modelling with the triangle sets also being stored as map data.

Geologic and mine modelling using volume components is central to Lynx. In order to facilitate 3D volume modelling, Lynx uses extremely productive display facilities in which the viewplane can be oriented to any scale, inclination and position. Any data type can be displayed in the viewplane, typically as an intersection. Once the background display has been set up, the viewplane can be rapidly shifted, rotated or cut at any angle. Considerable time can be spent in setting up a viewplane and background display. The currently active data, viewplane and background display can be saved as a session file which can be restored as needed. The productivity of these facilities has been greatly enhanced in Version 4, in which X-windows has been implemented for data visualization and geologic modelling.

Lynx's volume modelling method is probably the best available. While most packages use wire framing to move from a series of 2D cross-sectional outlines of a volume component to a 3D representation, Lynx has a unique methodology which lends itself to far greater accuracy than wire framing. The number and configuration of wires used to connect to sectional interpretations of a volume depends on the density and shape of the polygons defined on the sections. This can lead to missing or overlapping volumes when two volume components are placed in contact, as is typically the case during geologic modelling. This is because two wire framed volumes can either cross each other or gap between the sections on which they are defined. This problem is avoided in Lynx.

Triangulation and interactive volume modelling are the only methodologies available for surface modelling. Lynx's triangulation algorithm is highly developed and ensures that the original data is honored. Surfaces that have a thickness rather than only an offset or, equivalently, an elevation can be modelled by mapping two triangle sets on to each other and converting the combined triangle set to prisms which are then used as volume components. When there aren't enough hard data to form a satisfactory triangle set, the interactive volume interpretation facilities can be used to create very thin volume components. This technique is very effective for representing structures such as faults whose positions must be interpreted based on limited information.

As of Version 4, some gridding and engineering facilities are not fully converted to X-windows. Instead, the Version 3 programs are linked to the Version 4 menu system. Thus, some of the productivity enhancements of Version 4 are lost when working with these programs. Navigation in both of these versions is covered in the following section.

3D gridding, i.e. block model estimation is accomplished using either inverse distance or kriging. 2D grid estimation is only available by inverse distance map to grid estimation in which the primary purpose is to convert topographic map data to a grid that can be intersected with a block modelling for surface mine design. The ability to intersect block models with volume components using numeric integration allows highly accurate estimation of ore volumes by rock type to be included in the block model without the risk of overlap resulting in a total material volume exceeding the block volume or there being undefined material left in a block.

Surface mining facilities include a Lerches-Grossman pit optimizer and conical pit expansion. Pit optimization results are converted to map data which can be modified using the conical expansion facilities to meet any operational requirements. These same facilities can be used in top-down mode for dump design.

Lynx's underground mine design is excellent due to a number of specialized volume modelling facilities and its superb volume modelling capabilities. Highly inclined openings, such as shafts and raises, can be generated automatically by specifying their sectional dimensions, orientation and endpoints. Ramps can be rapidly designed by specifying their dimensions, slope, radius of curvature and digitizing their path. Drifts and crosscuts are treated similarly to ramps. Stopes are treated as irregular volumes that are modelled as per geologic volume components. Engineering volume components can be readily modified to reflect detailed design such as curved intersections, ore storage pockets or detailed survey information in an operational mine.

1.7.2 *Limitations of Lynx*

Lynx also has its limitations. As a product, Lynx has focussed on high-end Unix users interested in detailed geologic modelling and underground design. The database is no broader or more versatile that it needs to be to serve these facilities, but Lynx does make excellent use of map data to store a wide variety of data.

There are limited facilities for 2D gridding. While map data can be converted into a grid, the only available estimation algorithm is a limited version of inverse distance. Additionally, grid data structures are not integrated into the database eliminating the use of 2D grids for volumetric calculations or the ability to view 2D grids as part of the general background display facilities. As a result, surface modelling is limited to representation with triangle sets or interactive graphic interpretations. 3D grid estimation is limited to the two weighted average methods: inverse distance and kriging, but these methods cover the vast majority of users. The other estimation algorithms, such as trend surfaces and splines are usually applied to 2D gridding although they can be applied to 3D grids with valuable results, but not in Lynx or Techbase.

Lynx's statistical and geostatistical capabilities are limited to basic analysis and estimation. They include basic distribution analysis, scatter plots and good variography but little more. Data export to GSLIB is a major enhancement in Version 4.7 that enables the use of advanced geostatistical estimation methods. As noted earlier, a modeller should not expect to be able to rely solely on the statistical and geostatistical facilties included in geologic and mine modelling packages. Data can be imported and exported, but there is very little control over the format of the files. The expectation is that the user will export a standard Lynx ASCII file and then manipulate the file as needed using a more powerful file manipulation language in Unix such as AWK.

As noted earlier, the facilities for analysis, gridding and engineering have not been converted to X-windows as of Version 4. The older version does not support pull-down lists and pop-up menus and is significantly more difficult to navigate. This results in a slower learning curve and decreased productivity for the user. Full X-windows implementation is slated for Version 5.

1.7.3 *Navigating Lynx*

Several versions of Lynx are available for viewing data, geologic modelling and gridding and engineering. To run these versions enter <lynx -view>, <lynx -main> or <lynx-eng> from the terminal's command line. Only entering <lynx> will execute Lynx View. Unless the engineering facilities of Version 3 are needed, only execute Lynx Main, as this will conserve memory. This will result in the opening of Lynx's main menu shown in Figure 1.7.

Navigation in Version 4 is based on using the mouse to point and click on menu items, lists and toggles. For many tasks little or no keyboard use is needed. Figure 1.7 is an example of using the mouse and pull-down menus for selecting a project. From the main menu selecting **File** will pull down the File Applications menu from which **Project Data** is selected to pull down the Project Data Management Applications menu. Click on **Project Select** and the Project Select list will pop-up from which project TUTORIAL can be selected. Clicking on **OK** will make TUTORIAL the active project. Note that in any of the following examples in the text this procedure would be given as **File ⇒ Project Data ⇒ Project Select**.

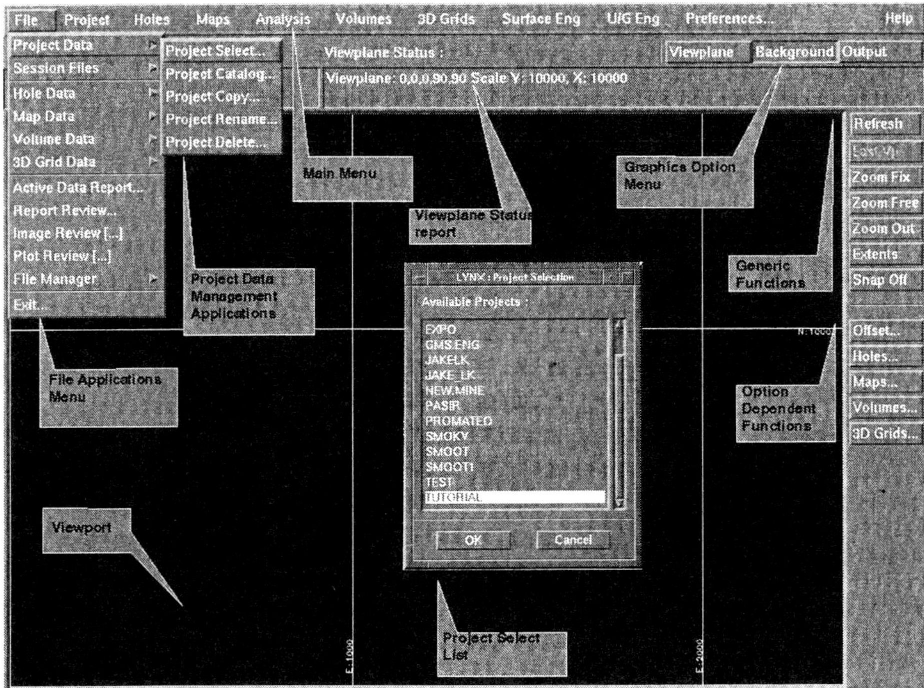

Figure 1.7. Lynx Version 4 main menu showing project selection.

The File Applications menu requires further consideration before continuing. Before project, hole, map, volume or grid data can be manipulated, the data source or destination that is going to be manipulated must be active. In the previous example, project TUTORIAL was activated, but prior to doing this the applications listed in the Project Data Applications menu (just to the right of the File Applications menu) would not even be available for use (indicated by a grey, low resolution type rather than a clear white type). This is true of the first six items on the File Applications menu. Selection of any of these items having a right facing arrow will pull down an associated menu with a pushbutton for data selection. In summary, when following any of the Lynx examples in this text, remember that the data associated with the example must be active. Otherwise, the pertinent menu items will not even be available for use.

Navigation through any of the menus in Version 4 will culminate in one or more pop-up entry forms or lists. These forms are the specific information needed by the program to execute a task is specified by the user. For example, consider the procedure for displaying drillholes in background (explained in more detail in Example

3.3). Again, remember that before continuing a drillhole subset must be selected. Selecting **Background** from the Graphics Option menu (see Fig. 1.7) will bring up the Background Options from which **Holes** can be selected. This will pop-up the Background Hole Data Selection entry form as shown in Figure 1.8 which illustrates entry form functionality. Clicking on pushbuttons, such as **Session**, will result in some action, such as displaying a report of active data, executing and closing the form (**OK**), executing but not closing the form (**Apply**), cancelling the task (**Cancel**), or opening up a dialog window (**Help**). Toggle buttons are switched on or off. In Figure 1.8 up to five displays can be active at one time, but only Display 1 is toggled on. The display parameters listing in Figure 1.8 are for Display 1 as indicated by the multiple pushbutton. Selection of similar pushbuttons, such as DISPLAY FORMAT and TRACE DISPLAY, require parameter values taken from a limited number of options selected from a pop-up menu. Other pushbuttons, such as **Subset Holes Select** display lists when selected or, alternatively, entry fields are provided for keyboard input as in the case of DISPLAY THICKNESS: FORE/BACK in which any reasonable numeric values can be entered.

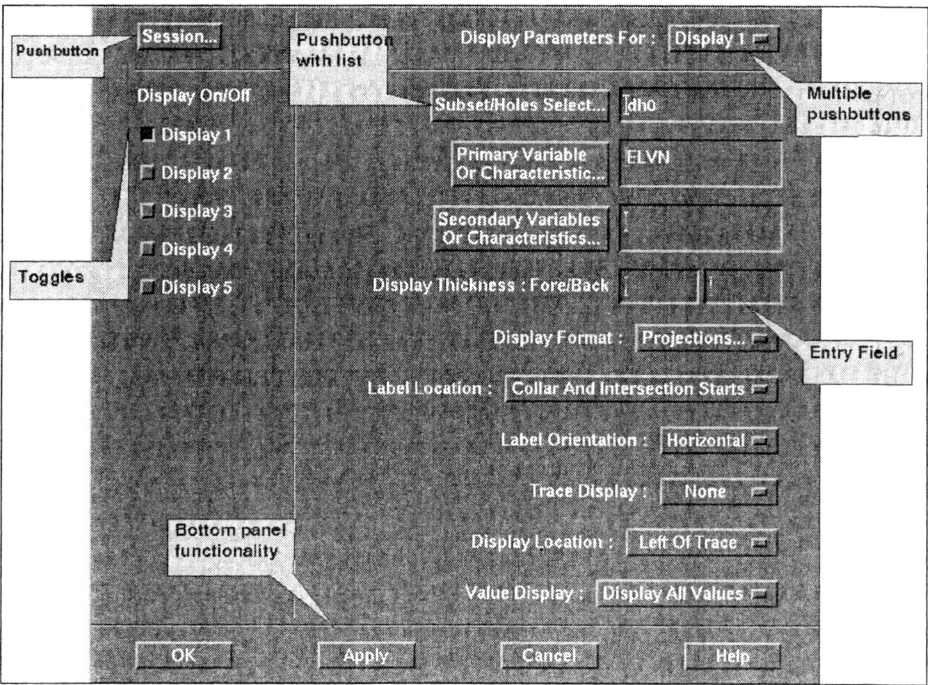

Figure 1.8. A typical data entry form.

As of Version 4.7 some facilities have not been converted to X-windows, specifically: 3D Grid Data, Open Pit Engineering and Strip Mine Engineering. In these instances facilities dating from Lynx Version 3.11 are included with this text. Navigation in Version 3.11 does not use lists, data entry forms or other X-windows devices.

Pushbuttons are only used to navigate the menus. Otherwise function keys (F1-F8 on your keyboard) or mouse selection from a menu are used to navigate the menu tree. Version 3 prompts the user for the type of input shown in Figure 1.8 either directly from the menu or from data entry forms of Lynx Geosystems own design. If a single menu level includes more than eight items, then the inclusion of **More** indicates that this function key must be selected to view the rest of the menu. **Return** is used to move up one level in the menu hierarchy. When selecting the menu items in Version 3, note that not all of the menu items are selected by clicking on them with the mouse. If the mouse doesn't work, use the associated function key (F1-F8). Also, some functions such as **Edit** and **Insert** in the design facilities and **X-sect** in the Version 3 sectional display facilities work by placing the mouse on the display where, for example, a point is to be entered and then using the function key to execute the desired function. Examples of these procedures are given for Lynx in Chapters 5 and 6.

1.8 A COMPARISON OF TECHBASE AND LYNX

There is an extremely wide variety of geologic structures that may have to be modelled by geologists and engineers involved in projects running the gamut of mineral extraction, groundwater studies and hazardous waste characterization. Any system that claims to be equally effective for modelling all geologic structures for any application is likely to in be in reality not very good at anything in particular. The market for geologic and mine modelling packages is very competitive and has forced the competitors to become increasingly specialized both in application and marketing. In this light it is meaningless to say which package is better than the other. They all have their strengths and weaknesses. Often where one system is lacking the other is strong. This is why two competing systems, Techbase and Lynx, were used in this text: neither has sufficient breadth or depth in all of the topics covered in this text to be used alone to demonstrate the theory and application of geologic and mine modelling. But they do complement each other remarkably well. This is not to suggest that someone considering the purchase of a similar system would have to get both Techbase and Lynx in order to have an effective system. No, not at all. It is highly unlikely that even the average consultant would need such a variety of modelling facilities, and in the case of an operating mine, the modelling requirements will be very specific so that having more than one system would likely be redundant.

The goal of this text is not to provide a tutorial to be used for two packages. If that was the goal it would be better to publish two separate manuals. Rather, it is the author's objective to present a comprehensive discussion of the process of geologic and mine modelling and to demonstrate modelling with two excellent commercial packages which have graciously been placed at the disposal of the academic community.

Anyway, how do Techbase and Lynx compare and how are they used in this text? Techbase has marketed itself more towards exploration and the mining of layered deposits. As such, it has excellent database facilities and its modelling algorithms are based on grids defined in the horizontal. Its statistical analysis is very good compared to the bulk of its competitors; this is added strength for its database facilities. The engineering design facilities are weak, especially for underground applications, but due to Techbase's high degree of flexibility a skilled user can handle almost any type of model even to the point of building customized routines for cumbersome tasks.

Lynx's strength is in volume modelling and the display of data. Techbase essentially lacks the ability to work directly with 3D models. On the other hand, Lynx essentially lacks the ability to work with 2D grids and is limited in its gridding algorithms, an area in which Techbase excels.

Neither program is exceptionally strong in surface mine design, although Techbase's gridding facilities and Lynx's surface handling make them strong contenders for strip mines. Still, by using different approaches to open pit design they are used to good purpose in this text to demonstrate the wide variety of methodologies that are available.

In terms of underground design, Lynx excells. Techbase does not have specific underground design facilities, but is flexible enough that it can be used to good effect in conjunction with a skilled CAD operator when mining layered deposits.

CHAPTER 2

Geologic database management

The geologic data used for project feasibility, ore reserve analysis, mine design and production scheduling typically consist of point and line data obtained from surveying and drilling. Survey data are provided as *xyz* coordinates from either surface mapping and underground surveys, while drillhole logs are based on intervals which are assumed to be continuous over their length. The first task in generating a geologic database is to transfer data from surveys and drillhole logs into a format that can be intrepreted by a geologic modelling package.

2.1 DATA FORMATS

ASCII is the lingua franca for data exchange between word processors, spreadsheets and more complex graphics-oriented programs. For the purposes of this discussion it is assumed that data is being provided in ASCII files.

2.1.1 *Survey or map data*

In general, survey data is used to represent information that would normally be presented on a map as either *xy* points in the plane of the map, such as property boundaries, or as *z* offsets from the map's surface, such as contours. Survey data can be provided as either point or transverse lines directly from a survey or from secondary sources, most typically contour maps. In either case, each record in the survey represents one point in space. A survey record consists of the *xyz* coordinates of the point and any additional relevant information. Two examples of this would be contour and polygon data. In the case of contour maps, each continuous contour will be digitized into a file where each record will have three fields: easting, northing, and a fixed elevation field which is entered at the beginning of each contour (see Table 2.1).

When the data file is used to define the contours for a map, additional information might be included to control the appearance of each contour such as the line width, style and color of major and minor contours, but the format for including these variables in the file are specific to the program used to interpretation and plotting.

Polygon files follow the same format as used for contours, but *x* and *y* are the adjacent vertices of the polygon moving either clockwise or counterclockwise. *Z* for the polygon file is the identification of the polygon and doesn't have a physical meaning. Polygons are often used to represent boundaries.

Survey data can also be used for point and tranverse data. Examples of point data

Table 2.1. Contour data.

Easting	Northing	Contour
10201.0	22096.9	5710
9996.6	21981.7	5700
9993.3	21832.1	5700
10239.9	21973.0	5700
10408.8	22093.0	5700
9998.2	22093.3	5700
10084.6	21861.8	5700
10812.8	22072.0	5650
10736.1	21964.6	5650
10608.9	21827.4	5650
10346.4	21641.0	5650
10122.5	21498.0	5650

are sampling sites such as well collars, trenches or channel samples. A transverse is a series of connected *xyz* points in which *z* can vary. A transverse would be used to represent the centerline of drift or ramp, or the proposed route of a road or powerline.

2.1.2 *Drillhole data*

Drillhole data files will have one record representing either a *xyz* location or an interval. Generally, the only *xyz* data provided will be the location of the collar of the hole. In a collar file there will be only one record per hole. Each record will include a hole identifier, easting, northing, elevation, and additional fields that provide information such as drill identification (rotary versus core), date, lease, total depth and any other pertinent data, such as contractor. Except for very shallow holes, drillhole traces are usually inclined. Survey files define the change in azimuth and dip over the length of each hole. In this case, there are as many records as there are recorded changes in direction. Each record in the file represents a continuous length of hole having the same inclination. Each of these records' files will have the following fields: hole identification, from, to (or depth interval), azimuth, dip and other derivative data such as length (see Table 2.2)

Two or more other files are usually provided that provide information on assays, geotechnical data and lithology. There will be as many files as are required by the variety of intervals used to record the data. Assays and geotechnical information (RQD, permeability) may all be analyzed on the same interval and can be included in the same file. The file format is the same as for the survey file: hole identification, sample identification, from, to (or depth interval), and all assays. Lithology will be reported in a separate file since the from/to intervals are fixed as being the contact boundaries between the lithologic units. There are advantages to including as much information as possible in the same file (see Table 2.3).

Table 2.2. Drillhole survey data.

Hole_id	Depth	Azimuth	Dip
84-90	110.0	150	85
84-91	0.0	–30	90
84-91	50.0	–30	88
84-91	100.0	–30	87
84-91	150.0	–30	86
84-91	200.0	–30	85
84-91	250.0	–30	84
84-91	300.0	–30	83
84-91	350.0	–30	82
84-91	400.0	–30	81
84-91	500.0	–30	78
85-93	0.0	150	90
85-93	20.0	150	89

Table 2.3. Drillhole assay and lithology data.

Hole_id	From	to	Lith	Au	Ag	As
83-100	20.0	24.0	OA	0.522	7.955	77.300
83-100	24.0	30.0	OA	0.122	3.310	179.100
83-100	30.0	40.0	OA	0.282	5.513	115.800
83-100	40.0	50.0	OA	0.193	2.719	183.700
83-100	50.0	60.0	OA	0.369	6.378	104.700
83-100	60.0	70.0	OA	0.346	5.751	119.100
83-100	70.0	80.0	OA	0.193	3.569	170.600
83-100	80.0	90.0	OA	0.289	5.642	87.500
83-100	90.0	100.0	OA	0.227	3.964	151.100
83-100	100.0	110.0	OA	0.315	5.453	98.700
83-100	110.0	120.0	OA	0.197	4.774	121.800
83-100	120.0	130.0	OA	0.494	8.339	73.100
83-100	130.0	140.0	OA	0.262	5.099	129.300
83-100	140.0	150.0	OA	0.152	3.600	146.200
83-100	150.0	151.3	OA	0.123	2.455	216.900

2.2 GEOLOGIC DATABASE STRUCTURE

The specifics of database structure and terminology are as varied as the programs that are in the market, but all of them must relate to the original data and hold estimation results of surfaces and volumes in much the same way. The structure of a database reflects its content. For geologic databases, content can be broken into two gross catagories: data and estimated values derived from the data. Geologic data comes primarily in the forms addressed above, either as survey or drillhole data. For most purposes, the raw data are not immediately useful since they represent isolated point values which must be used for estimating continuous structures as discrete sets of values. Therefore, geologic databases must be adaptable to the formats in which raw data are collected and in which estimated values are stored. The means of collecting and recording geologic data are predicated by the conventions followed in surveying and drilling. Likewise, geostatistical and conventional estimation methods have popularized a few formats for estimating spatial values. As a result, any comprehensive geologic database will have a structure matching the accepted formats for data and estimated value storage. This structure is most readily discussed by following the terminology used in one package. The following terminology originates from Techbase and will be used as an example.

A database is composed of tables which contain records of consistent format and content as defined by their fields. Fields can be thought of as variables whose values are found in each record of the table. In Techbase, a database consists of a collection of tables, and each type of table is designed to suit special formats of information. These table types include: flat, cell, layer, polygon and block tables. Flat, cell and block table types are common to all geologic databases and will be given the most attention herein.

Flat tables are used to contain data that are input from drillhole logs and surveys and derivative data such as composited sampling intervals. All other tables are based on flat table data. The structure of a flat table parallels the ASCII file data discussed in the previous section: there will be flat tables for topography, assays, lithology, etc. Flat tables are composed of records and fields where the records represent rows and the fields columns; they are used as a means of holding data that will be used for estimation in other table types and for graphical displays such as geologic sections. Flat table structure is illustrated in Figure 2.1. A flat table consists of one or more header

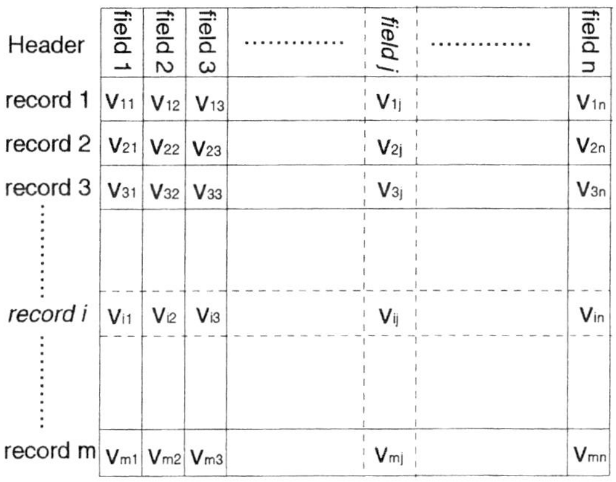

Header	field 1	field 2	field 3	....	*field j*	....	field n
record 1	v_{11}	v_{12}	v_{13}		v_{1j}		v_{1n}
record 2	v_{21}	v_{22}	v_{23}		v_{2j}		v_{2n}
record 3	v_{31}	v_{32}	v_{33}		v_{3j}		v_{3n}
⋮							
record i	v_{i1}	v_{i2}	v_{i3}		v_{ij}		v_{in}
⋮							
record m	v_{m1}	v_{m2}	v_{m3}		v_{mj}		v_{mn}

Figure 2.1. Flat table organization for storing input data.

lines followed by 1...m records. Each record records the 1...n field values, *v*. The identity of the fields is given in the header line such that field 1 would correspond to the first entry in any record. A typical example is given in Table 2.2 that has as a header 'holeid depth azimuth dip' as the field names for fields 1-4. The corresponding values in record one are: 84-90 110.0 150 85. In a flat table, a record corresponds to a sample point and the field values relate to a datum identification, its position in space, measured values and any other pertinent information. Flat tables are commonly used to contain raw input data or processed versions of that data.

Cell tables represent a two-dimensional matrix of cells which have an *xy* rectangular dimension and location; cell tables are used to hold estimated surfaces such as topography, top and bottom elevations of hanging walls and foot walls or seams, thickness or any other data which can be represented in two dimensions. The origin, orientation and extent of cell tables define their location in plan view while the fields in the table define the elevation or thickness of each cell. To understand how a cell table functions, imagine viewing the topology of a geologic structure in plan view and laying a grid over it and then assigning the value of that data to the cells that subdivided it so that each cell is assigned the average value of the area that it encompasses. Cell tables are commonly used to represent lithologic contacts, seam thicknesses, geochemical concentration levels and topography. Sedimentary deposits which are not too highly contorted are best represented by cell tables. Volumetric calculations made for coal and industrial minerals use cell tables. Often cell tables are referred to as gridded surfaces. Non-tablular geologic structures such as disseminated ore bodies or complex, high angle veins should not be estimated using a cell table.

It is possible to create cell tables which have variable sizes; there are advantages to using variable cell dimensions. For instance, areas of high data density should use smaller cell sizes. Also, regions having high angle faulting or steep dip should have a decreased cell dimension along dip.

Block tables are used for massive deposits which have characteristics which vary with depth and cannot be easily represented by cell tables. The definition of a block table is the same as for cell tables, except that a block table is a three-dimensional matrix and is also defined by a starting elevation and the number of blocks in the *z* direction. Blocks have *x*, *y* and *z* rectangular dimensions which are fixed for all blocks. Block tables are used for estimation of massive deposits. The extent of the table should be sufficient for enclosing not only the deposit but also, in the case of near surface deposits, the ultimate extent of the pit, including all stripping. The intersection of the block table with the cell table defining the surface topography delineates reserves and is used for volumetric calculations. The fields which are commonly kept in block tables include material density, assays and calculated fields for tonnage, cost and net mining value.

Layer tables are similar to a set of cell tables having the same definitions as to cell size, origin and orientation, but layer tables are listed in a vertical stack with each layer having its own name. The advantages of a layer table over a cell table are organizational. Each layer can be used to represent a specific layer in a deposit with its own list of fields; a set of cell tables could be used to represent individual layers, but there are difficulties that can arise in manipulating fields that are in different tables. In a layer table, all fields are still in the same table.

Polygon tables are used to store two- or three-dimensional polygons. One common use of polygons is to define lease or claim boundaries, but they are also very useful during production scheduling. Mining cuts are defined as polygons and are intersected with cell, block or layer tables in production volumetric calculations. This method is used to determine the average ore grade and tonnage of lifts and bench cuts. The polygon table can be used to keep track of production history. While all of the above tables can be found in comprehensive geologic and mining software, Techbase includes an additional table type which is peculiar to its database structure: the join table. Join tables are used to connect one table with another so that fields contained in both tables can be used simultaneously. Usually, there is a many-to-few relationship between records in the two tables which are to be joined (see Fig. 2.2). A

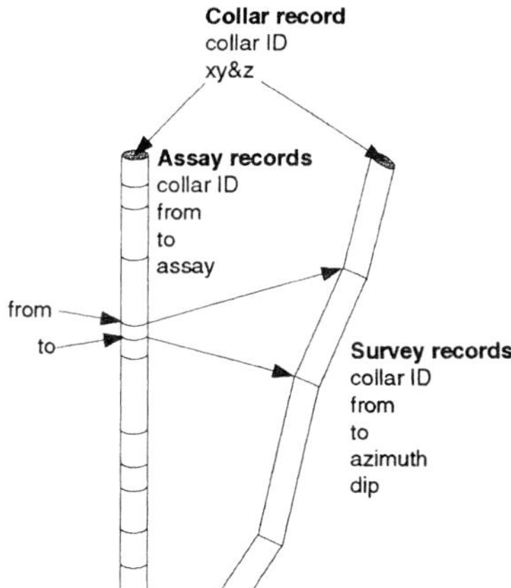

Figure 2.2. Cross referencing drillhole records taken from three tables.

join table is commonly used with the flat tables containing drillhole data. Drillhole log data is not commonly provided with the *xyz* coordinates of samples or lithologic boundaries; instead, it is necessary to calculate the *xyz* coordinates using the collar location, the from/to depth intervals down the hole and survey information on the inclination of the hole. Since the three flat tables which hold this information have different numbers of records, there must be a means of connecting them. This connectivity is accomplished through the join table, and will be discussed in depth under Compositing (see Section 4.1).

Techbase's join table highlights an essential component of any geologic database; there must be a means of cross referencing records in different but related tables. It was stated earlier that each record in a table lists the values that are relevant to a specific datum. For instance, in the case of drillhole data, this datum might be either the collar location of a hole, an assay value at some depth down the hole or a record that defines the deviation of the hole over some depth interval. As shown in Figure 2.2, all three record types are related: the collar location gives the starting coordinate of the hole; the assays are specified as being logged for a depth interval down the hole; and the survey table gives the deviation of the hole. All three sets of information are needed to define the assay at a specific location in space, but the number of records in each table won't be the same. For each hole there will be only one collar record, but in the hole there might be hundreds of assays and dozens of changes in inclination. All three tables are related to the collar file by the hole identification forming a one-to-many relation between the records of the collar table and the assay and survey tables. In order to determine the location of an assay, the depth interval of the assay must be compared against the depths of the inclination intervals in the survey file for a given hole.

2.3 CREATING A TECHBASE DATABASE

Before information can be loaded into a database via an ASCII file or by direct input, the structure of the database needs to be defined for the tables that are meant to be used to hold the raw data. Unless files containing estimated values from other programs are to be imported into Techbase, the tables to be defined will all be flat tables. Following the examples given above, at least the following tables and fields will be required:

1. Collars: hole_id, easting, northing, elevation, and any additional information which can be defined for the entire hole such as lease, depth and contractor.
2. Survey: hole_id, from, to, azimuth, dip.
3. Assays: hole_id, from, to, and a list of assays.
4. Lithology: hole_id, from, to, unit identifier.
5. Topography: x, y, z.

Flat tables must be defined for each of these along with the necessary fields. Note that it is important to use identical field names in separate tables for data that are consistent in their values between tables. For instance, the hole identifier will always be the same between tables and the same field should be used, but the from/to intervals will not have the same values in the survey, assay and lithology tables and their names in the three tables should be unique.

Example 2.1: Initializing a database in Techbase

From the Techbase main menu select **Techbase** ⇒ **Define** ⇒ **Database** ⇒ **Initialize** and create a new database called 'example' by entering a database name <example>, password and a descriptive title. Note that any alternative name is acceptable. Only the name is required. Care should be used when using a password. Like a login password, there is no way of determining what a password is if it is forgotten. Once the database is initialized, **eXit** out of **Database** in preparation for defining tables.

Exit from Techbase back to DOS and check the directory. There should be three new files: EXAMPLE.001, EXAMPLE.002 and EXAMPLE.003. Techbase subdivides the database into smaller files having the name of the database and a three-digit file extension. The database files are only accessible through Techbase and cannot otherwise be read.

Summary of procedure: Example 2.1

Initial data requirements: none, but create a project subdirectory and execute the following examples from the new project directory.

To start Techbase, enter **Techbase** from the command line.
From the main menu select **Techbase** ⇒ **Define** ⇒ **Database** ⇒ **Initialize**
NAME: <example> PASSWORD: <>
TITLE: <>
where <> indicates no entry
⇒ **eXit** ⇒ **eXit**

Example 2.2: Defining tables and fields in Techbase
From the **Define** main menu, select **Tables** ⇒ **Create**. Enter the names of four new tables: assays, collars, surveys and topo and make them flat tables.

Exit from **Create** and select **Fields**. To create new fields and attach them to a table, select **autotable** and enter the name of the table in which the fields belong. **autotable** will automatically default to the last table created. Next select **Create** and enter the name, type and class of the new field. The allowable types of fields are integer, real, date and text. The class is either actual or calculated. All data which is either loaded into Techbase or directly entered into the database is an 'actual' data field. A calculated field is one that is calculated from other actual fields in the same table. An example of a calculated field would be sample interval length, which would be the difference of the from and to depth fields.

Note that fields that are not attached to a table will still reside in the database but will not be able to contain data or be accessed. To see the contents of the database (but not the records) use Ctrl R. If it is necessary to **delete**, **modify**, **rename** or review (**show**) the definition of a field, this can be done in the **Fields** menu. To attach existing fields to a Table (whether they're currently attached to a table or not) return to the **Tables** menu and **Add field**. If the same field will be used on multiple tables, as in the case of the hole identifier, then only create it once and use **Add field** to include it in the other tables.

Repeat this procedure for every table that is required and for every field. Don't forget to use Ctrl R to review your progress to be sure that there are the right number of tables and that the required fields are attached to them. Look for any unattached fields.

Summary of procedure: Example 2.2
Initial data requirements: an open database (<example> as created in Example 2.1, for table and file structure see Boland database).

1. Creating tables: from the main Techbase menu **Techbase** ⇒ **Define** ⇒ **Tables** ⇒ **Create**
   NAME: <collars> TYPE: <flat>
   ⇒ **eXit**
2. Defining fields
   2.1 From the **Define** main menu select **Fields** ⇒ **Autotable**
   TABLE NAME: <collars>
   2.2 Create fields
   FIELD NAME: <hole_id> TYPE: <text> CLASS: <actual>
   LENGTH: <6> JUSTIFICATION: <left>

(Repeat Step 2.2 for each field in the collars.dat file: *x, y, z* and depth. See Boland database for details using **Define** ⇒ **Fields** ⇒ **Show** and the ^R option the review database structure.)

(Note: The prompt for integer and real numeric field types would be for the minimum and maximum values that the field can possess. This is optional. For real fields the **Display precision** will also be required. This is the number of decimal places.)

Repeat the process for survey, assay and topo tables. Refer to the data files survey.dat, assay.dat and topo.dat to determine the required field names and types.

2.4 KEY FIELDS AND JOIN TABLES

As discussed earlier, it will be necessary to have a mechanism that will make it possible to communicate between fields in different tables. This is done by defining key fields and creating join tables. A key field is a field which is common between two or more tables such as the hole identifier. It must have the same values in each table.

A join table can be created for two tables which share a key field and have a many-to-few relationship in their records. For instance, it might be of interest to determine the average assay for samples in a specific sediment. In Techbase, this would be difficult to determine since the fields for assays and lithology are commonly contained in two different tables, but both the assay table and lithology table will contain a field that identifies the drillhole which can serve as a key field. There is a many-to-few relationship between assays and lithology since the from/to interval for assayed samples can be expected to be much shorter than the length of intersection of the drillhole with the sediment type of interest.

A key field should be created before records are put in its table. This is done after the field has been defined by selecting **key field** in the **Tables** menu of **Define**. From the Techbase main menu the path is

Techbase ⇒ Define ⇒ Tables ⇒ Key field

Enter the Table in which the field resides and a list of fields which are to be treated as key fields. Repeat this for every table that is to be joined in a join table. Key fields are marked with an asterisk when the database is viewed with Ctrl R.

To create the join table, follow the same procedure as for creating any other table, but enter a table type of 'join'. You will be prompted for tables one and two. To maintain the many-to-one relationship between the records in the two tables, enter as table one the table that will have the larger number of records. Table two must be a flat table. The key field which is common in the two joined tables must at least be keyed in the second table. Now, fields in both tables can be accessed as if they were in one table, even though the structure of records is not the same in the tables.

Example 2.3: Loading ASCII files into a Techbase database and using runlogs
When working with an unfamiliar program, it is always a good idea to create a runlog of the session. A runlog records every menu item selection and keystroke and can be used to automate the procedure and to play back the exact sequence of events as they occurred while running the program. This provides an extremely useful record in that the runlog can be resurrected for later use either as an example of how to run a program or as a means of presenting a question to another user who might be more versed in the proper use of the program. Presenting a mentor with a runlog and output file with **Message output** toggled on is vastly more effective than the more common approach of explaining that, 'I tried to load in some data but it didn't work'.

To start a runlog from any program's main menu, select **Options ⇒ Runlog** and enter a file name. The file extension of *.rlg will be added by default. To put Techbase generated messages into the output file, select **Message output** in the same menu. Also, change the **Output name** to avoid overwriting this file later. The use of the runlog will be demonstrated following the remainder of this example.

For each of the flat tables created, there should be a corresponding ASCII file that was used as a guide to the fields needed and their format. At this point, the database should not have any records in any of its fields. Now load actual data into these tables. To execute the load program from the main menu, follow the path: **Techbase ⇒ Load ⇒ Setup**.

Make sure that the correct database opened. When a database is open, its name will be displayed in red highlight in the upper right corner. This corner also indicates how many filters are active by attaching a pipe symbol followed by the number of filters. For example: PROJECT | 2 would show that the database project is open and two filters are active. If there is no filter open, then select **Database** in the **Load** menu and **Open** the desired database. Note that if Techbase is not being executed from the project directory in which the database binary files reside, then the complete path will have to be provided as part of the database name.

Identify the ASCII file that contains the records that will be loaded and select **Data file**. If you're not sure which file you are loading, you do not have to exit to DOS or to the DOS **Command** line in the **System** menu. Instead, **List Files** can be used to list all files in the directory with the extension *.dat. To view the file, select **View file** and enter the file name. Upon viewing the file, note how many rows proceed the first data record and the order of the fields that are to be read. Select **Data file** and enter the file's name <collars.dat> and how many records are to be skipped in the file before the first data record is read.

Next, enter the list of fields that are to be loaded from the records contained in the ASCII file by selecting **Fields**. Enter a list of field names in the same order that they appear in the ASCII file. Techbase loads records one at a time and equates the order of appearance from left to right in a record to the order given in the field list. Note that Techbase cannot differentiate between a missing field value and a simple blank between fields. Thus, if the field list was given as: hole_id *x y z* type lease, and the data record in the ASCII file collars.dat appeared as: 84-102 10555 20300 3011 54, with a missing value for the 'type' of hole, then Techbase would read type = 54 and lease = null. Therefore, it is very important to browse the ASCII file before loading it to see that there is a complete set of records for each field. This can be done using either the operating system's text editor or with **View file**.

It is very common that there will be missing fields in a data file. In order to avoid the problems this creates in Techbase, a format can be used for reading the fields in each record. Both **Field list** and **View file** provide scales that can be used to locate each field in a record according to the starting column and length. If every record isn't consistent in the position and length of its fields for all of the records in the file, then use an editor to make sure they are consistent. Return to the **Field list** and enter each field name in its order of appearance, but include after each field in parentheses the starting column and length. In the example given above from collars.dat, enter:

hole_id (1,6) *x* (8,5) *y* (14,5) *z* (20,4) type (25,2) lease (28,2)

Note that it is only necessary to enter the positions of the fields that are going to be affected by null field entries which would include the field itself and all subsequent fields for the record. In this case, only type and lease would have to be identified as to position and length if no earlier record value were missing in any subsequent record.

After the field list has been entered in either free or fixed format, select **Parameters** in the main load menu. Here specify the **Update type**, **Silent?** and **Delimiters**. For an initial loading of a table, the default values for these parameters will be fine, but later, if the records in the tables have to be modified by loading new information, you will have to understand what these options do.

Update type controls how new records are added to a table. While loading, records in a database can be **Replace**d, **Append**ed or **Overlay**ed. **Replace** will first delete all existing records in the table before adding new records from the data file. Only use this option when you are starting out fresh. **Append** is used to add additional records to a table that already has records. For instance, additional new drill-hole logs might be added. **Overlay** will add new field values to existing records. In this way, the existing records could be overwritten to add new information that was forgotten in an earlier load. As long as the field names being loaded into the table are new, the previous field values will be left intact. Use of **Overlay** with the same field list as used in the previous update will result in overwritten field values.

If Techbase encounters any errors in the correspondence between the field list and the ASCII file, it will report a message as to the nature of the error. In some cases, Techbase reports warnings that are unimportant to the task that you are doing. Setting **Silent?** to 'yes' will suppress all messages. In general, this should set to 'no' so that you will be aware that there is a problem and can kill the run and start over.

Delimiter is the type of character that indicates a break between fields in a record. Usually, this is **whitespace** or a blank between field values in the ASCII file, but commas or other characters are acceptable alternatives.

To load the ASCII file into the database, return to the **Load** main menu and select **Load**. If all goes well there will be a message that a certain number of records were written into the table with zero exceptions. The number of loaded records should correspond to the number of records in the ASCII file.

Now there is a runlog available for use <test.rlg>. To run Load with the runlog enter <load-itest.rlg> from the operating system command line. Use the spacebar and enter keys to move through the sequence of commands.

Summary of procedure: Example 2.3
Initial data requirements: ASCII data file of collar records, a Techbase database with a flat table containing fields for hole identity, location and depth. Data files for assays, surveys and topography are also provided.

Procedure:
1. Place a runlog in the project directory:
   From the Techbase main menu **Techbase** ⇒ **Load** ⇒ **Options Runlog**
      NEW FILE NAME: <test.rlg>
   ⇒ **Output name**
      FILE NAME: <test.out>
   ⇒ **Message output** <yes>
   ⇒ **eXit**
2. Setup for loading data from ASCII **Setup** ⇒ **Data file**
      FILE NAME: <collars>   SKIP: <5>

⇒ **Fields**
<hole_id *x y z* depth>
⇒ **Parameters**
UPDATE TYPE: <append> (or replace)
3. Load data **Load**
(Repeat Steps 2 and 3 for each ASCII data file.)

2.5 CREATING A LYNX DATABASE

In Lynx, the process of defining a database is much simpler than in Techbase due to a more structured set of default project variables. Only those variables which are not among the default key words (not to be confused with Techbase Key fields) need to be defined. The disadvantage of Lynx's highly structured approach to geologic data is a loss of flexibility in importing and manipulating data within the database. Lynx relies on the import and export of ASCII files and the availability of Unix editor facilities such as AWK, ED, VI and SED to process data files. Both Techbase's and Lynx's data management strategies are well suited to their environments, DOS versus Unix.

Lynx database management will be illustrated by importing a data set into a new project. The data consist of a net of vertical wells which have been logged during drilling and casing for sediment type and subsequently measured for soil moisture content using a neutron probe with a 30 cm sampling interval. The data has been **Reported** from Techbase into a set of three ASCII files: collars, counts and lithology. The collars file contains the well id and *xyz* coordinates of the hole collar. Both the counts and lithology files are data files consisting of the well id, from and to intervals and corresponding log value for either neutron count or sediment classification as a string description and as a two digit code assigning a sediment class and horizon. ftp was used to transfer the ASCII file from the Techbase project directory to the home directory on the Unix workstation where they were converted to UNIX files using the dos2ux command and then modified and combined into a single data file that follows the required Lynx format for ASCII file drillhole import.

ASCII drillhole files that are to be imported into Lynx must follow a strict format. The file must consist of two types of records. Type 1 records define the format of subsequent Type 2 data records and must include certain predefined variable names. There are potentially three different Type 1 records, those defining either collar (header), survey or data records.

2.5.1 *Type 1: Header records*

1 [classname] Hole Category North East Elev Length Region (any other variables)

The first three entries must follow this example with 1 indicating that this is a format or Type 1 record, classname taking on any name such as collar or header, and HOLE being used to store the well id. The remaining names in the list can follow any order but must include at least NORTH, EAST and ELEV. If additional information is available on the type of drill, then CATEGORY is used to store a field consisting of up to two characters to represent a value such as DDH or reverse rotary. If the drilling was done in a specific area or time, then this field can be stored as REGION. Any other variables which are relevant to collar data (hole specific) can be included on this HEADER line.

2.5.2 *Type 1: Survey records*

1 [classname] HOLE DEPTH AZIM DIP (any other variables)

Again, the first three entries must be included as shown with classname being any suitable name such as survey or inclination. All capitalized entries must be entered as shown but DEPTH, AZIM and DIP can be in any order and can include additional survey specific field names. If the holes are vertical, this record class can be disregarded.

2.5.3 *Type 1: Data records*

1 [classname] HOLE FROM TO (list of assay and/or geologic variables)

There can be as many classnames and corresponding Type 1 data format records as there are intervals over which data has been recorded. Generally, these records will contain assay and geologic rock unit data. The first three entries are given as shown with a classname which corresponds to the type of data contained in the subsequent data record (assay, lithology, counts, etc.). FROM and TO must be included or the default variable names FROM, THICK or DIST can be used if appropriate. Note that Techbase would require a different field name for any from-to field associated with a different interval, but in Lynx FROM and TO are used for all record class types, either for survey or for data records regardless of interval length.

Type 2 records contain data with the variable values entered in the same order as given in the corresponding Type 1 records. The first entry on the line is a 1 followed by the same classname used in the Type 1 record. All subsequent data values follow the same order as given in the Type 1 record. Consider the example of the well data. The Type 1 records would be as follows and consist of a header record and two data format records:

```
1 HEADER HOLE EAST NORTH ELEV
2 HEADER  A-1   853.40    922.89    21895.15
2 HEADER  A-3   853.40   1122.83    21895.15
2 HEADER  A-5   853.40   1322.77    21895.15
1 COUNT HOLE FROM TO counts id
2 COUNT   A-1   0         30         1019   11
2 COUNT   A-1   30        60          858   11
2 COUNT   A-1   60        90          901   11
1 LITH HOLE FROM TO sed id
2 LITH    A-1   0.00      160.00     ss-ms  11
2 LITH    A-1   160.00    510.00     ms-cs  21
2 LITH    A-1   510.00    610.00     cs-gr  71
```

The default variables must be entered as shown, but the three classnames (HEADER, COUNT and LITH) can be any descriptive name. Note that the Type 2 records use exactly the same classnames, are preceded by a 2 and contain any additional variable values other than the capitalized default variables. Type 1 and 2 records can appear in any order in the data file. Value entry is free format using spaces as delimiters.

Example 2.4: Lynx Project Create, Define, Select and Copy

– ***Project Create***: Execute Lynx by entering <lynx -main> from your home directory. From the main window select **Project ⇒ Project Create** using the pull down menus to display the Project Create entry form as shown in Figure 2.3. Enter the project name, description, units of measurement and dip convention as shown in Figure 2.4. Selecting <OK> will result in Lynx generating a set of standard project sub-

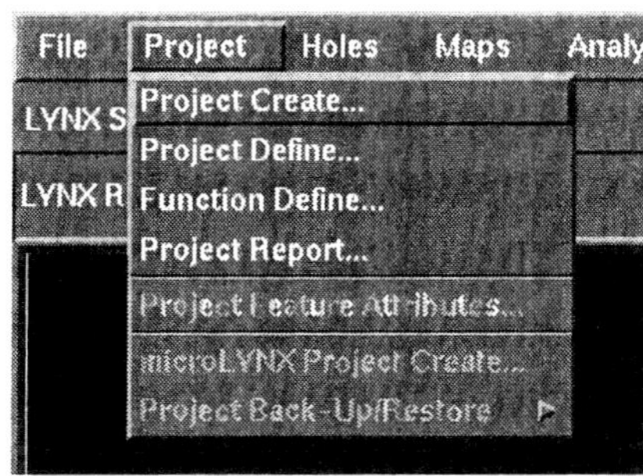

Figure 2.3. Project pull down menu.

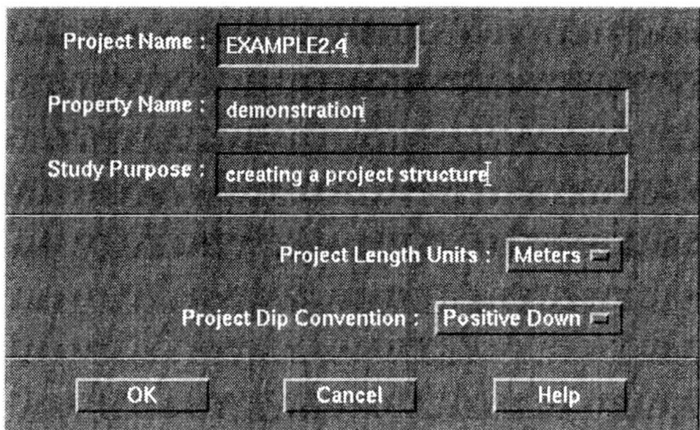

Figure 2.4. Project create entry form.

directories off of the path: /apps/lynx/projects/EXAMPLE 2.4 where EXAMPLE 2.4 is the project name entered during project definition. The directory structure is always as follows under EXAMPLE 2.4:

3d/	dh_prj.INDX	maps/	overlays/
designs/	dholes/	misc/	wave/
dh_prj	geostats/	models/	

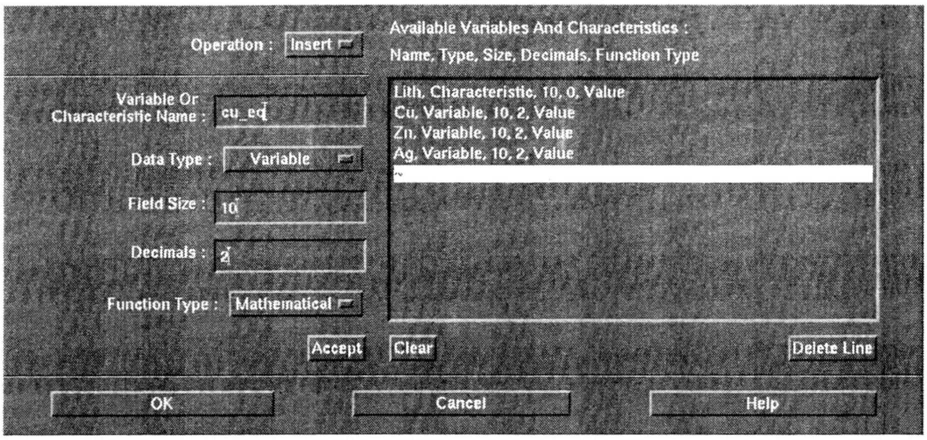

Figure 2.5. Project define entry form (TUTORIAL).

– ***Project Define***: To enter variable definitions select **Project ⇒ Project Define** and the Project Define menu will appear as in Figure 2.5 which shows the various user defined variables that are used in the TUTORIAL project. Each project variable or characteristic is defined in terms of a name, type (variable or characteristic), field size, decimals and function type (value, mathematical or logical). The functions or logical relationships are not defined in this window, but under the Function Define window which is explained next. Note that only non-default variables should be defined, such as assay or geologic measurements. The types of default variables created with every project were discussed in the preceding section on Type 1 data records. Selecting <OK> will result in the display of the Project Define Confirm menu. Select <OK> to include these variable definitions in the new project.

– ***Function Define***: The list of variables in Figure 2.5 include a total metal equivalent value <cu_eq> that will be used later for calculating net mining block values (see Example 6.5 for more details). To define a function select **Project Function ⇒ Define** (see Fig. 2.3) and the Function Define window will display. Select cu_eq from the list of available functions followed by **Define** to display the Mathematical Function Define window as in Figure 2.6 which also shows the function definition:

cu_eq=cu+.48zn+.009ag

Select <OK> and confirm the function definition. Function definition can always be changed at a later date as would be necessary for cu_eq as metal prices change.

– ***Project Select***: Project EXAMPLE 2.4 now exists, even though it has no data. Generally, the first step in a Lynx session will be to open an existing project. To open an existing project select **File ⇒ Project Data ⇒ Project Select** as shown in Figure 2.7 and the Project Selection window will be opened, listing all of the project subdirectories under the Lynx projects directory.

– ***Project Copy***: Its always good practice to work from a copy of the example databases rather than risk corruption of the original files and the resulting confusion. Alternatively, a new project can be created that contains or is based on a selection of data from another project. Before proceeding with copying a project, the project into

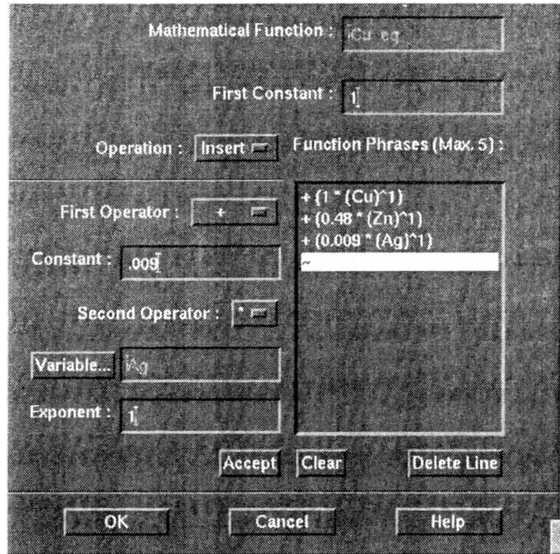

Figure 2.6. Math function define for cu_eq (TUTORIAL).

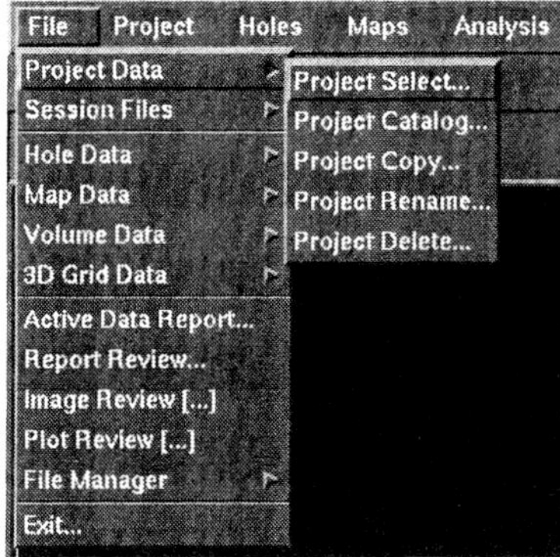

Figure 2.7. Project data pull down menu.

which it is to be copied must have already been created along with variable definitions matching the data to be copied over from the original source project. To copy a project select **File ⇒ Project Data ⇒ Project Copy** as shown in Figure 2.7 to display the Project Copy entry form (see Fig. 2.8).

The Project Copy entry window shows the range of file data that is contained within a Lynx project directory. For the initial stages of a project, only map and hole data will be needed. Selection of **Source Project** will display a list of available projects. Selection of **Variable and Characteristic Matching** opens the Project and Variables and Characteristics Matching entry form shown in Figure 2.9 with the selection process completed for copying the principle variables from the TUTORIAL project. When the variable names used in both projects are the same, Auto Match can be used.

The same procedure can be used to copy drill hole and map data. If all map and

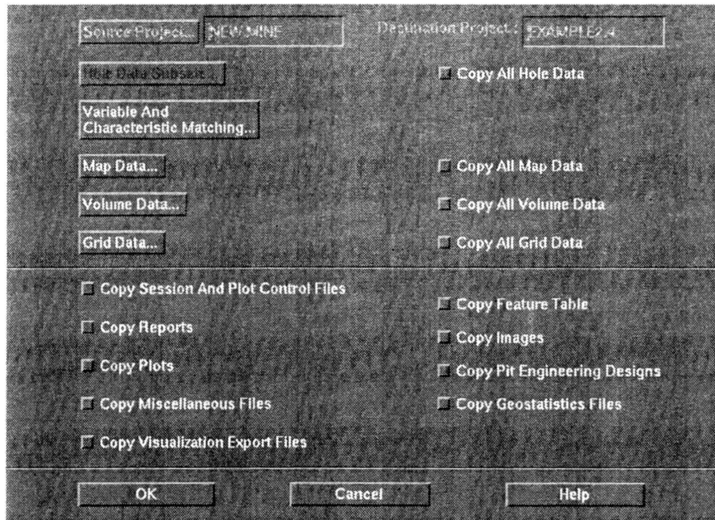

Figure 2.8. Project copy entry form.

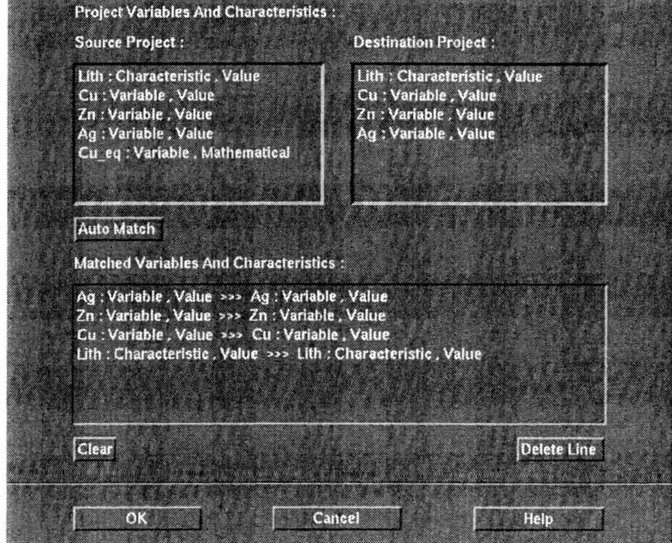

Figure 2.9. Project variables and characteristics matching entry form.

drillhole data are to be copied, then Copy All Map Data and Copy All Hole Data can be toggled on, but this is generally not the case. Selection of the **Map Data** and **Hole Data** push-button will display forms for the selection of specific maps and subsets of hole data. The Copy Feature Table should also be toggled on if map data are being copied since the Feature Table contains the characteristics of the objects that the map is composed of, for instance, contour line type, color and weight. When copying hole data, the subset dh0 generally represents the full data set with other values such as dh1 or dh2 being subsets of dh0. Selecting <OK> completes Project Copy.

Summary of procedure: Example 2.4
Initial data requirements: for project creation and definition, none; for project select, a project to select; and for project copy, an existing source project and a project definition with matching variables.

Procedure:
1. Project creation **Project ⇒ Project Create**
   PROJECT NAME: <EXAMPLE 2.4>
   PROJECT LENGTH UNITS: <meters>
   PROJECT DIP CONVENTION: <positive down>
   ⇒ **OK**
2. Project definition **Project ⇒ Project Define**
   Enter definitions as shown in Figure 2.5.
   ⇒ **OK** (confirm project definition)
   ⇒ **OK**
   Project ⇒ Function Define
   FUNCTION VARIABLE OR CHARACTERISTIC: <cu_eq>
   ⇒ **Define Function** (set up function as shown in Figure 2.6)
   ⇒ **OK**
   ⇒ **OK**
3. Project selection **File ⇒ Project Data ⇒ Project Select** <TUTORIAL>
4. Project copying **File ⇒ Project Data ⇒ Project Copy**
   SOURCE PROJECT: <TUTORIAL>
   HOLE DATA SUBSETS: <dh0>
   VARIABLE AND CHARACTERISTIC MATCHING: (see Fig. 2.9)
   MAP DATA: <topo, sp1, sp2> Copy Feature Table <on>
   ⇒ **OK**

Example 2.5: Exporting and importing ASCII drillhole data from and to Lynx
ASCII drillholes files can be imported into Lynx using the drillhole import/export facilities. As shown in Example 2.4, drillhole data subsets can be copied directly into a new Lynx project, but it may be necessary to copy drillholes across projects at a later date, or to generate an ASCII drillhole file that can be imported into some other application such as Techbase. Before proceeding with export and import, a project must have been created with variables defined that can hold the drillhole data as demonstrated in Example 2.4. Drillhole export occurs from an existing project that contains drillhole data. To export, select an existing project: select **File ⇒ Project Data ⇒ Project Select** and pick the project <TUTORIAL> from the Project Selection List. Before any operations can be done using drillhole data, a drillhole data subset will have to be active. Select **File ⇒ Hole Data ⇒ Subset Select** as shown in Figure 2.10 and select <dh0> from the listing. Subsets allow the data to be defined in the drillhole database as different subsets, that, for instance, might represent annual drill log entries or data with different ownerships. In any case, this initial set of holes can be considered to be raw data or subset 0. Now that a subset is active, the data can be exported by selecting **Holes ⇒ Hole Data Import/Export ⇒ Hole Data ASCII Export** and completing the ASCII Hole Data Export form including the drillhole subset (assumed to be the active subset unless changed at this point), a list of variables to be exported, the destination file name (default path is ./misc under the project

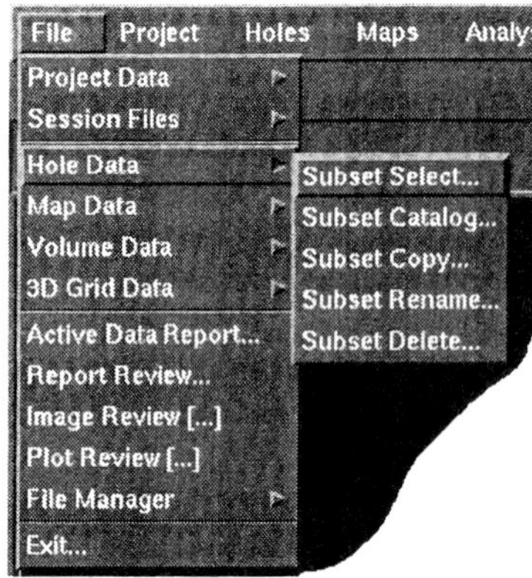

Figure 2.10. Hole subset selection procedure.

subdirectory) and toggle switches for including assay and survey records. The completed form is shown in Figure 2.12.

To import the drillhole data, select the project into which the data are to be imported and create a new active drillhole data subset. If there are no existing data this can be <dh0> which will have to first be created by selecting **Holes ⇒ Subset ⇒ Create** (see Fig. 2.11). The new subset will now be active and available to import data into following the same procedure as for drillhole data export, i.e **Holes ⇒ Hole**

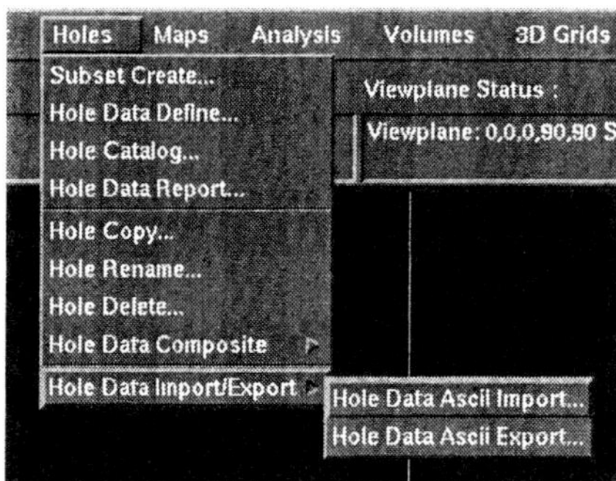

Figure 2.11. Hole data ASCII import/export selection.

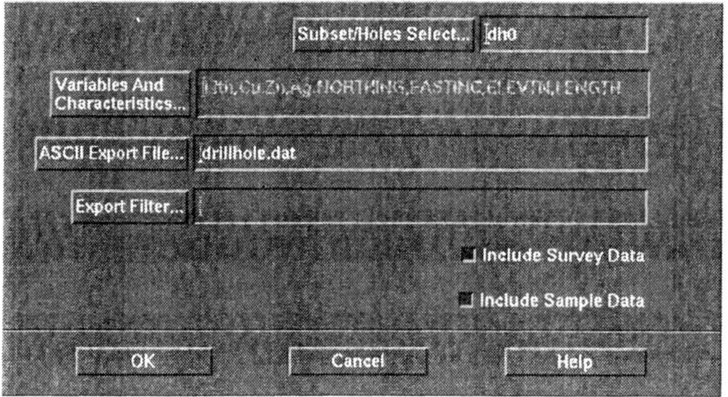

Figure 2.12. ASCII hole data export from project TUTORIAL.

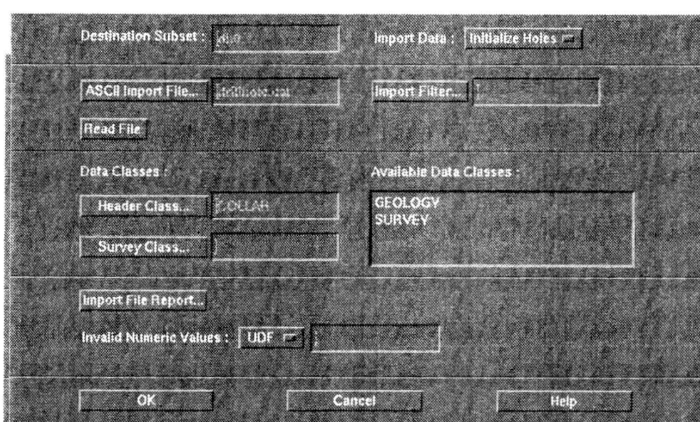

Figure 2.13. Hole data ASCII import for Example 2.4 from dh0 in TUTORIAL.

Data Import/Export ⇒ Hole Data ASCII Import and complete the Hole Data ASCII Import form as shown in Figure 2.13. The drillhole destination subset is the currently active subset <dh0> and can be either Updated or Initialized. In the case of a new subset, either mode will work, but in general, updating is used to overwrite or add new information to existing records, while initialization is used to replace records or read new information. The import file is assumed to be in the misc subdirectory of the destination project and may have to be copied there before it can read. Once read, the Header and Survey classes can be selected from the listing of available data classes. Invalid numeric values can be left undefined <UDF> or replaced. For example, in geochemical assays, measurements below the analytic methods tolerance may be reported as <.001> or as 'trace'. These should be replaced either with a zero or a null entry, '--'. Upon completion of this form, the drillholes will be imported into the drillhole subset.

When lynx has finished importing drillholes, it will generate an error file, import_err, in your projects misc directory. This should be viewed as a check on the import process. When errors occur during import, it may be more helpful to examine the Lynx generated file /tmp/data.no than the ASCII drillhole file to determine the source of import errors. Its also a good idea to check the veracity of the database by examining the drillhole listings in **Holes ⇒ Hole Data Report**.

Summary of procedure: Example 2.5

Initial data requirements: a project containing drillhole data or an ASCII file in the proper format for importing into Lynx and another project for importing data with appropriate variable definitions.

Procedure:

1. Exporting Drillhole Data: (project TUTORIAL)
   1.1 Select a drillhole subset as active **File ⇒ Project Data ⇒ Subset Select** <dh0>
   1.2 Export to ASCII **Holes ⇒ Hole Data Import/Export ⇒ Hole Data ASCII Export**
   SUBJET/HOLES SELECT: <dh0>
   VARIABLES AND CHARACTERISTICS: <select all> or <lith,cu,zn,ag>
   ASCII FILE EXPORT: <drillhole.dat> (created in ./lynx/projects/ TUTORIAL/misc/)
   INCLUDE SURVEY DATA: <on> INCLUDE SAMPLE DATA: <on>
2. Importing Drillhole Data: (project EXAMPLE 2.4)

2.1 Create or select a drillhole subset as active **Holes ⇒ Subset Create** <dh0>
2.2 Import to Lynx **Holes ⇒ Hole Data Import/Export ⇒ Hole Data ASCII Import**
DESTINATION SUBSET: <dh0> IMPORT DATA: <initialize holes>
ASCII IMPORT FILE: <drillhole.dat> (imported from ./lynx/projects/EXAMPLE 2.4/misc/)
⇒ **Read File**
HEADER CLASS: <collars>
SURVEY CLASS: <survey>

2.6 SOME HINTS ON UNIX, TEXT EDITING AND GENERATING LYNX ASCII DRILLHOLE DATA FILES

Source data is often in ASCII format but cannot be expected to arrive in the format required for importing into Lynx projects. The normal format should be similar to the examples given for loading data into a Techbase Database, i.e. three or more files which separately contain collar, survey and assay/geologic data records. It is also likely that the data will arrive on a DOS formatted disk that will have to be transferred to Unix. This discussion starts with using ftp to transfer files across a network.

FTP

1. Connect with a destination Unix server from a PC.
   > ftp 129.101.108.35 (this is an example IP address, check your network map for all other addresses)
   Next, exit to the DOS shell. This can vary. In Unix this is accomplished by entering an '!'. Go to the directory or drive containing the ASCII files and then return to ftp by entering the command 'exit.'
2. Change to binary file transfer mode.
   > binary
3. Change to the desired destination directory on the server. For instance, to move down from your home directory to a data directory called pasir.
   > cd pasir
4. Transfer files.
   > put (ASCII filename)
   The command 'mput' can be used with wildcards such as *.dat to transfer multiple files.

Converting DOS to Unix
ASCII files transferred from DOS will have end-of-line markers such as ^M which need to be removed. The easiest approach is to convert the file to unix format.
> dos2ux oldfile newfile

Inserting the first and second columns of Type 2 data records
Lynx requires that the first two columns of the data file be a 2 followed by the classname. This can be done line by line in the text editor, but assay files can easily consist of thousands of records. Hand entry is not only tedious, but is a sure source of errors. One way to approach this problem is to use the AWK editor as follows (see Aho et al. 1988):

1. Insert the classname (say, HEADER) as the first column of the collars.dat file and output the resulting file as collars.dat.lx.
   > awk '{print 'HEADER', $0}' < collars.dat > collars.dat.lx
   Repeat this for each of the data files with the corresponding classname.
2. Insert 2 into the data file containing the classname.
   > awk '{print '2', $0}' < collars.dat.lx > collars.lynx
   Repeat for each of the data files.

Concatenating data files
Put three data files together into one file and place them into drillhole.dat.
> cat collars.lynx survey.lynx assay.lynx > drillhole.dat
Finally, edit the combined file, drillhole.dat, by entering Type 1 data records as described above.

2.7 WORKING WITH MAP DATA

Data which are best defined with reference to a plane can be stored as a map in Lynx or as a flat or polygon table in Techbase. The distinction between drillhole and map data is a convention used in Lynx around which this discussion will focus.

Topographic data is the most common form of map data in which 2D features called contours are used to represent a line of constant elevation in plan view. But map data is not limited to the horizontal or to 2D data: the map plane is defined by a lower left origin, an azimuth and an inclination so that the plane to which the map data is referenced can be in any position. Features which are part of the map can be points, lines or contours which are defined by their xy position on the plane or features such as survey traverses which are defined both by their xy positions and by offsets taken normal to the plane. Since a map can take any orientation in Lynx (unlike Techbase), maps are extremely useful for working with underground structures such as veins, faults or stopes. There will be ample opportunities to manipulate map data in later exercises.

Techbase does not specifically work with 'map' data, but with one exception treats map data as it would any data. Point or contour type data are loaded into a flat table and all z data manipulation or estimation is based on an xy coordinate in plan view. Constructs such as lines, boundaries or more complex 3D polygons can be stored in a polygon table. Two-dimensional estimated data are stored in cell tables, but are also based on a strictly horizontal coordinate system.

The remainder of this section will focus on Lynx map definition and importing map data into Lynx from ASCII files.

2.8 LYNX MAP DEFINITION

The map plane is defined by its origin's (lower left hand corner) northing, easting and elevation (NEL), azimuth (clockwise from north) and inclination (positive, counterclockwise). Thus, a plan view map would be defined as having the orientation: 0,0,0,90,90. Two coordinate systems are used in Lynx: local and global. The global coordinate of a map is based on an origin of 0,0, whereas a local coordinate system can be defined based on a relative position from some important design feature of current interest, such as a shaft centerline. All data in the map are referenced to the map origin by local *xy* coordinates in the map plane. Offset data are taken as being normal to the plane, centered over *xy* and positive out of the plane towards the viewer.

An example of a Lynx map is given in Figure 2.14 in which the map file <sitemap> is displayed. Display of a map in background is discussed in Chapter 1. Note that a map has an orientation, origin and scale as shown in the **Viewplane** ⇒

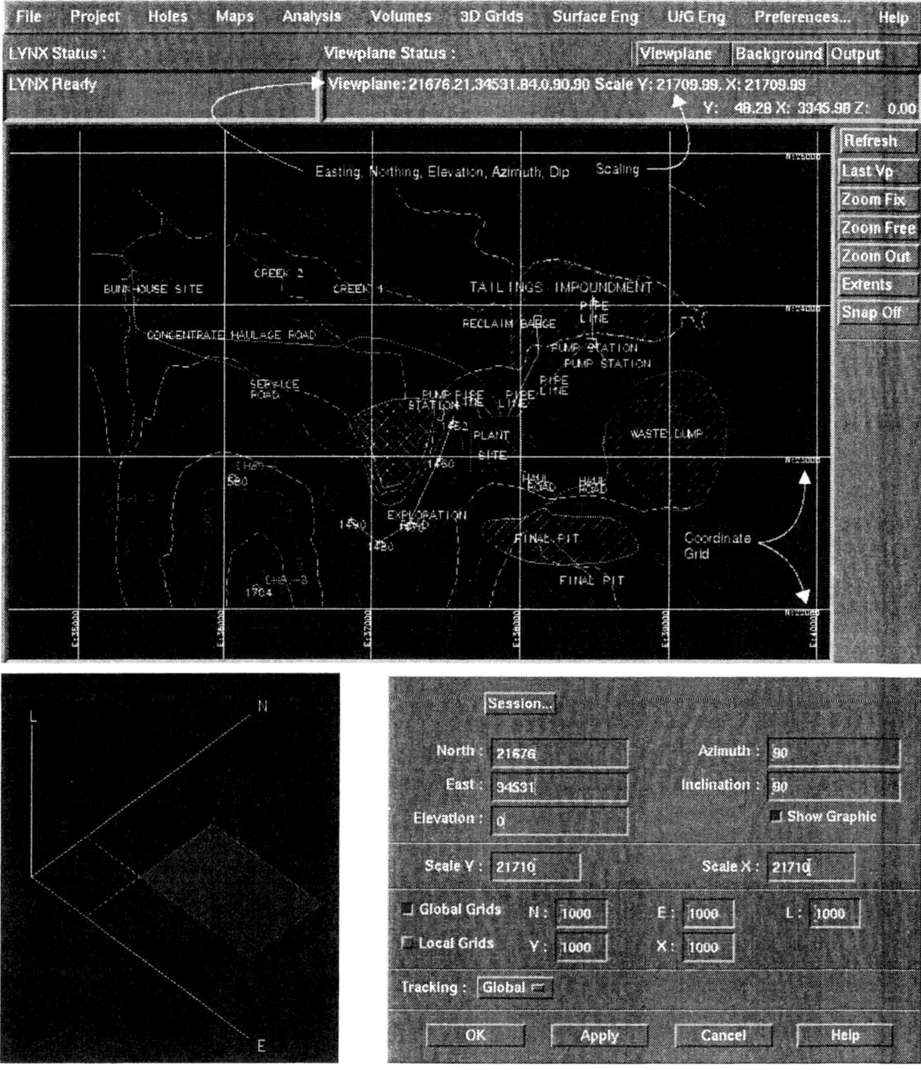

Figure 2.14. Map viewplane setup and orientation. (Figure in colour, see opposite page 54.)

Setup window of Figure 2.14, as well as other characteristics in display such as scale, global and local grids. All data stored in a map are associated with map features. Lynx includes a wide variety of map features having predefined attributes. A number of features are used in the map shown in Figure 2.14 including major and minor contours, text, traverses, points and lines.

2.9 IMPORTING ASCII MAP DATA

To illustrate the nature of map data, Lynx's ASCII Map Data Export has been used to generate an ASCII data file of the surface topography for the TUTORIAL project. The layout of the ASCII map is similar to that used for drillhole data with Type 1 and 2 records being used for format definition and data, respectively, in the first column. The second column is reserved for the class type which includes map header (HEADER), map origin (ORIGIN), group table header formats (HFORMAT) and map data which can have group names which are either default feature types or user defined. HFORMAT records are not necessary unless the defaults for displaying a data feature are not to be used. The following records are taken from the exported TUTORIAL ASCII map file TOP.

1 HEADER NAME DESCRIPTION
2 HEADER TOP 'Thinned topographic map for pit design'
1 ORIGIN NORTH EAST ELVN AZIM INC
2 ORIGIN 0 0 0 90 90
1 HFORMAT GROUP PEN LINE FLABEL FILL SYMBOL PLABEL LDIR CSIZE PDIM FDIM FVALUE F FFLAG FDESC
2 HFORMAT cmin 3 2 1 0 0 0 0 20 2 1 1 0 'minor contour *x,y* + var. (required)'
2 HFORMAT cmaj 1 1 1 0 0 0 0 25 2 1 1 0 'major contour *x,y* + var. (required)'

The HEADER lines require a map NAME and DESCRIPTION enclosed in quotes. The ORIGIN includes the attributes: NORTH, EAST, ELVN, AZIM, and INC to be provided on the corresponding Type 2 record. The HFORMAT Type 1 record can be included if the pen colors, line types, or any other default display attribute is to be changed for any of the data groups. In this example, the major (cmaj) and minor (cmin) contour group format parameters are being redefined in the data file. HFORMAT records are then followed by data records for various groups. Examples of these groups are as follows for the default group names cmin and cmaj:

1 cmin FEATURE NUMBER *x y* Elvn
2 cmin 1 1 37599.398 22683.33 1540
2 cmin 1 2 37635 22709.75 1540
1 cmaj FEATURE NUMBER *x y* Elvn
2 cmaj 1 1 37599.141 22653.221 1550
2 cmaj 1 2 37620.32 22671.209 1550

The attributes FEATURE and NUMBER give the line and point number for the contour. *x*, *y* and Elvn are the coordinate and offset values on the map plane. Note that the group name defines the manner in which the map feature will be displayed as well as the data itself. The format of map features can be modified later within Lynx if desired. Predefined group types include the following:

Group attributes table catalog.

#	Name	Description
0	–	
1	cmin	minor contour x,y + var. (required)
2	cmaj	major contour x,y + var. (required)
3	trav	traverse x,y,z (required)
4	line	line x,y (required)
5	point	3D point (x,y,z) + vars. (required)
6	pnt2d	2D point (x,y) + variables
7	ptxy	2D point (x,y) no variables
8	ptxyz	3D point (x,y,z) no variables
9	isogr	isograd (x,y,z) + variables
10	txt1	text Type 1
11	txt2	text Type 2
12	txt3	text Type 3
13	wall	u/g survey features wall shot (required)
14	sill	u/g survey feature sill shot (required)
15	pcrst	conical pit crest line (required)
16	crest	conical pit bench crest line (required)
17	toe	conical pit bench toe line (required)
18	road	in pit haul road (required)

Example 2.6: Export/import of ASCII file map data to and from Lynx
Map data will be exported as an ASCII file from project TUTORIAL and then imported into EXAMPLE 2.4. From the main Lynx menu have the TUTORIAL project active and any map data also selected as being active so that the map menu will be available. Select **Maps ⇒ Map Data Import/Export ⇒ Map Data ASCII Export** as shown in Figure 2.15 to bring up the Map Data ASCII Export form of Figure 2.16. Enter the source map that is to be exported <SiteTopo> and the ASCII export file name <sitetopo.asc>. Additionally, export the map feature's attributes if the file is to contain a record of the attributes of the features such as the text fonts and line types. The exported text file will be in the misc directory.

Next the ASCII map data will be imported into project EXAMPLE 2.4. After opening this project select **Maps ⇒ Map Data Import/Export ⇒ Map Data ASCII Import** as shown in Figure 2.15 to bring up the Map Data ASCII Import form of Figure 2.17. As in the case of exporting data, the file to be imported is assumed to reside in the misc directory of the host project. Enter the source map <SiteTopo>, select the Overwrite Existing Maps toggle if an existing map in the project is being replaced. If the feature attributes are to be taken from the project FAT select project, but if the features in the source project where the map was created were different and these attributes were exported with the ASCII file and are to be retained, then select ASCII file.

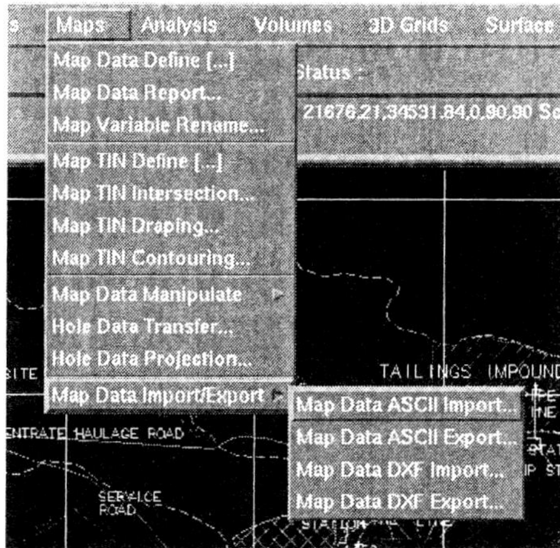

Figure 2.15. Map data ASCII import/ export selection.

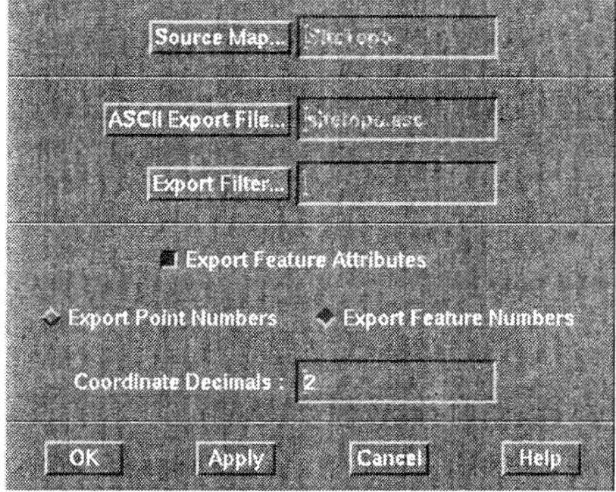

Figure 2.16. Map data ASCII export form for TUTORIAL site map.

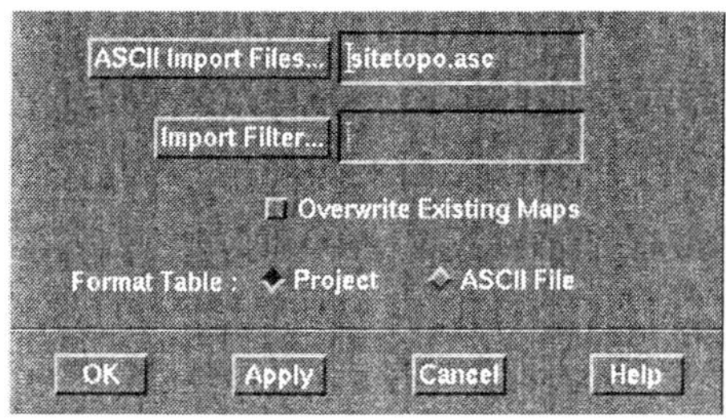

Figure 2.17. Map data ASCII import form for TUTORIAL site map.

Map data can also be imported and exported in the DXF format common to many CAD packages by following the same procedure as above (see Fig. 2.15).
Check to see that the map was loaded by looking for its name in the map catalogue listing by selecting **File ⇒ Map Data ⇒ Map Data Catalogue**. You can also use **Map Data Report** to check the map's contents by selecting **Maps ⇒ Map Data Report** as shown in Figure 2.15 and picking the map from the selection list. You should see something similar to Figure 2.18.

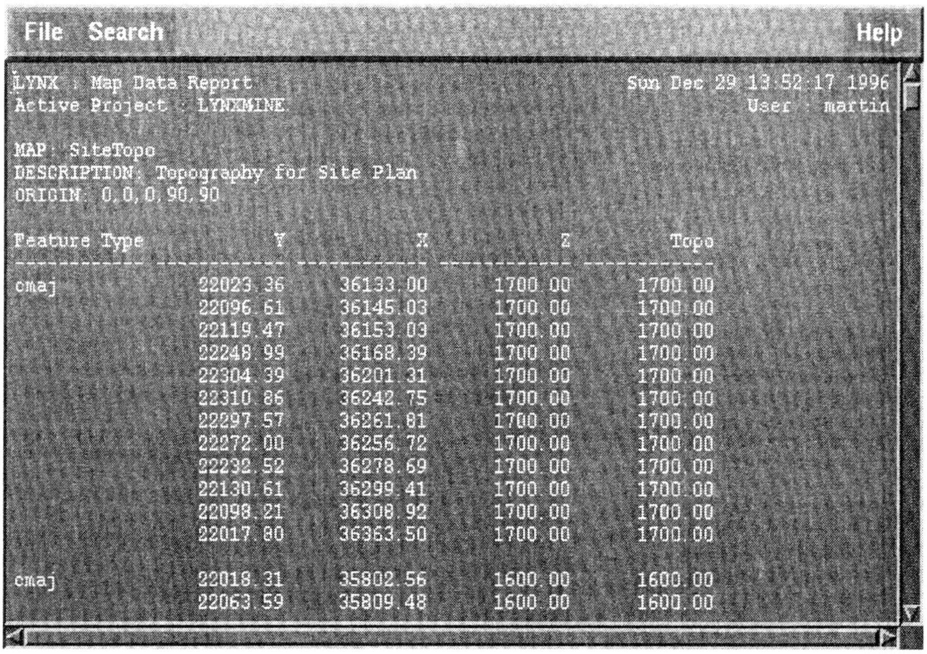

File Search Help

```
LYNX : Map Data Report                              Sun Dec 29 13:52:17 1996
Active Project : LYNXMINE                                      User : martin

MAP: SiteTopo
DESCRIPTION: Topography for Site Plan
ORIGIN: 0,0,0,90,90

Feature Type           Y           X           Z        Topo
------------ ----------- ----------- ----------- -----------
cmaj            22023.36    36133.00     1700.00     1700.00
                22096.61    36145.03     1700.00     1700.00
                22119.47    36153.03     1700.00     1700.00
                22248.99    36168.39     1700.00     1700.00
                22304.39    36201.31     1700.00     1700.00
                22310.86    36242.75     1700.00     1700.00
                22297.57    36261.81     1700.00     1700.00
                22272.00    36256.72     1700.00     1700.00
                22232.52    36278.69     1700.00     1700.00
                22130.61    36299.41     1700.00     1700.00
                22098.21    36308.92     1700.00     1700.00
                22017.80    36363.50     1700.00     1700.00

cmaj            22018.31    35802.56     1600.00     1600.00
                22063.59    35809.48     1600.00     1600.00
```

Figure 2.18. Report of map <sitetopo> from project TUTORIAL.

Summary of procedure: Example 2.6
Initial data requirements: A project containing a map to export and another project in which the map can be imported.

Procedure:
1. ASCII Map Export from the source project <TUTORIAL>
   Select Maps ⇒ Map Data Import/Export ⇒ Map Data ASCII Export
   SOURCE MAP: <SiteTopo>
   ASCII EXPORT FILE: <sitetopo.asc>
   EXPORT FEATURE ATTRIBUTES: <on> (optional)
   EXPORT FEATURE NUMBERS: <on>
   (either feature or point numbers can be toggled on)
2. ASCII Map Import into the destination project <EXAMPLE 2.4>
   Select Maps ⇒ Map Data Import/Export ⇒ Map Data ASCII Import
   ASCII IMPORT FILE: <sitetopo.asc>
   PROJECT: <on> (alternatively, ASCII FILE)
   OVERWRITE EXISTING MAPS: <on> (optional)
3. Map Reporting: select **Maps ⇒ Map Data Report** and select <SiteTopo>

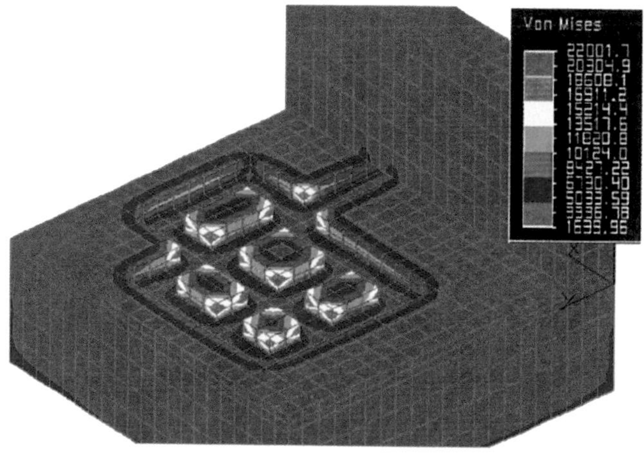

Figure 1.1. Finite element modelling of stresses about a 3D mine opening.

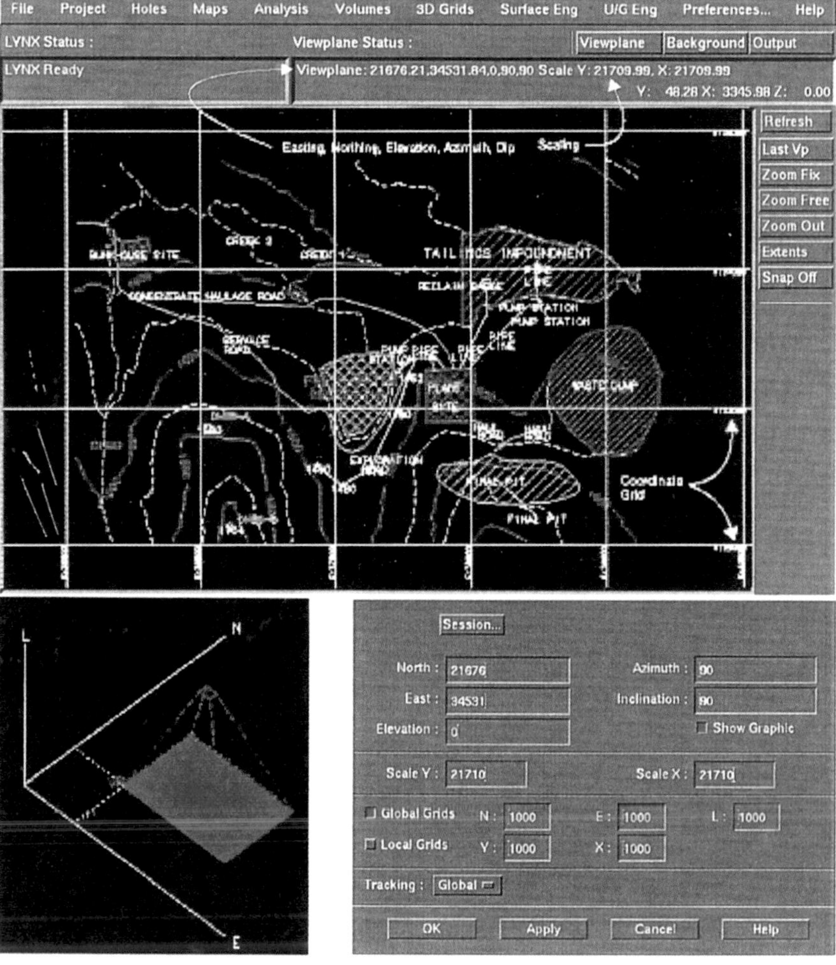

Figure 2.14. Map viewplane setup and orientation.

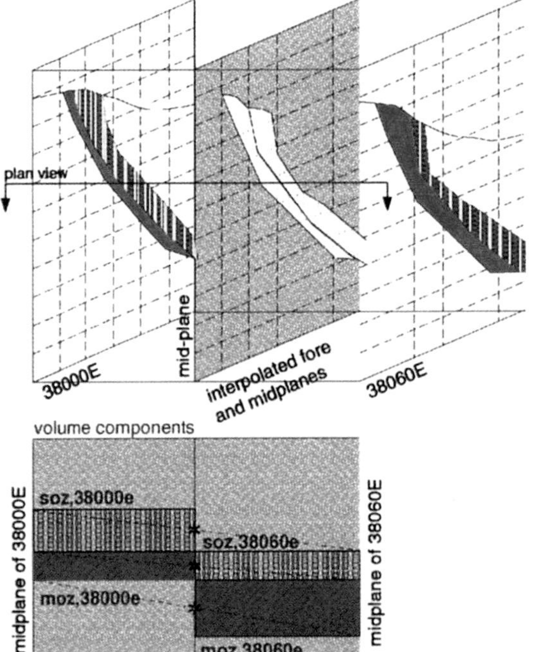

Figure 4.33. Linear interpolation of fore and back plane polygons of adjacent volume components.

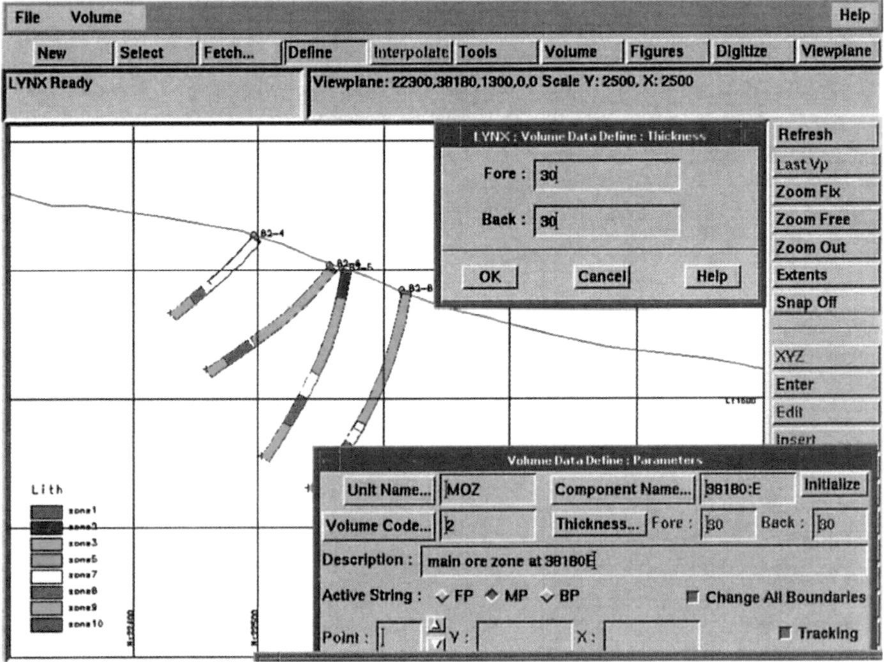

Figure 4.35. Volume data define menu and parameter entry form with color map display of drillhole lithology at section 38180:E.

CHAPTER 3

Data analysis

3.1 INTRODUCTION

Geologic data differ greatly from data generated in the experimental sciences in that geologic data are fragmentary manifestations of processes which are hidden or, at best, imperfectly understood. A laboratory experiment can be controlled to generate data values which correspond to known conditions; the experiment can be statistically designed to evaluate the response of a variable to changes in parameters such as heat and pressure. By comparison, geologic data are the realization of a multitude of past events and gradual processes which have come and gone over the eons. It is the geologist's task to examine the evidence left in the rock to uncover what these geologic events and processes might have been, especially in relation to the extent, distribution and concentration of valuable minerals.

Statistics can provide guidance at an early stage of exploration to determine how sampling should proceed, what data should be collected, how the data should be collected, how many samples are needed and from where. Unfortunately, those responsible for data collection and reserve estimation are rarely the same individuals. While the estimator must have a keen understanding of statistical methods, the geologist often does not, and leans heavily towards a qualitative description of a deposit and its setting. But only quantitative data are passed from the geologist to the estimator so that at the critical stage of reserve estimation the intuitive understanding of the deposit gained by the field geologist is not passed on to the estimator. Often, these two professionals will never even meet. As a result, the data collected are rarely adequate and may even be misleading to an estimator who may commence a reserve evaluation without the vaguest idea of what the deposit 'looks' like. For this reason, statistical inference and graphical displays of the data must be used to familiarize the estimator with the geology of the deposit and to help avoid gross errors during interpretation and estimation.

Another complication arises in geologic data which are not present in the experimental sciences and are therefore ignored in classical statistical modelling and inference: Mining and other earth science data are by their nature spatial. In the previous chapter, all of the data that were loaded into the geologic database were associated with position either as an offset from a plane or along a drillhole trace. The spatial nature of geologic data presents special challenges that are not considered in classical statistical analysis. The most common characteristics of spatial data that require specialized methods of analysis are that the data are not random, independent or normally distributed: the assumptions that most statistical modelling and testing algorithms are based on. Although methods for working with spatially dependent data have been developed within the classical statistical community, an offshoot of statistics, termed geostatistics, is currently the dominant method of data analysis, geoscience modelling and ore reserve estimation. Mining and geological modelling software are not heavily oriented towards statistical estimation, but generally do provide for basic statistical analysis and geostatistical estimation. Even when the software being used is inadequate for statistical analysis, the data of interest can be exported to a statistical package. It is not the purpose of this text to enter into a detailed review of either exploratory data analysis, statistics or geostatistics. Instead, a few useful analytic methods will be presented with the context of geologic modelling and estimation.

A comprehensive data analysis must be carried out as part of any new project. Failure to do so will inevitably result in poor estimation and lost time at a later date as previous data entry errors are being resolved. Preliminary data analysis should include the components reviewed in Sections 3.1.1-3.16.

3.1.1 *Data validation*

Survey, map and drillhole data for a single project can include tens of thousand of records that originate from numerous sources and were recorded by many different individuals. There are numerous opportunities for errors, such as incorrect recording of the data in the field, analytical error during assaying, and keypunching or digitizing mistakes when data are transferred from paper to computer files. A goal of exploratory data analysis is to identify data records whose field values are suspicious in their magnitude and to eliminate duplicate records.

3.1.2 *Exploratory analysis*

More often than not, the person responsible for building a geologic model is not the geologist who collected much of the data and has field experience with the deposit, yet a certain degree of familiarity with the data is a necessary part of geologic modelling. This must go beyond the basic database building requirements of the types of fields, maps or tables which need to be included in the structure of the geologic database. Additionally, it is necessary to be able to relate the distribution of the data to the underlying geology. For instance, what are the average grades, thicknesses, assay variability and continuity of the variables within the different lithologic units; where are areas that have been intensely sampled or that are under sampled; are there trends that are apparent in the data; where's the high and low grade material? In summary, do not embark upon a new project blindfolded. Don't expect the computer – no matter how elaborate the software – to do your job for you. Have an intrinsic understanding of the data so that you will be in a position to recognize errors as they occur in modelling and estimation.

3.1.3 *Outliers and population classes*

The characteristics of geologic units and the variables being measured are often highly correlated. While drillholes may be sampled and assayed for their entire length, high grade ore may only be found in one or two lithologic units. For instance, when modelling ground water flow through soils, the size distribution of different sediments is often the controlling factor in transmissivity. In some cases, what is and is not 'ore' may be obvious, as would be the case for coal seams or a metallic vein with distinct boundaries, but when the distinction between barren rock and an ore zone is not distinct, basic statistical analysis of the distribution of assays (or any other variable of interest) can reveal distinct population classes which are tied to lithologic units. Failure to distinguish between these classes during modelling and estimation will result in high estimation error and confusing results.

Outliers are data values that lie well outside of the distribution of the variable, typically at values in excess of three standard deviations from the mean. Outliers can be identified during the process of checking the population distributions and should be noted. Outliers are important for two reasons: they may be a data error or an isolated and undersampled zone of anomalous grade. In terms of assigning population classes, a large number of outliers may indicate that the population is actually composed of more than one population. Closer examination of the geologic log may reveal that these outliers belong to samples which were placed in the wrong lithologic classification. If there is no way of explaining the origin of an outlier, additional nearby samples are called for. Outliers should never be discarded without further investigation. This is especially true of populations which are highly skewed such as in the case of a lognormal distribution. Lognormal data should be normalized first before a check for outliers can be made.

3.1.4 *Distributions*

The empirical distributions of the variables of interest should be examined for a number of practical reasons. As part of exploratory analysis, examination of the histograms and cumulative frequency plots provides the best means of becoming familiar with the data. In terms of data validation and identification of population classes, the existence of outliers or mixed population classes can be identified. The distributions of values taken from two potentially different geologic zones can be compared to reevaluate the need to treat the zones as separate populations during estimation. Also, the shape of the distribution can have important consequences as to the methods which should be used for global reserve calculation and geostatistical estimation.

3.1.5 *Correlations*

Geochemical/geophysical data sets commonly have sets of variables which display significant correlation. Correlation typically reflects underlying chemical or physical relationships. These relationships may be obvious, such as a positive correlation between lead, zinc and silver assays in a sphalerite/galena deposit, but the mechanics of the correlation might be less obvious and in their study may lead to a deeper understanding of factors controlling the spatial distribution of the variables, an example of which would be a positive correlation between pre-deposition placer topography and gold concentration. Additionally, when the primary variable of interest is difficult or expensive to measure, there may be correlated variables that are readily available and inexpensive to measure. Inclusion of the correlated variables as part of the estimation of a more sparsely sampled primary variable can greatly improve estimation quality and actually reduce the cost of exploration.

3.1.6 *Spatial covariance*

Earth science data are inherently spatial and must be sampled, analyzed and estimated using methods which account for the influence of location on the variable. For this reason, variables taken from geologic data sets are referred to as being 'regionalized' variables. Regionalized variables are autocorrelated in space, i.e. the value of samples which are taken in closer proximity will be more similar over shorter distances of separation. The underlying principal of geostatistics is that there is an underlying covariance function which describes the covariance of a regionalized variable as a function of distance and direction of separation (see Matheron 1971). Geostatistics attempts to minimize the estimation error by including a model of this covariance function in the analysis and estimation process. The most popular proxy for this covariance function is called the variogram: more on this in Chapter 5.

The remainder of this chapter will focus of the practice of data analysis as outlined above and within the limitations imposed by Techbase and Lynx. A thorough discussion of these topics would require the use of additional statistical and geostatistical software such as Splus (Venables & Ripley 1994), GeoEas (Englund & Sparks 1991) and customized programs. Mining software packages tend to be canned and have only the most rudimentary statistical applications.

3.2 VISUAL DATA ANALYSIS

There are numerous effective ways in which data can be displayed graphically that can greatly speed data analysis. In the following exercises, simple graphical devices will be used to address some of the issues raised in Section 3.1.

3.2.1 *Scatter plots*

Scatter plots are simple *xy* plots of paired data values. Two common uses of scatter plots are for examining data patterns and correlation.

A scatter plot of *x* versus *y* from collars data is a good way of getting a handle on the spatial distribution of drillhole data. Before maps of data locations and contours of topography can be generated, an appropriate map scale and paper size must be determined. Simple summary statistics of the easting and northing can provide use with the minimum, maximum and range of the hole coordinates, but tell very little about where the bulk of the values are located. A simple scatter plot gives all this information at a glance, showing isolated drillholes that might be disregarded for the sake of using a larger scale and smaller plot size. An isolated drillhole location might also represent a data entry error which needs to be corrected.

Example 3.1: Generating a scatter plot in Techbase

A collars table consisting of 105 drillholes is to be displayed as a poster. Preliminary to doing this, the scale, range and map size must be selected.

From the Techbase main menu: **Statistics ⇒ Scatter ⇒ Field Names**.

Enter the field names for the easting and northing as the x and y axis and select **Plot scatter**. Enter the **Axis** labels and type, the style of **Markers** used to post data points in the plot and whether you want a grid, correlation line and statistics included on the plot under **Parameters. Draw** generates the scatter plot shown in Figure 3.1.

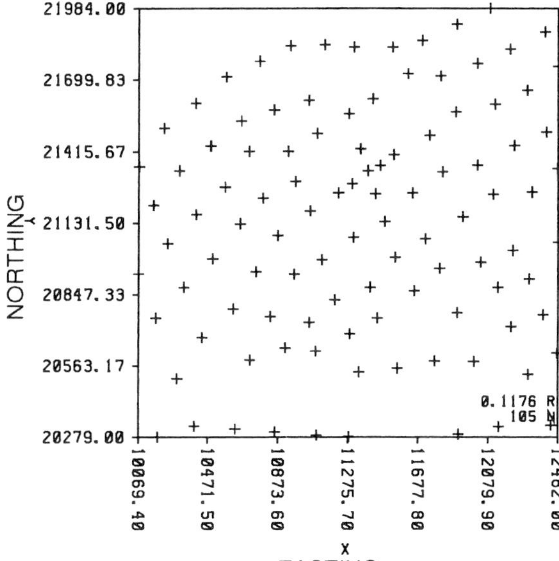

Figure 3.1. Scatter plot of drillhole collar coordinates for the Boland Banya project.

Scatter plots are also a quick way to check for correlation and can even be used to indicate the presence of multiple population classes or changes in correlation as a function of the magnitude of the assay for one or both variables. A correlation coefficient is calculated based on the entire data set and can be strongly influenced by a few outliers. From the scatter plot of Figure 3.2, it is apparent that the assays for gold and silver have a high positive correlation, but that they come from two different sources: in this particular case, from two separate lenses which have been sampled using the same set of drillholes. Note how the two populations skew the regression line. Inclusion of a regression line in the scatter plot is a good means of visually evaluating the impact of potential outliers. In this example, no outliers are apparent, but consider the impression that would be made if the plot was supposed to be of samples taken from only one of the two ore bodies and one of the samples from the other ore body was included due to missclassification of its lithologic unit. In this example, the missclassified sample would appear to be an outlier. Examination of the drillhole logs would lead to correction of this mistake.

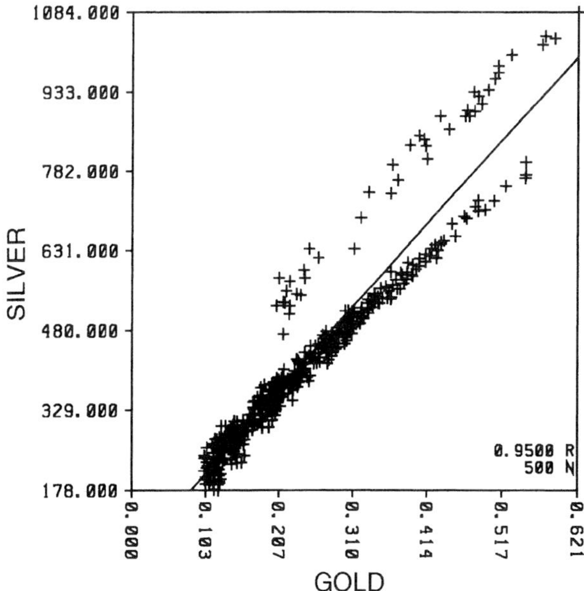

Figure 3.2. Scatter plot of gold and silver assays from the two ore zones of the Boland Banya deposit.

Summary of procedure: Example 3.1
Initial data requirements: a flat table containing collar coordinates.

Procedure: (Boland project)
From the main Techbase menu select **Statistics** ⇒ **Scatter** ⇒ **Graphics** ⇒ **Metafile Name**
FILE NAME: <holes.met> (entering .met optional)
⇒ **eXit** ⇒ **Field names**
X FIELD NAME: <x> Y FIELD NAME: <y>
⇒ **Plot Scatter** ⇒ **Parameters**
GRID: <no>
CORRELATION LINE STYLE: <blank>
⇒ **Draw** ⇒ **Graphics** ⇒ **Metafile Review**

3.2.2 *Histograms*

The shape of the samples' distribution can provide important insights as to the nature of the variable being measured and is a good starting point for data validation, checking for outliers and population classification. Since most statistical methods are based on the assumption of a normal distribution, the validity of this assumption must be checked before calculating even such basic measures as the mean and variance of an assayed grade. It is common for assay data to be highly positively skewed. Gold assays are often lognormally distributed, and use of normal statistical procedures for calculating statistics and grade estimation can result in serious bias (cf., Sichel 1966).

As an example, consider the summary statistics of the 4618 au assays from the Boland Banya project.

Number 4618	
Mean	0.03501
Std Dev	0.08149
Variance	0.007
Maximum	0.621
Minimum	0.000
Range	0.621
Coef Var	232.7494
Std Err	0.0012
Median	0.0050
Mode	0.000
Skewness	3.3453
Kurtosis	11.9386

Note that while the mean is .035 oz/st, the standard deviation is .08. For a normal distribution, most of the sample data values should fall within the mean assay + or – three standard deviations (–.046 to .116), which in this case would make no sense at all. Obviously, the distribution is highly positively skewed; otherwise, there would be negative grades. This impression is confirmed by observing the difference between the mean and median. The median is the middle value if the assays are ranked from lowest to highest. For normally distributed data, the median and mean should coincide, but in this example, the median is far less than the mean. The mode is the most probable value that will be observed. Here the mode is nearly zero, implying that the vast majority of assays are near zero even though the maximum assay is .06.

Skewness and kurtosis are measures of the shape of the distribution. If the distribution is perfectly normal, then the mean, median and mode will be equal. Otherwise, the distribution will be either positively or negatively skewed. In this example, the distribution has a high positive skewness of 3.3, implying that the distribution has a long positive tail of infrequent but comparatively large values. Kurtosis is a measure of the 'peakedness' of the distribution, i.e. how high the the proportion of values near the median (at the high point of the distribution) are in comparison to the rest of the higher and lower values. With a kurtosis of 11, a mode of zero and skewness of 3.3, we can expect a distribution which appears very lognormal.

The distribution of au assays taken from all lithologic units is highly skewed to the right with the vast majority of values are near zero (Fig. 3.3a). It should be obvious that the majority of assays are taken from barren samples and that the histogram should be limited to the ore zone (OA). Restricting the histogram to the 465 samples of the main ore zone results in a much higher mean grade and a distribution that while still positively skewed, is not lognormal (Fig. 3.3b). A $\log_e$ transformation of au results in a distribution which is effectively normal (Fig. 3.3c).

Since lognormally distributed data is so common in geologic data, distributions should be checked to see if they are very positively skewed. An easy way to check is to see if the data is actually lognormal is to plot the values on probability paper. Data plotting on a straight line on normal probability paper (in which the *x* axis uses a probability scale, 1-100%, and the *y* axis is linear) is considered to be normally distributed. Data which plots as a straight line on lognormal probability paper (the log of the same *x* and *y* axis) is lognormal.

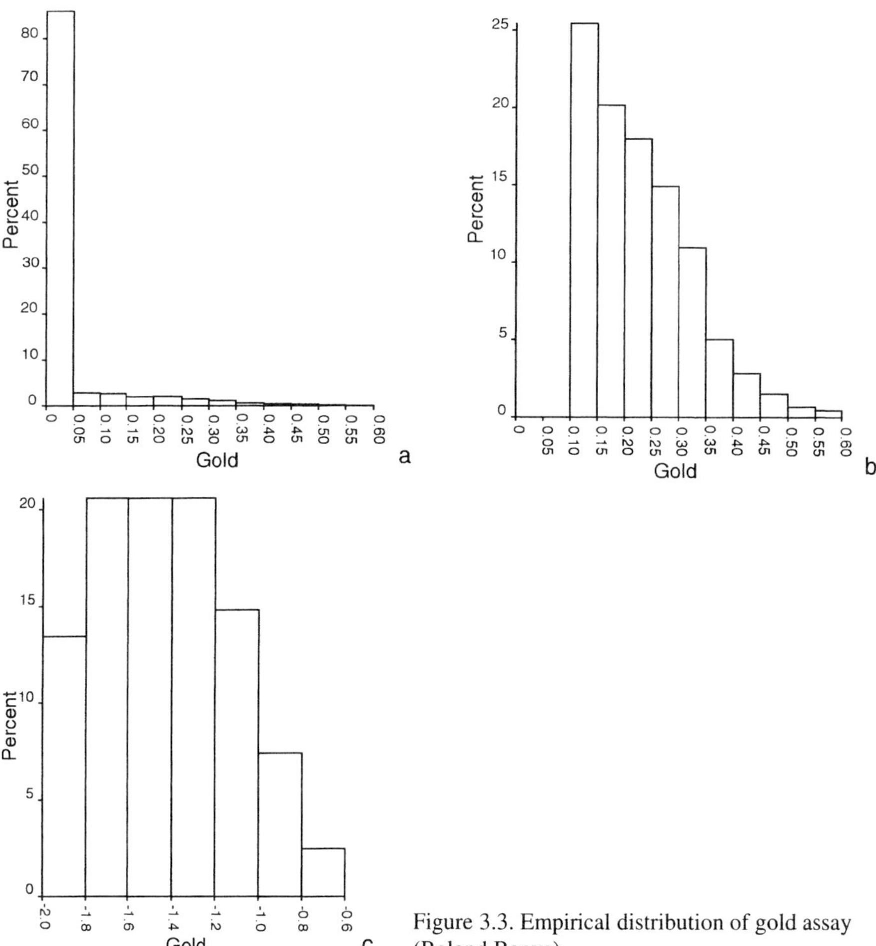

Figure 3.3. Empirical distribution of gold assay (Boland Banya).

Example 3.2: Exploring variable distributions in Lynx

Selected variable distributions from the Expo project will be explored to illustrate Lynx's Data Analysis facilities. The data set consists of shallow soil samples assayed for industrial contaminants. Figures 3.4a and 3.4b show the site (near the GM Stadium in Vancouver, BC) and the drillhole locations.

The soil consists of three types: clean fill, industrial fill and bedrock as shown in the NS section of Figure 3.5.

Before starting the analysis example, select Expo as the active project (**File ⇒ Project Data ⇒ Project Select**). Once the Expo project is active, select **Analysis ⇒ Data Analysis** to open the Data Analysis menu (see Fig. 3.6). Before proceeding with data analysis the data to be analyzed must be selected: **File ⇒ Data Select ⇒ Select hole + Map Data**. This brings up the Data Select: Hole Data & Map Data Entry Form. If a data selection file already exists (with a *.sh or *.dat extension in the /geostatistics project subdirectory), it can be retrieved by selecting **Selection File**.

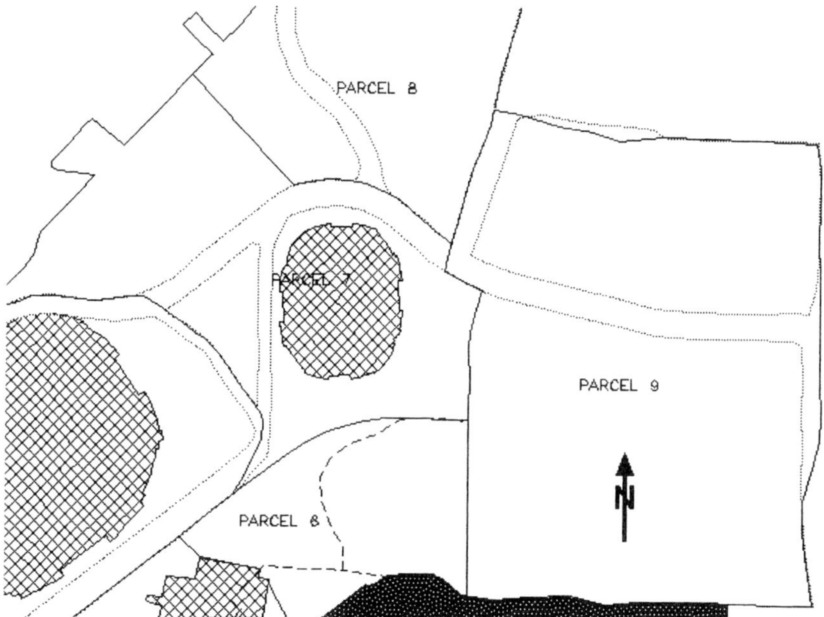

Figure 3.4a. EXPO project site between GM stadium and river.

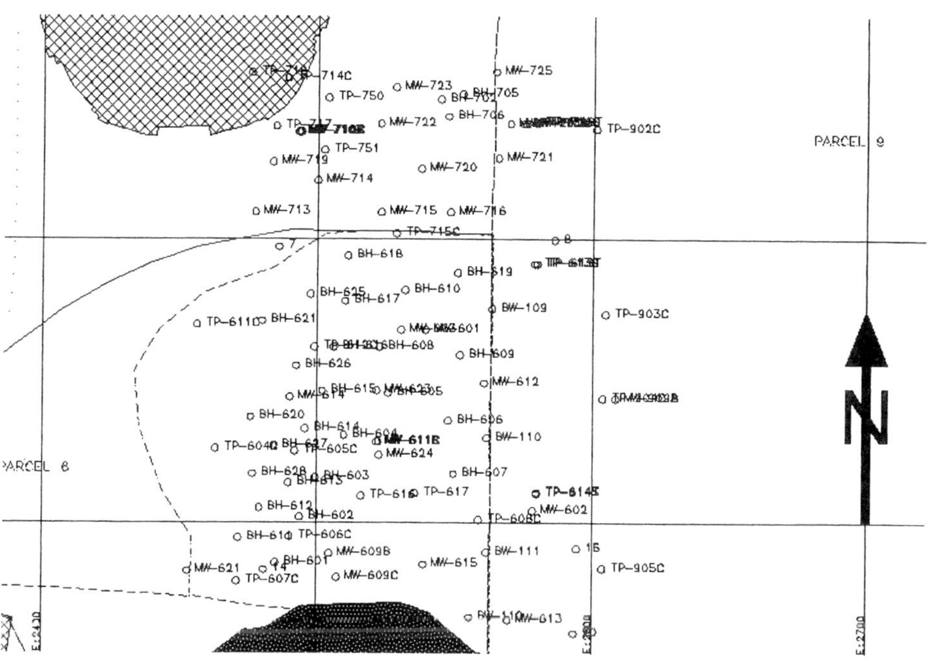

Figure 3.4b. Drillhole locations.

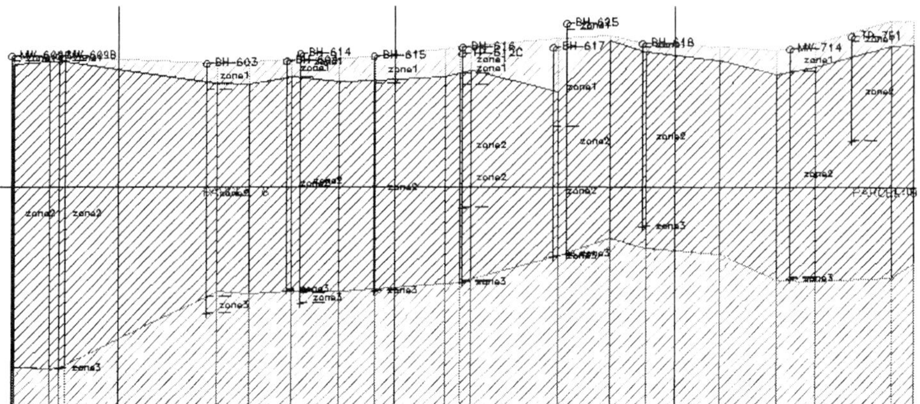

Figure 3.5. NS section of EXPO site: zone1 = clean fill, zone2 = industrial fill, zone3 = bedrock.

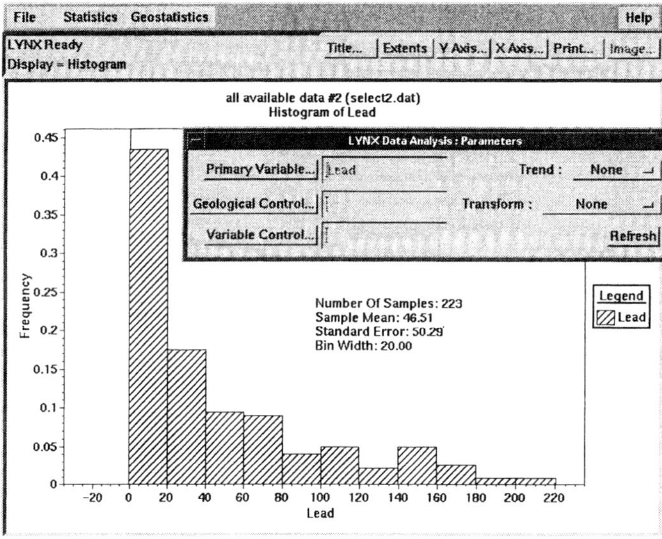

Figure 3.6. Lead histrogram from EXPO project.

In the Data Select Entry Form (see Fig. 3.7) click on the **Selection File** push-button to enter a name for the data selection file, toggle on **Create Data File**, provide a **Data File** name and **Data Select Title** to be displayed with the results of all statistics and geostatistics applications.

Click on the **Variable/Characteristic Select** push-button to bring up the Variable/Characteristic Sequence Select Entry Form. Here, click on the **Hole Data Select** push-button bringing up the Hole Data select Entry Form and select drillhole subset dh0. Select Lead as the variable to be analyzed. Variable minimum and maximums can also be set as data filters on numeric variables. Characteristics values, such as the soil type (Zone) can also be used to filter out selected records into the data selection file. In this example **Map Data Select** will be left empty, but remember that map

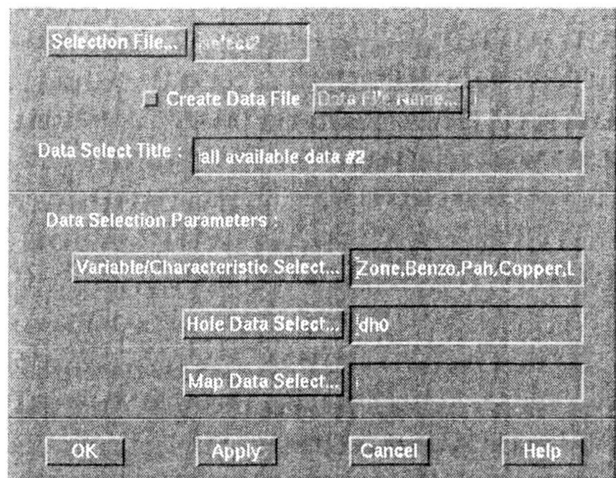

Figure 3.7. Data Select Entry Form setup for analysis all drillhole data in the EXPO project.

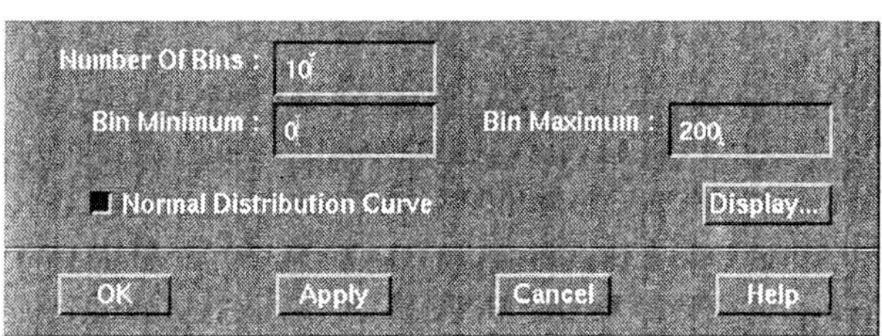

Figure 3.8. Parameter values for a histogram of untransformed lead levels in drillhole data taken from the EXPO project.

data structures are using in Lynx to hold a variety of information with 2D locations such as surface geochemical surveys. Once data selection is completed, the Data Analysis Parameter Form will become active. Note that there are a number of different drillhole classifications that might be related to different sampling protocols and therefore may be correlated with any subsequent analysis results. Since dh0 contains all drillholes, it might be worthwhile to create different drillhole subsets for analysis using the **Hole ⇒ Hole Data Define** facility in the main Lynx window.

Once the data file has been selected, analysis can continue with an examination of the distribution of lead contamination. Select **Statistics ⇒ Primary Statistics** to generate a report of the distribution's statistics. Use this information, especially the mean, standard deviation, minimum, maximum and coefficient of variation, to estimate a suitable range and bin size for the histogram. Next, display the Histogram Analysis Entry Form by selecting **Statistics ⇒ Histogram Analysis**. A suitable set of parameters for a Lead histogram is shown in Figure 3.8. The resulting histogram is shown in Figure 3.6. The **Normal Distribution Curve** checkbox will include a fitted normal distribution curve on the histogram. This can be used to evaluate the goodness of fit of the data to a normal distribution.

The type of distribution has important consequences when estimating distribution statistics and variograms. Estimates of the mean, variance and confidence intervals taken from relatively small data sets that have a lognormal distribution will be incorrect when calculated using the gaussian distribution function. Similarly, variogram analysis assumes that data values have a normal distribution.

A plot of the cumulative frequency distribution provides another visual aid to checking the type of distribution. Select **Statistics ⇒ Probability Analysis** to bring up the Probability Analysis Entry Form which will default to the same parameter values as for histogram analysis. Selecting **Apply** will generate a plot of the cumulative frequency distribution for lead as shown in Figure 3.9. This can be use to determine the proportion of samples that have a lead level below a given value and, by inference, the proportion of the soil in the area sampled that contaminated above and below a given cutoff. For instance, about 90% of the sampled values are below 115 ppm lead. Note that in Figure 3.9 that the normal distribution curve is not even close to the cumulative frequency plot of the data. The summary statistics and histogram indicated a data set that has a strong positive skewness and that might be lognormally distributed. To check this, the data can be lognormally transformed. The cumulative frequence plot might then match a normal cumulative frequency.

To check for a lognormal lead distribution, select **Transform** <log> in the Data Analysis Parameters Entry Form. Generate a primary statistics report as for the original lead data and note that the log transformed lead values range from about –8 to 6 with a relatively low coefficient of variation. Generate a cumulative frequency plot using this range and 12 bins. The resulting plot (Fig. 3.10) shows that the transformed data is much closer to a normal distribution. This indicates a lognormal lead distribution.

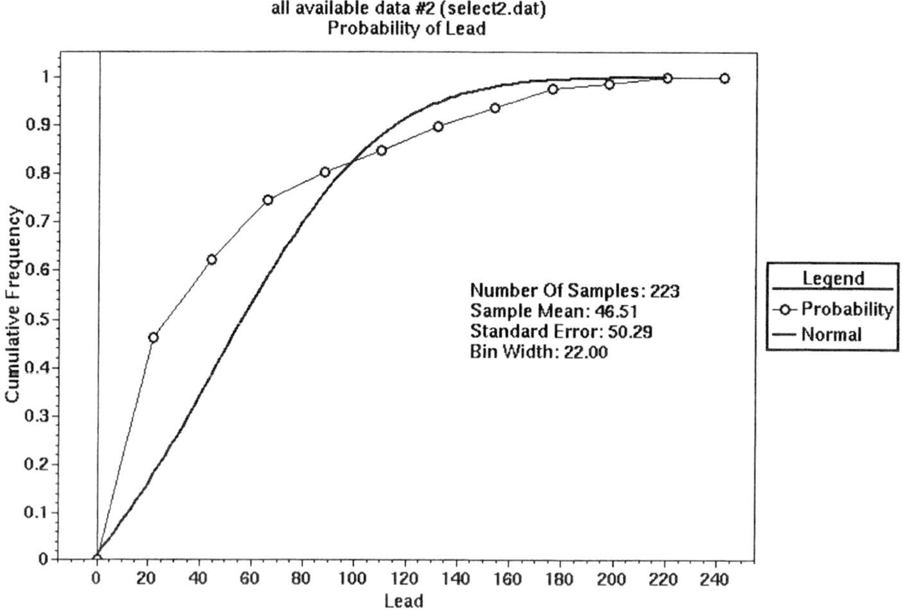

Figure 3.9. Cumulative distribution of untransformed lead levels from EXPO project drillhole data.

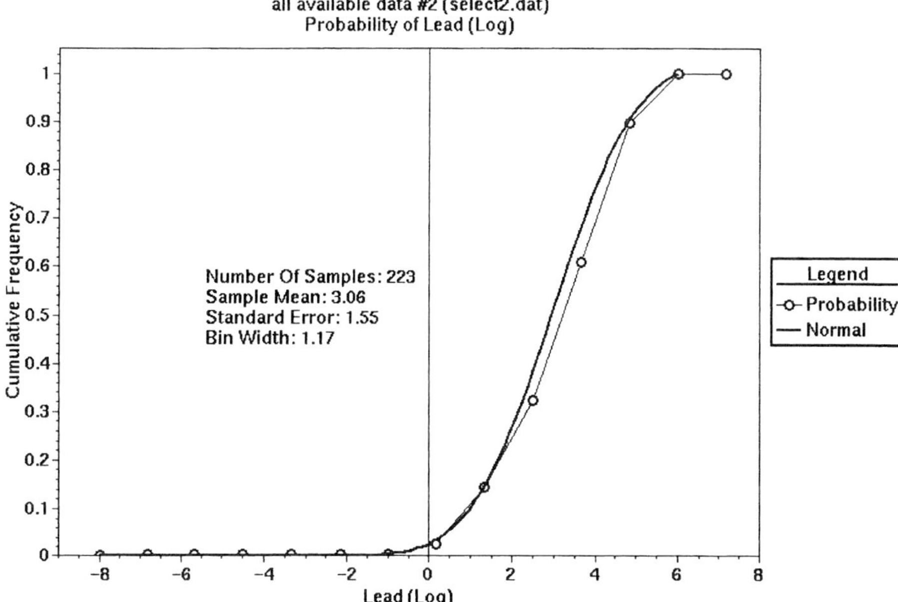

Figure 3.10. Cumulative distribution of log-transformed lead levels from EXPO project drillhole data.

Summary of procedure: Example 3.2
Initial data requirements: EXPO project including Lead assays from drillhole subset dh0.

Procedure:
From the main Lynx window select **Analysis ⇒ Data Analysis** (Data Analysis window).

1. Create data selection and data files for subsequent analysis **File ⇒ Data Select ⇒ Select Hole + Map Data** (Hole Data & Map Data Entry Form):
   SELECTION FILE: <select2>
   CREATE DATA FILE: <on> DATA FILE NAME: <select2>
   DATA SELECT TITLE: <optional>
   VARIABLE/CHARACTERISTIC SELECT: <Lead>
   HOLE DATA SELECT: <dh0>
2. Examine primary statistics **Statistics ⇒ Primary Statistics**
3. Generate a histogram **Statistics ⇒ Histogram Analysis** (see Fig. 3.8 for input values)
4. Generate a cumulative distribution function **Statistics ⇒ Probability Analysis** (same input values as for Step 3)
5. Use a log transform of the input data to check for lognormality:
   TRANSFORM: <lognormal> (Data Analysis Parameters Entry Form)
   Repeat Step 4 with a range of -8 to 6 and 12 bins (see Step 2).

Example 3.3: Data posting in Techbase

Use data postings to display data locations, selected information from each hole, polygons and additional physical features of interest. Creating a poster of collar locations is used as an example of this procedure.

The **Statistics** program can first be used to determine the extent of data before defining map limits and scaling. The max and min easting and northing values must be known for the area that is to be displayed before creating a graphics file in **Poster**. From the main menu, select **Statistics** ⇒ **Summary** ⇒ **Fields** and list the field names for the easting and northing coordinates of the drillhole collars.

From the main **Techbase** menu, select **Graphics** ⇒ **Poster** ⇒ **Graphics** ⇒ **Metafile name** and provide an original name so that your metafile will not be overwritten at a later date.

Often, the density of data in some areas of the map will make it difficult to post values on your poster without the field values overwriting each other because of the scale of the map and size of the paper. Say that we want to display collars contained in a specific lease which might be crowded on the scale used to display all data from all leases. In this case, a **Filter** can be defined under the **Database** menu option in most TECHBASE routines. Select **Database** ⇒ **Filter** ⇒ **Add filter** and add the field relations that are needed to filter out data records that are not of interest, eg., enter LS = 8 to only work with records in lease 8. Filters are retained for all TECHBASE executables after they have been set, so remember to **Delete filter** once you're done with them.

A map must first be provided with a physical dimension and limits by defining the range of coordinate values to be displayed along with margins, scale and sheet size. Select **Scaling** ⇒ **Scale** and enter the coordinate units (feet or meters) per inch of physical plot. You should have considered the size of paper that the plot will be produced on and the range and density of data before you get to this point. The scale used should also conform to some standard scale such as 50 ft/inch.

Next select **Range** and enter the coordinates of the bottom left and top right corners of the area that contains the data that you're interested in displaying. Provide sufficient margins so that any annotation that is to be placed outside of the border of the map will not be cut during plotting.

Select **Sheets**. Here you have a second opportunity to select margins. You can also rotate the plot if your coordinate system is relative and you want to use plot rotation so that the true sense of direction is shown. Enter a SHEETSIZE that is large enough to contain your plot. A size plots are: 8.5X11, B size are: 11X18, C, D and E are larger. When finished with SHEETS select BORDER. Before the border is drawn, TECHBASE will provide a message detailing the physical size of the plot and whether the plot will fit on the sheet size selected. Use this message to determine if the sheet size selected is appropriate.

Select **Grid** to put a coordinate grid on your plot. Normally, you should use the same coordinate ranges for grid as you used when defining the map. Use F1 for help on the various display options. Also, don't forget to use **Graphics** ⇒ **Review** to check each layer of the metafile. **Undraw** in the **Graphics** menu can be used to delete individual layers of the graphics file. To change the scaling or extent of the file use **New frame**.

Select **Point** for point posting. Here you will define the type of marker that is to be used to identify collar locations and the name of the **Fields** that are to be displayed with the marker. Select **Markers** and then enter the numeric code for the TYPE of marker (see the introduction section of the user manual for a list of the types of markers available). Under VALUE STYLE specify where and how the field that is to be plotted beside the collar is to be displayed.

Select **Draw** to post the markers and field values. Don't forget to use **Graphics ⇒ Review** to check the visibility of the markers and text on the plot.

Polygons can be added to your metafile to display lease lines, property boundaries, rivers, etc. The polygon file is created in ASCII usually by using a text editor. Each record in the polygon file defines a polygon vertex in the same coordinates as used for the map. Select **Polygon file** and enter the name of the polygon file. FILL STYLE and CLIP should be hollow and no, respectively. Otherwise, graphics within the polygon will be overwritten or items outside of the polygon will be clipped from the plot.

Use **Annotation** to add picture files, text and lines to the poster. When detailed annotation is needed, it is best to export a techbase metalfile as a DXF file and then import the DXF file into a CAD package

Summary of procedure: Example 3.3
Initial data requirements: same as for Example 3.1 using Boland database

Procedure: From the main Techbase menu select **Graphics ⇒ Poster**
1. Name the metafile <collars.met> (see Summary of procedure: Example 3.1)
2. Set up scaling, range, and grid: (parameters not specified in the following can be left at default values)
   2.1 **Scaling ⇒ Scale**
   X-SCALE: <350>   Y-SCALE: <350>
   2.2 **⇒ Range**
   LEFT X: <9900>   RIGHT X: <13140>   LEFT MARGIN: <.5>
   BOTTOM Y: < 19850>   TOP Y: <22330>   RIGHT MARGIN: <.5>
   2.3 **⇒ Sheets**
   SHEET SIZE: <A>   ORIENTATION: <landscape>
   2.4 **⇒ Border ⇒ Grid**
   X MINIMUM: <9900>   X MAXIMUM: <13140>   INCREMENT: <200>
   Y MINIMUM: <19850>   Y MAXIMUM: <22330>   INCREMENT: <200>
   GRID STYLE: <ticks>
   TEXT SIZE: <.10>   TICKS: <in>   TEXT: <in>
   ⇒ eXit ⇒ Graphics ⇒ Review (**Undraw** if unacceptable)
3. Post drillhole collars
   3.1 **Point ⇒ Fields**
   X-COORD: <x>   Y-COORD: <y>
   VALUE 1: <hole_id>
   3.2 **Marker**
   TYPE: <22>   SIZE: <.10>
   3.3 **Value Style**
   LOCATION: <top>   SIZE: <.10>
   3.4 **Draw⇒ eXit ⇒ Graphics ⇒ Review**

Example 3.4: Creating and displaying a poster of collar locations in Lynx
Drillhole collars from the TUTORIAL project will be displayed in plan view and projected into a map. To create a map that will contain the drillhole data select **File**

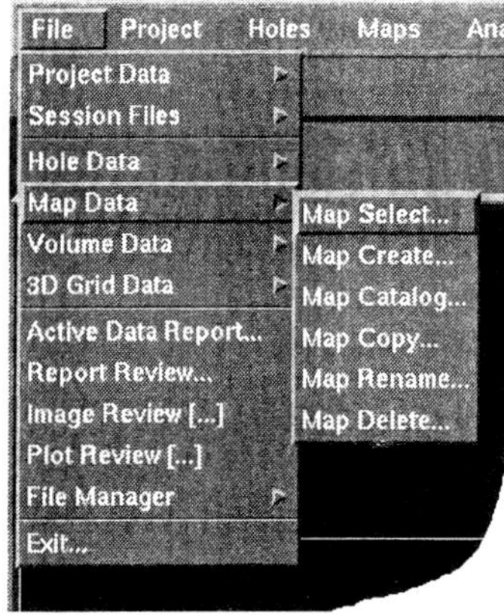

Figure 3.11. Map create selection procedure.

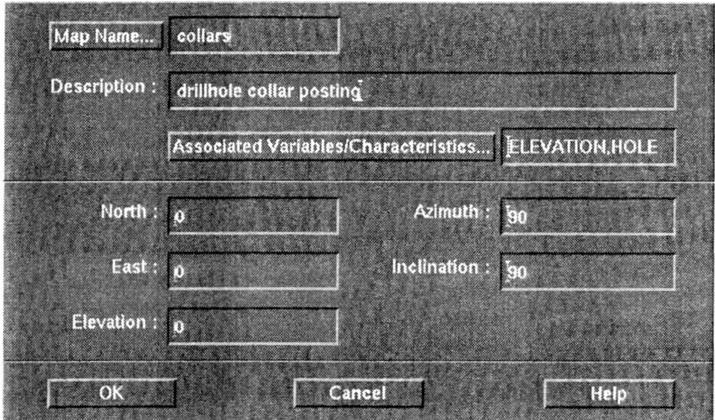

Figure 3.12. Map create entry form for map collars.

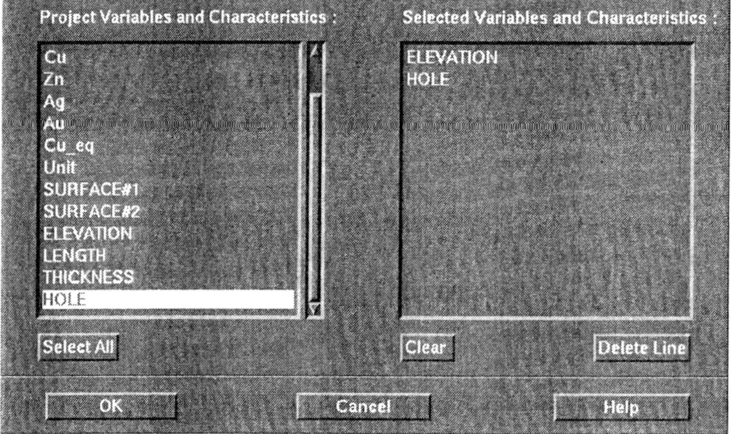

Figure 3.13. Variable/characteristic sequence selection entry form for map collars.

⇒ **Map Data** ⇒ **Map Create** as shown in Figure 3.11 to display the Map Create entry form of Figure 3.12. Enter the map name and description, the variables and characteristics that are to be included in the map and the map origin and inclination. Clicking on Associated Variables/Characteristics will display the Variable/Characteristic Sequence Selection entry form as shown in Figure 3.13. In this example the map will contain data for the hole identity which is a system characteristic and the collar elevation, a system variable. Variables such as Cu are user defined. Map origins are typically 0,0,0 but can be more localized. Plan view is 90,90.

Before placing drillhole information into the map, it's a good idea to use the background display facilities to examine the data first. From the main window select **Background** ⇒ **Holes** to display the Background Hole Data Selection entry form shown in Figure 3.14, which has already been set up for plan view display of hole collars. This menu will be revisted in later chapters for generating sectional views of hole data. Note that there are up to five different hole displays that can be defined at any one time. In this example only one display, Display 1, is toggled on. Click on **Subset/Holes Select** to display the Subset/Holes Select entry form and select drillhole subset <dh0> which by convention is the full raw drillhole data set (see Fig. 3.15). Choose <ELVN> as the **Primary Variable or Characterisitic** and set **Display Format** to <projections>. Selecting 'projections' brings up the Background Hole Data Projection Display entry form in which **Hole Name** should be toggled on to include the system characteristic HOLE in the display (see Fig. 3.16).

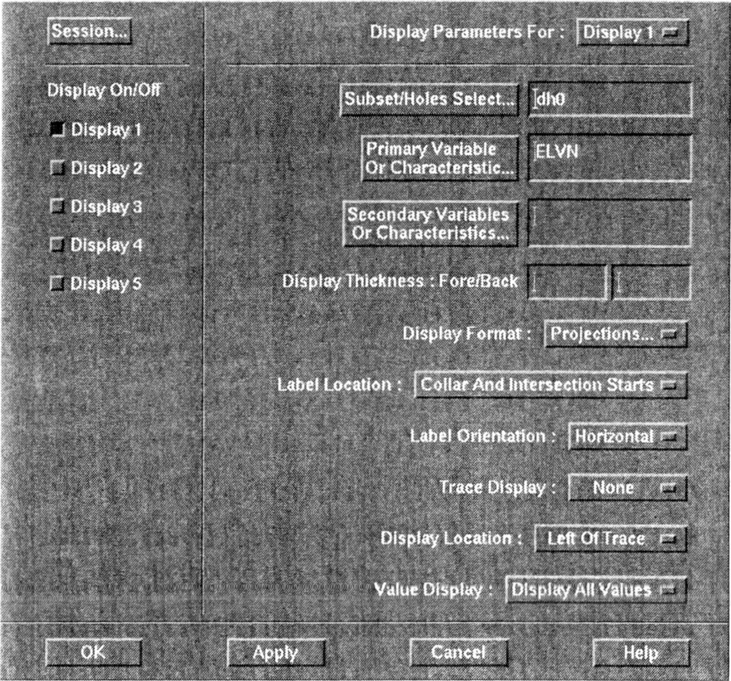

Figure 3.14. Background hole data selection entry form for displaying collars in plan.

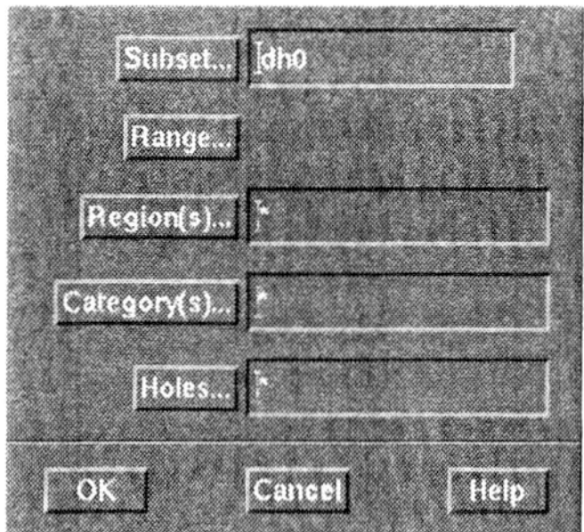

Figure 3.15. Subset/holes select entry form.

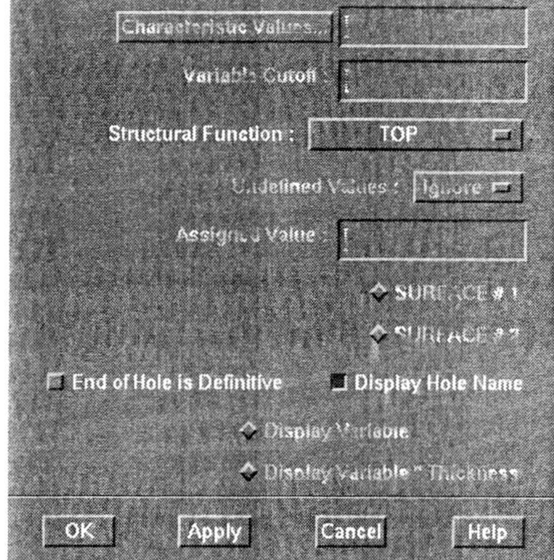

Figure 3.16. Background hole data projection display entry form for collars posting.

Returning to the Background Hole Data Selection entry form (Fig. 3.16), set **Label Location** to <collar and intersection starts> and **Trace Display** to <none>. Select Apply or OK and then **Extents** in the main window to display the collars as in Figure 3.17. Note that at this scale (about 4200:1) that many of the hole labels overwrite each other, a common problem for many drillhole data sets since more than one angled hole is often drilled from the same drilling platform.

Finally, this example will be concluded by placing the map data projections displayed in Figure 3.17 into the map <collars> that was previously created and still resides in active memory. From the main menu select **Maps ⇒ Hole Data Projection** to display the Hole Data Projection Parameters entry form (Fig. 3.18). Hole data projections onto a map plane are commonly used to determine downhole surface intersections with a given lithology, fault or cutoff and then project that as a z offset value onto the map plane where they are stored as point features. These point features can then be triangulated to form a surface. More on this topic in Chapter 4. In this example only the hole name and its elevation (offset from sealevel) is needed. Click on **Subset/Holes Select** and enter <dh0> (see Fig. 3.15). For **Variable of Characteristic** enter <ELVN>. Toggle on **Display Hole Name** and set the **Destina-**

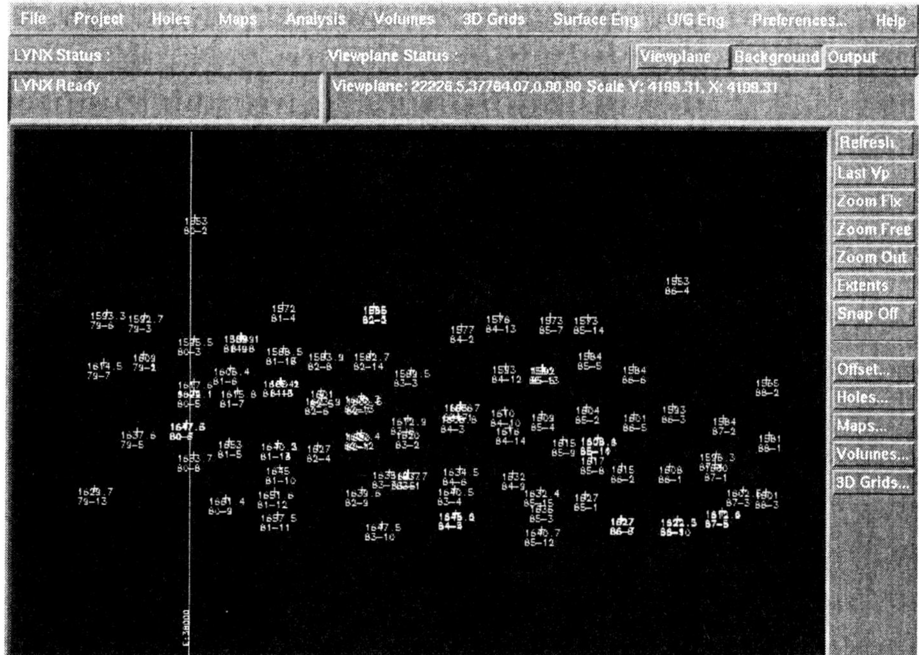

Figure 3.17. Background display of hole collars for TUTORIAL project.

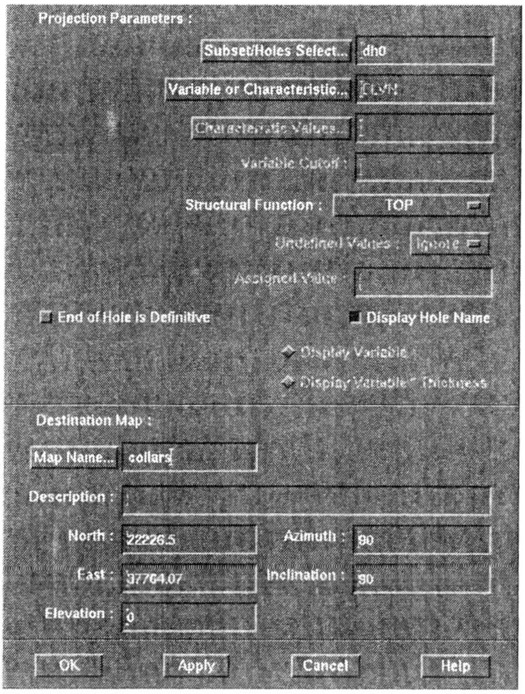

Projection Parameters :

Subset/Holes Select... dh0

Variable or Characteristic... LYN

Characteristic Values...

Variable Cutoff :

Structural Function : TOP

Undefined Values : Ignore

Assigned Value :

End of Hole is Definitive

Display Hole Name

Display Variable

Display Variable * Thickness

Destination Map :

Map Name... collars

Description :

North : 22226.5 Azimuth : 90

East : 37764.07 Inclination : 90

Elevation : 0

OK Apply Cancel Help

Figure 3.18. Hole data projection parameters entry form for displaying drillhole collars in plan.

tion Map to <collars>. Selecting **OK** will result in the overwriting of this empty map file.

The map can now be displayed in background by selecting **Background** ⇒ **Holes** and turning Display 1 off so that the holes display won't interfere with the display of the map. Actually, the two background displays should be indistinguishable. To display the map select **Background** ⇒ **Maps** and the Background Map Display Selection entry form will be displayed as in Figure 3.19. Toggle the display on and select <collars> from the map listing. Select **OK** and **Refresh** to get the map displayed as in Figure 3.17.

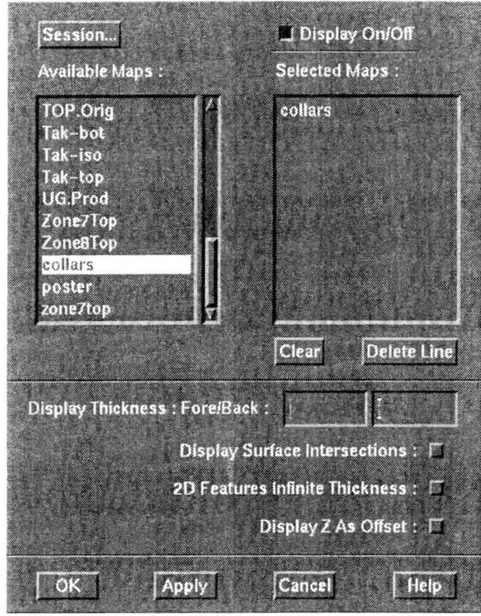

Figure 3.19. Background map display entry form.

Summary of procedure: Example 3.4
Initial data requirements: a project containing a drillhole data set.

Procedure:

1. Create a map file to contain collars **File** ⇒ **Map Data** ⇒ **Map Create** (Map Create entry form)
   MAP NAME: <collars>
   ASSOCIATED VARIABLES/CHARACTERISTICS: <HOLE> <ELVN>
   ⇒ **OK**
2. Display collars in background **Background** ⇒ **Holes** (Background Hole Data Selection entry form)
   SUBSET/HOLES SELECT: <dh0>
   PRIMARY VARIABLE/CHARACTERISTIC: <ELVN>
   DISPLAY FORMAT: <projections> (Background Hole Data Selection Display entry form)
   DISPLAY HOLE NAME: <on>
   LABEL LOCATION: <collar and intersection starts>

TRACE DISPLAY: <none>
DISPLAY PARAMETERS FOR: <Display 1>
DISPLAY ON/OFF: <on>
⇒ **OK**
⇒ **Refresh**

3. Putting hole projection data into a map **Maps** ⇒ **Hole Data Projection** (Hole Data Projection Parameters entry form)
SUBSET HOLES SELECT: <dh0>
VARIABLE OR CHARACTERISTIC: <ELVN>
DISPLAY HOLE NAME: <on>
DESTINATION MAP: <collars>
4. Display map in background **Background** ⇒ **Maps** (Background Map Data Selection entry form)
SELECTED MAP: <collars>
DISPLAY ON/OFF: <on>
⇒ **OK**
⇒ **Refresh**

3.2.3 *Contour Maps*

Contour maps are the most common visual aid used to understand three-dimensional relationships. Like a scatter plot, variables x and y are plotted against each other on two axes, but a third dimension is added by using the value of a third variable to generate contours or isolines of constant value. The most common application of contouring is where the xy axes represent location and the z value is the elevation or some other variable of interest, such as an assay or thickness. In earth science data, there is no underlying function available from which the isolines can be generated. Instead, a finite number of data locations must be used to estimate the path traveled by each isoline. The two most popular approaches to contouring are triangulation and gridding.

Triangulation algorithms attempt to form triangle sets from adjacent data. Since there are many possible triangle combinations, the algorithms attempt to achieve the best aereal coverage with the fewest triangles by forming near equilateral triangles. An example of a triangle set based on digitized elevation contours is given in Figure 3.20. A plane is then fit to the three xyz data points and the equation for the plane is then used to estimate the location of any isolines that fall within the range z values covered by the three vertices of the triangle.

In Techbase, triangulation can be applied either to random contouring (**Randcont**) or to grid estimation (**triGrid**). Both approaches have their strong points. Topographic data is often available from aerial photographic surveys or from paper contour maps. These sources provide a very high density of data when scanned or hand digitized. Contours based on triangulating a high density of data can be very detailed and as accurate as the original source data (Fig. 3.21). When the density of xyz data points is low, as in the case of drillhole thickness data, then triangulation is not a good choice for contouring. Since triangulation uses the equation of a plane to fit z values, the overall effect is to smooth the contours. When the triangles span individual digitized contours, this is a desirable side effect, but for sparse drillhole data, the

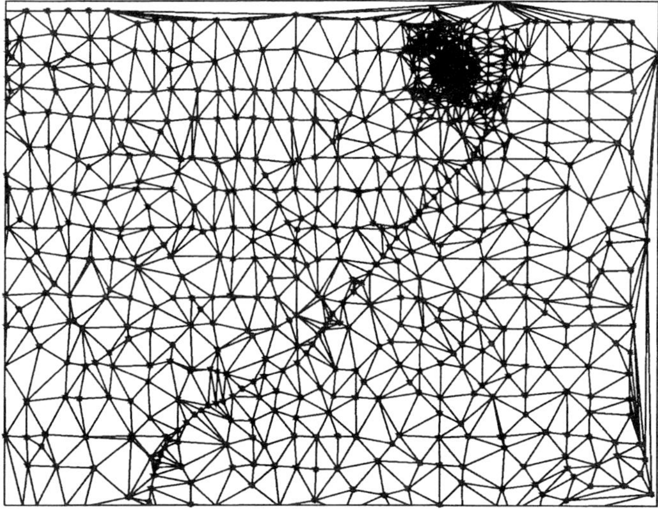

Figure 3.20. Triangle set for random contouring of topography in the neighborhood of the Boland Banya project.

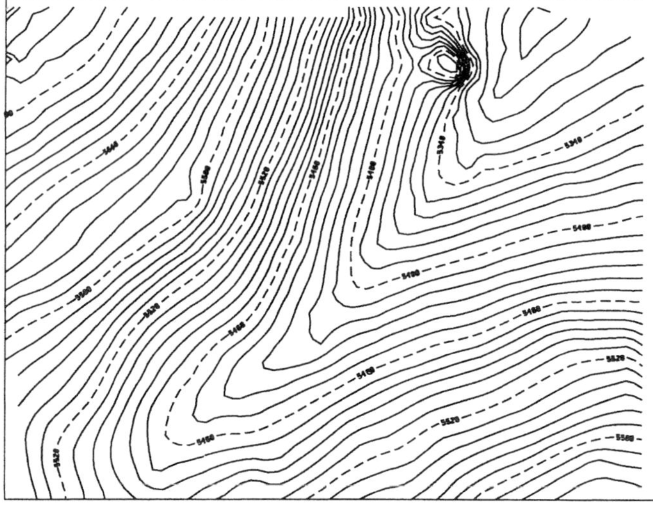

Figure 3.21. Random contouring of topography based on the triangle set of Figure 3.20.

smoothing can be very deceptive: a fault passing through a triangle would be reduced to a flat incline between the nearest drillholes.

Another consideration is the availability of using triangles as stored surfaces. Techbase only uses triangulation for generating graphics or as a method of grid estimation. Triangle sets are not stored for later use.

Lynx makes heavy use of triangles as a main method of surface definition. Triangles are stored and can be used at a later date to generate map data. The differences between gridded and triangulated surfaces will be covered more fully under geologic modelling.

3.3 SPATIAL COVARIANCE

As mentioned at the beginning of this chapter, earth science data are spatial in their nature, and there commonly exists a spatial correlation between 'nearby' samples. Subsequent analysis of the data must account for the fact that the data are regionalized, i.e. they are autocorrelated in space. In recent years, classical statistical methods have been developed around time series and regression to account for spatial data, but from its French and South African origins, geostatistics has developed into a largely separate discipline which dominates statistics in the earth sciences. Geostatistics is based on incorporating a measure of spatial continuity into the statistical modelling process. The most popular measure of spatial continuity is the variogram.

The experimental variogram is a representation of the average increase in variance between sample values as a function of the distance and direction of separation. A model variogram (henceforth referred to as the variogram) results from fitting one or more functions to the experimental variogram. In this way, the variogram value can be calculated for any vector of separation and used for grid and block model estimation. For more details on variograms see Isaaks & Srivastava (1989).

The experimental variogram is generated by searching for pairs data values that are separated by a set of separation distances, referred to as lags, either in a specific direction or irrespective of direction – a global search. For all pairs of data values found to be separated by the lag distance, their difference is squared, then all these squared differences are summed, and the summation is averaged by dividing by twice the number of data pairs found for that lag. The resulting experimental variogram value is the average variance of samples which are separated by the lag distance in the direction of the search. This is repeated for each lag and the resulting set of average variances is plotted against the lag distances to produce the experimental variogram for the direction of interest.

Several parameters are used to describe the shape of the variogram. The nugget represents the variance of points at zero distance of separation ($h = 0$). Theoretically, it is assumed that there will be a certain amount of variability between two points even at very short distances, if for no other reason than measurement and analytical error. But in practice, the shortest distance between samples will be significantly large, and the nugget must be estimated by extrapolating towards $h = 0$ using the average slope of the first couple of lags. As the lag distance increases, the correlation between the data pairs can be expected to decrease, i.e. the variance as represented by the variogram value will increase. Eventually, the distance will be so great that there is no correlation between the data pairs and the variogram value will stabilize about a constant value referred to as the sill of the variogram. The distance at which the variance stabilizes to the sill is the range of correlation (Fig. 3.22).

Note that experimental variograms are often non-transitional in that they never reach a sill over the range of the data. There are several possible explanations for this. Several covariance structures might be acting at different ranges of correlation, one of which is very long. Another more worrisome possibility is that the data have trend, or drift, in their magnitude. Drift is a directional dependence of the mean and is often modeled as a trend surface. When an experimental variogram is apparently

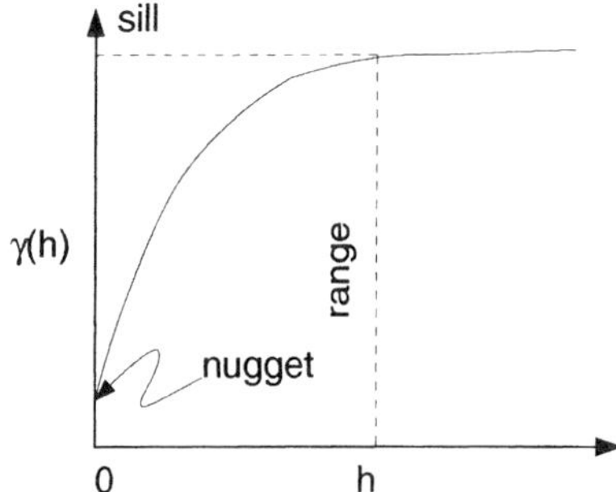

Figure 3.22. Variogram terminology.

non-transitional, the possibility of there being drift in the mean value should be explored, since this phenomenon can have dire consequences during estimation when using geostatistical methods such as kriging. These topics will be covered in greater depth both at the end of this chapter and in Chapter 5.

Example 3.5: Generating an experimental variogram in Techbase

Using the drillhole assay data from the Smoot PNL soil moisture data, we will generate a global (nondirectional) experimental variogram of neutron probe count. The PNL data set consists of wells drilled through a number of soil lenses on which the moisture level is dependent. For this reason, the distribution of moisture as measured by the field 'count' should be examined as a function of sediment class. If the distribution of the moisture content is a function of sediment type, then all statistics, including variograms, should be generated separately for each sediment type. Figure 3.23 shows a preliminary analysis that consists of generating a scatter plot of the population class versus moisture content. The procedure is the same as for Example 3.1 except that the sediment class (id) is an integer variable that is being used to numerically represent the original sediment descriptions. Such integer conversions of text field are frequently useful in statistical analysis or for use in data filters (cf., Smith 1992). For now, it will be assumed that moisture content is dependent on sediment type and the variogram analysis will be based on class 21.

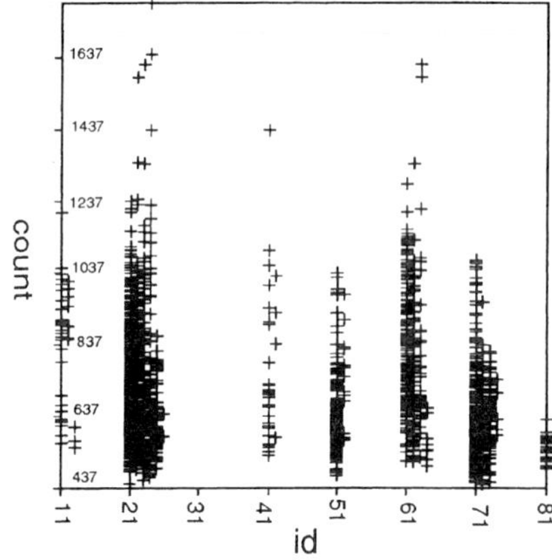

Figure 3.23. Scatter plot histogram of neutron probe count versus sediment category.

From the main menu select **Model ⇒ Vario ⇒Database ⇒ Add filter** and limit the selection of records by specifying <id = 21>. Assign a **metafile name** in **Graphics** and then select **Setup**. Under **Fields** enter the **Value** to be analyzed <count> and its *xyz* coordinates. Just enter *x* and *y* for two-dimensional data.

The **Directions** menu can be used to specify the vector's direction when acumulating pairs for each lag distance. For a global search, these default to azimuth, dip and tolerance angle of 0, 0, 90, respectively. **Parameters** for the lag distance and number of lags must be given. The lag distance refers to the initial lag. In this data set, samples were taken down the wells at 30 cm intervals. A good starting point for the first lag distance is the smallest commonly occuring sample interval, in this case 30 cm. Each subsequent lag will double the previous lag distance. So, if we entered an initial lag of 30 cm and 20 lags, then the total distance for which the variogram will be generated will be h = {30,60,90,...,600}. The horizontal distance spanned by these wells is roughly 18 m. Generally, the maximum lag distance shouldn't exceed half the span of the data. Techbase provides the option of excluding a given direction from the search, but this should not be used unless there is good reason. Finally, enter a suitable **title** in the **style** menu **eXit** and select **Variogram** to generate the experimental variogram which can be viewed in the **Graphics** menu or in more detail under **Model.**

The resulting experimental variogram based on a 30 cm lag interval displays high variability between lags and is difficult to interpret (Fig. 3.24a). Often, by increasing the lag distance, the resulting experimental variogram will display less variability between lags and is easier to interpret. By doubling the initial lag distance to 60 cm, the experimental variogram does become less erratic in its variability (Fig. 3.24b).

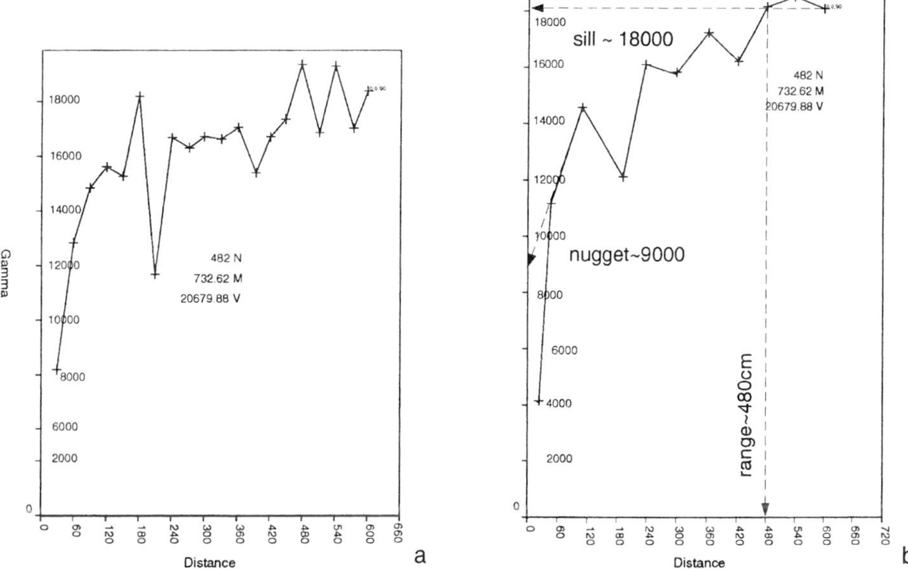

Figure 3.24. Experimental variogram of neutron probe counts at: a) 30 cm lag intervals, and b) 60 cm lag intervals.

Note that there are no quantitative rules for interpreting the values of the sill, range and nugget for an experimental variogram. Trial and error attempts at finding a model using different functions and parameter values are the best guide. Note that in either case, that the magnitude of the nugget effect is difficult to specify due to the sharp change in slope at 200 cm. The accepted rule-of-thumb for estimating the nugget effect is to extrapolate the slope along the first few lags to h = 0. Variogram interpretation and modelling will be discussed at greater length in Chapter 5.

Summary of procedure: Example 3.5
Initial data requirements: a spatially dependent variable, in this case <count> in the geology table of the Smoot database.

Procedure:
1. Open the database and define the filter. From the main Techbase menu select
   Model ⇒ Vario ⇒ Database ⇒ Open
   NAME: <smoot>
   ⇒ **Add filter**
   FIELD: <id>   RELATION: <=>   FIELD/VALUE: <21>
   ⇒ **eXit**
2. Define the fields to be analized **Setup ⇒ Fields**
   VALUE: <count>
   X: <easting>   Y: <northing>   Z: <elev>
3. Variogram parameters ⇒ **Directions**
   AZIMUTH: <0>   DIP: <0>   TOLERANCE: <90>
   ⇒ **Parameters**
   LAG DISTANCE: <30>   NUMBER: <20>
4. Generate the experimental variogram ⇒ **Variogram**
5. **Graphics ⇒ Review**

3.4 DRIFT AND TREND SURFACES

As was mentioned earlier, data in which there is a trend will have a non-transitional experimental variogram, i.e. the mean value is non-stationary. Stationarity is one of the key assumptions of geostatistical modelling: the moments of a regionalized variable's distribution should not be spatially dependent. When there is a trend, the average value will have a strong relationship with location. An example of this would be the elevation value for a survey taken on the side of a hill. Since values showing a trend will be increasingly dissimilar as the separation distance between them increases (because of the changing average magnitude of their values), the average variogram value will also increase with the lag distance. The resulting experimental variogram will display a drift in increasing variance with lag distance. The implication is that the true underlying variogram is hidden under this dominant trend, which should be removed by fitting a trend surface. The residuals of the trend surface fit should have a mean of zero. Otherwise, the estimate is spatially biased. Even if the average residual is zero, a map of the residuals might reveal concentrations of positive residuals in some localities and negative residuals in others, in which case there exists local bias in the estimate.

When working with an apparent trend in a regionalized variable, variography should proceed in two stages. First, the drift in the mean should be fitted using a simple surface that reflects the obvious overall trend. This is commonly done using a trend surface, a multivariate regression model in which the independent variables are combinations of the *x* and *y* coordinates of the dependent variable, typically elevation. More on trend surfaces in Chapter 4. If the trend surface has been correctly specified, then the remaining error terms, commonly referred to as residuals, should be free of drift. Variography can then be conducted on the residuals with an ultimate goal of estimation as discussed in Chapter 5.

Example 3.6: Experimental variogram analysis of piezometric data in Lynx
Directional variograms will be generated for piezometric map data in the EXPO project.

From the main menu select **Analysis ⇒ Data Analysis** to bring up the Data Analysis main menu.

The piezometric data exists in the form of a map. Note that *xyz*, referred to as point type feature in Lynx, is commonly stored in map form and can be displayed, contoured or analyzed as in the following examples. As in the case of drillhole data analysis, the first step for map data analysis is to read an existing data file or to initialize a data selection file. To initialize both the data file and selection, file click on **File ⇒ Select Hole and Map Data** bringing up the Data Select Hole & Map Data Entry Form. At this point the map data selection process is similar to drillhole data selection (see Example 3.2) except that the **Map Data Select** push-button is selected instead of **Hole Data Select** bringing up the Map Data Select Entry Form. Here click on **Map Select** <piezo> ⇒ **Variable/Characteristic Select** <MapElevation> and default the **minimum** and **maximum** *Z* values to be extracted from the map to the full range of data by simply clicking on each of the two push-buttons. Once the data has been selected, its content can be confirmed by the activation of the parameters form and by using **File ⇒ Report**.

For initial variogram analysis click on **Geostatistics ⇒ Semi-variogram Analysis** bringing up the Semi-variogram Analysis Entry Form (see Fig. 3.25). The type of variogram selected can be either Normal or Relative. A 'normal' variogram, as

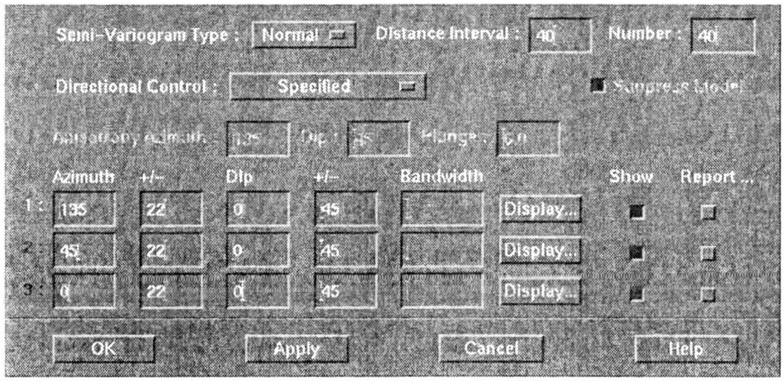

Figure 3.25. Semi-variogram Analysis Entry Form parameter settings for examining anisotropy in EXPO project piezometric map data.

discussed previously in this chapter, is the average squared difference of samples separated by a vector **h**. In practice the vector **h** is modified to have a length (lag) tolerance and angular tolerance which commonly default to half the lag distance and 45°. The formula for calculating the variogram value for a specific vector of separation, **h**, is:

$$\gamma(\mathbf{h}) = \frac{1}{2N(\mathbf{h})} \sum_{(i,j)|\, h_{ij} \approx \mathbf{h}} \left(z_i - z_j\right)^2$$

The denominator 2N(**h**) is twice the number of pairs (*i,j*) found within an approximate vector of separation **h**. The 2 originates from having the search for pairs being performed both from the location of sample *i* and *j*, which have values z_i and z_j. This results in the pair (*i,j*) and (*j,i*) being included in the average and is the origin of the term semi-variogram, which has been retained in the Lynx menus but is more commonly referred to as an experimental variogram or simply as a variogram.

Relative variograms are used to adjust the variogram value for each lag by the mean of the variable being measured. The idea here is to adjust the variogram for the effects of non-stationarity, i.e. spatially depended fluctuations in the mean. The use of the variogram assumes local stationarity so that when pairs (*i,j*) are found at **h***ij* they come from populations that have the same distribution, specifically, the same mean. There are several types of relative variograms (see Isaaks & Srivastava 1989), but all do essentially the same as Lynx which '...divides the difference in a pair of values by their mean value'.

The **Distance Interval** and **Number** refer to the initial lag distance and number of lags to be used in calculating a variogram plot. For example, a lag distance of 40 m and 40 intervals will result in values being calculated and plotted for **h** = {40, 80, 120, ..., 1600}. As a general guideline, set the initial lag to the shortest common distance of separation and limit the number of intervals to no more than half the extent of the data field.

Have **Suppress Model** toggled on. Variogram modelling will be discussed along with kriging estimation in Chapter 5.

Directional Control is used to generate a global variogram (no directional control) or to check for anisotropy (specifying multiple variograms in specific directions). Choose azimuths of 45° starting at due north (0,45,135) and click on **Apply** to display the three resulting experimental variograms (Fig. 3.26). From the directional experimental variograms, it is very apparent that there is a strong directional drift in the piezometric level to the north and northeast. Perpendicular to this direction (135°), the variogram is virtually flat. It would be hard to explain this behavior without considering the possibility of there being a strong trend in *z*. Otherwise, one would have to say that there is almost no nugget effect (unlikely) and that there is no spatial correlation at 135°, while at the same time there is a strong long-range correlation at 45° (even harder to believe). A closer look at the piezometric data is called for.

Examination of the contours for *z* in the piezometric data reveals that there is a strong trend in increasing *z* from the northeast to the southwest, and fairly uniform elevation perpendicular to this direction (Fig. 3.27). Fairly uniform surfaces such as

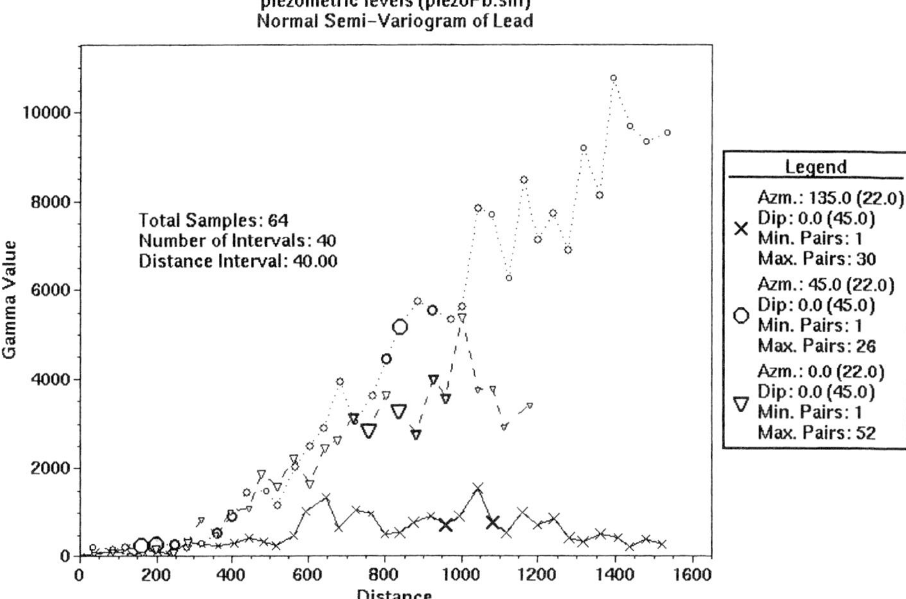

Figure 3.26. Directional variograms of raw piezometric data from the Expo site.

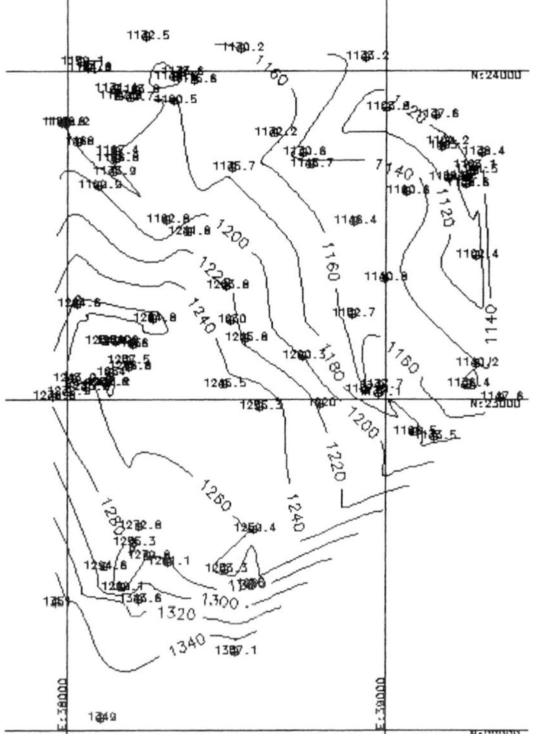

Figure 3.27. Contours of piezometric level and data locations.

this are well suited to trend surface fitting. Trend surface fits are accomplished using linear regression with polynomial equations. Linear terms will fit a plane, quadratic terms will fit a convex or concave surface and cubic terms will fit a saddle. Note that one polynomial will be fit to the entire data set used as input, so the resulting surface will be very smooth. As a result, the residual variation at the data locations will be high. For this reason, trend surfaces are not popular as a primary method of surface estimation. Usually, trend surfaces are used for trend removal and the trend surface residuals are used in some other method, kriging in particular.

Summary of procedure: Example 3.6
Initial data requirements: a regionalized variable (in this example the map piezo included in the EXPO project)

Procedure:

1. Data selection **File** ⇒ **Select Hole and Map Data** (Data Select Hole and Map Data Entry Form)
   Selection File <piezo> ⇒ **Variable/Characteristic Select** <MapElevation>
   ⇒ **Map Data Select** (Map Data Select Entry Form)
   ⇒ **Map Select** <piezo> ⇒ **Variable/Characteristic Select** <MapElevation> ⇒ **Accept** ⇒ **OK**
2. Variogram analysis **Geostatistics** ⇒ **Semi-variogram Analysis** (Semi-variogram Analysis Entry Form). Enter variogram parameters as shown in Figure 3.25.

Example 3.7: Trend surface Analysis and residual variograms in Lynx
The directional experimental variograms for the piezometric level display drift in the northeast direction. Examination of a contour map of z (Fig. 3.27) confirmed that elevation in the piezometric surface increases to the southwest in a linear fashion. We will fit a trend surface to z, save the residuals and generate a residual variogram to see if the drift in the mean value of z has been removed.

While a trend surface can be automatically fit to the data and the residuals used for the variogram analysis, let's first examine the trend surface equations. From the Data Analysis menu select **Geostatistics** ⇒ **Trend Analysis**. Linear and quadratic trend surface fitted to the piezometric data are displayed (as in Fig. 3.28) and can be saved as an ASCII report file in the project's ./report subdirectory.

Two polynomial equations for the linear and quadratic surfaces are displayed (Fig. 3.28). Note the form of the two equations and the near equivalence of the negative x and y coefficients in the linear model. The coefficients in the linear model reflect the fact that the piezometric surface sinks to the northeast. The slight difference in the magnitude of the x and y coefficients is mostly due to the slightly greater magnitude of the x coordinate values.

It is apparent from the contours of z that a linear trend will suffice, but in three-dimensional data sets, the nature of the trend cannot be as easily explored as is the case for a two-dimensional surfaces. As an aid in evaluating the performance of higher degree polynomials, an Analysis of Variance (ANOVA) can be conducted as in Figure 3.28. In the linear model, the sum of the squared residuals is 21680. If adding more terms to the polynomial is used to improve the fit of the trend surface to

```
Trend Analysis Report                     Mon Jun 29 12:59:52 1998
Active Project : EXPO                          User : martin
piezometric levels (piezoPb.slh)
Primary Variable Lead:

Linear Trend:   const = 756, x = -0.0783, y = -0.0837, z = 5.41e+04

Quadratic Trend: const = -1.55e+03, x = 0.262, y = -0.126, z = 4.22e+04
            x^2 = -1.55e-07, y^2 = -3.05e-07, z^2 = -1.81e+05,
            xy = -5.43e-06, xz = -2.01, yz = 2.66

Source        Sum of Squares    D.f.    Mean Square       F-Ratio
-----------------------------------------------------------------
Linear:        115168.00          3      38389.33          106.24
Residual:       21680.00         60        361.33
-----------------------------------------------------------------
Quadratic:     113984.00          9      12664.89           29.91
Difference:     -1184.00          6       -197.33           -0.47
Residual:       22864.00         54        423.41
-----------------------------------------------------------------
Total:         136848.00         63

Percentage of Total Sum Of Squares:
Linear Component:     84.16
Quadratic Component:  83.29
```

Figure 3.28. Report of trend surface analysis for piezometric level in the EXPO project.

the data, then the values of the squared residuals should markedly decrease. In the example of piezometric data, this is not so, and we can conclude that there is no significant improvement by using a quadratic equation. Care should be taken when using tests such as the F or t statistic which are based on the assumptions of normality and independence of the data. z may not be normal and is certainly not independent.

The very high F ratio demonstrates that the piezometric level is definitely not stationary and that this predictable fluctuation in the mean elevation dominates spatial variability. Select **Trend ⇒ Linear** in the Parameters menu and click on **Refresh**. As shown in Figure 3.29, the residual variogram from the linear trend surface no longer exhibits drift, nor is there any noticeable difference in variance between the 45° and 135° directional variograms in which case the directional variograms, should be replaced with a single global variogram which can be more easily interpreted (Fig. 3.30).

An additional check on the adequacy of this procedure is to generate a map of the residuals and examining it for any remaining trends or spatial bias by looking for concentrations of positive or negative residuals. This topic is covered in Chapter 5 and Example 5.11.

Summary of procedure: Example 3.7

Initial data requirements: any two-dimensional data set, for this example the data selection file of Example 3.6 (piezo.dat).

Procedure:
1. Either select an existing data file, (from the Data Analysis menu) **File** ⇒ **Data File** <piezo> or create a new selection file (see Summary of procedure: Example 3.6, Step 1).
2. Trend surface analysis **Geostatistcs** ⇒ **Trend Analysis**. Save report file if needed.
3. Plot residual variogram: In Data Analysis Parameter Entry Form select **Trend** <linear> ⇒ **Refresh**.

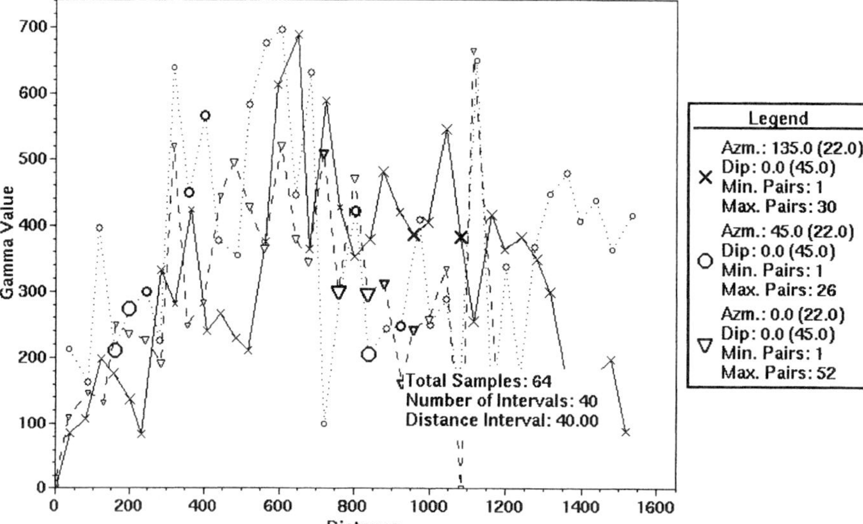

Figure 3.29. Residual directional variograms (same parameters as Fig. 3.26).

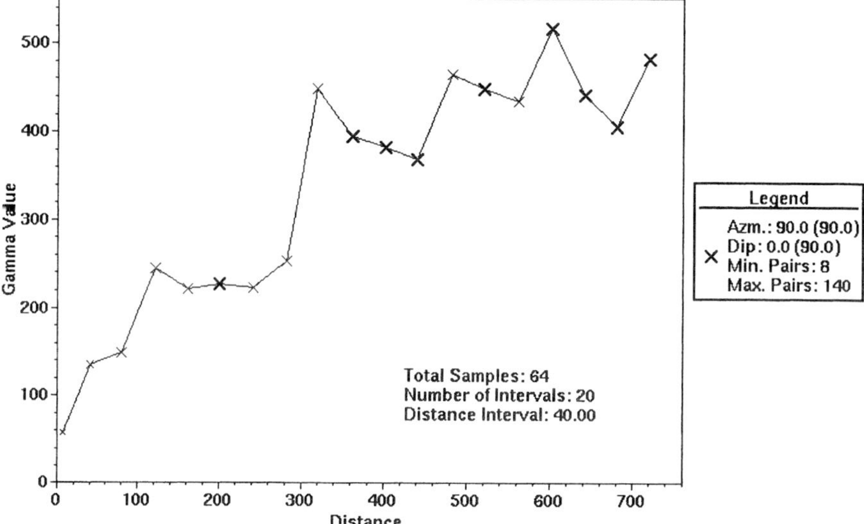

Figure 3.30. Global residual variogram of piezometric level.

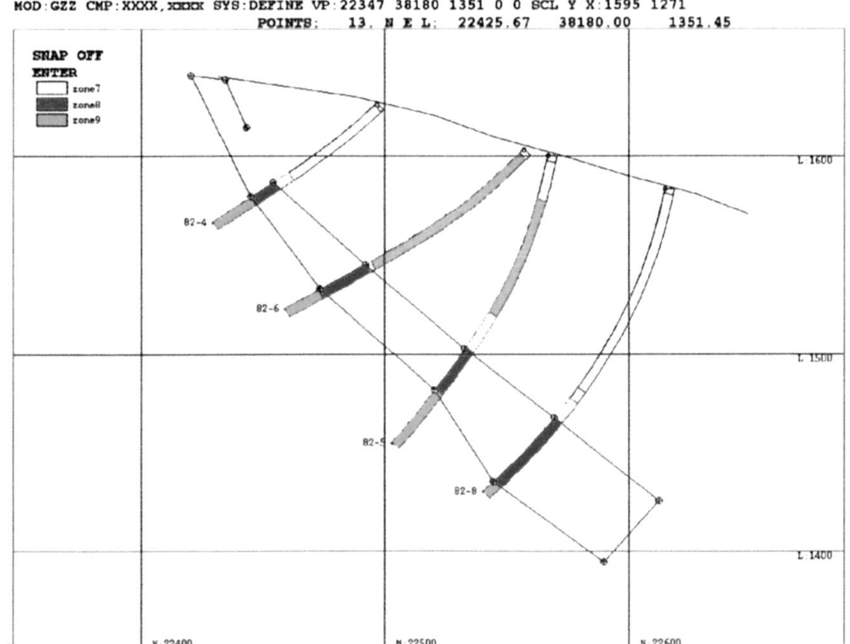

Figure 4.36. Interpretation of main ore zone limits on drillhole sections displaying rock type.

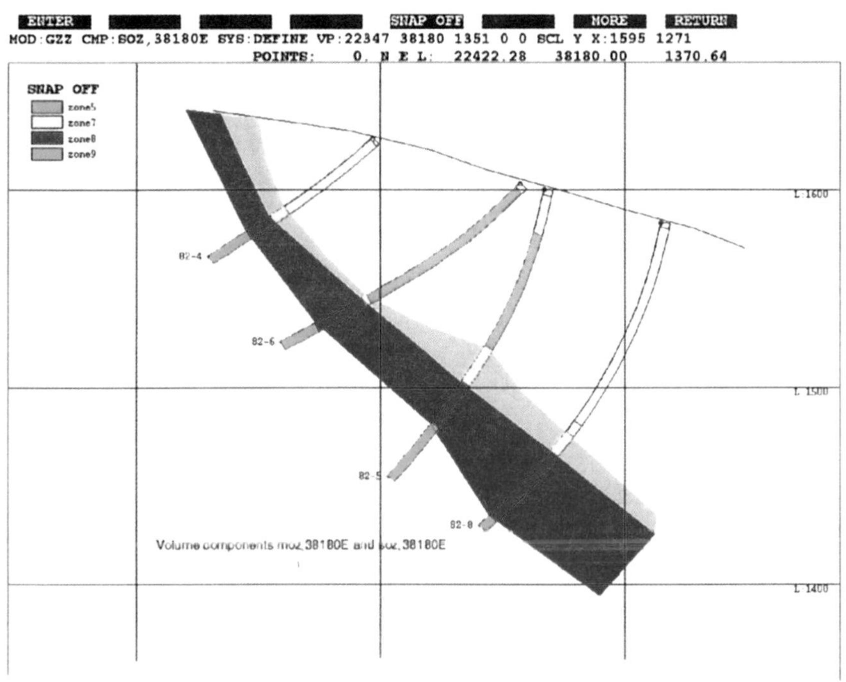

Figure 4.37. Primary and secondary ore zone volume components.

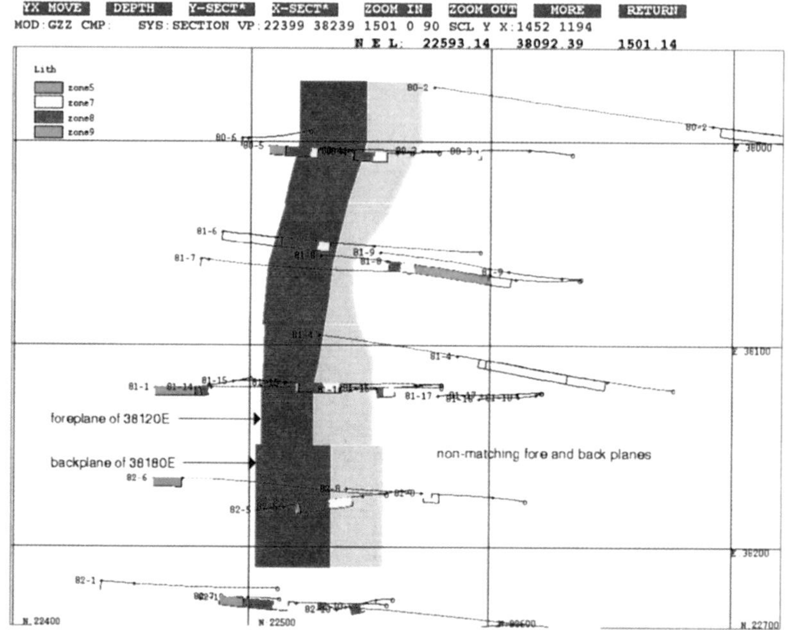

Figure 4.38. Linear interpolation of components on sections at 38120E, 38060E and 38000E. Component at 38180E not interpolated with 38120E.

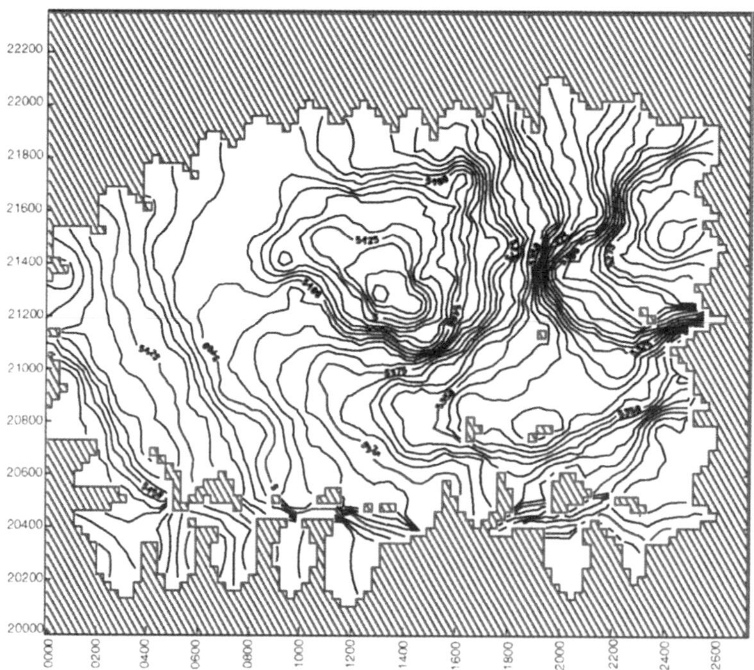

Figure 5.11. Inverse distance estimation of the top structure of the main (OA) ore zone showing the extent of cell assignments (Boland Banya project).

CHAPTER 4

Modelling geologic surfaces

Earth science data are not only spatial, they are also controlled by geologic structures. The analysis and estimation of regionalized variables can rarely be separated from the geology in which they're found: assay grades are a function of the ore zone, coal quality characteristics are a function of the seam, piezometric head is dependent on the soil type, and rock strength varies from one side of a fault to the other. Geologic structures must be modeled to provide a means of control during the analysis and modelling of sample data and as a means of estimating volumes. Two-dimensional surfaces can be used to represent faults, hanging and foot walls, and top and bottom structures during the estimation of volumes and thicknesses. Three-dimensional volume models are used for more complex geologic structures which cannot be represented in a plane.

Geologic modelling is a highly intuitive process. The geologist or engineer is faced with the task of transforming a sparse set of data values into a continuous three-dimensional model of the geologic structures which are only hinted at by field sampling. There are no empirical or statistical methods which can be effectively used to estimate the geometry of the irregular surfaces and volumes, only a few drillhole intercepts, outcrops and shadowy geophysical data which provide a starting point for what, in reality, is an educated guess.

Geologic software such as Lynx and Techbase do not really provide any new tools for building geologic models. What they do provide is a highly automated means of working with and manipulating the traditional tools of maps and sections. The real power behind the software is their ability to manipulate geologic models once they have been created, and their ability to rapidly estimate values and volumes as a function of the shape, size and position of the geologic model. This chapter concentrates on the methodology of geologic modelling without dwelling on the algorithms. The modelling process can be broken down into three steps: compositing of geologic data to obtain surface intercept coordinates, exploration of structures using maps and sections, and interpretation/estimation of section and map data into continuous surface or volume models.

4.1 DRILLHOLE DATA COMPOSITING

Drillhole and well data are not logged in a fashion which can be readily used for geologic interpretation. As was noted in Chapter 2, sample *xyz* coordinates must be reconstructed using the down hole sample depth, the hole inclination and the coordinate of the hole collar. Compositing methods are used to transform separate records for collar location, sample depth and inclination into a single record that includes the sample value, its lithology and coordinates. There are several common uses for composited data. Among these, the most common applications are regularization of the sample length and the assignment of coordinates to the drillhole intercepts of geologic boundaries.

4.1.1 *Regularization*

A long recognized pitfall of ore reserves estimation is the dependence between sample size and assay distribution, often referred to as the volume/variance relationship. Mathematically, samples are treated as point values without dimensions, but in reality, samples are taken at many different support sizes. It has been observed that as the support size increases, the variance of the assay will decrease. This has important consequences. For one, samples of differing support should not be carelessly mixed during estimation or statistical evaluation, since they probably have different distributions. A second issue is the change in the variability of assay from estimation when an estimate is based on point support data and the units being estimated are ore blocks consisting of thousands of tons. This volume/variance relationship is the source of continuous contention between the mine and mill, in that planned mine production grade is based on volumes sufficient for several days' production, while the mill must work with volumes which vary by the truckload. Invariably, the variance in grade of the production units used for production scheduling will be far less than the variance which is actually experienced between truckloads of ore.

Fluctuation of grade levels as a function of support size can be a function of the physical characteristics of the material being sampled. For instance, drill core samples are consistent in diameter, but the sample length can vary from a few broken fragments that are assigned to an interval of a few centimeters to cores many meters in length. In many deposits, high grade zones are associated with lower rock strengths. Portions of the core that are associated with higher grade will tend to lack cohesion in the core while barren zones will produce longer cores. Since cores are logged and sampled largely in the same intervals as they are physically produced, there can be an inverse relationship between sample length and grade. In the case of gouge and mud seams, a positive correlation between assay and length could also be discovered.

In the case of core length, the relationship between assay and length is only a consequence of the geology. Even the difference between a few centimeters and a couple of meters of core does not lead to that great of a difference in support size, but in many cases geochemical data is based on initial sample volumes which are different by several orders of magnitude as in the case of soil samples taken from trenches and small diameter split spoons.

4.1.2 *Boundary intercepts*

In order to estimate surfaces from drillholes, coordinate data for the intercept point of the hole through the boundary must be made available in a useful form. In Lynx, the locations of surfaces are treated as offsets from a plane. For a reasonably tabular ore zone, a map is defined wherein the azimuth and inclination correspond to the strike and dip of the formation with an origin which places the map plane within the major portion of the ore body. Lynx then automatically generates offset data for projections of drillhole intercepts with the hanging and footwall which can be stored as part of the map and subsequent triangulation. The triangle set can subsequently be used for contouring of surfaces and volumetric calculation of thicknesses and intersection volumes between different geologic and engineering models.

Techbase takes a different approach in that drillhole intercept data is retained as drillhole data, rather than as a map. Drillhole compositing is used to reduce the entire drillhole intercept with the unit to a single composite and then assign the composites' location to either the top or bottom of the composited sample length. The results are stored in two new flat tables, one for the top surface and the other for the bottom surface. These composite tables will have one record for each drillhole that was found to intercept the unit of interest and will include fields for the *xyz* coordinates, the hole identifier, a description of the lithologic unit and composited values for any sample data of interest. Using this approach, the original set of drillhole tables (collar, survey and assorted assay and geologic data tables) are translated into two tables for each lithologic unit of interest: one table to hold the top intercept and the other to hold the bottom. The records in these tables can then be used for two-dimensional estimation of grid values such as the top and bottom surfaces and the composited data associated with the lithologic unit. Grids of the top and bottom surfaces are also used for calculation of thicknesses and indirectly for the calculation of intersection volumes with other surfaces, such as pits and block models.

4.1.3 *Compositing methods*

Drillhole data is commonly composited to a fixed interval length as a proxy for regularization of the sample support size as a means of producing a smaller data set that is easier to work with in certain applications. The basic procedure is to combine samples into length composites and assign a single assay to the composite based on a weighted average of the core lengths of the constituent raw sample data. As an example, consider a drillhole which is to be composited to 10 m intervals starting at the collar of the hole and having as its first four samples assays of .4, .35, .5 and .3 oz/t having core lengths of 2, 4, .5 and 5 m, respectively as shown in Figure 4.1. The first composited value would be calculated as:

$$(.4 \times 2 + .35 \times 4 + .5 \times .5 + .3 \times 3.5) / (2 + 4 + .5 + 3.5) = .35$$

Note that the final 5 m interval extends 1.5 m beyond the 10 m composite. 3.5 m are assigned to the first composited sample while the remaining 1.5 m will be included in the next composite. This calculation is continued down the hole until the final interval, which will probably be truncated to less than 10 m.

Techbase provides several methods of compositing: value, length, bench, code and interval. Value compositing simply assigns an *xyz* coordinate to each of the records in a data table which contain either assay, geologic or other down hole interval based data. The composited values will be overlaid in the same data table. No new

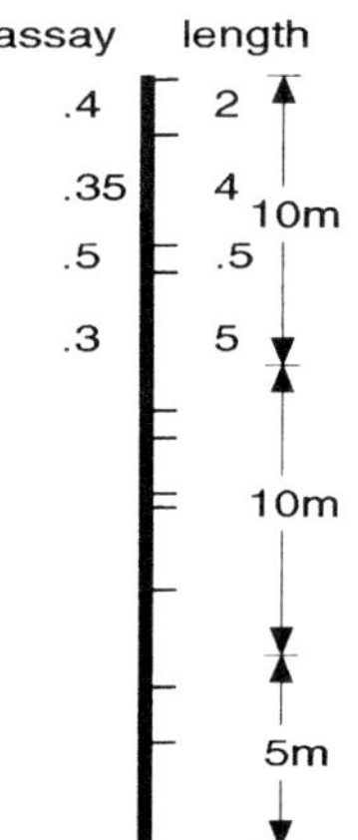

Figure 4.1. 10 m interval composites.

composited data are created beyond the coordinates each of the data records. This is useful for three-dimensional analysis and estimation when geologic control and a standard support size are not of interest. Length composites are based on a fixed compositing interval which starts at the hole collar. The composited assays are stored in a new composite table. Bench composites are the same as length composites except that the starting elevation for compositing is the same for all holes. The idea here is to generate a data set in which there is only one sample for each mining bench in a hole. The composite starting elevation would be the crest of the uppermost bench in the pit. Bench compositing simplifies surface mine planning and production scheduling by allowing two-dimensional estimation and displays to be based on benches. Like bench and length compositing, code composites use fixed composite intervals (starting at the collar) and assign coordinates and average assays to records in a composite table. The difference is that a code field can be used to control the break points in composites down the hole. The typical application for code compositing would be to use a lithology field as the code, so that composites would break with a change in the lithology. Length, bench and code compositing all result in a composite table based on a fixed interval which is broken either by the end of the hole or by a change in a code field. Interval compositing enables the compositing interval to vary in accordance with the intervals specified in a separate Interval table containing fields to contain the required from/to interval values. The resulting irregular interval composites can be written to the Interval table.

Example 4.1: Using length compositing to standardize sample length in Techbase
Gold assays from the Boland Banya database will be composited to 10 m intervals in the OA ore zone.

Under **Techbase** ⇒ **Define** ⇒ **Tables**, **Create** a new flat table called compOA and **Add** the hole identification field to this table. Then **eXit** from **Tables** and in **Fields** set **Autotable** to compOA and **Create** new fields for the coordinates of the composited assays and the average gold assay value, say *x*_10, *y*_10, *z*_10 and au_10. You may also want to add new fields for other assays and to store the composite length.

From the main menu, select: **Modelling ⇒ Composite ⇒ Database ⇒Add filter** and add the filter: < lith = OA> where lith is a field in the assays table which defines the lithologic unit in which each assay record belongs. In this example, only records in the OA main ore zone will be composited. **eXit** and in **Setup ⇒ Fields** enter the field names that correspond to the collar location, hole inclination, and assay from/to intervals. The Azimuth, Dip and Depth inclination fields are left blank if all the holes are vertical. The hole_id, easting, northing and elevation are taken from the collar table; bearing, dip and depth are from the survey table and from and to are from the assays table.

In the **Coordinates** menu, enter the output coordinate field names that were added and created for the new composite table, compOA to identify the output field names for the composited records.

In **Assignment,** identify the field in the composite table in which the composited assay is to be recorded and then under **Parameters** set TYPE to length, LOCATION to middle, and MAXIMUM LENGTH to 10 m. The other compositing parameters refer to other compositing methods and should be left blank. Unless this is the first time in which composited records are being added to the composite table, you should be sure that the UPDATE TYPE is set to the appropriate method. Selection of **Composite** will place new records into the composite table. Note the number of composites created and the number of records entered into the composite table.

Compositing illustrates the volume/variance relationship in the case of drillhole data. The average length of the raw assay data is 6.35 m with cores ranging in length between .1 and 10 m. Prior to compositing, there are 456 samples having an assay variance of .008 with a Coefficient of Variation of 41 (Fig. 4.2); following compositing to 391 10 m samples, the assay variance falls to .006 and CV = 35 (Fig. 4.3). In both cases the average assay is 0.22 oz/t, but variance of the composited data decreases and the shape of the distribution becomes less skewed as the higher assays are absorbed into lower grade composites.

It is interesting to note that while the mean assay is not effected by compositing,

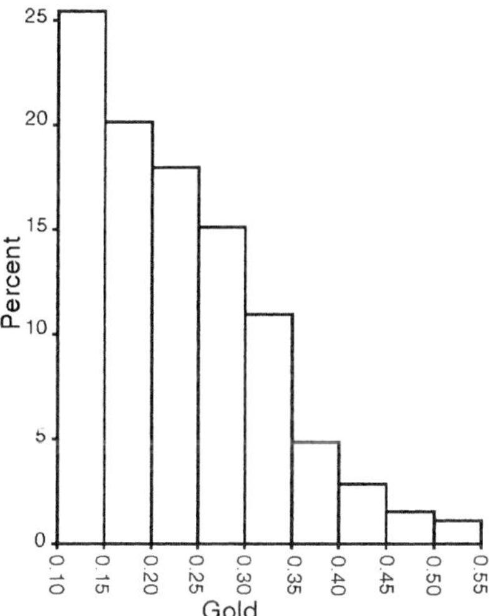

Figure 4.2. Distribution of original gold assay data.

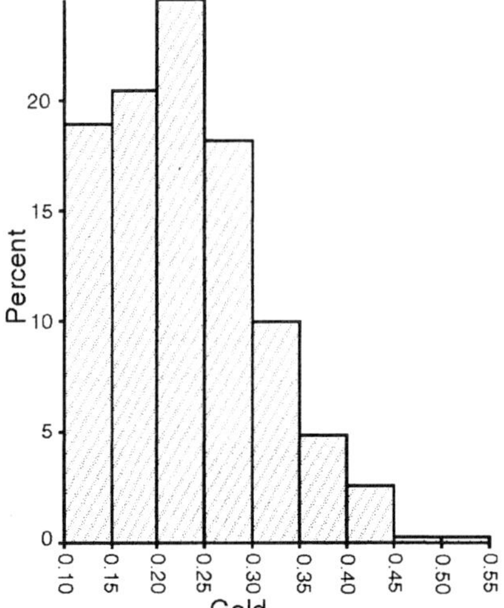

Figure 4.3. Distribution of gold assay for 10 m composites.

the variance and shape of the distribution does. In fact, if the compositing process was continued to even greater lengths, say 20 or 30 m in this example, the composited distribution would rapidly approach a normal distribution. This is a consequence of the Central Limit Theorem which states that as the sample size drawn from a distribution of any size increases, the mean of the sample size will tend toward a normal distribution. In compositing, the sample size is the number of samples averaged into a composite, which is directly proportional to the composite length. The sample mean is the composited value. Therefore, as the composite length increases, the distribution of the composites will tend towards normality.

Summary of procedure: Example 4.1

Initial data requirements: A drillhole data set loaded into Techbase including at least collar locations and assays. This example uses the Boland database.

Procedure:

1. Create a flat table for composited assays **Techbase ⇒ Define ⇒ Tables ⇒ Create**
   TABLE NAME: <compOA> TYPE: <flat>
   ⇒ **Add field**
   TABLE NAME: <compOA> FIELD LIST: <hole_id>
   ⇒ **eXit**
2. Create fields to hold composites **Fields ⇒ auto Table**
   TABLE NAME: <compOA>
   ⇒ **Create**
   FIELD NAME: <x_10> TYPE: <real> CLASS: <actual> DISPLAY PRECISION: <0>
3. Repeat Step 3 for *y*_10, *z*_10 and au_10, except that DISPLAY PRECISION should be three decimal places for au_10.
4. Compositing
   4.1 From the main Techbase menu **Modelling ⇒ Composite ⇒ Database ⇒ Add filter** <lith = OA>
   ⇒ **eXit**

4.2 Input fields **Setup** ⇒ **Fields**
HOLE ID: <hole_id> TOTAL DEPTH: <depth>
X: <x> Y: <y> Z: <z>
AZIMUTH: <bearing> DIP: <dip>
FROM: <from> TO: <to>
4.3 Output fields ⇒ **Coordinates**
X: <x_10> Y: <y_10> Z: <z_10>
HOLE_ID: <hole_id>
4.4 ⇒ **Assignment**
COMPOSITE FIELD: <au_10> ASSAY FIELD: <Au>
(repeat for each assay field to be composited)
4.5 ⇒ **Parameters**
TYPE: <length> LOCATION: <middle> MAXIMUM LENGTH: <10>
⇒ **eXit** ⇒ **Composite**

Example 4.2: Using length compositing to generate surface intercepts in Techbase
Length compositing has another application: it can be used to assign coordinates to the points at which drillholes intercept the boundary between lithologic units. This is accomplished using the same procedure as Example 4.1. The difference in this example is that the length interval will be made so long that it will be possible for only one composite sample to be formed within the geologic unit of interest. Also, instead of assigning the composite's coordinate **Location** to the middle of the composite, the *xyz* coordinate will be alternately assigned to the top and bottom of the composite in two separate runs with value assignments being made into two different composite tables, one for the top intercepts and the other for the bottom intercepts.

Before leaving **Define**, create a new field, called length, in the assays table that is to be composited. Create length as a calculated field equal to the difference of the fields from and to that are used in the assays table (length = to from –). Run **Statistics** ⇒ **Summary** for the field length to determine the maximum length of the drillhole intercepts with the lithologic unit of interest. This will be used as an approximation of the maximum thickness of the rock unit.

Following the same procedure used in Example 4.1, create two flat tables with fields to contain the composite samples' top and bottom coordinates. Note that since the number of records in these two tables will be equivalent in number and identity, one table can be used to hold both the top and bottom intercepts, but be sure that there are separate *xyz* coordinate fields for the two surfaces.

Still following the steps outlined in Example 4.1, assign a filter to the lithologic unit of interest (e.g. lith = OA) and under **Setup** enter the collar, survey and assay table fields. Since only the coordinates of the top and bottom intercepts are of interest, it is not necessary to assign raw assay data to a composited field unless you will want to work with two-dimensional composited data later.

In the **Parameters** menu, set **Type** to length, **Location** to top, and **Maximum length** to the maximum difference found between the from/to fields in that lithologic unit. Set **Update type** to replace and then **Composite**. Techbase will generate a warning message for each drillhole that intercepts the lithologic unit since the composite length exceeds the actual length of the available samples. Don't worry. Set DISPLAY MESSAGE to 'off' and proceed. Note that the number of composites and the number of records added to the table should be no more than the number of drillholes in the collar table. It's good practice to know in advance of compositing by lithologic unit how many of the drillholes intercept the rock unit. This provides a quick means of double checking composite results. For instance, if the number of composite records created exceeds the number of drillholes that intercept the rock unit of interest, it might indicate that multiple composites are being created in some of the longer holes and that the composite length needs to be increased.

Next, change the **Assignment** fields to the fields that will hold the bottom intercept coordinates, and under **Parameters** change **Location** to bottom and **Update type** to overlay (if the top and bottom intercept fields are to be held in the same flat table) and **Composite**. If all of the holes being composited are vertical, then the *xy* coordinates of the top and bottom structure intercept will be identical and one composite table can be used to hold the results, but in the case of inclined holes, the intercept location will be different, requiring different field names for *xy* coordinates of the top and bottom intercepts.

The resulting composite table(s) should be reviewed in **tbEdit**. There should be only one composite record per hole. Later in this chapter, the composite top and bottom intercepts will be used for generating gridded surfaces of the top and bottom structures.

Summary of procedure: Example 4.2

Initial data requirements: The same as for Example 4.1 with the addition of a lithology field in the assays table or a separate geology table. When the ore zone is based on an ore cutoff, a data filter restricting sample grades to be higher than cutoff can also be used, but only if assays are continuously above cutoff in the ore zone.

Procedure:

1. Create a flat table for the top of composited lithology **Techbase ⇒ Define ⇒ Tables ⇒ Create**
   TABLE NAME: <OAtop> TYPE: <flat>
   ⇒ **Add field**
   TABLE NAME: <OAtop> FIELD LIST: <hole_id>
   ⇒ **eXit**
2. Create fields to hold composites **Fields ⇒ auto Table**
   TABLE NAME: <OAtop>
   ⇒ **Create**
   FIELD NAME: <oaxt> TYPE: <real> CLASS: <actual> DISPLAY PRECISION: <0>
3. Repeat Step 2 for the *y* <oayt> and *z* <oazt> coordinates.
4. Repeat Steps 1-3 for the bottom of composited lithology (see table OAbotm)
5. Compositing
   5.1 From the main Techbase menu **Modelling ⇒ Composite ⇒ Database ⇒ Add filter** <lith = OA>
   ⇒ **eXit**
   5.2 Input fields **Setup ⇒ Fields**
   HOLE_ID: <hole_id> TOTAL DEPTH: <depth>
   X: <x> Y: <y> Z: <z>
   AZIMUTH: <bearing> DIP: <dip>
   FROM: <from> TO: <to>
   5.3 Output fields ⇒ **Coordinates**
   X: <oaxt> Y: <oayt> Z: <oazt>

HOLE_ID: <hole_id>
5.4 ⇒ **Parameters**
TYPE: <length> LOCATION: <top> MAXIMUM LENGTH: <1000>
⇒ **eXit** ⇒ **Composite**
6. Repeat Step 5 for the bottom surface.

4.2 TWO-DIMENSIONAL SURFACE MODELLING

A large variety of geologic structures can be represented as surfaces: seam top and bottom structures, soil horizon thickness, faults, phreatic surfaces, or any other topological data which can be represented by a single z value at a xy coordinate on a plane. When a geologic structure becomes so irregular in shape that a single z value at x and y is no longer sufficient to describe the surface, then volume modelling should be used. A example of this would be a lithologic boundary that is so highly folded that the surface folds over itself so that a line normal to any plane would puncture the surface in two or more places as demonstrated in Figure 4.4. Two surfaces with similar orientations can be used to store z values such as top and bottom structure elevations of a seam, which can be used to estimate thickness as $z_{top} - z_{bottom}$. However, when two surfaces are not sufficient to model a structure, as would be the case for a cylindrical ore body, volume modelling must be used.

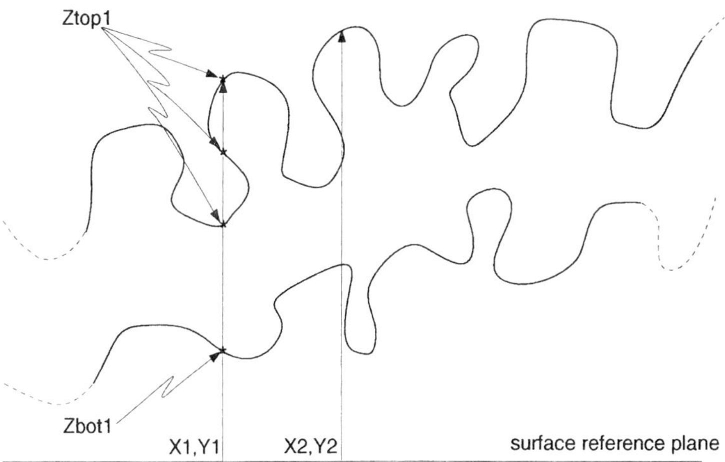

Figure 4.4. Highly convoluted geologic structures cannot be projected onto a plane as a single z coordinate position xy in the plane.

In general, structures should be modeled as surfaces whenever possible. Surfaces are easier to generate, manipulate and display. Volume models should only be resorted to when the geometry is too complex or when surface intercept data is so sparse that a high degree of interactive interpretation is necessary in order to define the surface.

Either grids or triangle sets can be used to store surface information. The choice between these two methods determines the options which are available for modelling and the limitations of the model. Grids are used for surfaces in CAD packages. In CAD software, grids can be used to lay a surface over a wire frame. Color and lighting algorithms can then be applied to the cells of the grid and the grid lines hidden to give the wire frame the appearance of a solid object. CAD grids are defined by the *xyz* coordinates of their vertices and can be either rectangular or irregular polygon meshes. Meshes can be fit to point data using splines or bezier curves.

The primary objective of using surfaces in CAD software is to display the surface as a visual representation of a design which is composed of regular geometric primitives, such as spheres, cubes, prisms and the like. This is because the primary application of CAD is aimed at manufacturing and design. Geologic structures are irregular in shape and are poorly represented by fitting geometric functions or by using primitives. Also, the objective of geologic modelling is not simply for display, but is aimed at calculating volumes of intersection. Gridding can be used to represent two surfaces, say a top and a bottom structure, but the method of defining the elements in those two gridded surfaces must lend itself to making calculations of the relative positions of the grids so that thicknesses and volumes can be calculated. By adding a third surface, such as a pit wall, which intersects the top and bottom structure grids of an ore zone, the three meshes can be algebraically manipulated to determine the volume or ore contained in the pit. In the following, two powerful methods of representing surfaces will be discussed more fully: 2D matrices (Techbase) and triangle sets (Lynx).

4.3 REPRESENTING A SURFACE AS A MATRIX

A two-dimensional matrix is an array of rows and columns used for storing data. The rows and columns may represent records and variable values and simply be a convenient means of storing related data, but a more powerful application of matrix structures occurs when the row and column indices are used to represent the descretization of two continuous independent variables x and y, whose values determine the value of a dependent variable $z = f(x,y)$. $f(x,y)$ may be a continuous function which is defined. In this case the matrix $\mathbf{Z}$ is only needed to hold values of z for fixed intervals of x and y, but when $f(x,y)$ is unknown, then $f(x,y)$ can be approximated and the matrix $\mathbf{Z}$ holds the values for that approximation.

Geologic surface modelling using grids is aimed at representing an unknown surface location z by an estimate of the function, i.e. $\hat{z} = f(x,y)$. Since we are working with spatial data, the independent variables x and y become a coordinate in a reference plane and z becomes an offset from that plane. The two-dimensional matrix $\mathbf{Z}$ is used to define the position of a geologic surface in three-dimensional space by translating the 2D xy matrix index, the z offset and the orientation of the reference plane into global coordinates. Surfaces can be represented as 2D matrices by storing the estimated value $\hat{z}$ by coordinates xy, which correspond to the indexing of the matrix, and by the orientation of the plane. This is illustrated by Figure 4.5 which shows a contoured map generated from a grid of 49 estimated values stored in a matrix with

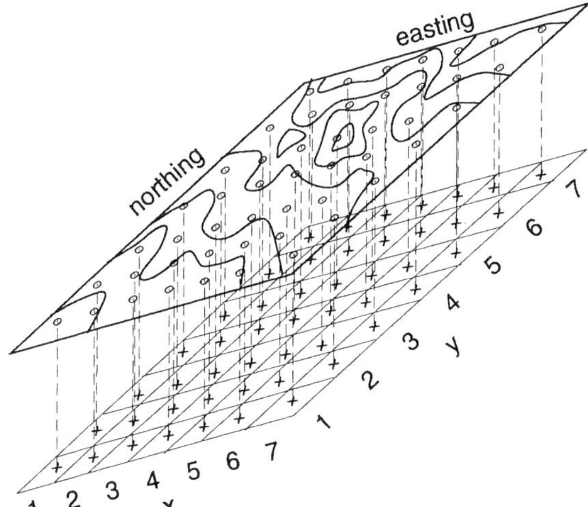

Figure 4.5. Representing surface offsets as a matrix.

elements at indices $x = 1...7$ and $y = 1...7$ that correspond to map coordinate eastings and northings, respectively.

There is a critical gap between actual data and the matrix representation of a surface. The actual data that is available does not exist on a regular grid whose elements can be translated directly into a matrix. Rather, data is available at irregular positions in x and y. These randomly located data values, z, must be used to estimate a value, $\hat{z}$, for every cell in the matrix. Estimation of matrices is the primary thrust of geostatistics and will be reviewed in Chapter 6.

When gridded surfaces are defined as matrices, matrix algebra provides a computationally efficient means of manipulating the $\hat{z}$ values stored in one or more related grids. Typically, the only actual datum which will be estimated into a matrix is the estimated $\hat{z}$ position of a surface. All other data are then calculated by performing matrix algebra on the $\hat{z}$ values of two or more surfaces. An example of this procedure follows.

Example 4.3: Matrix computations

Consider the case of two flat lying coal seams, A and B, for which drillhole data and a contour map of the topography is available over an area of 1000' on a side and wherein a local coordinate system is used. These two sources of data are converted into *xyz* surface coordinates by digitizing the contours of the topography and by compositing the drillhole data to yield the top and bottom drillhole intercepts with the two seams. This provides random point data for five surfaces: topography, top of A, bottom of A, top of B and bottom of B. These data are to be used to model these five surfaces and then from these models to estimate the thickness, volume and tonnage of: (1) seam A; (2) seam B; (3) the AB interburden; (4) the overburden; (5) total waste; (6) total coal; and (7) a stripping ratio. Note that these models are not to be merely used for drawing contour maps. Rather, these must be true spatial models which can be used to calculate thickness, volume and tonnage to be calculated for any area of the deposit.

The first step is to model the five surfaces onto five corresponding matrices: **A1**, **A2**, **B1**, **B2** and **T** for the tops and bottom elevations of seams A and B, and the topography. Each of the five matrices will be defined as having the same number of rows, *i*, and columns, *j*, and it will be assumed that the matrixes' cells correspond to a location and area of the property. Let the area covered by the property be 1000' × 1000'. The elevation of the surface topography of the property is to be contained by a 10 × 10 matrix. In this way, a cell $a1_{ij}$ has a position and a fixed areal dimension, 100' × 100'. Also, assume that **A1**'s orientation is in the horizontal plane, that its origin is coincident with the origin of the property to be modeled, and that the span of the cells of the matrix cover the entire property, or at least that portion of the resource that is of interest. Following this methodology, the corner of the first cell of the matrix, $a1_{11}$, will correspond to a location (0,0). Given a cell dimension of 100 × 100, the centroid of the cell will be at (50,50). Likewise, cell $a1_{12}$ will be centered on (50,150) while cell $a1_{21}$ will be centered on (150,50). The bottom right corner of the final cell, $a1_{10.10}$, will be located at (1000,1000). Thus, the matrix **A1** covers the full extent of the property. By defining **A2, B1, B2** and **T** in the same way, any location in a surface in the property can be represented by a cell in one of the three matrixes. Note that each of the matrices must be equivalent in terms of their origin, extent, orientation and number of cells or else there will not be a one-to-one correspondence between their cells, i.e. cells $a1_{ij}$, $a2_{ij}$, $b1_{ij}$, $b2_{ij}$ and t_{ij} must cover the same area on the property and have their centroids aligned at the same easting and northing.

The next step is to use the random point *xyz* data to estimate values into each of the cells of the corresponding matrix. Numerous algorithms are available for estimating from random point data to cells, the details of which will be covered in Chapter 5. Triangulation has already been introduced in Chapter 3 as a means of contouring. It can also be used for estimating grid values. In this case, the triangle set is overlaid on the matrix, such that for any cell there is one triangle whose extent includes the cell centroid. The value of the cell is then estimated by solving the equation of the plane defined by the three *xyz* vertices of the triangle using the xy centroid of the cell. Thus, each cell is assigned a *z* value.

The equation of a plane is $z = ax + by + c$. Consider the triangle shown in Figure 4.6 which has as its vertices the points (561,220,643), (620,230,670) and (580,310,700). The equation for the plane defined by these three points is found by solving for the parameters *a*, *b* and *c* in the system of equations

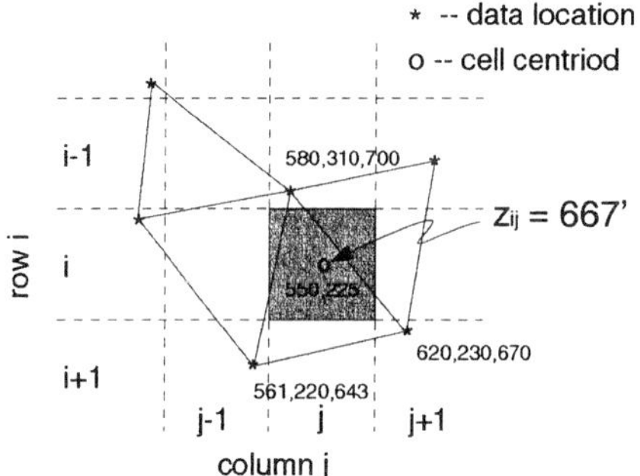

Figure 4.6. Using triangulation to estimate *z* values into a matrix.

$$561a + 220b + c = 643$$
$$620a + 230b + c = 670$$
$$580a + 310b + c = 700$$

By defining a matrix **A** to the coefficients of the equation, a vector **x** to the variables a, b and c and a vector **b** to the RHS, **Ax** = **b** can be solved as **x** = $\mathbf{A}^{-1}\mathbf{b}$.

```
>> A =
   561        200        1
   620        230        1
   580        310        1
>> b =
   643        670        700
>> inv(A) =
     -0.0135     0.0186     -0.0051
     -0.0068    -0.0032      0.0100
      9.9324    -9.7821      0.8497
>> inv(A)*b' =
      0.2128
      0.4814
   427.3142
```

Thus, the equation of the plan is

$$z = 0.2128x + 0.4814y + 427.3142$$

The centroid for the cell which falls within this triangle is (550,255). Plugging this coordinate into the equation of the plane, we get a $\hat{z}$ value of 667' for the cell. This process is continued for every cell in the matrix by locating the triangles which contain the cell centroid. A more detailed discussion of triangulation will be given in Chapter 5.

Once the matrices for all five surfaces have been estimated, matrix addition and subtraction can be used to obtain thicknesses. Note that the value estimated into all five matrices is the elevation of the respective surface. Therefore, to calculate the thickness of seam A, subtract **A2** from **A1** (**A1** as the top of seam A has the higher $\hat{z}$ value). For example, if the estimate z values for the A seams top and bottom structure are (Figs 4.7-4.9):

A1 =

400	420	427	450	452	460	457	490	480	484
403	423	433	440	455	465	460	488	485	485
410	415	430	460	462	467	460	482	481	480
411	412	433	440	470	475	470	480	485	478
413	420	430	435	454	470	468	480	480	475
417	422	433	440	450	460	465	470	475	470
420	420	425	430	440	450	460	465	470	465
425	422	430	435	445	440	445	460	465	460
430	427	425	435	450	445	450	455	460	460
440	430	430	440	460	465	470	465	465	455

A2 =

700	730	734	750	755	760	757	790	780	785
703	722	733	740	765	765	760	788	785	787
710	715	730	760	762	767	760	782	781	780
711	712	733	740	770	775	770	780	785	775
713	720	730	735	755	778	768	780	780	786
717	722	733	740	750	760	768	770	757	760
720	720	725	730	740	750	760	765	770	760
725	722	730	735	745	740	745	760	765	760
730	745	724	735	750	745	750	755	760	766
740	730	744	750	760	765	770	766	766	756

>> **A2-A1** =

300	310	307	300	303	300	300	300	300	301
300	299	300	300	310	300	300	300	300	302
300	300	300	300	300	300	300	300	300	300
300	300	300	300	300	300	300	300	300	297
300	300	300	300	301	308	300	300	300	311
300	300	300	300	300	300	303	300	282	290
300	300	300	300	300	300	300	300	300	295
300	300	300	300	300	300	300	300	300	300
300	318	299	300	300	300	300	300	300	306
300	300	314	310	300	300	300	301	301	301

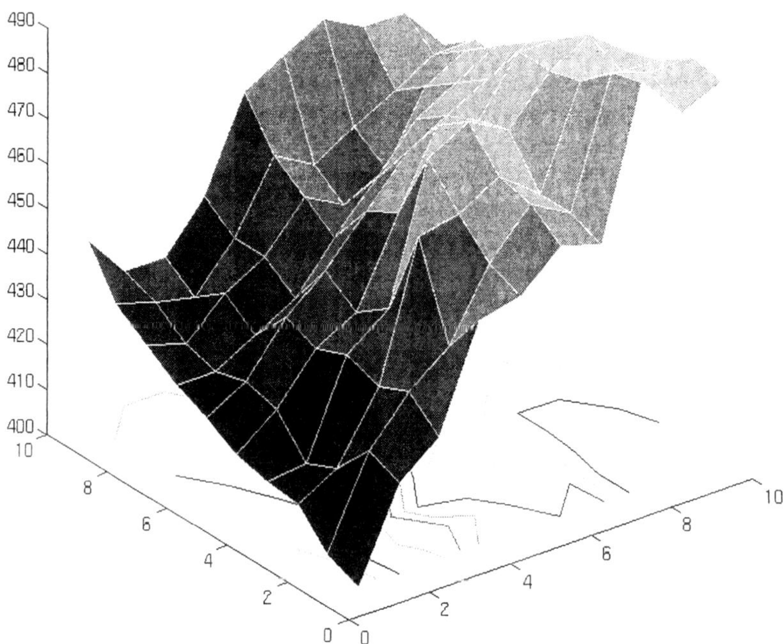

Figure 4.7. Top structure of seam A as stored in **A1**.

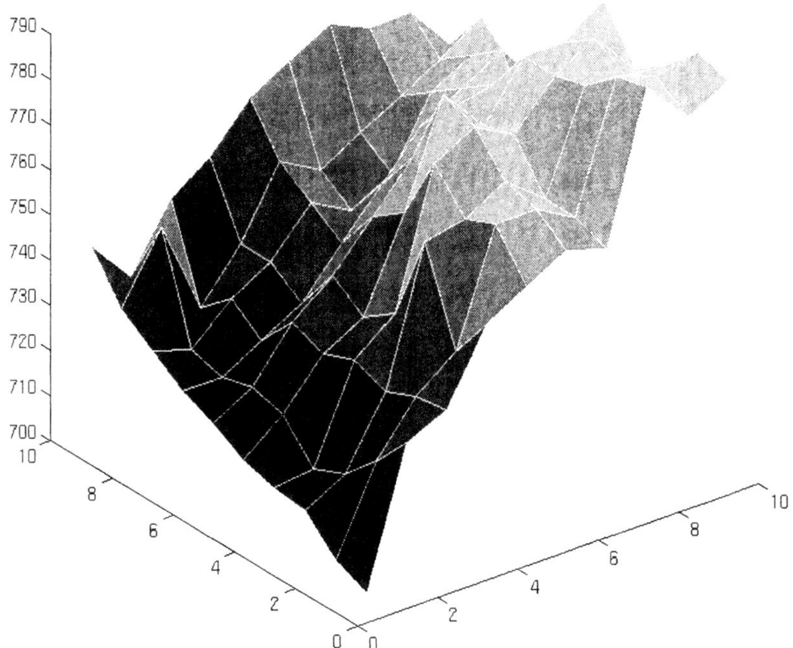

Figure 4.8. Bottom structure of seam A as stored in **A2**.

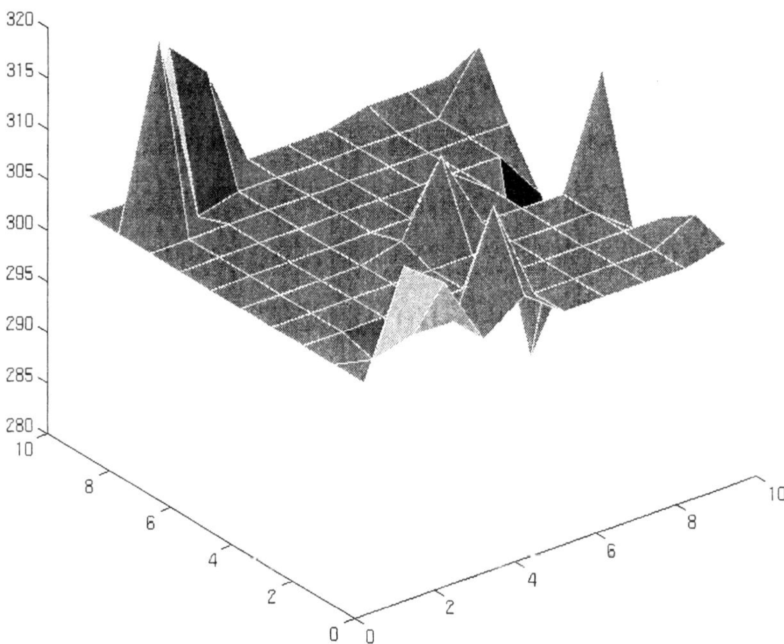

Figure 4.9. Seam A thickness (**A1**-**A2**).

Matrix to scalar operations can now be used to calculate the volume of coal in seam A as the product of the area of each cell and the thickness value of the cell, 100*100* (**A2-A1**). The cell volume times a density factor (say 60 lb/cu ft) yields the tons of coal in each cell, and the sum of the cells provides the total tons of seam A coal in the 1000' × 1000' property.

>> sum (sum (**A2-A1**))*60*100*100
ans = 1.8041*e*+10

To determine the thickness, volume and tonnage of seam B, the same procedure would be used. Once the surfaces for the bottom of A and the top of B are estimated, the interburden thickness can be found as: **A2-B1**. Likewise, the overburden is calculated as: **T-A1**. Total coal thickness is: (**A1-A2**) + (**B1-B2**), and total waste thickness is: (**T-A1**) + (**A2-B1**). The resulting stripping ratio (feet of waste to feet of coal) is found as: [(**T-A1**) + (**A2-B1**)] / [(**A1-A2**) + (**B1-B2**)]. Scalar operations and summations of cells can be used to translate thicknesses into volumes and tonnages. Quality characteristics can also be estimated as surfaces and used to calculate average characteristics such as Btu/ton.

4.4 CELL TABLES IN TECHBASE

Random contouring is an excellent way of generating contour maps of high density data, but is only suitable for generating graphics: it does not generate a surface that can be used for calculations. In Techbase, surface models are stored in cell tables, which are matrix structures. The main differences between a cell table and the matrices described in Example 4.3 are that a cell table can be used to store many estimated values and the cell dimension can vary in size across rows and columns.

The first step in defining a cell table is to determine appropriate values for its defining parameters by examination of the data which are going to be used for estimating its cell values, i.e. the origin, orientation, extent and density of the actual random point data. By calculating the **Summary** statistics for the coordinates of the data, as in Chapter 3, the minimum x and y coordinates that are to be contained in the table can be determined. Likewise, by examining a **Scatter** plot of x and y the extent, orientation and spatial density of the data set can be examined. The column and row size used in defining the cell table correspond to the dimension of the cell and are the most difficult parameters to determine. The cell dimensions should be based on the average spacing of the actual data. As the cell size is decreased, greater graphical resolution and improved volumetric accuracy will be possible, but the number of records (cells) in the table will increase rapidly, consuming more memory and slowing all calculations involving the table. The variogram and statistical evaluation of the estimation error for alternative cell sizes can be used to determine a statistically near optimal cell dimension, but this is a tedious iterative process. As a rule of thumb, a cell size between a half to a quarter of the average data spacing should be used. The average data spacing doesn't correspond to the average x and y coordinates, but to the average triangle side length resulting from a Delaney triangulation. Since this information is not easily obtained, it is probably best to eyeball a data posting and estimate an average spacing between neighboring data points. Note that the density of data may vary drastically from one area to another. If areas of drastically different sampling density constitute large or especially critical areas of the property, then the cell dimensions in that area can be varied in order to take advantage of increased data density by decreasing cell size and improving local estimation accuracy.

Another consideration when defining a cell table is that more than one value may be estimated into it. In fact, volumetric calculation will require multiple surfaces estimated into the same cell table or into equivalently aligned cell tables. For example, surfaces for both the topography, and composited drillhole *xyz* lithologic intercepts will be needed to calculate overburden. Digitized topographic data will have a much greater data density than drillhole data and will probably extend over a larger area. Thus, the cell table origin and extent should be determined using the topographic data, but while a small cell size might be desirable for the topography, a much coarser grid would be appropriate for the drillhole data. Matrix algebra involving two different cell tables having different cell sizes is not possible, since matrix operations between the two tables require that the cell tables have the same dimensions both in terms of the numbers of rows and columns as well as the cell sizes.

Example 4.4: Defining a cell table

Consider the Boland Banya gold deposit for which the topography was gridded using random contouring in Chapter 3. Execute the **Define** program, then select **Tables** and **Create**. Give the table a descriptive name (structures or assays) and define the table as being a CELL table. BASELINE AZIMUTH can be used to rotate the table so that a relative coordinate system can be aligned to true north if desired. The default is 90°, which aligns the x axis with an azimuth of 90°. Assign the table's lower left corner as X MINIMUM = 9900 and Y MINIMUM = 19800 (minimum XY as found for generating a poster). The upper left corner of the cell table is defined by the XY dimensions of the cells (ROW SIZE = x grid interval length, COLUMN SIZE = y grid interval length). Assign a fixed cell size of 40 × 40 ft. The NUMBER of cells times the cell size in that dimension should correspond to the upper right corner of the property or deposit. 83 cells of 40' in the x dimension will span easting from 9900 to 11,500, while 64 cells of 40' in the y dimension will span northing from 19,800 to 22,360. The resulting table will be defined as shown in Figure 4.10. **Exit** from the TABLE menu and select **Fields** followed by **Auto table** as per usual. Select **Create** and add new fields to the cell table to hold the values that are to be gridded. Note that it is not necessary to add coordinate fields; TECHBASE creates default fields for the coordinates of the cells when the table is defined. These default fields will be called: {table name}_rec for holding the table record number, {table name}_ nul is a null entry, {table name}_row is the matrix row number, {table name}_col is the matrix column number, {table name}*_xc* is the x coordinate of the cell centroid, {table name}*_yc* is the y coordinate of the cell centroid, {table name}*_csz* is the column cell dimension and {table name}*_rsz* is the row cell dimension. The default fields for the matrix index (_row and _col) can be used to export matrices in or out of a Techbase database. This has important applications for transferring matrices between Techbase and other programs which are used for advanced geostatistical modelling.

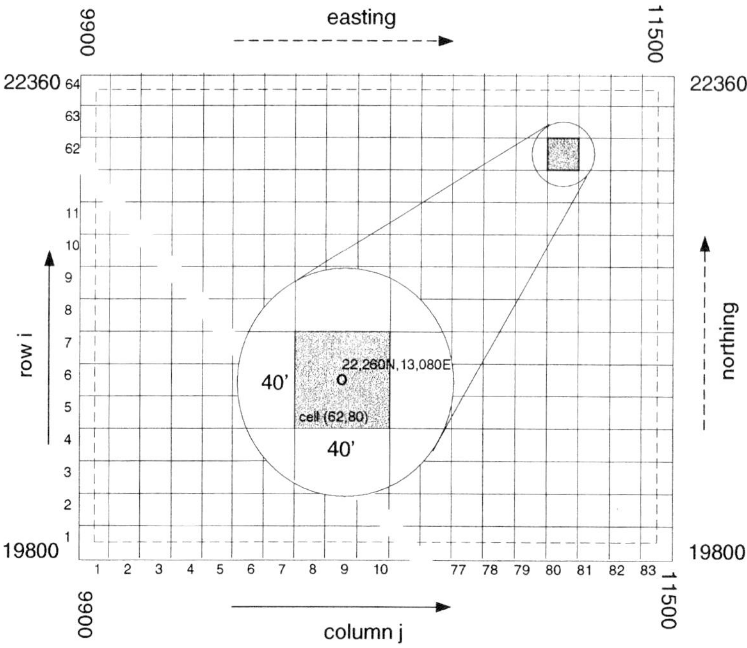

Figure 4.10. The coincidence of a cell table matrix and the location and extent of the area stored in it.

Summary of procedure: Example 4.4
Initial data requirements: An initialized database.

Procedure:

1. **Techbase ⇒ Define ⇒ Tables ⇒ Create**
   TABLE NAME: <surfaces> TYPE: <cell>
   X-COORD: <9900> COLUMN SIZE: <40> NUMBER: <83>
   Y-COORD: <19800> ROW SIZE: <40> NUMBER: <64>
   ⇒ **eXit**
2. Add fields (other than coordinates) to cell table:
   2.1 **Fields ⇒ auto Table**
   TABLE NAME: <surfaces>
   2.2 ⇒ **Create** (create fields as needed for gridding as per Examples 4.1 and 4.2)

4.5 TRIANGULATED SURFACES

Matrix representations of surfaces are extremely well suited to laminar geologic structures for which the top and bottom structures are reasonably parallel to each other and the reference plane, but when the structure to be modeled is steep, non-laminar or has major changes in strike and dip, a matrix coordinate system which is fixed to a plane becomes incapable of representing the surface since the orientation of the reference plane should be based on the strike and dip. Even the use of multiple reference planes cannot solve this problem, since the orientation of these planes will not be in alignment with each other. As a result of the misalignment of the reference planes, there will be no way to apply matrix algebra for calculating thicknesses and volumes. If a set of aligned matrixes can no longer be used to represent top and bottom surfaces, then the justification for using a rectangular cell for gridding is no more valid than any other polygonal mesh, since the opportunity to use matrix algebra is lost. In this case, the vertices of the polygon represent an *xyz* coordinate on the surface being modeled. This differs from a cell table definition where the corners of the cell only have an *xy* position. Of the possible polygonal shapes that could be used on a surface, Delaney triangles will provide the greatest surface resolution since a triangle defined by three nearest neighbors data points will have the smallest area of any planar geometry which could be defined by the data.

Lynx uses triangle sets as the basis for constructing volume models based on prisms. Volume models based on surfaces can be generated much more rapidly through this method than through pure interactive definition of surfaces on sectional views. Thickness calculations can be performed between subparallel and contorted triangle sets, but the basic limitations of surface modelling still apply.

Example 4.5: Generating surface offset data in Lynx

Unlike Techbase's use of drillhole compositing to obtain *xyz* intercept data, Lynx calculates top and bottom structures as offsets from a map and then stores the offset data in the map. The map viewplane is defined as being roughly parallel to strike and dip and, when possible, is positioned between the structures from which offset data are to be stored. The *xyz* points of penetration of the drillholes with the hangingwall and footwall are then projected normal to the reference plane with *z* being the vertical offset distance from the plane and *xy* being coordinates in the plane as in Figure 4.11. These offset values are then stored as map data and can be triangulated. The triangle sets for the offsets of the hangingwall and footwall can then be used to calculate thicknesses and form volume components.

Using the Lynx TUTORIAL project in the Lynx main menu select **Viewplane ⇒ Setup** and set the viewplane orientation to: 22700, 38850, 1400, 270, 40. Set the scale to 5000:1 and use a global grid of 200, 200, 200. Select **Refresh** to refresh the display.

Create a new map that will hold the offset data from the hangingwall of zone 7 to the viewplane. Select **File ⇒ Map Data ⇒ Map Create** and enter the **Map Name** <z7t> and description. If an existing map is selected from the Map Selection list, it will be overwritten with the new definition and its data will be lost. Note that this should be a new map since it will be assigned the same orientation that has already been provided for the viewplane. Selecting **Associated Variables** brings up the Variable/Characteristic Selection list from which THICKNESS should be selected. This will allow both offsets and thickness data to be stored in the map.

Before continuing with drillhole projection, make sure that the raw drillhole subset (dh0) is active. From the main menu select **Maps ⇒ Hole Data Projection** to display the Hole Data Projection Parameters entry form. Again, select the drillhole subset to be used. The posting variable selected must contain the lithologic data being used for surface definition, in this case, the variable 'lith'. The CHARAC-

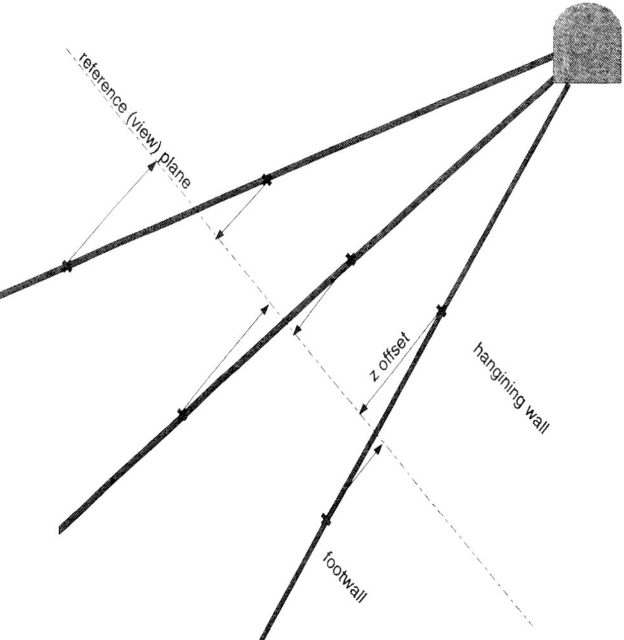

Figure 4.11. Projections of hangingwall and footwall drillhole intercepts (**x**) as z offsets on a map reference plane.

TERISTIC values are the names of the lithologic units for which offset data are to be calculated, for instance 'zone 7.' Enter the type of offset to be calculated: top, bottom, or top + thickness. Select top. Enter the destination map's name. Note that once any map is active, access to this entry form can be used to create new maps having the origin and orientation of the current viewplane. In fact, selection of **OK** or **Apply** will result in the overwriting of the map. The offset will be calculated for each drillhole trace that penetrates zone 7 (Fig. 4.12) and written to the map (z7t).

Repeat this exercise for the bottom of zone 8, writing the results into another map (z8b). The background display facilities and map report can be used to review the offset data.

Summary of procedure: Example 4.5
Initial data requirements: a drillhole data set with structural or assay data.

Procedure:
1. Orient viewplane **Viewplane ⇒ Setup**
   NORTH: <22700> AZIMUTH: <270>
   EAST: <38850> INCLINATION: <40>
   ELEVATION: <1400> SHOW GRAPHIC: <on>
   SCALE Y: <5000> SCALE X: <5000>
   GLOBAL GRIDS: <on> N: <200> E: <200> L: <200>
   LOCAL GRIDS: <no> TRACKING: <global>
   ⇒ **OK**

⇒ **Refresh**

2. Create a new map **File** ⇒ **Map Data** ⇒ **Map Create** (Map Create entry form)
   ⇒ **Map Name** <z7t> (Map Selection list)
   ⇒ **Associated Variables** <thickness> (Variable/Characteristic Sequence Selection)
   NORTH: <22700> AZIMUTH: <270>
   EAST: <38850> INCLINATION: <40>
   ⇒ **OK**
3. Project drillhole data onto map **Maps** ⇒ **Hole Data Projection** (Hole Data Projection Parameters)
   SUBSET/HOLES SELECT: <dh0>
   VARIABLE OR CHARACTERISTIC: <lith>
   CHARACTERISTIC VALUES: <zone7>
   STRUCTURAL FUNCTION: <top>
   DISPLAY HOLE NAME: <on>
   MAP NAME: <z7t>
   ⇒ **OK**
4. Check results with map report and background display
   4.1 **Maps** ⇒ **Map Data Report**
   MAP SELECTION: <z7t> (Map Data Report)
   ⇒ **OK**
   4.2 **Background** ⇒ **Maps** (Background Map Data Selection)
   DISPLAY ON/OFF: <on>
   AVAILABLE MAPS: <z7t>
   ⇒ **OK**
5. Repeat Steps 3 and 4 creating a duplicate map (z8b) to hold the footwall offset distance for zone 8 to be used in surface handling.

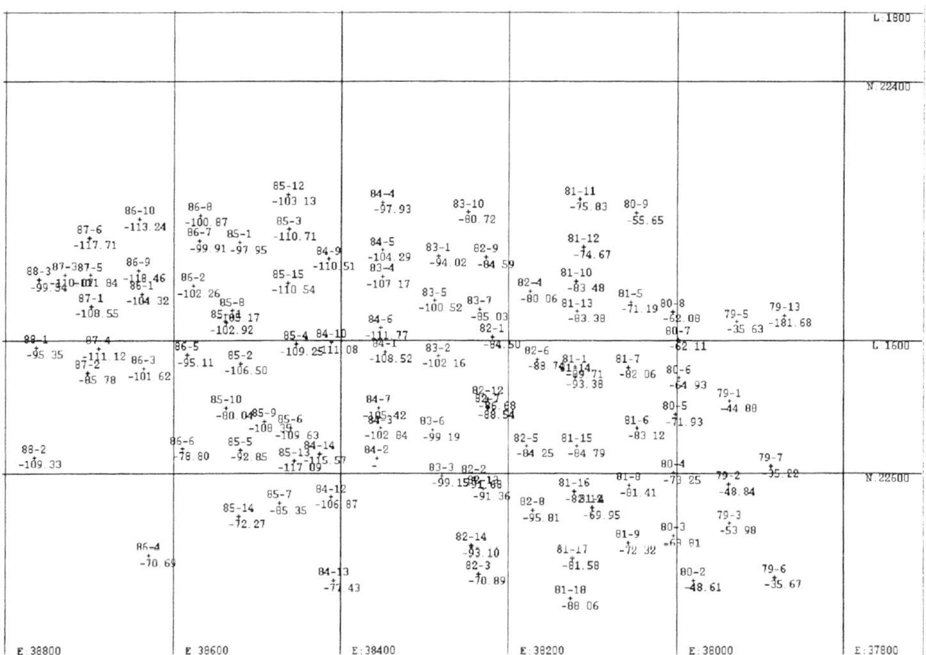

Figure 4.12. *Z* offset map data posted showing drillhole ID.

4.6 CALCULATING VOLUMES FROM TRIANGLE SETS

Grids are not directly based on the *xyz* data which are used in estimation of their *z* values. In fact, some grid estimation algorithms, such as trend surfaces, do not necessarily honor the *xyz* data; the residual error at the points on the grid which correspond to the original data locations will be nonzero. There are other problems with gridded surfaces.

It was noted earlier that matrix calculations between grids required that the dimensions of the matrices be equivalent. Actually, this is not exactly true. It is always possible to perform calculations between submatrices in which one of the two matrices is contained entirely within the limits of the other. The overriding criterion for relating two or more matrices is that the rows and columns must all be aligned so that a *z* value in the cell of one matrix can be related to another by referring to its index *ij*. As two surfaces move further away from being near parallel, it becomes increasingly difficult to project both surfaces onto the same reference plane.

Discrepancies in data density for two surfaces present another problem for gridding. The cell dimension should be based on the average separation distance between adjacent data, but topographic data are usually available at a much greater density than geologic drillhole data. With digitized contours available at a spacing of around 10 m and drillhole data taken at intervals of 500 m, the topography cells should be around 3 m on a side, but the coal seam grid should have cells as large as 150 m square. However, if these two grids are to be used to calculate overburden and stripping ratios, the same grid size must be used. Applying the 150 m grid to the topography would be ridiculous unless the surface was very even. Using a 10 m grid for the seam structures would result in a huge database which would be very slow to process. Because of the nature of matrices, a compromise must be made between efficiency and accuracy.

Triangulated surfaces avoid these problems by being directly based on the data locations and not being tied to matrix manipulations. Since the vertices of the triangles are formed from the *xyz* data, the original data are honored exactly. Even though the efficiency of using matrix operations is lost, this liberates this method of surface representation from the artificial requirements of cell alignment which are imposed by matrix algebra (see Holding 1994).

The object of forming a triangle set is to combine the set with another triangulated surface in order to create a volume model. Whereas a cell is associated with a fixed estimated *z* value, a triangle's *z* value varies depending on the *xy* location on the triangle surface. A triangle is used to form a prism which becomes a volume component of a larger volume model based on two triangle sets. The simplest case would be that of forming a volume model based on a triangle set and a fixed thickness as in Figure 4.13. In this case, a thickness would be assigned to each vertex in a direction normal to the map plane of the original offset data. Similarly, a surface of constant elevation could be used with a triangulated surface to generate volume components. In this case, topography may have been triangulated, and this is to be used to determine the volume of material to be removed as part of digging a foundation to a constant elevation for a plant site. To form prisms, the vertices of the triangles are projected onto a secondary surface of constant elevation. The edges of the prisms are

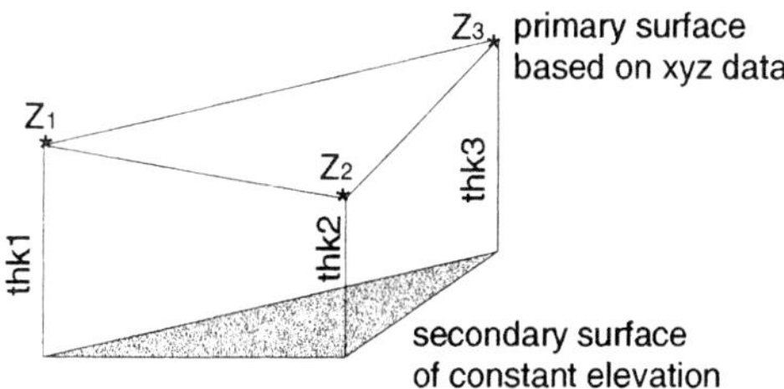

Figure 4.13. Prism volume component with a planar secondary surface.

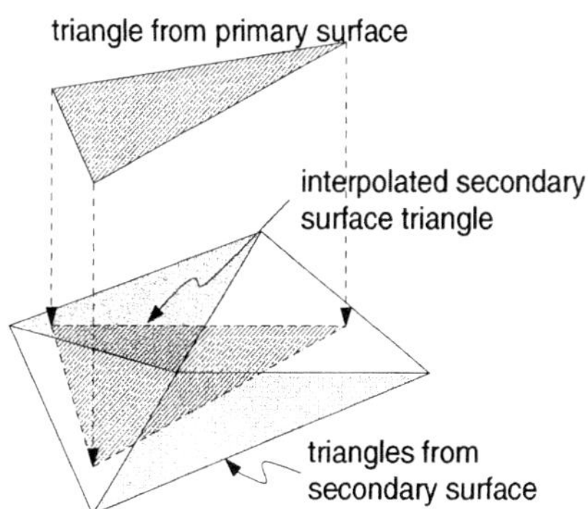

Figure 4.14. Intersected triangles of secondary surface.

parallel to each other and normal to the orientation of the viewplane.

A more complex situation arises when volume components must be formed from two triangle sets. In this case, it is not apparent how the prisms are to be formed since there isn't a one-to-one correspondence between the triangle vertices of the two sets (Fig. 4.14). In order to understand how this is done, it is important to note that final volume components must honor the original *xyz* data for both surfaces and that the triangles which were formed from the *xyz* data are artificial constructs which need not be retained in the final volume model. Prism-based volume model components can be formed as follows:

1. Select one surface to be the primary surface and the other to be the secondary surface.

2. Map the data from the primary surface onto the secondary surface and discard the original triangle set of the secondary surface. The *z* value of the primary points mapped onto the secondary surface will be determined from the original secondary triangle set by identifying the secondary surface triangle which is intersected by the projected ray, calculating the *xy* point of intersection, and then finding the *z* value by solving the equation of the plane associated with the secondary triangle. This interpolated *z* value is assigned as the thickness of the primary surface vertex (Fig. 4.15). Thickness estimation can end at this point by using the interpolated thickness, but this method will not honor the data of the secondary surface and can result in gross error if the secondary surface is very irregular and does not closely parallel the primary surface. To ensure that the data for both surfaces are honored, two more steps can be added.

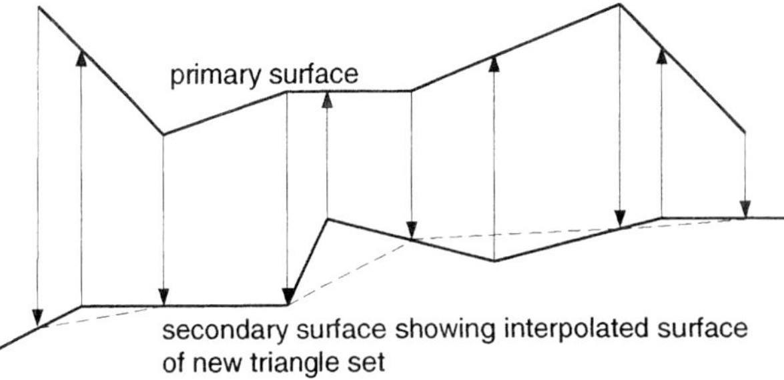

Figure 4.15. Generating prism volume components by projecting the triangle set of the primary surface onto the secondary triangle set. Vertices of the secondary surface are not honored.

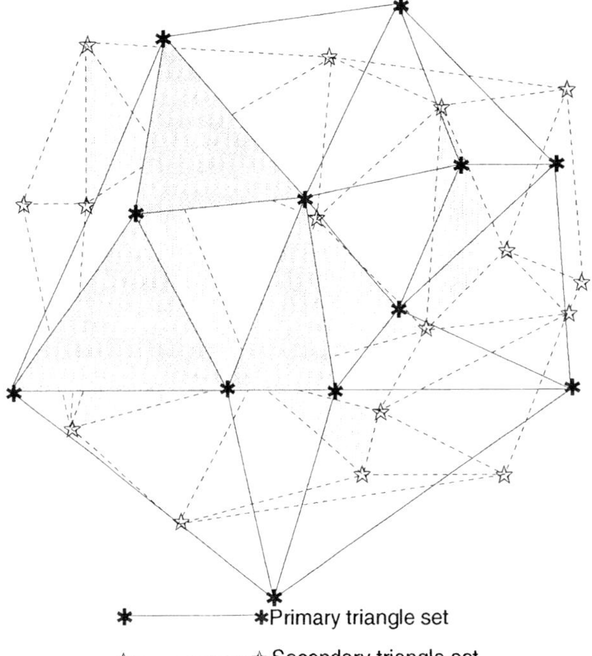

Figure 4.16. Primary and secondary triangle sets with non-coincident verticies.

3. Map the data from the secondary surface onto the primary surface, determining the z values from the primary triangle set, then subdivide the primary triangle set to honor the data from the secondary surface (Fig. 4.16).

4. Form prisms based on the subdivided triangle set in the primary surface (Fig. 4.17).

Note that there must be a means of controlling the direction in which the vertices from one surface are mapped onto the other. Lynx projects offset data normal to the current viewplane. When the offset data belongs to maps having different orientations, a viewplane can be used which splits the difference between the orientations of the two maps. This ensures that the projections from the two surfaces will come into alignment even if the maps's azimuths or dips are at a high angle to each other.

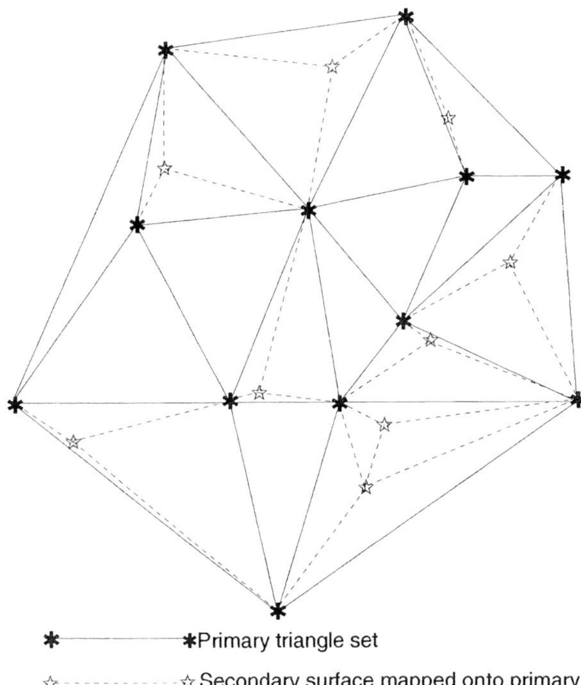

Figure 4.17. Both primary and secondary triangle sets honored.

Example 4.6: Generating triangle sets and volume components

Since a geologic volume model is going to be created or modified, the name of that model must be entered at this point before proceeding. The volume model will consist of components, in this case prisms, which will be stored in a file G.x, where *x* is a file extension of up to 10 characters of your choice. Enter any unique name and provide a description of the component file's contents.

Under **Viewplane** ⇒ **Setup** orient the viewplane to: 22300,38850,1400,270,40 (N, E, L, Az., Dip). This viewplane parallels the orientation of the offset map data of the top of zone 7 and bottom of zone 8 in the TUTORIAL project that was stored in maps z7t and z8b, respectively. Set the viewplane scale to 5000:1 and use a 200 m grid. Remember to **Refresh** the screen. **Return** from the viewplane menu and select **Maps** to use z7t offset data as an overlay. The same data should be displayed that were created in Example 4.5.

Create a volume model that will be based on triangle sets (TINs) for the hangingwall of zone 7 and the footwall of zone 8. The resulting volume model will encompass the thickness and extent of the primary and secondary ore zones. From the main menu select **File** ⇒ **Volume Data** ⇒ **Volume Create** to bring up the Volume Create entry form and enter the new volume name (G.zone78). The volume model must be active for subsequent volume operations.

Before map TIN manipulation can proceed, the map containing the offset data must be active. From the main menu select **File** ⇒ **Map Data** ⇒ **Map Select** and enter the zone 7 map from Example 4.5 (z7t).

TINs must be generated from offset data in the active map to generate prism-based volume components using surface handling. From the main menu select **Map ⇒ Map TIN Define** bringing up the Map TIN Define menu and the TIN Define Parameters entry form as shown in Figure 4.18. You will be prompted for the removal

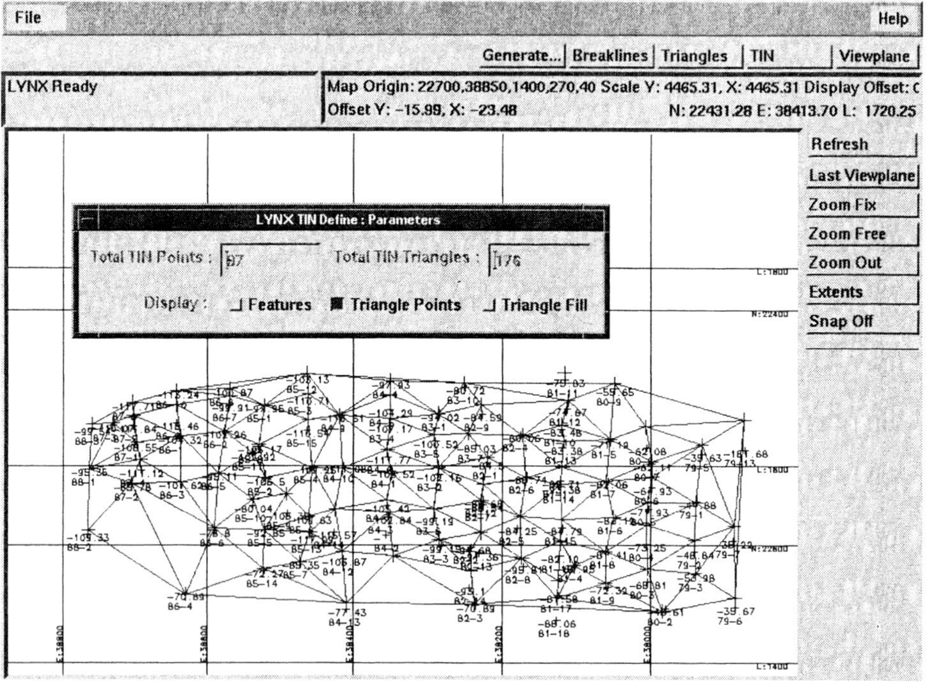

Figure 4.18. TIN used for modelling the surface as an offset for the top of zone 7.

of duplicate *xyz* data. Take notice if any points are removed. Duplicate points indicate a problem: either the surface folds on itself or more than one lithologic unit uses the same name. These problems must be corrected before continuing or interactive volume component definition on sections should be used. In the TIN Define Parameters entry form FEATURES should be toggled on. This will force triangle generation to honor map features such as traverse lines along creek bottoms or pit toe and crest lines. TRAINGLE POINTS and TRIANGLE FILL can be toggled on to display triangle apexes and highlight with fill the individual triangles in the display. In the Map TIN Define menu select **Generate** to generate the triangle set for the active map (z7t) and select **File ⇒ Save & Exit** to return to the main Lynx menu. Next repeat the TIN generation process for the second surface that will be used to define the volume model (z8b).

Volumes ⇒ Surface Handling brings up the Surface Handling menu and the Surface Handling Parameters entry form. One set of triangles will be designated as the primary set of *z* values and stored in one of three memory registers: a, b or c. z7t map data have been already loaded. Selecting **Primary** and entering z7t will result in defining *z* data in the map z7t as the primary surface. Select **Secondary** and enter z8b as being the secondary surface. Select **Manipulate** bringing up the Surface Manipulate entry form. At this point, the calculation that is to be used in defining the volume components is defined using three registers: a, b and c. From the Surface Manipulate entry form (see Fig. 4.19), select Define to bring up the Manipulate De-

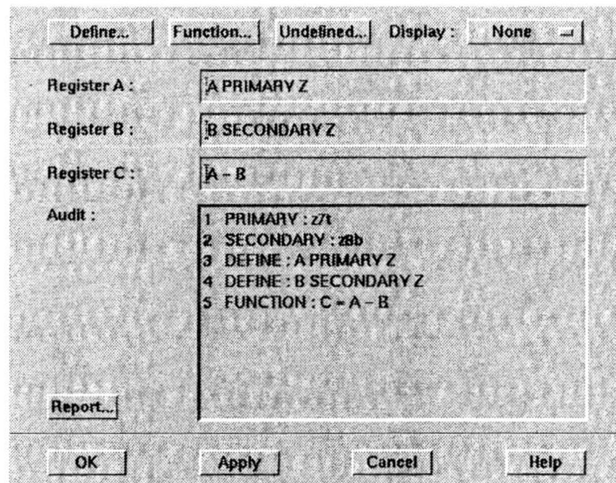

Figure 4.19. Surface Manipulate entry form for volume modelling of zones 7 and 8.

fine entry form from which the contents of the three registers is defined. Define registers a and b as being *z* offsets associated with the primary and secondary surfaces, respectively (i.e. the hangingwall (a) and footwall (b) offset from the viewplane). Register c will be defined as a thickness (c = a – b). From the Surface Manipulate entry form select **Function** to bring up the Manipulate Function entry form and enter REGISTER: <c> as being SUBJECT: <a>, OPERATOR: <->, and OBJECT: <b>.

What Lynx has done is to project the *z* values contained in a primary surface (the *xyz* map data of z7t) onto a secondary surface (the triangles of z8b). The projection has been normal to the viewplane (22300,38850,1400,270,40). The calculated thickness is based on the intersection point of the projected z7t data onto the triangles of z8b. This represents an interpolated thickness since the z7t triangles do not honor the data of z8b. Now that the primary and secondary surfaces have been defined along with the formula for calculating thickness, the volume model can be generated.

From the Surface Handling menu, select **Generate** to display the Volume Generate entry form and enter a name for the geologic model, the unit, prefix, and a pen color that will be attached as an attribute of the components and a description. The components will be generated and displayed as shown in Figure 4.20. Note that components will not be generated for primary triangles that aren't completely covered by corresponding triangles in the secondary surface. Missing prisms can present a major stumbling block to applying triangulation to geologic modelling when the top and bottom surfaces aren't of the same extent. Figure 4.21 shows the volume model in cross section with the drillholes shown with zones 7 and 8 in the same color. Both the holes and volume model are shown with a 20 m fore and back plane projection. The off section volume components are shown as dashed lines. Note that it is difficult to verify the volume thickness by visual comparison with the drillhole data.

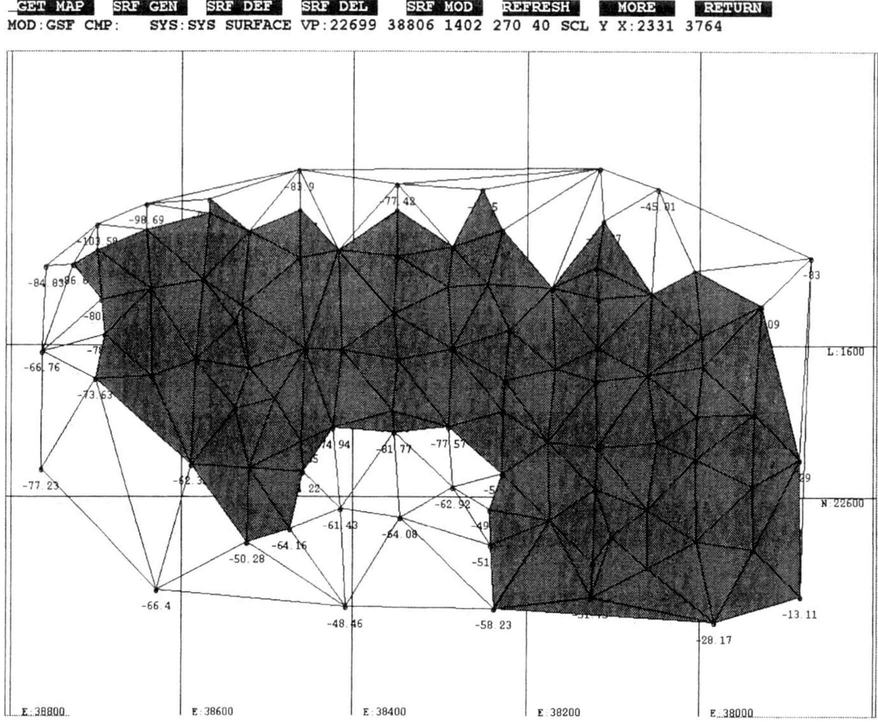

Figure 4.20. Prism volume components of zones 7 and 8 for the TUTORIAL database.

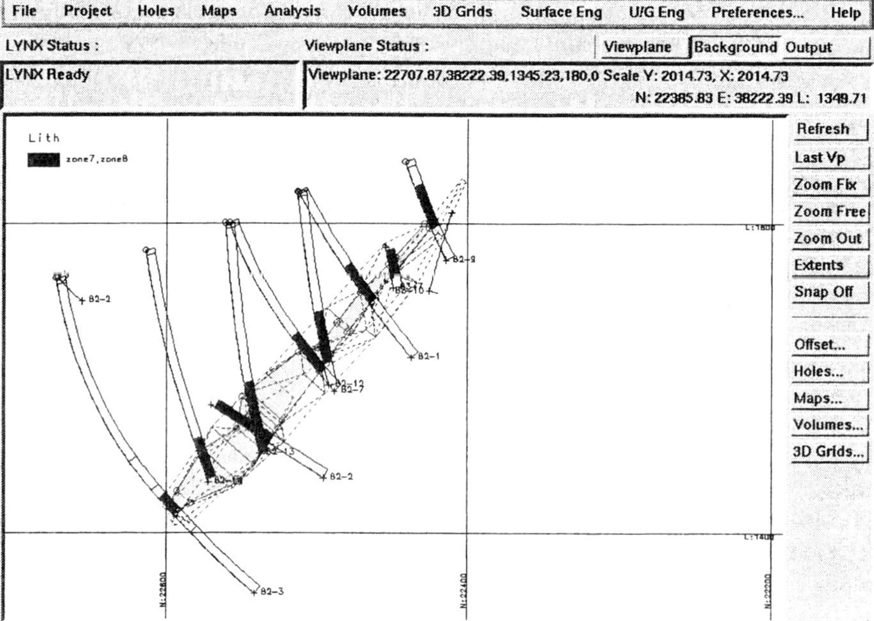

Figure 4.21. Zones 7 and 8 TIN-based volume model in X-section showing defining drillholes.

Summary of procedure: Example 4.6

Initial data requirements: Offset map data for a pair of subparallel, non-intersecting surfaces (see Example 4.5). In this case zone 7 hangingwall (z7t) and zone 8 footwall (z8b) maps from project TUTORIAL.

Procedure:

1. Make one of the offset maps active **File ⇒ Map Data ⇒ Map Select** <z7t>
2. Create the volume model **File ⇒ Volume Data ⇒ Volume Create** (Volume Create entry form)
   ⇒ **Volume**
   NEW VOLUME NAME: <G.zone78>
   ⇒ **OK**
3. Triangulate offset data **Map ⇒ Map TIN Define** (Map TIN Define menu and LYNX TIN Define: Parameters entry form). In Parameters entry form:
   FEATURES: <on>
   TRIANGLE POINTS: <on>
   TRIANGLE FILL: <off>
   In Map TIN Define:
   Generate ⇒ Generate (ignore points with z undefined)
   ⇒ **File ⇒ Save & Exit**
4. Repeat Step 1 for map z8b.
5. Repeat Step 3 for map z8b.
6. Create a volume model **Volumes ⇒ Surface Handling** (Surface Handling menu and Surface Handling: Parameters entry form)
   ⇒ **Primary** <z7t> (from Primary TIN Select listing)
   ⇒ **Secondary** <z8b> (from Secondary TIN Select list)
   ⇒ **Manipulate** (Surface Manipulate entry from)
   ⇒ **Define** (Manipulate Define entry form)
   REGISTER: <A> SURFACE: <primary>
   VARIABLE: <z offset>
   ⇒ **OK**
   ⇒ **Define**
   REGISTER: <B> SURFACE: <secondary>
   VARIABLE: <z offset>
   ⇒ **Function** (Manipulate Function entry form)
   REGISTER: <C>
   SUBJECT: <A>
   OPERATION: <->
   OBJECT: <B>
   ⇒ **OK**
   ⇒ **Generate** (Volume Generate entry form)
   VOLUME: <G.SURF78>
   UNIT: <> PREFIX: <>
   CODE: <3>
   ⇒ **OK ⇒ File ⇒ Save & Exit**
   ⇒ **Viewplane ⇒ X-section ⇒ Extents** (Background Volume Data Selection entry form must have display on)

Example 4.7: Map data contouring in Lynx

Hole data projection can also be used to generate map data as thicknesses that can be contoured. The same data as used for Example 4.6 will be used, but the procedure will be carried out entirely within the format of map data. Volume component data cannot be directly converted into a contoured display. Note that in Lynx, contours are features (cmaj or cmin) that are intrinsically connected to maps. Two approaches will be demonstrated in the following, display and storage of thickness data. In the first case, the background facilities will be used to display the combined thickness of the adjacent ore zones 7 and 8. This background display will then be output to a map as text features. In the second case, the hole data projection facilities will be used to generate thickness data in a map, which can be triangulated and contoured.

In order to view the thickness of the ore zones, the viewplane must be oriented approximately parallel to their strike and dip as per Example 4.6. From the main menu select **Background** ⇒ **Holes** bringing up the Background Hole Data Display entry form. Select the PRIMARY CHARACTERISTIC to be <lith> and set DISPLAY FORMAT to <projections> bringing up the Background Hole Data Projection Display entry form in which the CHARACTERISTIC VALUES are selected as <zone7, zone8> and STRUCTURAL FUNCTION is selected as <top + thickness>. All other hole display parameters are as in the previous examples. This selection of parameter values will cause the total thickness of zone 7 plus zone 8 to be projected normal to the viewplane. As long as the viewplane is reasonable parallel to the strike and dip of the ore zones, the ore thickness will be reasonably accurate. After refreshing the display, use **Extents** and **Zoom Fix** to center the display as shown in Figure 4.22. This display can be output to a map by selecting **Output** ⇒ **Map** in the main menu. The Output Map entry form will be displayed in which an existing map can be selected.

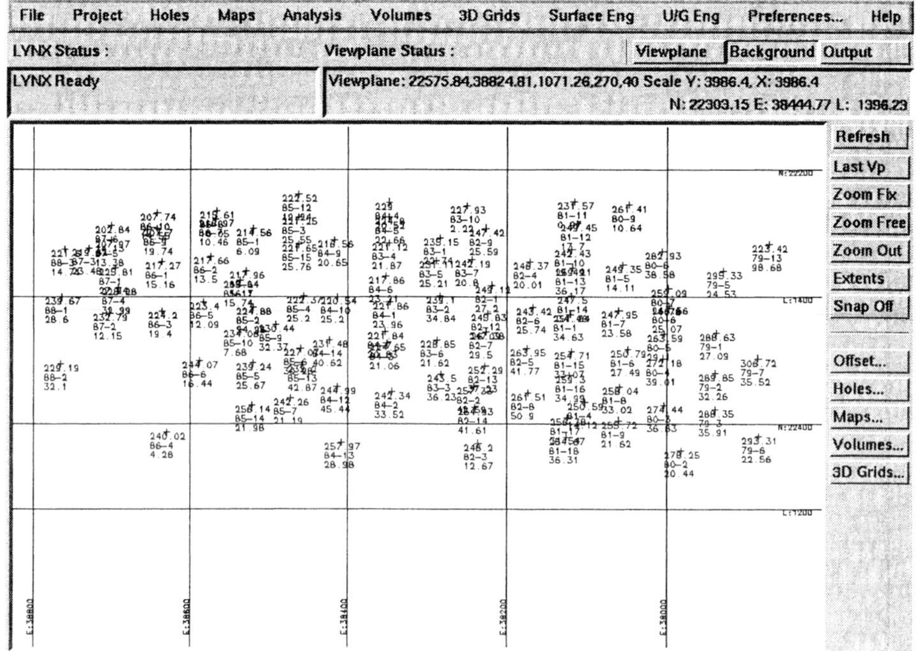

Figure 4.22. Hole projection display of thickness for zones 7 and 8.

The display will be either added to the selected map data or will overwrite the map's contents.

Map thickness data generated from a background display cannot be used as data for three-dimensional modelling as in surface handling. In fact, examination of the resulting map using the Map Report facilities will show that the thickness data has been stored as text features at a 2D *xy* coordinate. In other words, the map only contains 2D information and cannot be triangulated. In order to create useful 3D map data, the hole data projection facilities must be used. From the main menu select **Maps ⇒ Hole Data Projection** bringing up the Hole Data Projection Parameters entry form. This form is filled out as in Example 3.4 except that the CHARACTERISTIC is <lith>, the CHARACTERISTIC VALUES are <zone7, zone8> and the STRUCTURAL FUNCTION is selected as <top + thickness> as above. The name of the map in which the data is to be stored is entered in MAP NAME. Examination of a report of the resulting map will show that the data consists of *xyz* point features with an associated thickness. Map background display can be used to produce a display very similar to Figure 4.22.

The hole projection data contained in the map can be triangulated. Make the map part of the active data set and select **Maps ⇒ Map TIN Define** from the main menu bringing up the Map TIN Define menu and select **Generate** to create the TINs. Since the same set of drillhole projections are being used as in Example 4.6, the triangle set should appear as in Figure 4.20. Select **File ⇒ Save & Exit** to store the TIN to the map and return to the main menu. To contour the map select **Maps ⇒ Map TIN Contouring** bringing up the Map TIN Contouring entry form (see Fig. 4.23). Enter

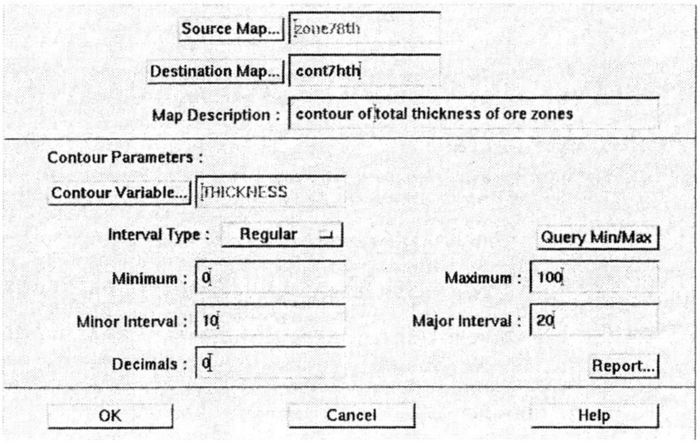

Source Map... zone78th
Destination Map... cont7hth
Map Description : contour of total thickness of ore zones
Contour Parameters :
Contour Variable... THICKNESS
Interval Type : Regular
Query Min/Max
Minimum : 0
Maximum : 100
Minor Interval : 10
Major Interval : 20
Decimals : 0
Report...
OK
Cancel
Help

Figure 4.23. Map contouring entry form.

the source map containing the thickness TIN and a destination map that will contain the contours. Using the same map for input and output will result in the overwriting of the original TIN data.

The results of map TIN contouring can be observed through map reporting and background map display. Note that in the contoured map file the features will consist of cmaj and cmin as defined in Figure 4.23. Figure 4.24 shows the thickness contours for the two ore zones.

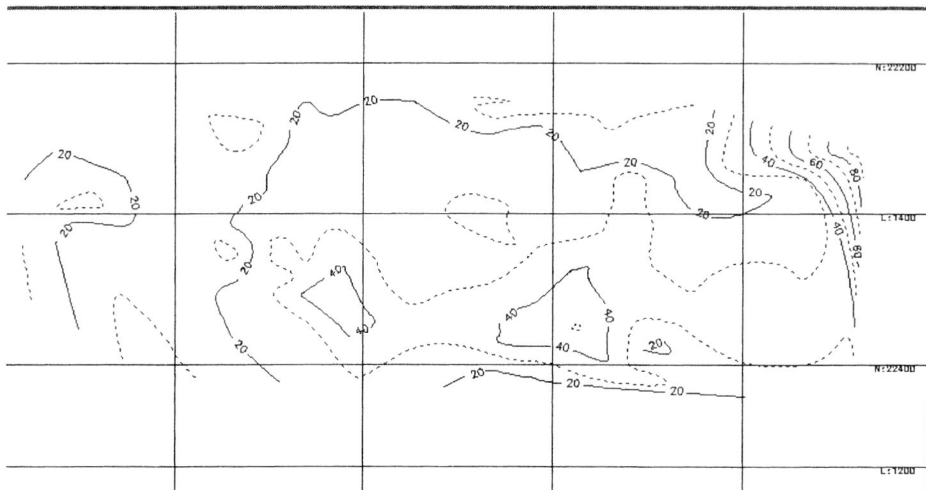

Figure 4.24. Contour plot of total thickness of zones 7 and 8.

Summary of procedure: Example 4.7
Initial data requirements: drillhole data containing lithology

Procedure:
1. Background display of thickness data (optional)
   1.1 Orient the viewplane **Viewplane ⇒ Setup** (Viewplane Set-up entry form)
      (Orient viewplane as per Summary of procedure: Example 4.5, Step 1)
   1.2 Display total thickness **Background ⇒ Holes** (Background Hole Data Display entry form)
      Display Format: <projections> (Background Hole Data Projections Display entry form)
      Characteristic Values: <zone7, zone8>
      Structural Function: <top + thickness>
      ⇒ **OK**
      ⇒ **OK**
      ⇒ **Refresh** ⇒ **Extents** ⇒ **Zoom Fix**
   1.3 Output background display to map (optional) **Output ⇒ Map** (Output Map entry form)
      Map Select: <zone78th>
      ⇒ **OK**
2. Generate map of hole data projection thickness (viewplane already set as per Step 1.1) **Maps ⇒ Hole Data Projection** (Hole Data Projection Parameters entry form)
   Subset/Holes Select: <dh0>
   Variable Or Character: <lith>
   Characteristic Values: <zone7,zone8>
   Structural Function: <top + thickness>
   Map Name: <zone78th>
   ⇒ **OK**
3. Generate TIN (map zone78th must be active) **Maps ⇒ Map TIN Define** (Map TIN Define menu)
   ⇒ **Generate ⇒ File ⇒ Save & Exit**
4. Contour thickness **Maps ⇒ Map TIN Contouring** (Map Contour entry form)
   Source Map: <zone78th>

DESTINATION MAP: <cont78th>
CONTOUR VARIABLE: <thickness>
⇒ **OK**

4.7 INTERACTIVE GEOLOGICAL INTERPRETATION

The previous sections have shown how thickness and volumes can be calculated using surfaces. Laminar deposits can be modeled as grids and thicknesses and volumes calculated using matrix algebra. More complex laminar deposits can be triangulated and the triangle sets from two surfaces can be used to generate volume components in the form of prisms. But when the geologic structure is non-laminar, its geometry must be directly interpreted.

Geology has been traditionally interpreted on cross sections using structural data from exploration drifts and drillholes. Drifts and drillholes are typically very widely separated. Since the actual structural complexity is great in comparison to the density of data, automated fitting routines, such as splines, cannot be used with much reliability. For this reason, the traditional methods of using hand drawn geologic boundaries on sectional views is still used for creating volume models of complex structures. The process is roughly as follows:

1. Generate a section (usually vertical) and display on the section drillhole lithology any relevant surfaces such as the topography and any other source of structural data such as exploration drifts.
2. Using the available structural data, digitize on the section an interpretation of the boundary of the structure. Save this interpretation as a polygon of *xyz* points. Usually, one out of the set of polygon vertex coordinates will be fixed to the orientation of the section, i.e. constant *x*, *y* or *z*.
3. Repeat the process of digitizing the boundary for a set of parallel sections which spans the structure to be modeled including an interpretation of boundaries for the extremes of the deposit where there are no available data.
4. Connect the boundaries for adjacent sections to form a volume model.
5. Repeat Steps 1-4 for each deposit/geologic structure, making sure that the connect boundaries between adjacent lithologic units are free of undefined gaps.

This seemingly simple procedure can be as complex and time consuming as the structural complexity and the importance of accuracy warrants. The following sections of this chapter will demonstrate by example how volume modelling is practiced in Lynx (Holding 1994).

4.8 CROSS SECTIONAL DISPLAYS

Sections are used to view surfaces, *xyz* data points and volumes as an intersection with a plane. Sections can be thought of as a means of viewing 3D data on a 2D surface. The 3D data can be either projected onto the plane of the section or can be shown as a trace line where the volume or surface is sliced by the viewplane associated with the section. Sectional displays are most commonly associated with drillhole data and are therefore vertical so as to be in rough alignment with the drillholes' inclinations. Surfaces are displayed on section as traces indicating where the sectional plane intersects the surface. Volume models of underground openings are displayed as polygons showing the boundary of the opening in the sections. Since most *xyz* type data, such as drillhole traces, will only intersect a plane at one point, data can be projected onto a section. For this reason, sections are assigned a fore and back thickness. Thus, instead of displaying values which intersect the plane, any information which falls within the rectangle defined by the extent, orientation and thickness of the section can be projected onto the viewplane.

Example 4.8: Creating sections in Techbase
A vertical section through the Boland Banya deposit will be created to display the following features: topography, top and bottom structures for the main ore body, and a posting of drillholes showing gold assays displayed as a down hole histogram. The section will run diagonally across the property from the SW at 10300E, 20200N to 12200E, 22000N. The section elevation change will be from 5050' to 5600'. The horizontal run of the section will be 2617' with a height of 550'. To print this on a B size sheet (18' × 11') with one inch margins (plot area of 16' × 9') the *x* scale will have to be 2617/16:1, say 175 ft/in. The vertical scale is normally exaggerated. To fit the page, the scale will be about 550/9.5:1, say 75 ft/in.

From the Techbase main menu select **Graphics ⇒ Section**. Assign a new metafile name, make sure that no filters are active, and proceed to **Scaling** to assign the horizontal and vertical scales determined above. Next assign the **End Points**. Under **Projection** assign a fore and back plane thickness. Don't use too great a thickness since the number of hole traces projected onto the section plane will become confusing. Try 75'. Select front and top views. If you also want a side view the *x* scaling factor will have to be increased for the section to fit on a B sheet. The top view is a handy way of checking the deviation of the holes away from the section line. Under **Sheets**, select a B size sheet and a landscape orientation. If you didn't assign margins as part of the **End Points** definition you can do so now. Next select **Border**. Before creating the metafile by drawing the border, Techbase will check to see if the plot can fit on the assigned sheet size and display the physical dimensions of the plot area which will have the border drawn around it. If the plot seems to be a good fit, then continue and use **Graphics ⇒ Review** to double check the extent of the graphic. Otherwise, abort using ^X and change either the scaling or the paper size. Add a **Grid** using the same section extents defined in **End Points**, but note that this is a diagonal section across the property, so the front and top views will include both easting and northing grids on the *x* axis. This might make the plot somewhat confusing if a small grid spacing is used.

The **Points** option can be used to display *xyz* data such as composited surface intercepts on the section. First, the intersection of the section plane and surfaces will be displayed. Select **Surface** and enter the **Field** name corresponding to the gridded topography. Note that all surfaces are by definition field names in cell tables in Techbase. Assign the intersection lines color and line **Style** and then **Draw**. Check the results be using metafile review and then continue by drawing the surfaces for top and bottom structures of the two ore zones. When you attempt to draw a cell field which does not intercept the section, Techbase will report that there are no values to draw. This is the case for the lower deposit on this section line.

Next plot the drillhole traces on the section. Select **Holes** and enter the **Fields** used for establishing the trace of the hole and the data to be posted down the hole. This field list will be the same as used in Example 4.1 to composite holes. **Hole Labels** should be assigned to the collar of the hole and if the holes are highly inclined, many of their traces may penetrate the front and back thickness limits of the section. These puncture points can also be labeled with the hole name. **Linestyle** controls the appearance of the hole trace. **Markers** controls the type, color and size of marker used for the hole collar, bottom and points of entry or exit from the fore and back plane. **Values** controls how the assay value that was listed in the Fields menu will be displayed down the hole. In this example a down hole histogram will be used. To get the down hole histogram, enter a histogram scale of 1 and TYPE as Bar. The scale is relative to the magnitude of the data being plotted. A scale of one results in bar heights equal to the magnitude of the assay. Thus, a histogram bar for an assay value of 0.02 will be 0.02 inches high. You will probably need to experiment with different scales. Selection of **Draw** will include the holes on the section. The resulting cross section is shown in Figure 4.25.

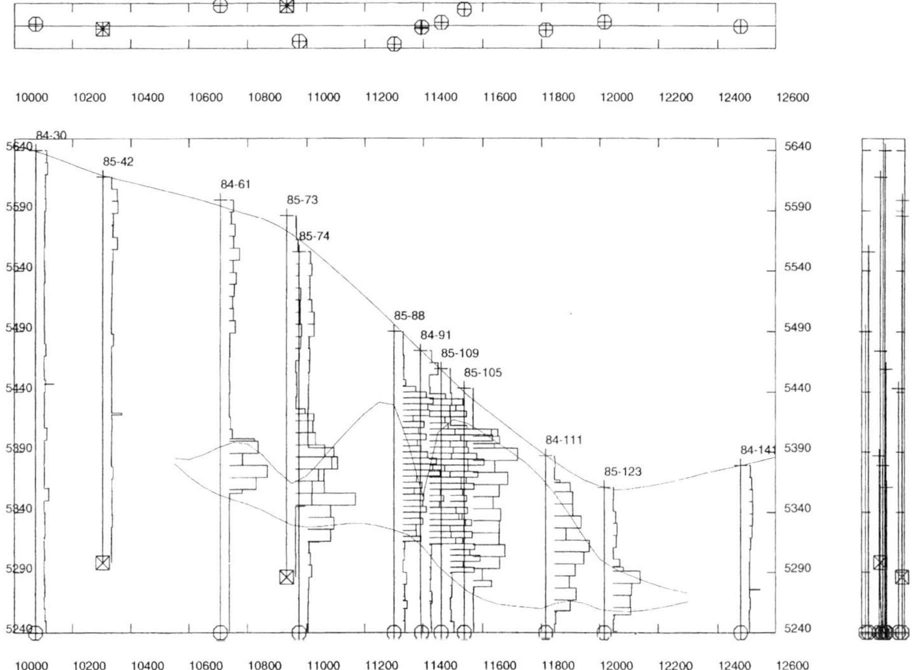

Figure 4.25. Sectional display of the estimated top and bottom structures for the main ore zone of the Boland Banya deposit with drillhole traces and gold histograms.

Summary of procedure: Example 4.8

Initial data requirements: drillhole data including assay and survey data and gridded surfaces of topography and lithologic contacts.

Procedure:

1. From the main menu **Graphics** ⇒ **Section** ⇒ **Graphics** ⇒ **Metafile name**
   File Name: <section>
   ⇒ **eXit**
2. **Scaling** ⇒ **Scales**
   Horizontal Scale: <300>
   Vertical Scale: <75>
   2.1 **End points**
   Left Endpoint X: <10000> Y: <21350>
   Right Endpoint X: <12600> Y: <21350>
   Bottom: <5240> Top: <5650>
   2.2 **Projection**
   Projection Distance
   Front: <100> Back: <100>
   Front View?: <yes>
   Top View?: <yes>
   Right View?: <yes>
   2.3 **Sheets**
   Sheetsize: <A> Orientation: <L>
   2.4 **Border**
   2.5 **Grid**
   X Minimum: <10000> Maximum: <12600> Increment: <200>
   Y Minimum: <21350> Maximum: <21350> Increment: <200>
   Z Minimum: <5240> Maximum: <5650> Increment: <50>
   Grid Style: <ticks> Text Size: <.1>
   Ticks: <in> Text: <in> Labels: <PERP> Sides: <LT>
3. **surFace** ⇒ **Field**
   Field Name: <topo_c>
   3.1 **Style**
   Linestyle: <solid> Color: <green>
   3.2 **Draw**
   ⇒ **eXit**
4. Repeat Step 3 for the ore zone top (oat_c) and bottom (oab_c) using different colors.
5. **Holes** ⇒ **Fields**
   Hole Id: <hole_id> Depth: <depth>
   X: <x> Y: <y> Z: <z>
   Azimuth: <bearing> Dip: <dip> Depth: <total>
   From: <from> To: <to>
   Value Fields: <Au>
   5.1 **Hole labels** (accept defaults)
   5.2 **Line style** (accept defaults)
   5.3 **Markers** (accept defaults)
   5.4 **Values**
   Location: <right> Color: <blue>
   Size: <.1> Histogram Scale: <1>
   Type: <rbar> Offset: <.1>
   Fill Style: <hollow>
   5.5 **Draw**
   ⇒ **eXit**
   ⇒ **Graphics** ⇒ **Review**

Example 4.9: Creating sections in Lynx

Sectional displays in Lynx have the same basic components as in Techbase but with a major difference: the sections are oriented to a viewplane which can be at any inclination. Maps, drillhole data, and volume models can be displayed either as an intersection with the section or as a thickness in the fore and back planes of the section.

Drillhole data from the DOE soil moisture program will be used to illustrate the process of interpreting complex lithologic data in sectional view. First, display the drillhole collars in plan view to identify the orientation of subsequent vertical sections. Select the project SMOOT1 as being active and then select **Viewplane ⇒ Setup** from the Lynx main menu to bring up the Viewplane Setup entry form. Enter a view plane orientation of 0,0,20100,90,90 (N, E, L, Az, Dip), a scale of 10,000:1 and a global grid of 500 cm on a side (for details see Examples 4.5 and 4.6). First, only the collar locations will be posted in plan view. Select **Background ⇒ Holes** from the main menu to display the Background Hole Data Display Selection entry form. Define one of the drillhole displays for Display 1 using the default display parameter values, select all of the holes (subset dh0) to be displayed. Set DISPLAY FORMAT to <value>, LABEL LOCATION to <collar and intersection starts>, turn the trace display off and **Apply ⇒ Refresh** to see the collar posting. As shown in Figure 4.26, the wells are laid out in an eight sided star pattern covering an area 15 m^2.

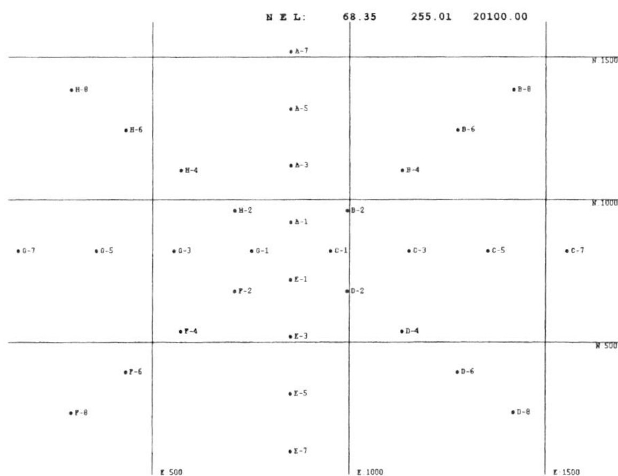

Figure 4.26. Well collar locations for the Smoot database.

Next, modify the drillhole posting so that both the original alpha and numeric codes for the sediment type are displayed along the drillhole trace. Enter <sed> as the PRIMARY VARIABLE and <Id> as the SECONDARY VARIABLE. Use 50 cm fore and back thicknesses for projection onto the viewplane, change the drillhole trace to <solid> and the VALUE DISPLAY to <composite values> which will cause adjacent samples having the same sediment description to be displayed together as a single interval.

To see a vertical section running from collar G-7 to C-7, return from the background display menu and select **Viewplane ⇒ T section**. Transverse sections can be run through each of the four axes of the pattern of wells. Digitize the two ends of the section line. **Shift**, **Extents** and **Zoom free** can be used to improve visibility (see Fig. 4.27).

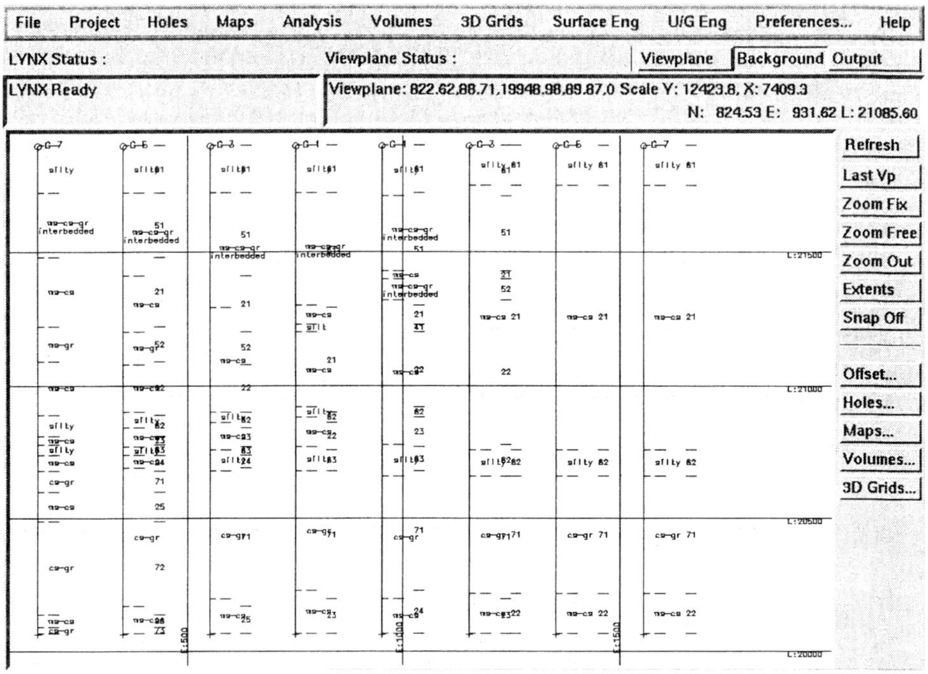

Figure 4.27. A section showing sediment classes defined for modelling the Smoot database.

Lynx allows up to 5 different cross-sectional displays to be defined. Drillhole data can be displayed as values posted along the drillhole trace as above or as histograms, color maps and contoured sections. When an alpha variable is being posted, the drillhole trace can be displayed as a color. In **Display format** in the Background Hole Data Display entry form, respond with <colormap> instead of <value> which will bring up the Background Hole Data Colormap Display entry form in which the sediment class is selected and associated with a pen color (see Fig. 4.28). For instance, <2,ss-ms> would display the hole trace in red for all from/two intervals of the variable Sed having the value ss-ms. All other parameters for hole background display can remain the same as in the previous examples. Displaying sediment types and lithologies is the most common means of displaying drillhole data during interactive volume modelling.

Numeric variables can be displayed as a histogram. In Lynx, the scaling factor for the histogram bar height is variable units/mm. A histogram scale of 100 for a value of 1000 would result in a bar 1 cm in height, i.e 1000/100 mm. In the Background Hole Data Display entry form change the PRIMARY VARIABLE to <count> and the DISPLAY FORMAT to <histogram> which will bring up the Background Hole Data Histogram Display entry form in which the SCALE is set to <500> units/mm.

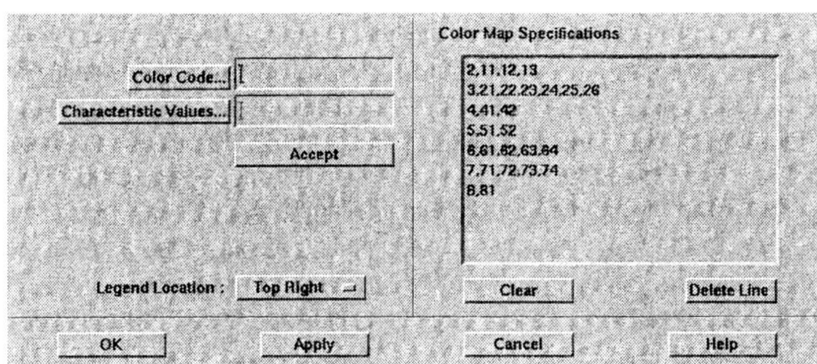

Figure 4.28. Background hole data colormap display entry form.

Numeric variables can also be contoured. With a three-dimensional data set, contours can be displayed in any section orientation. To do this, change DISPLAY FORMAT to <contour> bringing up the Background Hole Data Contour Display entry form. Contours in the plane of the section are based on an inverse distance grid. The inverse distance parameters for the search radii, skewness and inverse distance power must be entered. Inverse distance estimation of grids is covered in Chapter 5, but for now it should be noted that the search distance must be sufficiently great so that several samples are within the search radius of the grid nodes. This will be true if the search radius is at least the average spacing between drillholes. Two search radii are given so that when the major axis (R1) and minor axis are unequal, a search ellipse is being used which is oriented according to the skewness angle. SKEW = 0 (default)

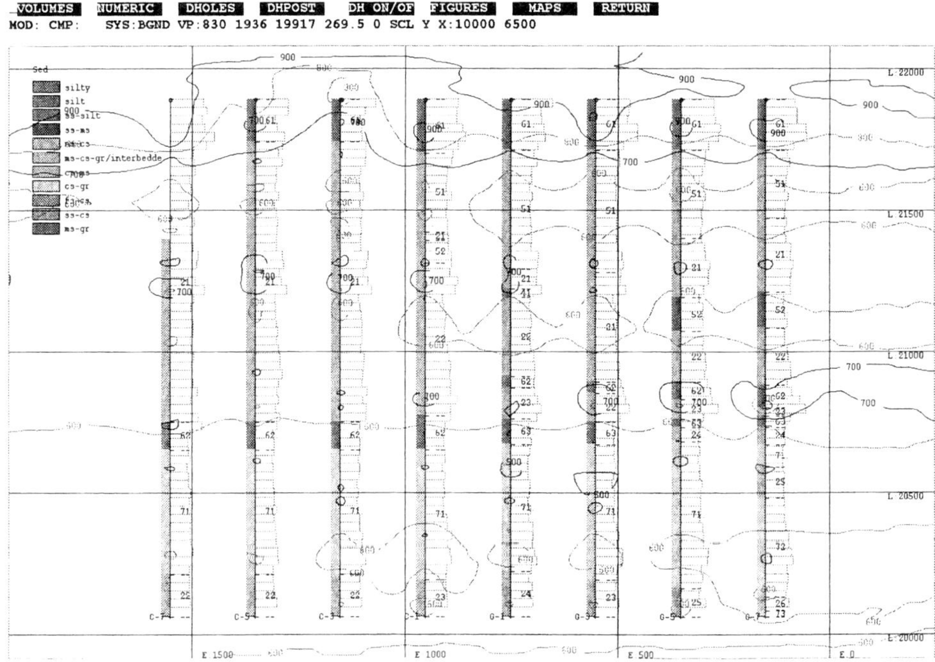

Figure 4.29. Histogram traces and contouring of moisture levels on section.

results in R1 being parallel to the *x* axis of the viewplane. The grid density is controlled by GRID SIZE. The default is a grid interval of 1/50 of the *x* axis range in the viewplane. Contouring is controlled by the contouring parameters. **Query Min/Max** can be used to report the minimum and maximum values of the numeric variable being contoured.

Sectioning and contouring of 3D estimates is extremely useful for checking the relationship between sediment type and moisture level for the Hanford well data. By examining sections having lithology posted as a color, and with downhole histograms and contours of the moisture level, it seems apparent that there is no strong relationship between moisture and soil type except for the possible exception of the silt lenses (in red). There is a trend in increasing moisture level towards the surface, and it seems as if some of this moisture as perched on the silt lenses. Below the lenses the neutron count runs around 500 to 600. Moving into the lenses from below, the count value increases to over 600 until just above the lenses where the level is above 700 and then decreases back down to under 600. This example demonstrates the importance of looking closely at the data in three dimensions using sections. At this point, it seems that only the silt horizons are important factors in controlling moisture levels and that instead of modelling as many as seven soil types, one may be sufficient. It should be noted that this data set was taken from very dry conditions and that the influence of other soil structures might become apparent if water were injected into the soils from the wells.

Summary of procedure: Example 4.9
Initial data requirements: drillhole data (dh0 in project SMOOT1) containing sediment type and numeric data.

Procedure:
1. Orient viewplane to section **Viewplane** ⇒ **Set-up** (Viewplane Set-up entry form)
   NORTH: <0> AZIMUTH: <90>
   EAST: <0> INCLINATION: <90>
   ELEVATION: <20100> SHOW GRAPHIC:<on>
   SCALE Y: <10000> SCALE X: <10000>
   GLOBAL GRIDS: <on>
   N: <500> E: <500> L: <500>
   TRACKING: <global>
   ⇒ **OK**
   ⇒ **Refresh**
2. Background collar posting **Background** ⇒ **Holes** (Background Hole Data Selection entry form)
   DISPLAY PARAMETERS FOR: <Display 1>
   DISPLAY ON/OFF: <on>
   ⇒ **Subset/Holes Select** (Subset/Holes Select : Background)
   SUBSET: <dh0> (all other parms *)
   ⇒ **OK**
   (skip entry of primary and secondary variables)
   DISPLAY FORMAT: <value>
   LABEL LOCATION: <collar & intersection starts>
   TRACE DISPLAY: <none>
   ⇒ **Apply**
   ⇒ **Refresh**
3. Change display for vertical section (remain in Background Hole Data Selection entry form)

Primary Variable: <sed>
Secondary Variable: <Id>
Display Thickness: Fore/Back <50> <50>
Trace Display: <solid>
Value Display: <composite values>
⇒ **OK**
⇒ **Refresh**

4. Change viewplane orientation to vertical (main menu) **Viewplane** ⇒ **T section** (use mouse to define section line from G7 to C7 and **Zoom Free** to center)
5. Vertical section displays **Background** ⇒ **Holes** (Background Hole Data Selection entry form)

5.1 Color map
Display Parameters For: <Display2>
⇒ **Subset/Hole Select** <dh0>
Primary Variable: <Id>
Display Thickness: Fore/Back <50> <50>
Display Format: <colormap> (Background Hole Data Color Map Display)
⇒ **Color Code** <2> (Color Selection list)
⇒ **Characteristic Values** <11,12,13> (Characteristic Value Selection)
⇒ **Accept** (repeat for each set of sediment types to be displayed with same color codes)
Legend Location: <top right>
⇒ **Apply**
⇒ **Refresh**

5.2 Histogram
Display Parameters For: <Display 3>
Display On/Off: <on> Display 3
Primary Variable/Characteristic: <count>
⇒ **Display Format** <histogram> (Background Hole Data Histogram Display)
Scale (units/mm): <500>
⇒ **OK**
Display Location: <right of trace>
(all other parameter settings the same)

5.3 Contours
Display Parameters For: <Display4>
Display On/Off: <on> Display: 4
Display Format: <contour> (Background Hole Data Contour Display)
Search Radii R1: <200> R2: <200>
Skew: <0> Order: <2>
Grid Size: <37>
⇒ **Query Min/Max** (437 - 1788)
Minimum: <500> Maximum: <1800>
Minor Interval: <50> Major Interval: <100>
⇒ **OK**
(all other values as in previous example)
⇒ **OK**
⇒ **Refresh**

4.9 INTERACTIVE GEOLOGICAL INTERPRETATION AND VOLUME MODELLING

Sectional displays, surface modelling and statistical analysis should be intensively used until the modeler has a three-dimensional picture of the important geologic structures in his mind's eye. He should be confident that he understands the geometry of the orebody and how it is related in space with respect to other structures such as faults. The apparent lithology may be much more complex than the level of detail required. A drillhole database consisting of a confusing array of rock types may be reduced to a few ore zones and a matrix of waste. Likewise, a hydrogeologic study of groundwater flow in a complex sequence of soil types can often be reduced to two basic categories of high and low porosity soils. The actual objectives of a project rarely require that the geology be reproduced in its full complexity, but the modeler should be aware of the major variables required for the completion of the project and include them in the model. For example, it is obvious that ore zones need to be modeled as part of a feasibility study as part of the reserve analysis, but ignoring the various rock types in the surrounding and overlying waste might be a critical mistake if the study advances to the level of design detail required for a bankable cost engineering study. Pit slopes and swell factors are all a function of rock type, and without a model which identifies waste by type of rock, it will be difficult to accurately determine stripping ratios and production fleet size during the life of the mine. Before proceeding with geologic modelling, both the complexity of the structures and the level of detail required by the project must be fully understood, or else the modelling process will have to be repeated at greater cost at a later date.

Complex geologic structures are best modeled interactively by displaying the available lithologic information on sections and digitizing the supposed outline of the structure as it would appear in the section. The outline is saved as a polygon of *xy* points on the section. The orientation of the plane of the section is used to translate the polygon vertices into the global coordinate system of the project. As many structures as needed are digitized on each section and are saved with a code that identifies the polygon in terms of the geologic structure that it represents. Once two or more of these boundaries have been generated on adjacent sections, these two-dimensional boundaries can be joined to form a volume component as shown in Figure 4.30.

Different methods are available for connecting polygon boundaries to form a vol-

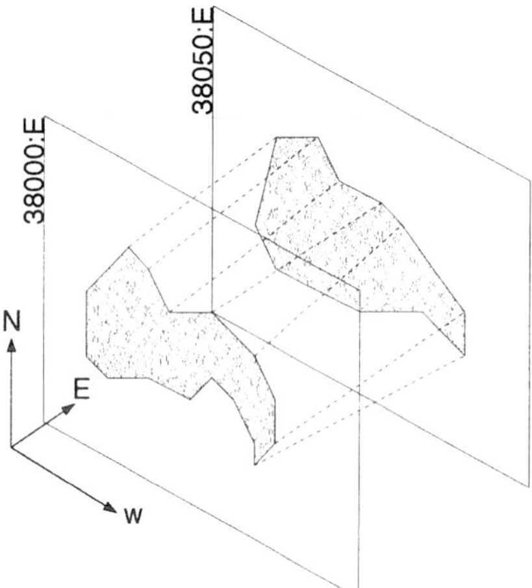

Figure 4.30. Relating boundary polygons in two sections.

ume component. It is important to consider the method used, since the interpolated surface between the two polygons is a representation of the geologic structure's surface topology. Some methods simply assign a front and back thickness to the boundary, and the resulting volume retains the same cross-section as the digitized polygon throughout its thickness (Fig. 4.31). For structures that are reasonably continuous

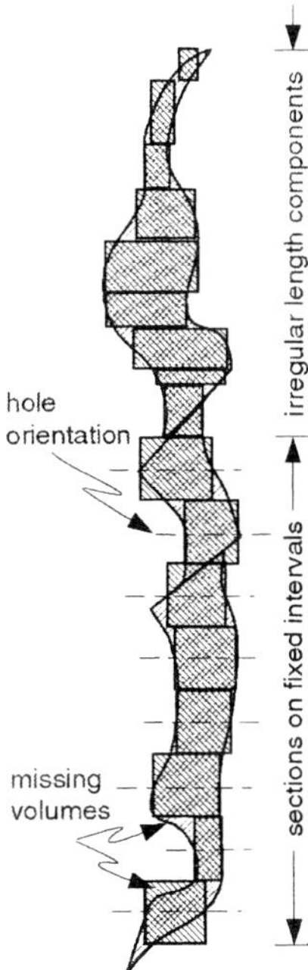

Figure 4.31. Missing and non-existent volumes resulting from using rectangular volume components based on fixed section intervals.

along one axis, this stepwise approach to volume modelling may be perfectly acceptable, but as the curvature of the axis along which the volume models are being generated increases, increasingly thinner volume components will be required if a glaring discontinuity between adjacent components is to be avoided. As the thickness of the volume components decreases, the number of components will increase, as will the time required to produce the model. For structures with a distinct strike and dip, holes are often drilled on predetermined section lines, and the volume components' lengths are essentially fixed to the section intervals. Since volume modelling is restricted to the available data, selecting a component thickness that will fit the complexity of the surface topology is not an option. Thus, when a component has a fixed cross-sectional area for its entire length, the possibility of lost or non-existent ore increases towards the extremities of the components.

Wire framing (Fig. 4.32) is another popular method of connecting polygon boundaries to form volume components. When using wire frames, two or more boundaries are displayed at an orientation in which the vertices of all the polygons are visible. Rays connecting topologically related vertices are defined, and then the vertices between the two adjacent polygons are connected while honoring the rays

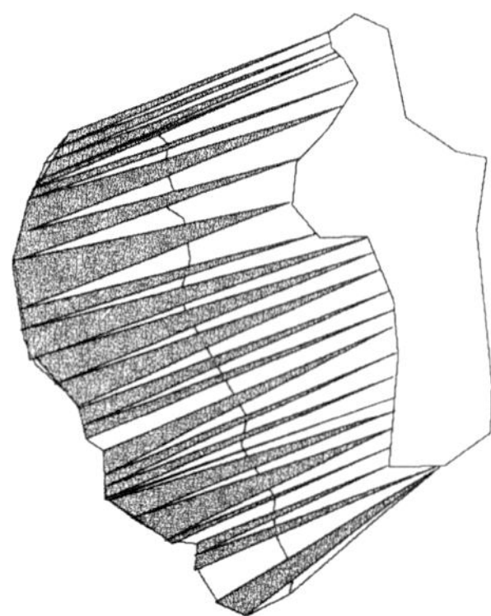

Figure 4.32. Wire framing between two boundaries.

that were selected. Since the polygons do not necessarily have the same number of vertices, a vertex in the less complex polygon may be connected to multiple vertices in the polygon which consists of more points. The faces defined by adjacent rays now define the volume component's topology, and the ends of the components are the two polygons' boundaries from the adjacent sections. As long as the polygons that form either end of the component are not too dissimilar in shape of complexity, the cross-sectional outline of the component at any point along its thickness should be a linear interpolation of the two defining polygons, but when the structure becomes increasingly complex and convoluted, so that the polygons' boundaries are excessively dissimilar, wire framing algorithms become increasingly unreliable.

Even when structures are being interactively digitized on sections, there has to be an adequate density of drillholes from which sections can be generated. If the geometry of the structure varies more rapidly than the average drillhole spacing, it will be increasingly difficult to use any of the previously mentioned volume modelling techniques. It should be noted that there is a statistical relationship between a surface's variability in space and the estimation error which can be expected for a given density of drillholes. This estimation error is a function of the variogram which describes the topological complexity at different scales, the drillhole pattern and the hole spacing. A regular rectangular or equilateral triangular grid will always result in the lowest estimation error and has the strong advantage of providing a set of parallel sections that consist of aligned holes. The grid should be oriented to the strike and dip of the main deposit. As the spacing between holes decreases, the estimation error will continue to rapidly decrease until there is little improvement in the accuracy achieved in the interpretation of the surface. The degree of accuracy needed for a given stage of the project or for the mining method that will probably be used must also be considered. A high-grade vein that is to be mined using cut and fill will require a much more accurate volume model during the feasibility and design stages than will a disseminated deposit that will be mined with a less selective method. From a practical viewpoint, the density of data available for any new project can be expected to be far less than what will be required for an accurate low-risk study.

In Lynx, volume model components are defined by a fore, mid and back plane (see Fig. 4.33). The mid-plane corresponds to the current viewplane. During the definition of the component's boundary, the mid-plane is oriented on the drillhole sec-

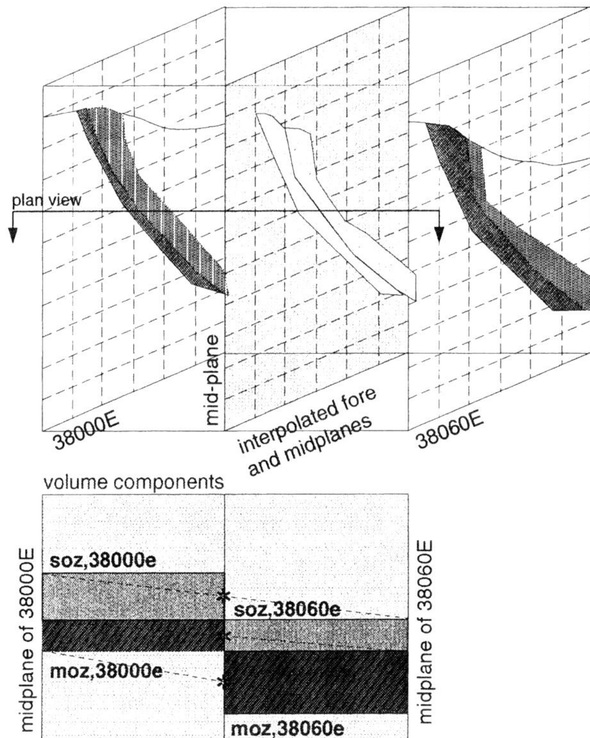

Figure 4.33. Linear interpolation of fore and back plane polygons of adjacent volume components. (Figure in colour, see opposite page 55.)

tion. The structure's boundary is defined on the midplane as an interpretation of the lithologic zones which are displayed on drillholes which fall on or near the section line. This boundary is then assigned a fore and back thickness. The resulting volume component has a constant sectional area and a thickness equal to half the distance between the two adjacent cross-sections. The structures' boundaries on adjacent sections will not be equivalent in either area or shape. Also, the locations of the *xy* centroids of the components will intercept the viewplane, and will diverge from each other as a function of the variability of cross-sectional area and curvature. Thus, the position of the adjacent component's fore and back planes will not match and there will be a major discontinuity at the fore plane/back plane contact between two adjacent components. This problem is resolved by generating a linear interpolation between the midplanes of the two components. Both boundary polygons are displayed offset from each other in section and the modeler selects vertices from the two boundaries which represent the equivalent topological features. These 'linked' vertices are then honored and form the basis for generating an interpolated boundary that replaces the two components' fore and back planes. The resulting volume components will no longer have a constant cross-section: the midplane of a component will retain the same boundary that was originally defined on-section, while the fore and back planes' boundaries will be a linear interpolation of the current component's midplane and the midplanes of the two adjacent volume components. This process is illustrated in the following example.

Example 4.10: Defining a volume model in Lynx
The minimum data that are required for interactive volume modelling are drillholes containing the pertinent structural or grade information and any bounding surfaces. In this example, ore zones 7 and 8 will be modelled based on lithology. The bounding surface will be the topography since the deposit outcrops. To view the topography in section, the map data features (cmaj and cmin) must be converted from contours to a triangulated surface. Since the resulting TIN is continuous, an intersection with it will be displayed as a polyline. Note that when the triangulating map contains contours, traverse lines, toes and crests, these map features should be honored to avoid smoothing of the surface. The procedures for triangulating surfaces are given in Examples 4.6 and 4.7.

Volume interpretation is done interactively on the screen based on the background display. So the first step is to select **Background ⇒ Maps**, bring up the Map Background Data Selection entry form and enter the map <TOP> and toggle DISPLAY SURFACE INTERSECTION on. The triangulated topography will now be visible only where it intersects the viewplane.

Next the drillholes are placed in background by selecting **Background ⇒ Holes** bringing up the Background Hole Data Selection entry form. As per previous examples, select subset <dh0>, use <lith> as the primary variable in the display, use a fore and back plane projection distance of 30 m, use a thick or solid hole trace and set the DISPLAY FORMAT to <color map> as shown in Figure 4.34. This will post the

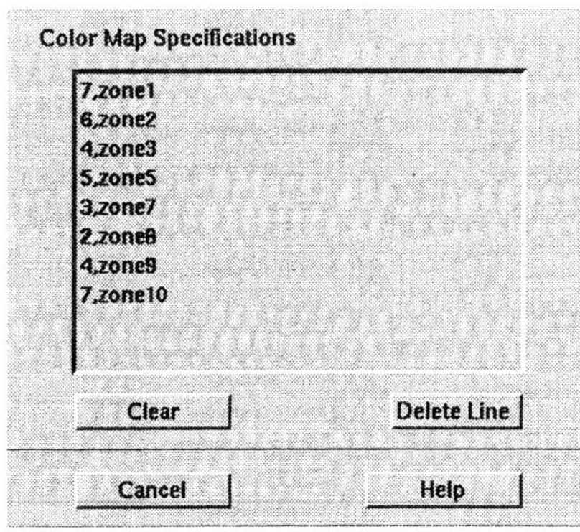

Figure 4.34. Color map specifications for lithology.

lithology on the holes as colors (5, zone5; 3, zone7; 2, zone8; 4, zone9). Note that these colors will be used as part of the definition of the components: the main ore zone (lith = zone8) will be displayed in red (pen = 2), while the secondary ore zone (lith = zone7) will be displayed in yellow (pen = 3). Select a set of colors for the different lithologic units and use these in all subsequent displays and component definitions.

In this example, four geology volume components are going to be created to represent the main and secondary ore zones between eastings of 38150E and 38270E. Drillholes in this project are aligned on sections perpendicular to strike at roughly 60 m intervals. The midplanes of the components will be oriented to these two sections and will have fore and back plane thicknesses of 30 m, respectively. Therefore, one component will have a midplane on the 38180E section, a foreplane at 38150E and backplane at 38210E. The other component will have its midplane at 38240E and cover from 38210E to 38270E. To start with 38180E, orient the viewplane to 22300,38180,1300,0,0 (N, E, Z, Azi, Inc), set the scale to 2500:1, and use a 100 m global grid. **Refresh** the screen to see the section.

Once the background display has been adjusted, the volume component can be defined. Select **Volumes ⇒ Volume Data Define** to open the Volume Data Define menu and the Volume Data Define Parameters entry form (see Fig. 4.35). Select **De-**

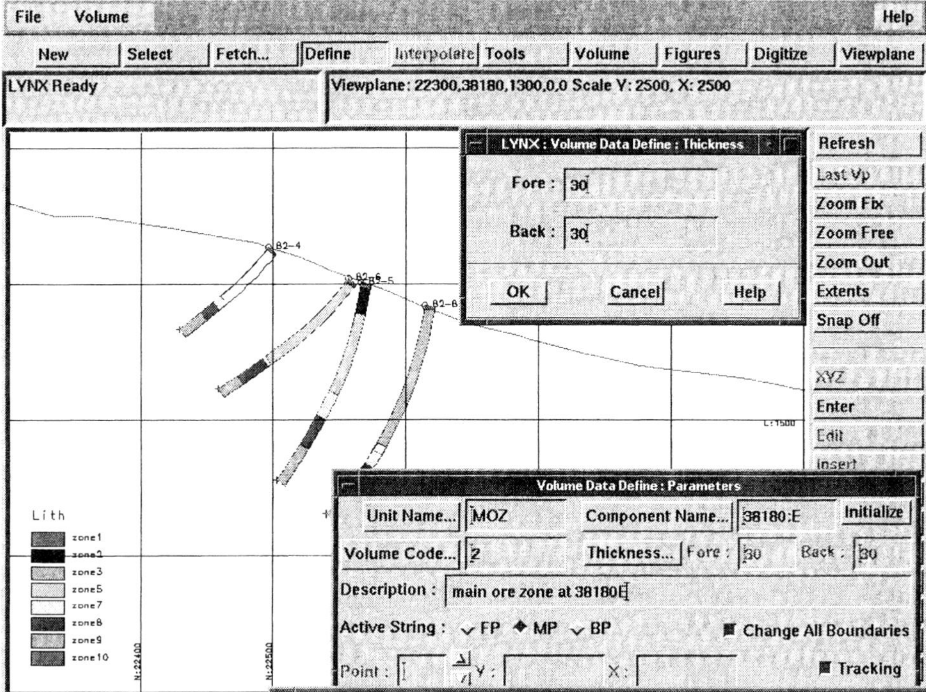

Figure 4.35. Volume data define menu and parameter entry form with color map display of drillhole lithology at section 38180:E. (Figure in colour, see opposite page 55.)

fine and in the Parameters entry form, enter the new component's Unit Name <MOZ>, Component Name <38180:E>, Volume Code <2> and fore and back thickness <30>. Toggle MP on so that the boundary being defined is the midplane of the component. Select **Enter**, and then digitize the top and bottom contact of the main ore zone on the drillholes (Fig. 4.36). Additional points can be added to give the boundary any reasonable shape between the drillhole intersection points. Additional points can be added between existing vertices using **Insert**. Vertices can be moved using **Edit** or deleted using **Del Point**. The last action can be undone with **Undo**. If the polygon is a hopeless mess, you can start fresh by selecting **Del all**. Practice by using these various volume editing features. When the boundary is satis-

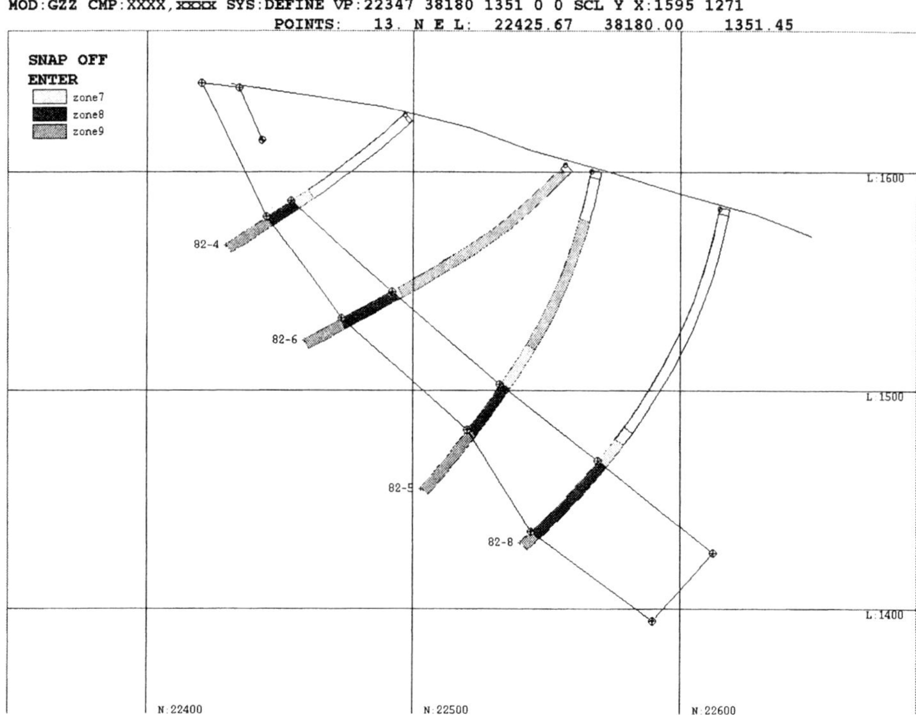

Figure 4.36. Interpretation of main ore zone limits on drillhole sections displaying rock type. (Figure in colour, see opposite page 86.)

factory, save it as a component by selecting **Save** or **Save & Clear**, which will clear the display of the midplane boundary. The stored component will be displayed in red. Next define a component for the secondary ore zone. The footwall of the secondary ore zone is the hangingwall of the main ore zone, so the two boundaries will have to be perfectly matched in order to avoid gaps between the two volume components. To ensure matching contacts, the vertices of MOZ,38180:E that are shared will be replicated on the secondary ore zone component. Select **Initialize** in the Parameters entry form and enter the definition of the zone 7 component with a thickness of 30,30, a color code of 3 and a name <SOZ,38180:E>. Select **Define** ⇒ **Snap on** ⇒ **Pick** and pick component MOZ,38180:E by clicking on its boundary with the mouse. Once selected, the boundary will be highlighted in cyan and the **Segment** option will become available. Select **Segment** and enter the first and last vertices on MOZ,38180:E that are to be shared with SOZ,38180:E, i.e. the contact between their hanging and footwalls, respectively. Next toggle **Snap off** and use **Enter** to complete the boundary, editing the polygon as needed. Both MOZ,38180:E and SOZ,38180:E are shown in vertical section in Figure 4.37.

To this point, the volume model components that have been generated are rectangular volumes in the *x* and *z* planes (normal to the midplane of the component) and have the boundary in the *y* plane that was digitized and stored. This can be seen by using **Viewplane** ⇒ **Y-sect** and running a plan view section through the components.

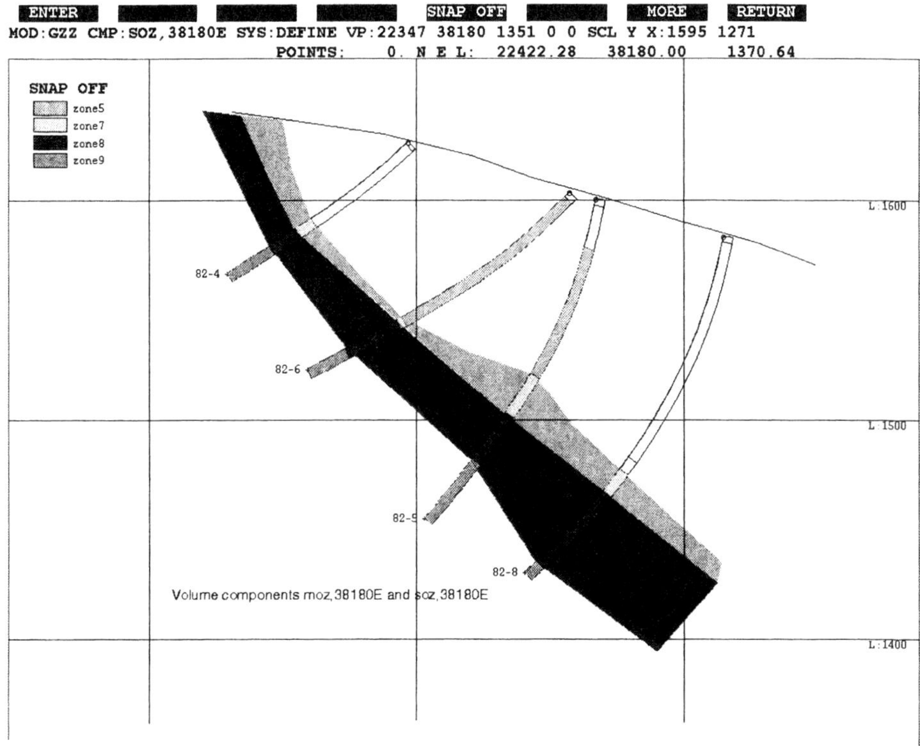

Figure 4.37. Primary and secondary ore zone volume components. (Figure in colour, see opposite page 86.)

The process of defining rectangular components is repeated on adjacent parallel sections until an adequate portion of the structure has been modeled. These components will not coincide on their fore and back planes with adjacent components. While defining components, it is desirable to be able to view adjacent components. To do this select **Volume ⇒ Replicate** and enter the name and the plane (MP, FP, or BP) of the component to be placed in memory and projected onto the current viewplane. This boundary can then be accessed by the **Define** functions **Pick** and **Segment**. This component will be displayed in red outline and is more than background display since it can be picked and used to define common vertices between different viewplanes or as in the contact between zones 7 and 8. Also, note that the volume background display can be toggled on and off from the Volume Data Define menu by selecting **Viewplane ⇒ Background** and toggling the volume display off. The display parameters for background display must be accessed from the main menu as per normal. When checking adjacent components for overlap, consider changing the volume display component fill to hatching.

The next step is to interpolate between the midplanes of adjacent components so that all coincident components match. Three sections of primary and secondary components have been previously defined and will now be displayed along with MOZ,381380:E and SOZ,38180:E. To display existing volume components, select

Background ⇒ **Volumes** in the main menu and select the geology model name followed by the names of the components to be viewed for the current geologic model. Enter a display thickness of 0,0 and run a plan view section to see all the components (Fig. 4.38). The components adjacent to the new components at section 38120:E do not coincide with the fore planes of the 38180:E components. Linear interpolation between component midplanes is used to correct this problem.

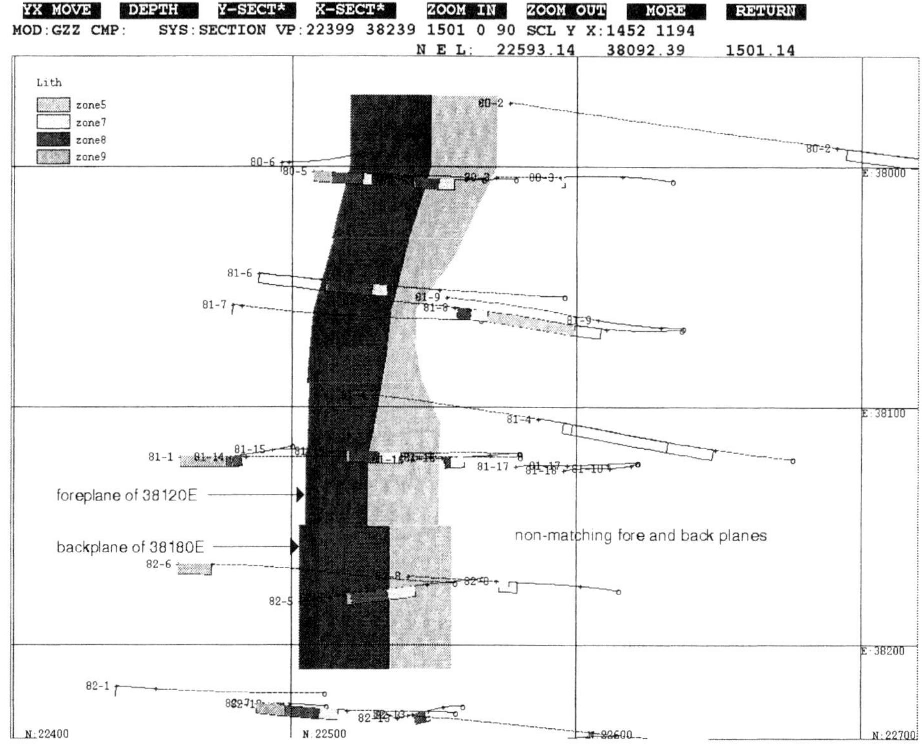

Figure 4.38. Linear interpolation of components on sections at 38120E, 38060E and 38000E. Component at 38180E not interpolated with 38120E. (Figure in colour, see opposite page 87.)

From the upper left corner of the Volume Data Define menu select **Volume** and toggle **Interpolate** on changing the menu to include the **Fetch both** and **Interpolate** options. Accessing **Interpolate** clears the active data. Select **Fetch** to bring up the Volume Data Interpolate Fetch Both entry form in which both the name of the component that is to be used as the primary surface (MOZ,38120:E) and whether that component's fore or back plane is to be used for interpolation (foreplane) or is entered along with the name of the secondary surface (MOZ,38180:E), for which the backplane is selected for interpolation (see Fig. 4.39). The primary surface will be dis- played as an unfilled polygon with its vertices posted with an 'o'. The secondary surface will be filled and its vertices will be posted with a '+'. Select **Interpolate** ⇒ **Link** and digitize rays from the primary surface to the secondary surface that link

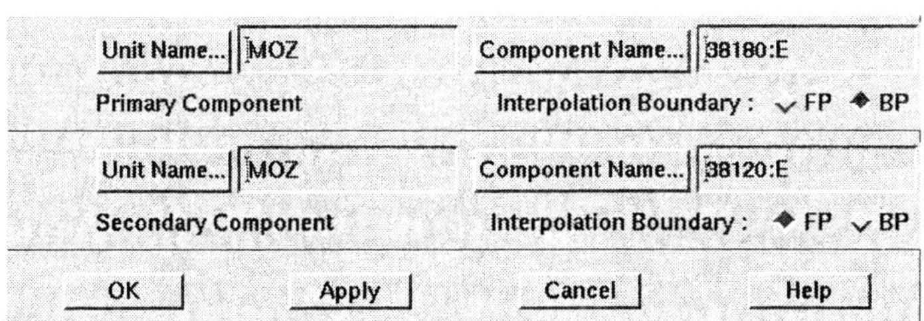

Figure 4.39. Interpolate's Fetch Both entry form.

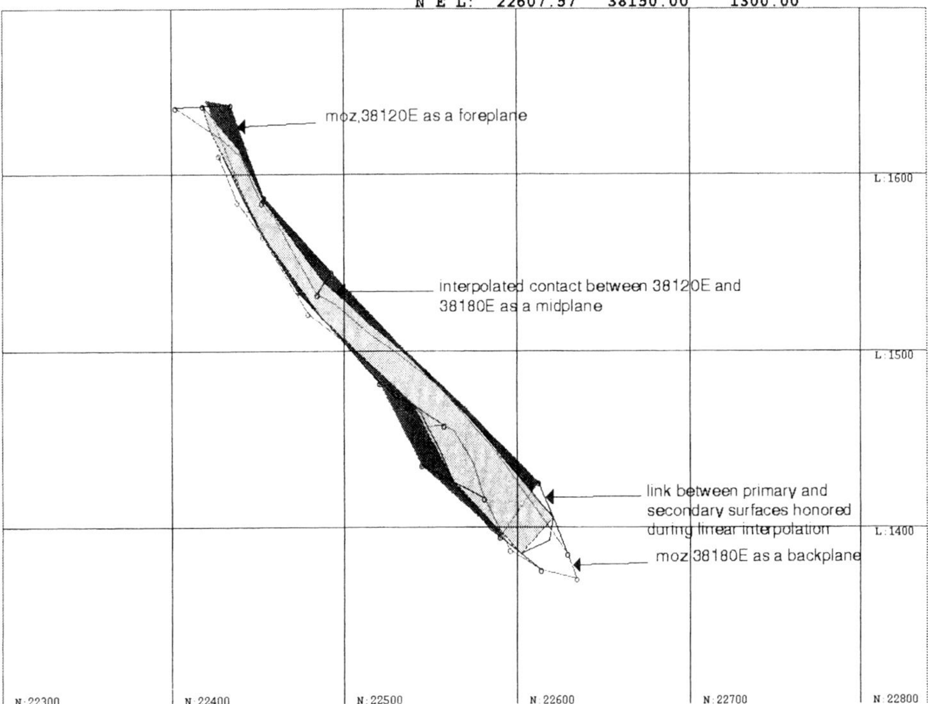

Figure 4.40. Fore, mid and back planes used for interpolation between adjacent components.

corresponding features between the fore and back planes. **Process** will interpolate a new midplane between the two polygons (Fig. 4.40). **Save both** the interpolated fore and back planes of the two components. If the two polygons are very dissimilar, the interpolated boundary might not be very satisfactory. If this is the case, you might want to reconsider the shapes of the two adjacent components, add a third component that is between them, or try a different set of links. Repeat the process for the secondary ore zones and toggle the volume's background display on to check to see that the interpolated surfaces are in alignment from several section views and depths. A common mistake that would result in a bizarre interpolation would be if the wrong fore and back planes were selected.

When the fore and back plans have been misspecified during interpolation, the easiest solution is to once again bring up the Fetch both entry form, correctly enter the combination of fore and back planes to be used for interpolation and repeat the process, but this will only correct the two adjacent halves of the components, leaving their opposing ends (for instance the backplane of MOZ,38120:E and the foreplane of MOZ,38180:E) incorrectly interpolated. While the addition of new components (say, MOZ,38060:E and MOZ,38240:E) can be used to correct this by subsequent interpolation, at times it is best to just reset the component to its original non-interpolated geometry and start over. To do this, toggle interpolation mode off (**Volume ⇒ Interpolate**). Make the component to be reset active by using either **Fetch**, **Select** or by Initializing it in the Parameters entry form in which the active plane must be the midplane. Use **Viewplane ⇒ Setup** to place the viewplane exactly on the midplane of the component. Finally, select **New** to clear the active memory followed by **Define ⇒ Snap on ⇒ Pick**. Pick the component's boundary which will become highlighted. Use **Segment** to select the entire boundary by selecting the first (displayed as a filled diamond) and last (adjacent to the first component but separated by a gap in the boundary) vertices on the component's midplane. By selecting **Save** or **Save & Clear** the component will be overwritten based on the midplane boundary.

Summary of procedure: Example 4.10
Initial data requirements: drillhole data of lithology (project TUTORIAL)

Procedure:

1. Make the volume model in which components will be added or modified active. From the main menu **Volume Data ⇒ Volume Select** (or **Create**): <G.EX4.10>
2. Setup a suitable background display.
   2.1 Triangulate the surface topography. Make <TOP> active and select **Maps ⇒ Map TIN Define** and in the Parameters entry form select
   FEATURES: <on> TRIANGLE POINTS: <on>
   ⇒ **Generate ⇒ File ⇒ Save & Exit**
   2.2 Display topography as an intersection **Background ⇒ Maps** (Background Map Data Selection entry form)
   SELECTED MAPS: <TOP>
   DISPLAY SURFACE INTERSECTION: <on>
   ⇒ **OK**
   2.3 Display lithology on the drillhole trace **Background ⇒ Holes**
   SUBSET/HOLES SELECT: <dh0>
   PRIMARY VARIABLE/CHARACTERISTIC: <lith>
   DISPLAY THICKNESS: FORE/BACK <30> <30>
   DISPLAY FORMAT: <color map> (see Fig. 4.34)
   TRACE DISPLAY: <solid>
   ⇒ **OK**
   2.4 Orient viewplane to (N,E,L,Az,Dip) (22300,38180,1300,0,0) **Viewplane ⇒ Setup**
3. Define a volume component in the main ore zone **Volumes ⇒ Volume Data Define** (Volume Data Define menu and Parameters entry form)
   Toggle off volume background display while in Define mode **Viewplane ⇒ Background**
   VOLUMES: <off>
   ⇒ **Define** (initialize component definition in Parameters entry form)
   UNIT NAME: <MOZ> COMPONENT NAME: <38180:E>
   VOLUME CODE: <2> THICKNESS: FORE <30> BACK <30>
   ACTIVE STRING: <MP>

⇒ **Enter** (digitize MP boundary of MOZ,38180:E)
⇒ **Save**

4. Define a secondary ore zone component sharing a common boundary with the main ore zone component
⇒ **New** ⇒ **Define** (initialize component definition in Parameters entry form)
Unit Name: <SOZ> Component Name: <38180:E>
Volume Code: <3> Thickness: Fore <30> Back <30>
Active String: <MP>
⇒ **Snap on** ⇒ **Pick** (pick MOZ,38180:E) ⇒ **Segment** (select shared boundary)
⇒ **Snap off** ⇒ **Enter** (complete SOZ,38180:E boundary)
⇒ **Save & Clear** (removed from active memory)
5. Repeat Steps 2.4, 3 and 4 creating components MOZ,38120:E and SOZ,38120:E with mid-planes at 38120:E.
6. Interpolate between components at 38180:E and 38120:E **Volume** ⇒ **Interpolate** (toggle on)
⇒ **Fetch** (Volume Data Interpolate Fetch Both entry form)
Primary Component:
Unit Name: <MOZ> Component Name: <38180:E> Interpolation Boundary: <BP>
Secondary Component:
Unit Name: <MOZ> Component Name: <38120:E> Interpolation Boundary: <FP>
⇒ **OK**
Interpolate ⇒ **Link** (link one or more vertices from the primary to secondary boundaries) ⇒ **Process** ⇒ **Save both**
7. Repeat Step 6 interpolating between SOZ,38120:E and SOZ,38180:E.

Example 4.11: Interactive editing of volume components in Lynx
Examination of adjoining components will inevitably uncover inconsistencies in the geologic model which need to be corrected. These inconsistencies commonly will be in the form of gaps and overlapping components in the extremities of the models, which can be corrected in Lynx by using the interactive volume model interpretation facilities that allow the modeler to select a component to be modified. For example, examination of the component created in Example 4.10 reveals that the foreplanes of MOZ,38180:E and SOZ,38180:E are overlapping. To correct this overlap, the foreplane of MOZ,38180:E will be modified. Before continuing you may want to change the display so that the components are not displayed in fill, are labeled and the drillholes that were in background are toggled off.

From the main menu select **Volumes** ⇒ **Volume Data Update** to open the Volume Data Update menu and Parameters entry form. With the viewplane in plan view, and cutting the volume components, use the **Select** function to select the component to be modified. Once selected, the component's definition will be displayed in the parameters form with either the fore (FP), mid (MP) or back plane (BP) toggled on. Select **Define** and the viewplane will orient itself to the selected plane of the component. This plane is then displayed in white with its vertices. The midplane of the component will be displayed as a dashed line in the color of the component (Fig. 4.41). There are different approaches that can be used to guide modification of the component. Volume background display can be used to bring up only the component of interest. For instance, if MOZ,38180:E overlaps SOZ,38180:E on their fore

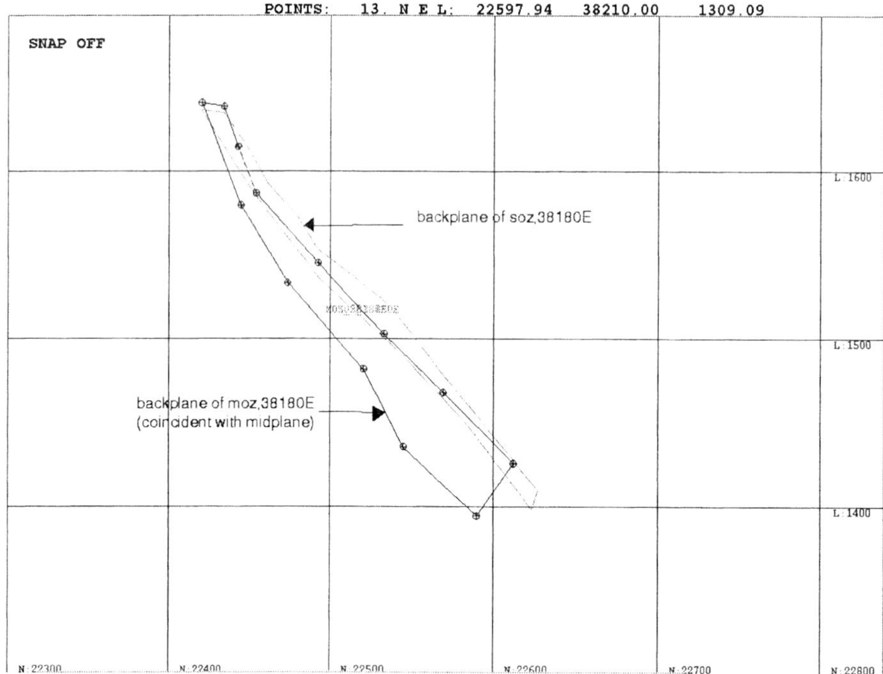

Figure 4.41. Interactive interpolation of a midplane.

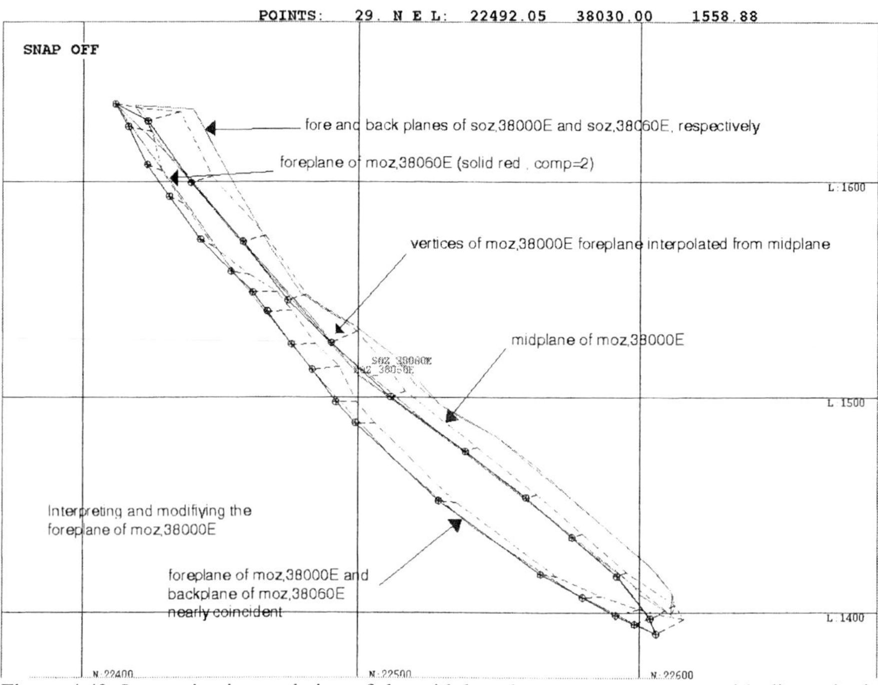

Figure 4.42. Interactive interpolation of the midplane between components with discontinuities at their extremities.

planes, then the select function could be used to bring MOZ,38180:E into active memory with the boundary displayed at 38240:E. SOZ,38180:E can also be displayed on this viewplane to serve as a guide, but the most accurate approach is to use **Volume ⇒ Replicate** to replicate SOZ,38180:E and **Snap On**, **Pick** and **Edit** to bring the two boundaries into alignment. If the two components are a poor match, **Insert** can be used to add new vertices that will provide a closer match. **Reset** can be used to return the component to its original boundary. When the modifications are finished, **Save** the modifications by overwriting the component (Fig. 4.42).

Summary of procedure: Example 4.11
Initial data requirements: adjacent volume components requiring modification

Procedure:
1. Make the volume model to be modified <G.EXP4.9> active.
2. Display the volume model in background as per Example 4.10 but in plan view.
3. Update the component's boundary. From the main menu select **Volumes ⇒ Volume Data Update** (Volume Data Update menu)
   3.1 Select the component to be modified
   ⇒ **Select** <MOZ,38180:E>
   Toggle on the desired boundary: <BP>
   3.2 Replicate the component to be used as a template **Volume ⇒ Replicate** (Replicate Selection entry form)
   UNIT: <SOZ> COMPONENT: <38180:E> BOUNDARIES: <BP>
   ⇒ **OK**
   3.3 Pick the boundary to be modified
   Define ⇒ Snap On ⇒ Pick: <MOZ,38180:E>
   3.4 **Edit ⇒ Save**

CHAPTER 5

Estimation of grids, areas and volumes of intersection

It has been demonstrated how grids, triangle sets, surface handling and interactive volume modelling can be used to represent the topology of geologic structures and how thicknesses and volumes can be estimated from these structural models. Once the limits are defined by the extent of the volume model, volumes can be estimated by methods such as triangulation, matrix algebra and integration. Thus, volumes are defined by the limits of one or more continuous surfaces. These continuous surfaces are in turn defined either by estimation or by interactive interpretation on geologic sections and extrapolation/interpolation between sections. Triangulation was briefly introduced in Chapter 4 as a means of estimating a surface $\hat{z}$ value for all locations x and y in the plane of the surface. For the purposes of estimating volumes, this $\hat{z}$ value represents either an elevation, offset distance or thickness, but in fact, there is no limitation on the definition of z which can just as readily be a coal seam's Btu value as its thickness. Whatever the definition of z, the modeler is faced with a troublesome problem of using a limited number of samples of relatively small volume to estimate a vastly greater area. This chapter explains the procedures used for estimation.

Grids are associated with two-dimensional estimation of a wide variety of geochemical and geotechnical variables. Estimation of surface boundaries was introduced in Chapters 3 and 4. Characteristics of soils and many bedded deposits can also be modeled as a two-dimensional grid due to their essentially laminar structure. Since these laminar or bedded deposits are commonly treated as being homogenous in character across their thickness, they are commonly treated as being two-dimensional surfaces, but as laminar deposits become thicker and are sampled in intervals along their thickness, it is often convenient to continue to model three-dimensional variability using a sequence of aligned or layered 2D grids, each layer being associated with a depth interval or bedding plane. As the thickness increases, it becomes more convenient to work with a three-dimensional grid, a matrix of three indices whose elements are associated with a volume. These three-dimensional grids are commonly referred to as block models. Thus, the discussion of two-dimensional estimation of surfaces will now be extended to estimation of any regionalized variable existing in either two or three dimensions.

While grids and block models can be used to assign an estimated value to an area or volume, other facilities are needed to determine areas and volumes of intersection. Thicknesses and volumes can be calculated between surfaces or by numeric integration of volume models, but reserve estimation and production planning must be based on average assays and tonnages within minable units. The minable unit is the volume associated with either a cell in a two-dimensional deposit or a block in a three-dimensional deposit model. Total reserve estimates are based on intersecting grids or volume models that represent the limits of an ore body with either a grid or block model of estimated grade. For a two-dimensional deposit, such as Appalachian coal, grids can be used to contain estimates of the top and bottom structures, btu, and other quality characteristics. A grid of the topography can be intersected with the seam's top and bottom structure grids and total reserves and average quality characteristics can be determined. For a three-dimensional deposit, a block model can be used to contain estimated grades. This block model can then be intersected with a volume model which represents the limit of the ore body and the average grade and tonnage of ore in the volume model can be calculated. For surface mines, minable reserves are determined by intersecting a surface representing the limits of mining with both a block model of ore grade and volume models of the various lithologic units to determine average grade and total waste and ore tonnage as a function of rock type. For estimating minable reserves of underground operations, volume models of stopes and ore bodies can be intersected with block models to provide estimates of tonnage and grade for each stope. The method used for estimating minable reserves can be extended to production scheduling. In the example of coal mining, the proposed limits of a cut associated with a mining period can be intersected with the grids representing seam limits and quality characteristics to determine tonnages of ore and waste and average coal quality. If the tonnages and quality do not match the production requirements for the mining period, then the limits of the cut can be modified or another cut evaluated until an acceptable production plan is found. This methodology for production scheduling can be extended to open pit and underground mining by determining volumes of intersection between block models of grades, volume models of geology and grids representing push-backs in pits or volume models of underground openings.

5.1 ESTIMATION METHODS

There are four basic classes of methods used for assigning values to grids and block models: splines, surface fitting, weighted averages and areas of influence. Of these, methods which base an estimate on a weighted average of the surrounding point data have received the most attention in the past and continue to hold the greatest potential for providing statistically optimal estimates, but all of these methods have advantages and are favored in certain circumstances and for some types of deposits or data sets. This discussion of estimation will start with the traditional, non-statistical algorithms and then delve more deeply into weighted averages, especially the geostatistical estimation methods jointly referred to as kriging.

5.2 AREAS OF INFLUENCE

When samples are widely scattered over the area of investigation, and there is an apparent absence of spatial correlation, a simple and popular method of assigning sample values to a volume is to assign the sample value to an area of influence around each hole. These methods are well suited to hand calculations and continue to be used for data which are taken on spacings which are so great as to negate the validity weighted average and spline methods which assume a correlation between nearby samples. Often, the justification for using this method is that the value of interest is very continuous within the distance of separation between neighboring samples and that minor variations in the estimated value have little economic consequence either due to the low value of the commodity or because of the use of a non-selective mining method.

Polygons of influence are formed using a triangle set generated from neighboring samples. The sides of the triangles are bisected at 90° with construction lines that extend until they intersect another bisector or until they reach a boundary which limits the estimation area. The intersection of two adjacent bisectors form a vertex for the polygons of influence of the three samples which were the basis of a triangle. Polygons formed around samples at the edge of the data set will have polygons which extend to the boundary of the area being evaluated. When sampling is on a rectangular grid, the resulting areas of influence will be rectangular. Irregular sample locations will result in an irregular set of polygons.

Figure 5.1 illustrates the construction of polygons for assigning thickness to areas of influence. A triangle set has been formed by selecting vertices which result in near equilateral triangles. Polygons were then formed by bisecting each side of the triangles at their midpoint with a perpendicular construction line. The point at which two

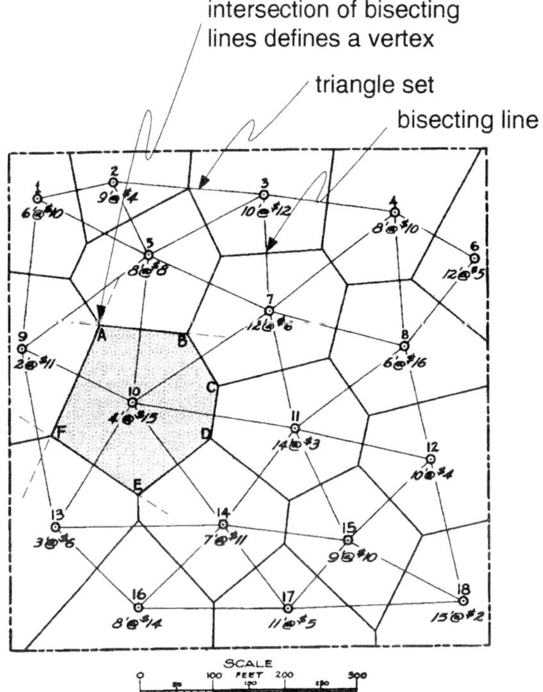

Figure 5.1. Construction of polygonal areas of influence (from Parks 1949).

adjacent construction lines cross establishes a polygon vertex. In the case of the data point 10, the hexagon ABCDEF is established and its area computed. Since deposit thickness at 10 was measured as 4 ft, the volume is based on the hexagon's area and a thickness of 4 ft. At sample 10 the materials value was determined to be \$15/bcy which is then applied to the volume of ABCDEF.

The polygons can be retained and used for estimation as would a triangle set. Since the polygons cover the entire area being evaluated, the value of any *xy* point of interest can be found by identifying the polygon that encompasses this location and referring to its assigned value. This same method can be used to assign values to grids from polygons by comparing the grid's centroid coordinate to the polygon set. Polygonal gridding is a handy means of transferring qualitative data into grids. In this case, the polygon can represent property boundaries, and production calculations from a grid can include a flag defining ownership on which the payment of royalties can be based.

Note that 'area of influence' methods are not commonly used for estimation since the underlying assumption is that the data points are spatially uncorrelated, i.e. the experimental variogram is a pure nugget effect. A pure nugget effect is uncommon and when observed is more likely to be the result of very widely separated samples or mixing of data taken from different sample populations. As shown in Figure 5.2, an estimate based on polygons will consist of planar surface with sharp discontinuities, a surface which is unlikely to be found in nature.

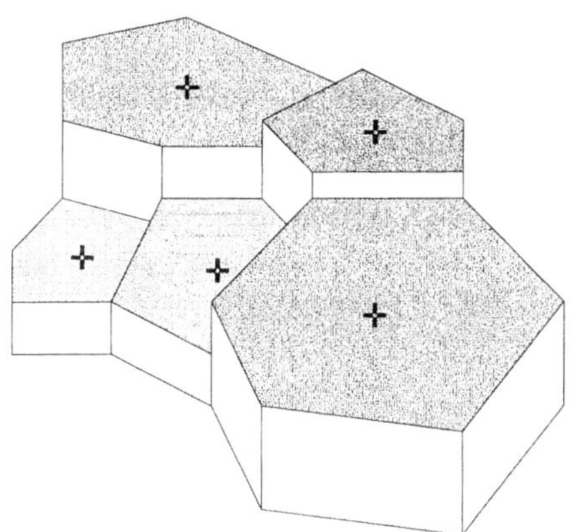

Figure 5.2. Polygons of influence showing discontinuities at edges.

Average quality characteristics and thicknesses can be estimated from polygons of influence by calculating the area of each polygon and then using the area as a weight for the polygon's sample value and calculating a weighted average. This method of using polygons to determine an area of influence for each sample has other valuable applications. Sample data are often concentrated around areas that are in some way anomalous. High grade or geologically complex areas will commonly be sampled more intensively than the rest of a property. Preferential sampling will introduce a bias into global reserve calculations and weighted average estimation methods. If the more intensively sampled areas are of substantially different grade, the average grade will be biased towards the grade found in the more intensely sampled areas. Since the intensively sampled area might represent a small area in comparison to the remainder of the study site, the magnitude and consequences of the bias can be great. Preferential sampling can be corrected by declustering the data. Declustering involves using the area of the polygon of influence around each sample as a weight. Instead of calculating a simple average of the sample data, a weighted average is calculated using the areas of the polygons.

Example 5.1: Using polygons of influence to decluster data in Techbase
Return to the topographic data for the Boland Banya project to illustrate the risk of bias from clustered data and the use of polygons of influence for declustering and correction of bias. Refer to the random contoured map and triangle set of topography that was used to illustrate triangulation and map contouring in Chapter 3. In the NE, jutting out of the left side of the valley is a volcanic neck which was densely surveyed. The densely sampled area comprising the neck is roughly bounded by the window 11900E to 12500E by 21800N to 22350N. By calculating **Summary** statistics for the surveyed topographic data over the entire property, we find that the average elevation is 5451' calculated from 855 values. Add a filter to restrict the calculation of the mean elevation to samples within the window bounding the neck and note that the average elevation is 5363' calculated from 293 samples. Gridded estimates and polygons of influence will be used to demonstrate that the average elevation of 5451' is an underestimate of the true average elevation due to having preferentially sampled 34% of the elevation data in an a much smaller area having 3.8% of the total area of the project.

The area from each polygon can be saved and stored in the same table which contains the coordinate *xyz* data used to generate the polygons. **Create** a new field in the topography table to contain the polygon areas. There will be one polygon for each record in this table. From the Techbase main menu select **Modelling** ⇒ **Polyest** ⇒ **Polygon** ⇒ **Fields** and enter as DATA POINTS the *xyz* values from the flat table containing the topographic data and the area field as the RESULT. Under **Parameters** provide a polygon radius of influence, the number of sides that a polygon which is outside of that radius will have, and a file name in which the polygons can be stored for later retrieval. When some data are separated by very long intervals, it might be preferable to limit the area assigned to a sample value. It is a good idea to use a reasonable area of influence that is based upon observation of the data density near the border of the data set. The default value is a large number, which without a boundary file to clip to the polygons would result in very large areas being assigned to boundary data; this would in itself cause biased results since boundary data would be more heavily weighted. The area of influence is provided to limit the radius of a polygon assigned to an isolated sample. Polygons which are not surrounded by samples, as in the case of isolated samples, will have a radius equal to the radius of influence and the assigned number of sides. If a file name is provided, the polygons will be written to an ASCII file with each record having an *x*, *y* and polygon id value. This file can be loaded into a polygon table to use for estimation or can be displayed in **Poster**. A **Boundary file** can be provided to limit the area in which polygons will be generated. The boundary file will be an ASCII file containing the vertices of a polygon that will be used to clip to the polygons calculated in Polyest. **Estimate** will generate the polygons, write their vertices to the ASCII file, and place values into the area field.

To display the polygons and the data from which they were generated execute **Poster**, set up the **Scaling** and then select **Polygon file** and enter the name of the polygon file generated by Polyest. The z values on which they are based can also be displayed by selecting **Points** ⇒ **Fields** and only entering the x and y coordinate of the elevation. Entering the elevation field will result in the value of the elevation being posted. Enter a **Marker** to indicate the data location in each polygon. The set of polygons of influence is shown in Figure 5.3.

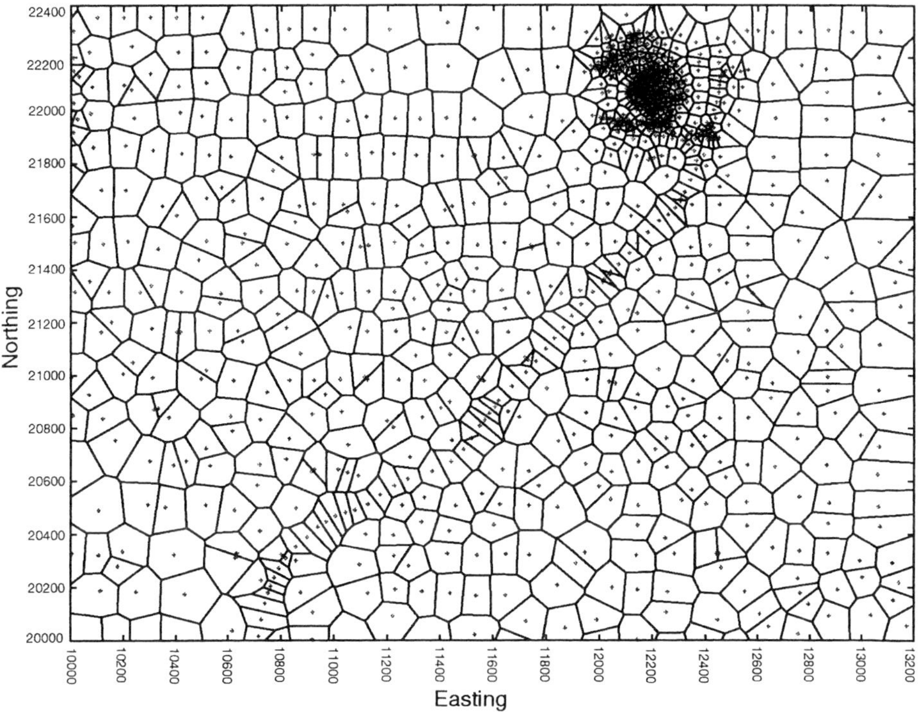

Figure 5.3. Polygons of influence for topographic data (Boland Banya).

To calculate the declustered weighted average elevation, create a new field in the flat table containing the topographic data. This will be a calculated field equal to the product of the field for area of the polygons and the elevation. The weighted average elevation is found by dividing the sum of this product (area × elevation). **Report** can be used to sum the values of the polygon areas and the weighted elevations.

The declustered weighted average elevation was found to be 5503'. This is much greater than the value of 5451' found by averaging all of the elevation data. An unbiased estimate of the average elevation can be found by estimating the elevation into a grid (cell table) and then finding the average of the estimated grid values. Cell table estimates are unbiased since each cell represents a fixed area, although individual cell values can be biased in the presence of clustered data. More on this later. Still, the average gridded estimate of the elevation should be closer to the true average elevation than the point data. The average gridded elevation was found to be 5495' which we find to be very close to the declustered elevation of 5503'.

Summary of procedure: Example 5.1
Initial data requirements: a flat table containing topographic data for the Boland database.

Procedure:
1. Create a field for polygon areas: **Techbase ⇒ Define ⇒ Fields ⇒ autoTable**
   TABLE NAME: <topo>
   ⇒ **Create**
   FIELD NAME: <parea> TYPE: <real> CLASS: <actual>
2. From the main menu generate polygons: **Modelling ⇒ Polyest ⇒ Polygon ⇒ Fields**
   DATA POINTS RESULTS
   VALUE : <elevation> AREA: <parea>
   X: <easting>
   Y: <northing>
   ⇒ **Parameters**
   RADIUS OF INFLUENCE: <200>
   NUMBER OF SIDES: <6>
   POLYGON FILE NAME: <parea.pol>
   ⇒ **Estimate**
3. View the results graphically: from the main menu **Graphics ⇒ Poster**
   3.1 Setup **Scaling** (metafile naming <topopoly.met>) (scaling X: 10000-13200, Y: 20000-22400; 250 ft/in; B sheet in landscape; plot **Border** and **Grid** (optional))
   3.2 **Display polygons ⇒ polygon File**
   POLYGON FILE: <parea.pol>
   3.3 Display calculated areas and values ⇒ **Point ⇒ Fields**
   X-COORD: <easting> Y-COORD: <northing> VALUE 1: <parea> (optional)
   ⇒ **Marker** (accept defaults)
   ⇒ **Value** style (optional)
   SIZE: <.05> BASELINE: <45>
   ⇒ **eXit**
   ⇒ **Graphics ⇒ Review**

5.3 TRIANGLES AND TRIANGULATION

A triangle set can be used in a number of ways. As was demonstrated in Chapter 4, triangle sets can be used to represent surfaces, calculate thicknesses and define prism-shaped volume components. They can be used as the basis for forming polygons of influence, or, as in the contouring example of Chapter 3, the equation of a plane defined by a triangle can be used for contouring either by using random contouring or by using the triangles for gridded estimates. Linear interpolation of z values along the legs of triangles used to be the primary method of creating hand-drawn contour maps from sample data (as in Fig. 5.4). Triangles can also be used to assign

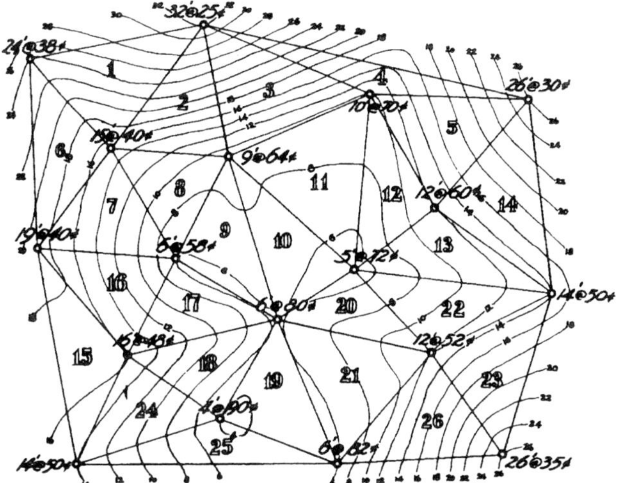

Figure 5.4. Triangulated areas of influence used to estimate contours (from Parks 1949).

estimated values as a function of the triangle in which the point is located. This method, while similar to polygons of influence, has the advantage of not having as sharp a discontinuity between estimated values that are on the edge of adjacent triangles. In the simplest approach, any point within the polygon can be assigned the average of the three values that form the triangle. Another version is to subdivide the triangle into three triangles having as vertices the three sample values and the point to be estimated. These three new triangles are used to estimate the unknown point as a weighted average of the three sample values of the original triangle with each of the samples being assigned as a weight the new triangle whose hypotenuse is opposite to the sample. As the unknown point gets closer to any one of the sample locations, the new triangle used as the weight for that sample will increase in area. Thus, this method has similarities both to polygons of influence and inverse distance methods of estimation. For example, if a triangle based on locations *ijk* (Fig. 5.5) and having sample values at those locations of z_i, z_j and z_k is used to estimate a value z_x at a location x that is inside of the triangle and subdivides the triangle area a_{ijk} into

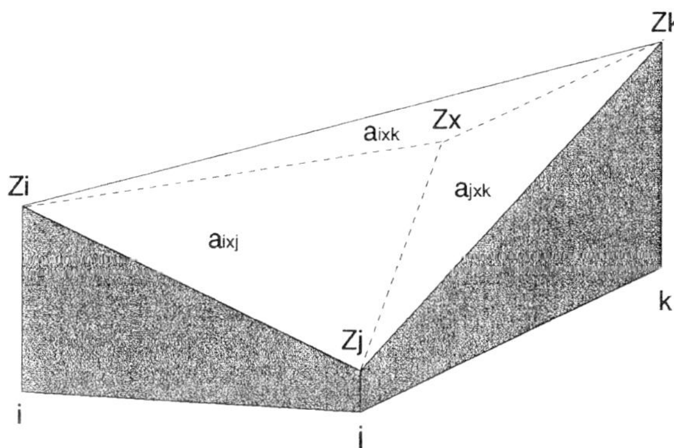

Figure 5.5. Estimating either random or gridded values using triangular areas of influence.

areas a_{ixj}, a_{ixk}, and a_{jxk}, then the assigned value will be found as

$$z_x = \frac{a_{jxk} z_i + a_{ixk} z_j + a_{ixj} z_k}{a_{ijk}}$$

Example 5.2: Gridding using triangle sets in Techbase

Techbase's triangulated gridding routine, **triGrid**, includes a surface fitting algorithm which can fit either a planer surface, as was described in Chapter 3, or a curved surface based on a maximum gradient calculation. The top structure of the OA main ore body in the Boland Banya deposit will be used to illustrate the application of this triangulation algorithm.

Before continuing with this exercise, *xyz* drillhole intersection data is required from the top and bottom contact of the ore zone OA. This can be generated from a length composite as was illustrated in Example 4.2. Additionally, there will have to be a cell table with a field to hold estimated values for the elevation of the OA top structure as was illustrated in Example 4.4.

From the main menu select **Modelling** ⇒ **triGrid** ⇒ **Grid** ⇒ **Fields**. Enter the *x*, *y* and *z* coordinate fields from the flat table holding the composited hole intercepts under DATA POINTS and the cell table field that will hold the gridded top structure under RESULTS. Select Parameters and enter the number of surrounding data points from which the slope (gradient) is to be derived and the maximum gradient magnitude.

If the gradient is set to zero, then:

1. Delanay triangles are constructed,

2. A planer patch is calculated from the three node points that surround the point that will be calculated,

3. The *z*-value is calculated from the three closest triangle nodes.

If the gradient is not zero (a value was entered or was left blank), then:

1. Delanay triangles are constructed,

2. The gradient is calculated at each triangle node. This is a five equation gradient – d*z*/d*x*, d*z*/d*y*, d*z*/d*xy*, etc. The user specifies the number of points or the program defaults to enough points to give an adequate answer,

3. Using the gradient equations at each triangle node location, the 'new' point is calculated from the equation. Three 'new' values are calculated from three node points and the results are averaged. The 'new' points coordinates can be an actual *xy*-node coordinate or the center of a cell.

Grid will estimate values into the cell table field using the composited data. In this example, there should be 105 data points used to create 109 triangles. Using the cell table defined in Example 4.4, there should be 4022 results (cells) estimated and 3167 cells not estimated. Note that triangulation does not extend beyond the limits of the data set. The 3167 cells are outside of the 109 triangles.

triGrid was carried out using two different sets of parameters. In one case, four data points were used and the gradient was set to zero. The resulting estimate is based on the planar equations. **Gridcont** was used to contour the resulting estimate of the OA top structure with a smoothing factor of zero. **Randcont** was used to generate a triangle set based on the 105 drillhole intersections with the top of OA. The resulting map (Fig. 5.6) clearly shows how the grid values are simply linear interpolations on each triangle. These are the same results that would be obtained by drawing the contours by hand.

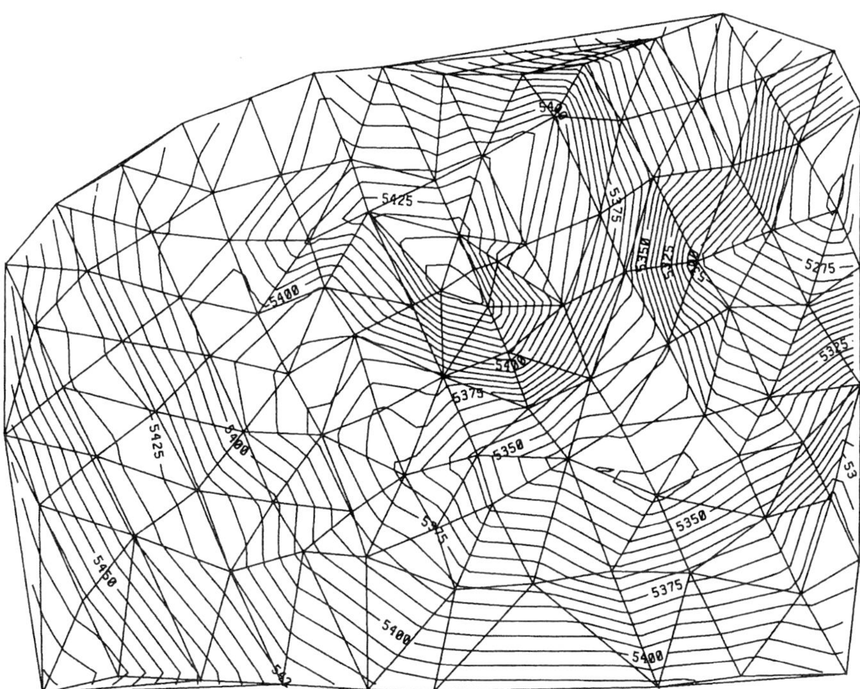

Figure 5.6. Results of using triGid with estimation based on planar surfaces (no contour smoothing).

Summary of procedure: Example 5.2
Initial data requirements: topographic *xyz* data from the Boland project

Procedure:

1. **Techbase ⇒ Define ⇒ Fields ⇒ autoTable**
   TABLE NAME: <surfaces>
   ⇒ **Create**
   FIELD NAME: <topo_t> TYPE: <real> CLASS: <actual>
2. From the main menu select **Modelling ⇒ triGrid ⇒ Grid ⇒ Fields**

DATA POINTS	RESULTS
VALUE: <elevation>	VALUE: <topo_t>
X: <easting>	X: <>
Y: <northing>	Y: <>

3. From level 2 ⇒ **Parameters** (accept defaults)
4. From level 2 ⇒ **Grid ⇒ eXit ⇒ eXit**

5.4 SPLINES

Splines are commonly used in one dimension to fit curves. In two dimensions they can be used to fit gridded surfaces. Splines use differential equations to fit random z values in 2D space while subject to boundary conditions acting at the sample data points, referred to as knots, and along the boundary of the grid. In the case of a continuous surface, the spline can be likened to a flexible sheet stretched over N observed elevations w_n (n = 1...N) having forces f_n acting at knots (x_n, y_n) (Briggs 1974). Fitting the sheet to the observations produces a displacement in the sheet f_n at w_n.

$$\frac{\partial^4 u}{\partial x^4} + 2\frac{\partial^4 u}{\partial x^2 \partial y^2} + \frac{\partial^4 u}{\partial y^4} = \left(f_n : x = x_n, y = y_n; \quad 0 \text{ otherwise}\right)$$

Various functions, u, can be used for fitting. Generally, the most common function used for fitting is a polynomial. A popular function is a cubic polynomial. For 2D surface fitting this results in the elevation, z, being a linear function of the independent variables x and y in first, second, third order single variable terms and cross-products. A spline differs from a cubic polynomial in that boundary conditions are enforced at the boundaries and knots. At the knots, the spline function u must be first and second order differentiable to ensure a smooth transition of the surface at the data points.

Along the edge of the sheet, the force and the bending moment about a tangential line are zero, yielding the following boundary conditions on the above equation.

$$\frac{\partial^2 u}{\partial x^2} = 0, \quad \frac{\partial}{\partial x}\left(\frac{\partial^2 u}{\partial x^2} + \frac{\partial^2 u}{\partial y^2}\right) = 0, \quad \text{and} \quad u\left(x_n, y_n\right) = w_n$$

Note that the boundary conditions require that the displacement function at a data point (x_n, y_n) must be equal its value w_n. The values used are based on the minimization of the total squared curvature.

$$C(u) = \int\int \left(\frac{\partial^2 u}{\partial x^2} + \frac{\partial^2 u}{\partial y^2}\right) \mathrm{d}x\mathrm{d}y$$

In practice, u (x, y) is not solved as a continuous surface but is approximated for each node in a grid, i.e. the grid estimates $u_{ij} \equiv u(x_i, y_j)$ such that $x_i = (i-1)h$, $y_j = (j-1)h$ for i = 1, ..., I, and j = 1, ..., J where h is the grid spacing. The discrete total squared curvature of the grid becomes

$$C = \sum_{i=1}^{I} \sum_{j=1}^{J} C_{ij}^2$$

where C_{ij} is the curvature at grid index ij and node location (x_i, y_j). This is a linear function of the u_{ij}. To determine this function, C is minimized by taking its partial derivative with respect to u_{ij} at each grid node, and setting it equal to zero.

$$\frac{\partial C}{\partial u_{ij}} = 0, i = 1, \ldots, I; \quad j = 1, \ldots, J$$

In two dimensions, the approximation of C at (x_n, y_n) is:

$$C_{ij} = (u_{i+1,j} + u_{i-1,j} + u_{i,j+1} + u_{i,j-1} - 4u_{ij}) / h^2$$

where the u_{ij}'s are the discrete displacement estimates at the grid nodes. Along the edge rows and corners, different expressions are used (see Briggs 1974). Using the three previous equations, the resulting difference equation for the interior of the grid is found to be

$$\begin{aligned} &u_{1+2,j} + u_{i,j+2} + u_{i-2,j} + u_{i,j-2} \\ &+ 2\left(u_{i+1,j+1} + u_{i-1,j+1} + u_{i+1,j-1} + u_{i-1,j-1}\right) \\ &- 8\left(u_{i+1,j} + u_{i-1,j} + u_{i,j-1} + u_{i,j+1}\right) + 20u_{i,j} = 0 \end{aligned}$$

while other equations are derived for the boundary. The various u_{ij} are solved for in these relationships by giving starting estimates for the neighboring grid nodes. If multiple w_n are within the limits of the cell, an initial assignment can be based on the average of the data values or on the nearest value. Otherwise, the sample average can be used or an inverse distance weighted average can be used as the starting estimate.

This procedure is followed to assign gridded estimates iteratively on successively finer grids with the estimate for each grid being based on the previous courser grid's estimates. The knots retain their initial values at each iteration, but cells without data values will change in each iteration. The solution procedure terminates when the solution converges so that the change between sucessive iterations meets an assigned tolerance value or when the number of iterations reaches a maximum.

Example 5.3: Minimum Quadratic Curvature (MINQ) estimation in Techbase
The topographic data of the Boland Banya deposit will be fit using the MINQ routine. Use **Techbase** ⇒ **Define** ⇒ **Fields** ⇒ **Create** to create a new field to hold the MINQ estimate <topo_m>. From the main menu select **Modelling** ⇒ **Minq** ⇒ **Minq** ⇒ **Fields** and input the source data from the topography table and the results field <topo_m> where the estimate will be stored. Note that the result must be stored in a cell table: the difference equations for spline estimation are based on gridding.

From the level 2 menu select **Parameters**. CONDITIONING refers to how initial displacement values are generated for the grid: CLOSEST assigns the elevation of the closest data point to the grid, while AVERAGE assigns the average elevation to the grid node. The default is AVERAGE. CONVERGENCE and MAX ITERATIONS are the two criteria that are used for stopping the iteration algorithm. CONVERGENCE is the tolerance between the data value(s) and estimate that must be satisfied in the grid before stopping. A CONVERGENCE of .01 to .1 times the contour interval is suggested for contouring; .01 to .001 is suggested if the grid is to be used for perspective viewing. The default is 1/300 of the elevation range. MAX ITERATIONS is the number of estimation iterations used. The default is 100. MAX ITERATIONS has precedence over CONVERGENCE. Select **Minq** to start estimation. The algorithm first initializes the grid with the available data and then commences iteration to convergence or the maximum number of allowed iterations.

The results of using MINQ for contouring topography using the closest data elevation for conditioning and 1 versus 100 iterations are shown in Figures 5.7a and 5.7b, respectively. For both of these estimates it is interesting to observe the degree of smoothing that occurs. Figure 5.7a has much more irregular and realistic contours than Figure 5.7b, yet, by definition, Figure 5.7a is 'less accurate' than Figure 5.7b in that the estimated surface is farther from the data. What is actually happening is that with one iteration the estimated surface is much more localized with the grid values being largely based on the closest datum value. In the initial iteration the maximum change can be quite large (23.48) since many of the cell estimates will have been based on the average elevation, but by the 100th iteration the estimates have been smoothed to the point that the maximum change is down to 0.02.

The advantages of MINQ are that it generates smooth contours and still honors the data within the cell. If there is only ony data value falling within the limit of the cell, it will be honored. If there is more than one datum value falling in the cell and AVERAGE is used for conditioning, then MINQ can be used to decluster the data much as in Example 5.1. But the use of minimum curvature criteria and boundary conditions presents hazards for some types of data sets. Smoothing of the data is generally undesirable in that local variations can be lost. The effect of smoothing is clearly visible in Figure 5.7, but in this case smoothing has actually resulted in bringing the estimated surface closer to the data points which are nothing more than strings of digitized contours, except in the area of the volcanic neck at the head of the valley. There, a high density of data points have been declustered and smoothed, resulting in a surface estimate that appears as a rounded hill rather than a sharp discontinuity. This suggests that splines should not be used for surfaces representing rugged terrain or faults even if the original data set contains a high degree of detail about sharp features. Methods which are much more localized and capable of honoring surface features, notably triangulation and not gridding, should be used in these cases. No surface estimate should be used for crossing a discontinuity like a fault. In the case of a fault displacement, the two sides of the faulted surface should be estimated separately.

Their are other potential probelms with using MINQ which are inherent in the nature of splines. Cells that are surrounded with data should converge to their final value in a few iterations, but areas that are isolated from data may require many iterations as the adjustments to the closest data shift across the interviening cells. Away from the knots, the only criteria for cell values are the principle of minimum curvature and the boundary conditions. These may act to yield unpredictable results: for instance, creating mountains where none exist. Additionally, the requirement that the surface be first and second order differentiable at the knots will force the surface to be a smooth curve at data locations. Thus, in the case of the volcanic neck of Figure 5.7, the top and sides of the neck would be rounded and smooth unless there was a very high density of data delineating the sharp change in topography. A final issue is the validity of the minimum curvature algorithm: who's to say if Figure 5.7a or b is the closer representation of reality? Splines found favor during the initial development of 'contouring' routines principally because they produced pleasingly smooth and continuous contours similar to those that would be produced by an experienced draftsman. Pleasing contours are not wise criteria to use for estimation.

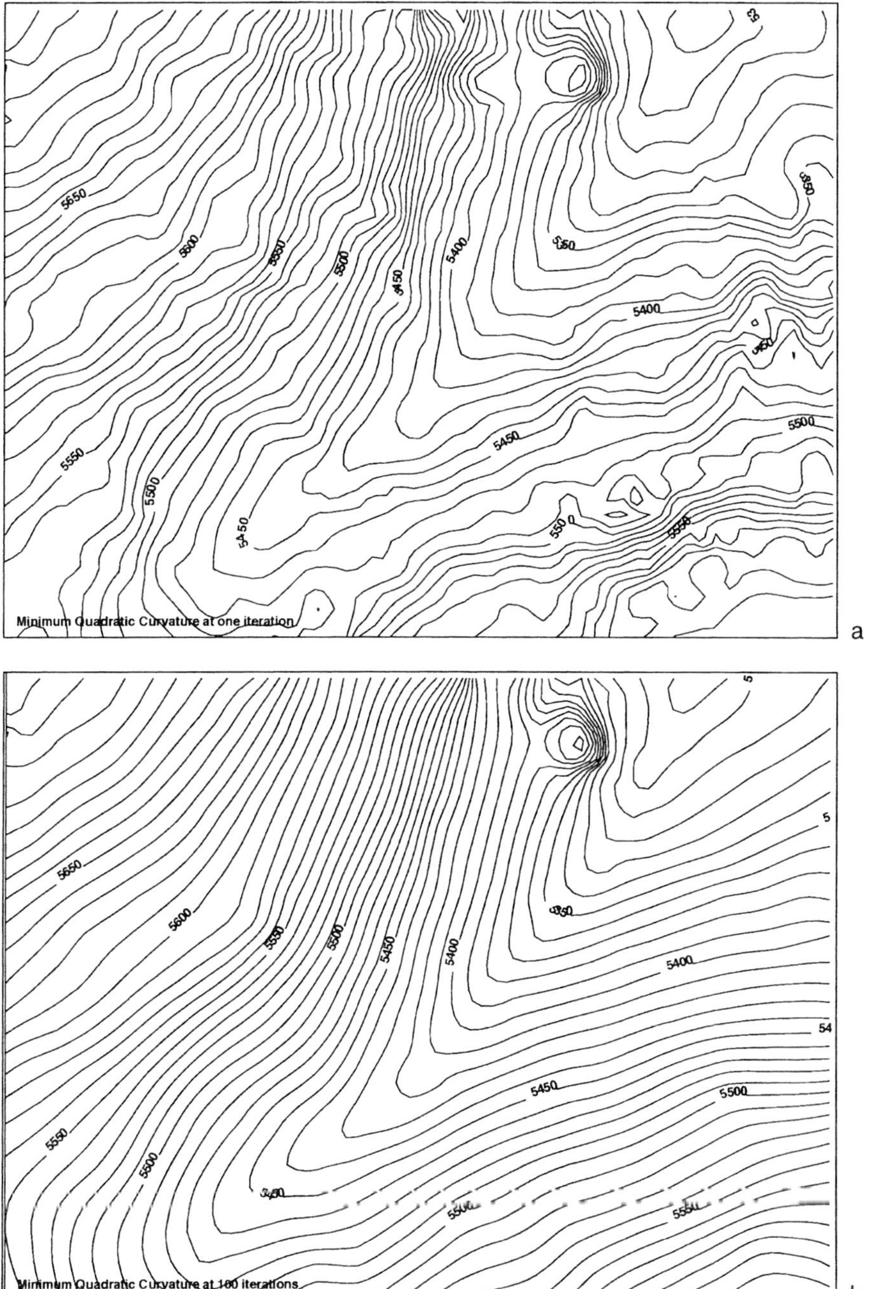

Figure 5.7. MINQ estimation of topographic contours with: a) 1 iteration, and b) 100 iterations.

Summary of procedure: Example 5.3
Initial data requirements: a flat table of topographic data or any other 2D data set.

Procedure:
1. **Create** a field <topo_m> in the surfaces cell table <surfaces> to hold the MINQ estimate.
2. From the main menu select **Modelling** ⇒ **Minq** ⇒ **Fields**
   DATA POINTS — RESULTS
   VALUE: <elevation> — VALUE: <topo_m>
   X: <easting>
   Y: <northing>
   2.1 **Parameters** (for generating Figure 5.7b's data)
      CONDITIONING: <closest> MAX ITERATIONS: <100>
      CONVERGENCE: <.01>
   2.2 **Minq**

5.5 WEIGHTED AVERAGES AND INVERSE DISTANCE

The concept of using a weighted average was introduced as an assignment method using triangles. Weighted average estimation is currently the most popular method of estimation. The estimation algorithms differ in how the weights are calculated.

Drillhole data-based estimation is a relatively recent development which became possible with the introduction of diamond drill bits following the 1880's. Prior to this, miners relied on channel samples from within the mine workings to estimate grades and tonnage. Channel samples were taken normal to the dip, when possible, from the hangingwall to footwall contacts so that the sample length was across the the true width of the vein. The grade could then be estimated from a weighted average of the channel assays using the vein width as the weight (see Fig. 5.8).

With the advent of drilling as the prime feature of deposit delineation, another method of weighted averages became popular: inverse distance. Inverse distance is based on the recognition of the spatial correlation of regionalized variables, in that as

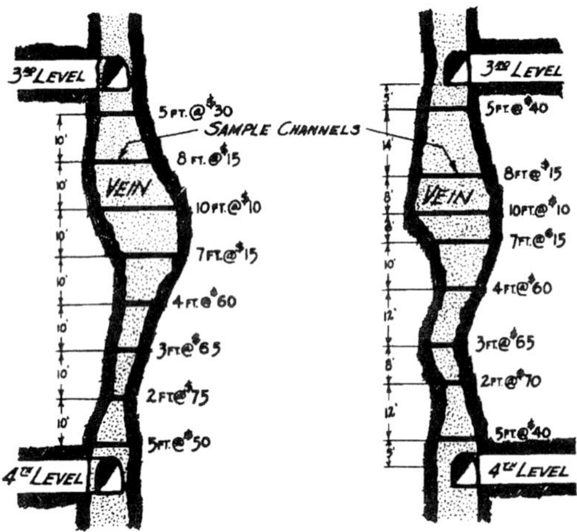

Figure 5.8. Channel samples used to calculate a weighted average grade by vein width (from Parks 1949).

the distance separating two locations increases, the probability that the values at those locations will be similar decreases. Inverse distance methods do not make use of a variogram or any other spatial function to define the relationship between distance and covariance. Instead, the inverse of the distance, d_i, separating the location to be estimated and nearby data locations is used as the basis of the weight in a weighted average of the sample values, z_i,

$$\hat{z} = \frac{\sum_{i=1}^{n} \frac{1}{d_i^p} z_i}{\sum_{i=1}^{n} \frac{1}{d^p}}$$

A power, p, is introduced as part of the weight calculation so that the importance of the difference in separation distances can be accounted for. The value of this power is commonly between zero and three. When $p = 0$, the inverse distance method is reduced to a simple average with all weights equal to one regardless of the distance. With $p = 1$, the weight increases linearly with distance so that a sample twice as distant to the point to be estimated as another will have half the weight. As p increases past 2 or 3, the nearest sample takes on increasingly greater importance until inverse distance is reduced to an area of influence method of assigning values based on the nearest sample. The choice of the power, p, is largely based on experience with the deposit or with the spatial phenomenon being modeled. Later in this chapter, jackknife estimation methods will be examined which can be used to evaluate the effectiveness of different estimation parameters based on their accuracy in estimating known values. Inverse distance estimation shares many similarities with kriging, which bases the calculation of its weights on the minimization of the error variance.

Estimation of grids using weighted average routines requires that some of the surrounding data are selected to estimate a cell or block centered point. Depending on the weighting scheme, the influence of data which is beyond a certain distance may be essentially null and should therefore be eliminated from the estimate for the sake of increased computational efficiency. A moderately sized grid can consist of tens of thousands of cells that are to be estimated with thousands of data points. As a result, estimation using the entire data set for each cell can be very slow. Even if speed of calculation is not an issue, the entire data set should not be used since this can result in an estimate which is much smoother than reality. One of the advantages of using inverse distance or kriging over fitted surface methods is that estimation can be based on only the local data and the estimate will therefore be more representative of the locality and less influenced by regional trends.

Data selection routines typically include the following elements: 1) the maximum and minimum samples to be used for estimation, 2) a search distance in up to three directions, and 3) the azimuth and inclination of the search axes.

One search radius is usually applied to the entire data set regardless of the location of the point to be estimated. Since the density of data can vary widely, in some areas the search routine may find far more data than is needed for a good estimate. Returning to the example of the surface topography of the Boland Banya project that was declustered in Example 5.1, a search radius of 100-200 ft would typically encompass no more than two or three data points in the western half of the property, but in the area of the volcanic neck, the same search radius would include nearly 38% of the data. By limiting estimation to only a couple of values, the closest data points from those found during the search will be used for estimation. So, where the average distance of separation between the point to be estimated and the three closest data values may have been 200' in the western portion of the property, in the area of the volcanic neck, the average distance to the nearest three data values may be only 10'. Thus, by assigning a relatively small value to the maximum number of data points that are to be used for estimation, inverse distance can retain its localized characteristics even when using clustered data.

When no data can be found within the search radius, the cell will not be estimated. The natural reaction to this is to keep increasing the search radius until all the cells in the grid are estimated. This is a grave mistake. Data are commonly too sparse to estimate all regions of a deposit. Forcing an estimate on a volume which is isolated only provides an illusion that there is ore of a certain value at a location when, in fact, there may be none. This practice is especially dangerous with a non-statistical method, such as inverse distance, since there is no way to evaluate the estimation error for the grid. Large blocks of null grid values are needed as a reminder that there are regions that are marked as terra incognita which require further exploration. Assigning a minimum number of data values for estimation provides a means of controlling the data requirements that must be met before an estimate will be made. Requiring that there be two or three samples within the search radius before an estimate will be made helps to ensure that the cell is not an isolated volume and probably has data surrounding it on several sides. Examination of the configuration of the data neighboring unestimated areas will confirm the need for either an increased search radius or additional sampling.

When there is no reason to believe that the continuity of the regionalized variables being estimated is a function of direction, then the search should be based on either a circle or sphere, but if the phenomenon varies more rapidly in one distance than in another, then the search should be based on either an ellipse or ellipsoid with the major axis oriented in the direction of greatest continuity. In a two-dimensional data set, the search ellipse is defined be the lengths of the major and minor axes. The major axis is assigned an azimuth of rotation from 0° and the minor axis is at 90° to the major axis (see Fig. 5.9). Ellipsoidal searches are defined as three angles of rotaoriented to strike, its second axis to dip and its third to plunge. This can be accomplished by defining a second coordinate system, uvw, which is rotated from the project coordinate axes, *xyz*, such that u corresponds to *x*, *v* to *y* and *w* to *z*. The main axis, *u*, of the search ellipse can be defined by its length and by a strike azimuth, θ, from *x* and plunge from the vertical, φ. The *w* and *v* search axes are also assigned lengths and a dip, δ, for *v*. Since *u*, *v* and *w* are orthogonal, *u* will be aligned with the strike and plunge, *w* will be parallel to dip and in the plane of the orebody and *v* will be normal to dip and aligned with the true thickness of the orebody. Figure 5.10 illustrates the relationship between search ellipse axes and the project coordinate system as a function of the three angles of rotation.

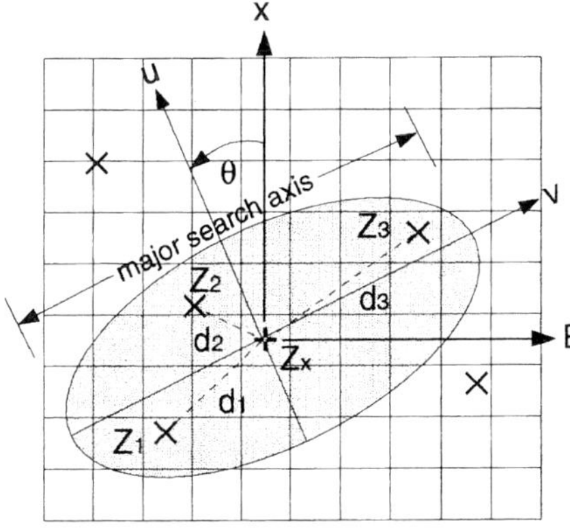

Figure 5.9. Using inverse distance to estimate grid centered values with a search ellipsoid with an orientation of N65°E.

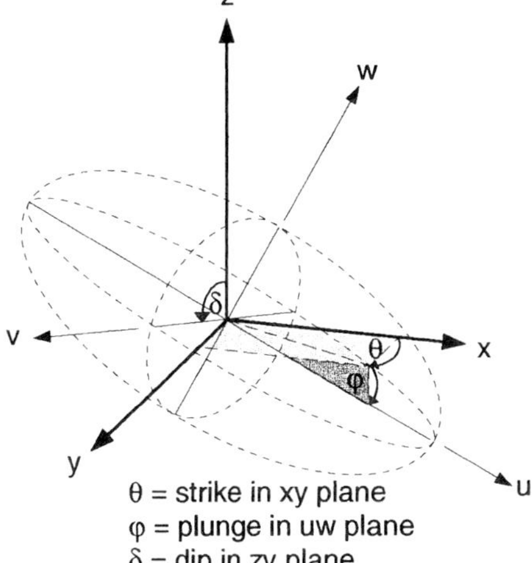

Figure 5.10. Search ellipsoid showing rotation from project *xyz* coordinate system.

Example 5.4: Inverse distance estimation of of a surface in Techbase
We will repeat the estimation of the top structure of the OA ore body that was carried out the Boland project in Example 5.3, but this time inverse distance will be used. Examination of the surface elevations of OA shows a horizontal, lense-shaped orebody with no particular strike, dip or plunge. The data values are about two to three hundred feet apart. Surface elevations do not change very rapidly. Even at the steepest gradients, the change in elevation is no more than 25'/100'. Note that a triangulated surface is a good way to first examine the data since it strictly honors the data, doesn't extrapolate and produces simple contours.

Before proceeding with this exercise, you will need to create three new fields in the cell table used to contain surface models: a field to contain the inverse distance estimate for the top of OA, an integer field to hold the number of samples used to estimate the cell, and a field to hold the average distance from the cell centroid to the data used for estimation.

From the Techbase main menu, select **Modelling** ⇒ **Inverse** ⇒ **Inverse** ⇒ **Fields** and enter the same *xyz* data fields that were used in Example 5.3. These are the composited intercept fields for the top of OA. The results of estimation will be stored in a cell table field which you have already created in the same cell table used in Example 5.3. Do not use the same field that was used to store the triangulated estimate: you will want to compare the inverse distance and triangulated surfaces later. Select **Search** and enter the minimum number of samples which must be found within the search radius for a cell to be estimated (2), the maximum number of samples that will be used to estimate a cell (6) , and define the u axis search distance as 200'. This will result in a circular search pattern out to 200' from the centroid of the cell in the horizontal plane. The azimuth and dip default to the horizontal, the *v* axis will be equal to the *u* axis (circular search) and the *w* axis has no meaning in the estimation of a two-dimensional data set. Under **Parameters**, enter the inverse distance power (2), respond negatively to the elliptical distance and jackknife prompts and enter the field names that were created to hold the sample count and average distance used in estimating each cell. If a non-circular search pattern was being used, the elliptical distance prompt would be set to 'Yes'. The use of jackknife methods will be addressed later in this chapter. Selection of **Estimate** implements inverse distance estimation.

Estimation with the above parameter settings should result in 4367 cells estimated from the 105 drillhole intercepts, leaving 2822 cells not estimated. When triGrid was used to estimate this same surface, 3167 cells were not estimated. Triangulation does not form estimates outside of the limits of the data, whereas inverse distance will. In this particular case, the search distance is 200'. As a result, estimation will extend 200' beyond the triangulated grid. Because inverse distance and other weighted average methods will extrapolate beyond the available data, there is a risk of estimating values into areas which are outside of the ore body. Since the OA ore zone is lens shaped, the tendency will be for extrapolated estimates at the edges of the deposit to increase in depth. When estimation is repeated for the bottom structure, the opposite will be true. The result is that the top and bottom surfaces may cross each other at the periphery of the deposit. If this cross-over occurs, it will be seen as a negative thickness which can be easily corrected by setting the top and bottom elevations for cells displaying a negative thickness to null values. Extrapolation and resetting of negative thickness to null values is actually a handy way of pinching out lens shaped deposits, but excessive extrapolation distances should always be avoided by limiting the search distances and setting a minimum value of at least two samples required for estimation.

A visual examination of the results of estimation is essential to ensure that the search and estimation parameter settings were reasonable and that those cells that should have been estimated were estimated and that extrapolation was not excessive. Two methods of displaying surfaces using Techbase will be demonstrated. Contouring will be used to provide a comparison with the surface estimated in Example 5.2, and then color posting of the cell table will be used to demonstrate the value of direct observation of the grid estimates without the confusion that can be caused by contours.

Use **Gridcont** to contour the cell table field containing the inverse distance estimate of the top of OA using the same contour intervals. Do not smooth the contours.

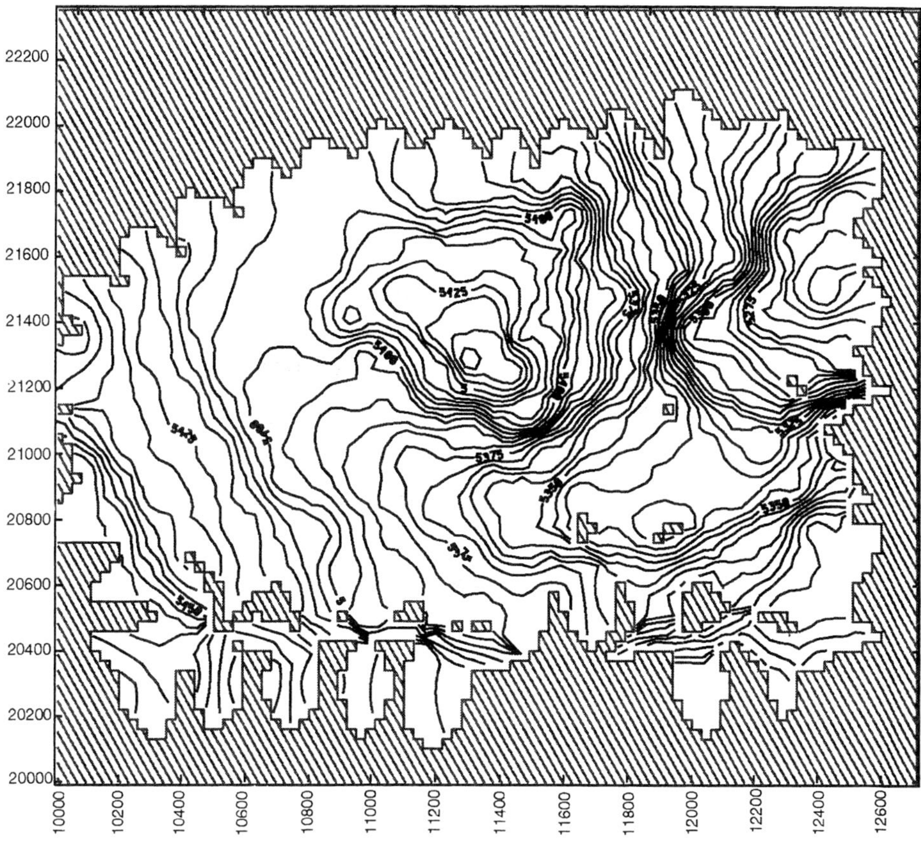

Figure 5.11. Inverse distance estimation of the top structure of the main (OA) ore zone showing the extent of cell assignments (Boland Banya project). (Figure in colour, see opposite page 87).

The resulting surface for OA is shown in Figure 5.11. The contours generated from triangulation (Example 5.2) and inverse distance are roughly the same, but there are some interesting differences. The triangulated surface is much smoother. Even though a planer estimate based on a triangle is restricted to the three points forming the triangle, it is still not as localized as an inverse distance squared estimate having the search parameters given in this example. In this case, the inverse distance estimate is much more variable and is probably a much closer representation of the true top structure. Also, it seems that the extent of the inverse distance surface is greater than for triangulation and highly irregular. To help us understand why, color cell posting of the fields related to the inverse distance estimate will be generated.

In **Poster**, select **Database** and **Add** a filter setting the inverse distance estimated field to null. For instance, if the top structure field was named oat_id the filter line would be: oat_id = ___ , where '___' is simply a blank, or null, entry. With this filter, only cells which were not estimated will be displayed. We will post all cells which have not been estimated. To do this select **Cell/poly** and for the field to be displayed alongside of each cell, enter the default null field for the cell table holding the surface estimate. If the cell table was named topo_grid, the null field would be topo_grid_nul. topo_grid_nul has null field values. Techbase requires that a value be posted in order to post the cells, but also posting a field value is not necessary for displaying the extent of estimation and will clutter the display. Entering topo_grid_nul allows the cell to be posted without having to display any cell related value. Select **Style** and enter a cell color (red) and fill style (19 is left slanting diagonal hatching). **Draw** will place the fill style in all cells which were not estimated. **Outline** will place a border around these cells. Figure 5.11 is an overlay of the contoured top structure of OA and the posting of the null values cells.

When the posting of unestimated cells is superimposed on the contours of OA's top structure, we can get a good idea of the extent and continuity of the estimate. It is now apparent that there are cells in the interior of the deposit which were not estimated. This is particularly true along the southern edge of the deposit where there are long fingers of alternating estimated and null cells. The entire border of the deposit has a ragged appearance which neither conforms with the data nor with the type of shape that might be expected for this type of deposit. It is likely that the fiord-like appearance of the boundary of the deposit is an artifact of the estimation method rather than a reflection of reality.

As part of the inverse distance estimation, the number of samples and the average distance to the samples used for estimation of each cell was saved into two optional fields in the cell table. These can also be displayed as color cell postings. The same procedure in **Poster** is used to display a range of values as is used for a single color: a filter is added, setting the field for the number of samples to either 2, 3, 4, 5 or 6

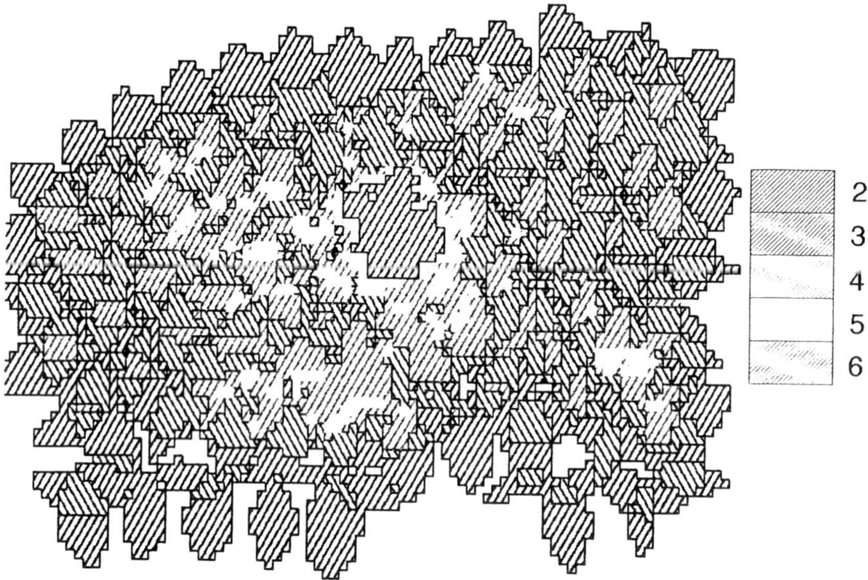

Figure 5.12. Color cell posting of number of data points used for estimation of the OA top structure using a search radius of 200'. (Figure in colour, see opposite page 214.)

(the minimum and maximum number of samples allowed during estimation), under **Cell/poly** a different color and style of hatching is selected for each filter setting and the cells are drawn and outlined. In this manner, the metafile is built in five layers of colors and hatching styles. A more elegant approach is to create a calculated field which is assigned pen colors acording to the value of the cell. An example of this is given in step 6.A of the following Summary of procedure. Examination of the results, as shown in Figure 5.12, shows that the irregular deposit boundary and unestimated cells correspond to areas which were remote from the data, areas in which only one sample could be found. This is a clue as to why the border has a fiord-like irregularity.

Poster's **Cell/poly** utility was used to color post the average estimation distance to the surrounding samples over 20' length intervals from 70 to 196 ft. Poster's **Points** facility was used to post the sample data on the same map (see Fig. 5.13). It is now apparent that the odd shape of the boundary is due to the use of a search radius of 200' and the requirement that there be at least two samples within the search radius before a cell is estimated. These results indicate that it might be beneficial to slightly increase the search radius (maybe to as high as 300') and also increase to a minimum number of samples required for estimation from 2 to 3.

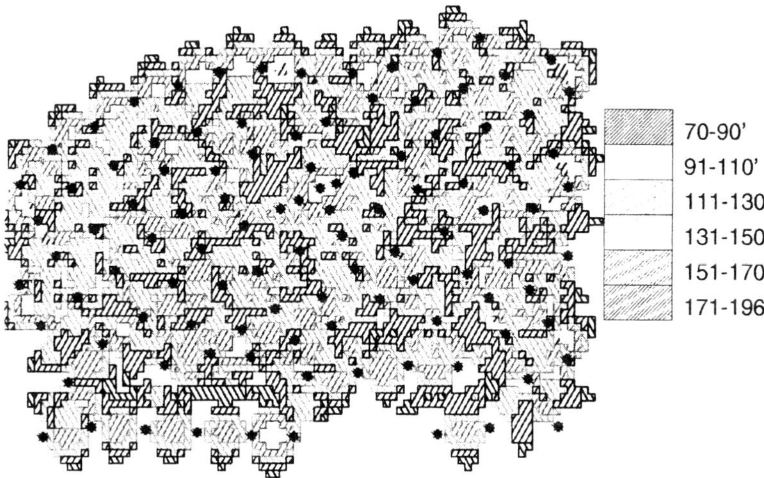

Figure 5.13. Colored cell posting of the average distance of estimation using inverse distance squared with a search radius of 200' and the sample data. (Figure in colour, see opposite page 214.)

Summary of procedure: Example 5.4
Initial data requirements: a 2D data set for surface modelling, in this example the length composites for the top structure of OA generated in Example 4.2.

Procedure:
1. Create three fields in the cell table <surfaces> to hold the inverse distance estimate <oat_id>, the number of samples used in the weighted average <oat_no> and the average weighting distance <oat_d>.
2. Generate the inverse distance estimate **Modelling ⇒ Inverse ⇒ Inverse**

2.1 Enter input and output fields ⇒ **Fields**
Data Points — Results
Value: <oazt> — Value: <oat_id>
X: <oaxt> (no coordinate fields are entered for cell tables)
Y: <oayt>
2.2 **Search** parameters
Max Samples: <6> Min Samples: <2> Search Length: <200>
2.3 **Parameters**
Inverse Distance Power: <2>
Sample Count: <oat_no> Distance: <oat_d>
2.4 **Estimate**

3. Contour the top of OA. From the main menu **Graphics** ⇒ **Gridcont** ⇒ **Graphics** ⇒ **Metafilename**: <oatidcont.met>
3.1 Set **Scaling** as per previous examples (see Examples 5.1 and 5.2).
3.2 **Contour** ⇒ **Field**
Contour Value: <oat_id>
3.3 **Line Style** (all others default)
Minor Contour Number: <4>
3.4 **Intervals**
Contour Interval: <5>
Low Contour Value: <5000>
3.5 **Parameters**
Contour Smoothing Factor: <0>
3.6 **Contour** ⇒ **Graphics** ⇒ **Review**

4. Post unestimated cells. From the main menu **Graphics** ⇒ **Poster**
4.1 Filter out all estimated cells ⇒ **Database** ⇒ **Add filter**
Field: <oat_id> Relation: <=> Field|Value: <> (null)
4.2 Name the metafile <oat_null>
4.3 Set **Scaling** to be the same as for the contour plot of Step 3.
4.4 Supress the plotting of values in the cell **Cell/Poly** ⇒ **Fields**
Value 1: <surfaces_nul>
4.5 **Style** (others default)
Cell Color: <red> Fill Style: <18> Fill Scale: <.5>
4.6 **Outline** ⇒ **Draw** ⇒ **Graphics** ⇒ **Review**

5. Overlay the two plots
Either from the Techbase main menu **System** ⇒ **Review metafile**
File Name: <oat_idnull oatidcont>
or from the operating system command line
> mftr -i oat_idnull oatidcont

6. Post cells in different colors according to the average distance of estimation (two methods: A and B)
A)
6.1 **Graphics** ⇒ **Poster** ⇒ **Database** ⇒ **Delete filter** (if still on from Step 4.1) ⇒ **Add filter**
Field: <oat_d> Relation: <>> Field|Value: <70>
Field: <oat_d> Relation: <<=> Field|Value: <90>
6.2 Repeat Step 4.3.
6.3 Repeat Steps 4.4-4.6.
6.4 Repeat Step 6.1 for a range of 91 to 110 ft (**Modify filter**) and Step 6.3 using a different color and or fill type.
6.5 Repeat Step 6.4 changing the filter's range to the color and type of fill as shown in the legend of Figure 5.13.

B)
6.1 Create a field to hold the cell color as a function of distance **Techbase ⇒ Define ⇒ Fields ⇒ Autotable**
TABLE NAME: <surfaces>
⇒ **Create**
FIELD NAME: <color_d> TYPE: <integer> CLASS: <actual>
⇒ **eXit** ⇒ **eXit** (to Techbase menu)
⇒ **tbCalc** ⇒ **Setup** ⇒ **Equation**
CALCULATION STEPS: <f,color_d.cal> (create a text file named color_d.cal containing the following lines)
2 oat_d 90 <= 25 skip
3 oat_d 110 <= 19 skip
4 oat_d 130 <= 13 skip
5 oat_d 150 <= 7 skip
6 oat_d 170 <= 1 skip
7 = color_d
⇒ **eXit** ⇒ **Calculate**
6.2 From the main menu **Graphics** ⇒ **Poster** (repeat Steps 4.2-4.4)
⇒ **Cell/poly** ⇒ **Fields**: <surfaces_nul>
⇒ **Style**
CELL COLOR: <color_d> (all oher entries as in step 4.5)
⇒ **Draw** ⇒ **Outline** ⇒ **eXit** ⇒ **Graphics** ⇒ **Review**

5.6 BLOCK MODELS

A block model is a three-dimensional equivalent to a grid which is used to contain estimates of volumes and volume-associated attributes. Where a grid is composed of cells which are associated with an area and a *xy* coordinate, a block model is composed of blocks that have a volume and are centered on *x*, *y* and *z*. Attributes of the volumes are estimated into each block to define the volume's composition. These attributes include the proportions of different classes of material and the quality characteristics of those materials. For example, a block model might be used to discretize the geology of a coal property, and individual blocks would contain estimates of the tonnage of each seam of coal, volumes of waste from the various interburdens and the overburden, tonnage factors, and quality characteristics for each seam including btu, sulfur, moisture and ash. Various methods can be used to estimate these values into a block model. Volumes of intersection with polygons, surfaces and other volume models of geology can be used to estimate the proportion of different classes of materials, and point to volume estimation methods, such as inverse distance and kriging, can be used to assign quality characteristics to the materials in the block. Once values for the variables needed for a feasibility study have been estimated into a block model, net mining values can be calculated for each block and the block model can be intersected with either stopes or a pit to determine mining revenues. These topics will be addressed in the following chapters.

The procedure followed for defining a block model parallels that of a grid. The origin of the block model will be the origin of the property, typically the SW corner. The only requirement is that the model should be positioned so that it can contain the full extent of the ore body and any mining works. In open pit operations, the model must be large enough to contain any reasonable pit expansion and pit depth. Additionally, the block model must be aligned with all structural grids with which it will intersect. The extent of a block model can be less than the extent of the topography, but the blocks must be aligned with the cells. This means that the block's *xy* dimension and orientation in the *xy* plane must be equivalent, and if their origins are not the same, then the difference in the positions of their origins must be an even multiple of the block/cell dimension. Figure 5.14 shows the shifting of position that is possible

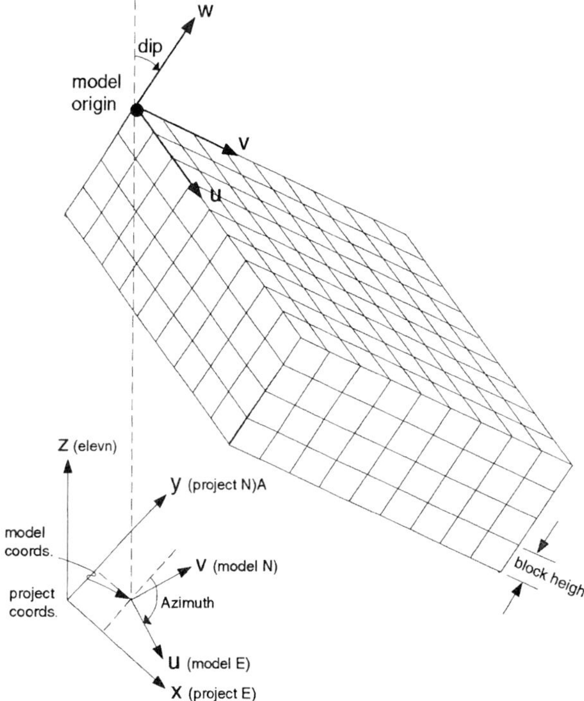

Figure 5.14. A block model rotated from project coordinates *xyz* to local coordinates *uvw* so as to be oriented in the plane of a deposit of corresponding strike, dip and plunge.

for a block model in Lynx. A block model has its own origin in the project coordinate system and can be rotated in the horizontal (azimuth) and vertical (dip). In this manner, the orientation of a block model can be made to correspond to the geometry of the deposit and thereby decrease the size of block model, increasing the accuracy and speed of volumetric calculations.

The extent of a block model is defined by the origin of the model, the dimensions of the blocks and the number of blocks along each axis. Selecting block dimensions presents the same problems as for a determining the cell size for a grid. Generally, block dimensions should be large enough to represent a volume that corresponds to-the volumes used in production planning. For this reason, the block height is usually set to the bench height for open pit applications. As for grids, the *xy* dimensions are based on the sample spacing and should be small enough to reflect the variability of the deposit, but not so small as to increase memory requirements and processing time without a significant improvement in estimation accuracy. The top of the block model should correspond to the crest of the uppermost bench in a pit or to the highest surface topography. Alignment of the blocks with the benches will allow layers of block values to be displayed as a bench and simplify production scheduling. Note that some effort should be made to select a reasonable ultimate pit crest limit since ultimate pit limit optimization algorithms are based on the mining of full block volumes in the block model. Therefore, the ultimate pit crest determined using pit optimization routines is dependent on the dimensions of the block model. Figure 5.15 illustrates how Lynx's background display facilities can be used to estimate the required block model dimensions.

Techbase allows for the definition of variable block dimension. Block dimensions can be specified by row, column and level, which makes it possible to have improved model definition in areas that have a high density of data or that are in more complex geology. Pit bench heights are determined from slope stability calculations which can change with depth. Variable bench heights can be built into the block model to correspond to changes in bench height.

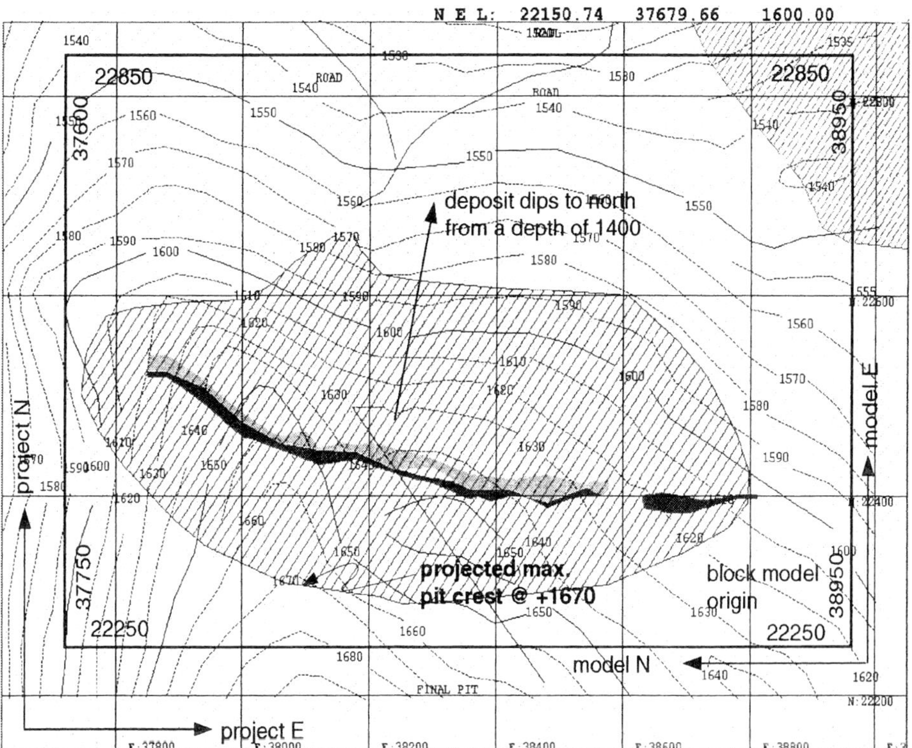

Figure 5.15. Displaying the maximum likely pit crest and deposit limits in order to determine block model origin, extent and orientation (Lynx Tutorial project).

Example 5.5: Defining a block model in Lynx

Block models in Lynx can have their own coordinate system comprised of an origin, azimuth and inclination. The orientation of the block model is relative to the project coordinate system: the model origin is stated in project coordinates and starts on the uppermost level, azimuth is entered as the positive clockwise angle between project north and model east, and the inclination is the clockwise positive rotation about model east axis.

We will define a block model that would be suitable for the Lynx TUTORIAL project. The complete volume model of the orebody (G.OR, zones 7 and 8) was displayed in plan view along with a map of surface topography which included an approximate ultimate pit outline. Sectional views of the orebody were used to identify the depth of the orebody. From these views, it is apparent that the highest pit crest elevation would be about 1670' and that the deepest minable ore would be at 1400 m. There is no plunge to the orebody, which strikes close to 90° for much of its length. A block model with an origin of 22250N, 38950E at 1700 m elevation with an extent of 1260 m (model north/project east) by 600 m (model east/project north) should be sufficiently large to contain both the ore body and either a pit or underground openings. The average spacing between section lines of drillholes is 60 m along model north which would imply a block dimension in this direction of around 20 m. The spacing between drillholes on section (model east) is closer, about 40-50 m, for which a dimension of 15 m would be reasonable. The bench height of any pit is going to be 20 m, which will be used for the block height. Using block dimensions of 20 × 15 × 15 will require 62.5 blocks along model north (round to 63), 40 blocks along model east and 20 blocks in depth (see Fig. 5.16). Note that the orientation of the block model is at a right angle to the project coordinates with increasing model north block coordinates corresponding to decreasing project east coordinates. The

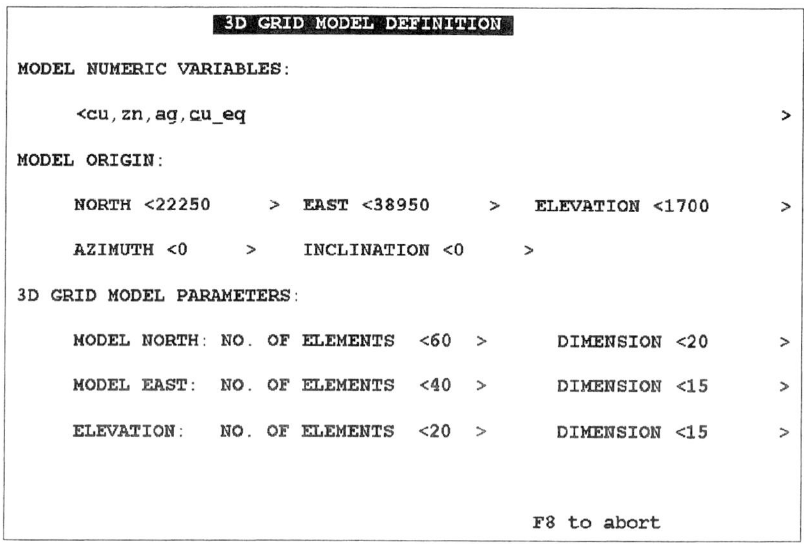

Figure 5.16. Block model definition (Lynx Tutorial project).

block model could have been aligned with the project coordinates, but model sections, which are defined as normal to the model north axis, are generally defined on the same viewplane that was used for generating geology volume components, hence the orientation of the block model.

From the Lynx main menu, select **3D Grids ⇒ V3 3D Grid Data ⇒ 3D Grid Model Definition** and enter a name for the new block model, say '3DBLOCK'. Since this is a new model, you will be prompted for a description and for the model parameters which includes the variables to be included in the model (enter a list of numeric and numeric calculated variables separated by commas), the model origin, azimuth, inclination, block dimensions and the number of blocks as determined above.

Summary of procedure: Example 5.5
Initial data requirements: none

Procedure: (grid definition remains entirely within Version 3 as of Version 4.7)

1. From the main menu select **3D Grids ⇒ V3 3D Grid Data** (opens Version 3 menu) ⇒ **Project Select** (Project, 3D Grid Model and Eng. Selection entry form)
   PROJECT: <TUTORIAL> (active data selection in Version 4 is not recognized in Version 3 and the process must be repeated)
2. Define a new 3D grid **3D Grid Model Definition**
   ENTER 3D GRID MODEL NAME: <3DBLOCK>
   ENTER 3D GRID MODEL DESCRIPTION: <TUTORIAL DEMO> (3D Grid Model Definition entry form)
   MODEL NUMERIC VARIABLES: <cu,zn,ag,cu_eq>
   MODEL ORIGIN: NORTH: <22250> EAST: <38950> ELEVATION: <1700>
   AZIMUTH: <0> ELEVATION: <0>
   3D GRID MODEL PARAMETERS:
   MODEL NORTH: NO. OF ELEMENTS: <60>DIMENSION: <20>
   MODEL EAST: NO. OF ELEMENTS: <40> DIMENSION: <15>
   ELEVATION: <20> DIMENSION: <15>
   (report of model definition) ⇒ **Continue**

Example 5.6: Estimating values into a Lynx block model using inverse distance
Once a block model has been defined, values can be estimated into the model. Two classes of values are commonly of interest: volumes of intersection and estimated values such as assays. Volumes of intersection are calculated by numeric intersection of the volumes of intersection between the block model elements and any geologic volume models. In this example, the block model created in Example 5.5 will be intersected with the orebody model G.OR. Volumes of intersection will be calculated for each element and stored in the model as the default variable 'volumes', which can then be used during the calculation of ore tonnages when the block model is intersected with a grid of pit limits or with underground mining volume models of stopes and other openings. Note that the block model which is used for estimation of ore grades can be defined with respect to the orientation and extent of the orebody. This block model may not be suitable for open pit mining which requires that the blocks be horizontal and correspond with the location of benches. Following the interpola-

tion of volumes, inverse distance will be used to estimate ore assays into the block model. The resulting model values will define the geologic reserves.

From the main lynx menu select **3d Grids ⇒ Grid Geology** to display the 3D Grid Geology entry form. Selecting the **Volume Data Select** push-button brings up the Volume Data Select entry form. Here click on **Volume** and select G.OR and accept the wildcard defaults (*) for **Unit**, **Component**, and **Volume Code**. Note that it is not desireable to process the entire volume model as in this example if only a few volume components have been added or modified, nor is it a good idea to reinitialize the geologic intersection results if only a few components are involved. Select **OK** to return to the 3D Grid Geology entry form and enter the **Integration Increment**. Narrower integration increments will result in increased estimation accuracy, but will also increase the processing time. Since the individual volume components in G.OR were defined on 60 m section widths, there is little justification in using a very narrow slice through the volume model. The thickness of the slice should be based on the complexity of the volume and should not be thicker than the dimensions of individual blocks. The orientation of these incremental volume slices is defined by the orientation of the block model with slices normal to the model north axis. For now, a five meter integration increment should be sufficient. The Geology Intersection File (SM_GEO) does not need to be initialized, since this was done during the creation of the block model, but if the volume model G.OR was modified following the definition of the block model, then re-intersection and initialization would be necessary. Clicking on **OK** starts integration and brings up the 3D Grid Geology Process report form which displays integration progress and the components being intersected. If the block and volume models have been correctly defined and exist in the same location in space, the volumes of intersection will be reported as they're calculated by block model cross-section (normal to model north) and plan section (normal to model inclination). These volumes are stored in the variable 'volumes' and can now be viewed in sections as part of a numeric background display.

As in the case of statisitcal analysis for the Lynx examples in Chapter 3, a data selection file containing the drillhole data to be used for estimation must be available. The procedure paralles that given in *Summary of procedure for Example 3.2, Step 1* and is given in the *Summary of procedure, Step 2* for this example. The main difference is that in this example a data selection file will be created out of all the copper assays taken from zone8, the main ore zone. Similarily, estimation will only be performed on blocks intersecting zone8. Note the decision to restrict both source samples and estimation to the same ore zone.

Note that before carrying out grid estimation, that the appropriate block model must have already have been defined (see Example 5.5) and must be in active memory (**File ⇒ 3D Grid Data ⇒ 3D Grid Select**).

From the Lynx main menu, select **3D Grids ⇒ Estimation Parameters** bringing up the 3D Grid Estimation Parameters entry form as shown in Figure 5.17. Click on **Parameter File** and either select an existing file (as in later examples) or enter a new file name <3dblock>. A parameter file **Description** can also be given. Click on **Source Select/Data File** to select the data file to be used for estimation <fulldh0.dat>. Choose one of the project variables included in the data selection file <Cu> and Click on the **Geologic Control** push-button bringing up the Geologic

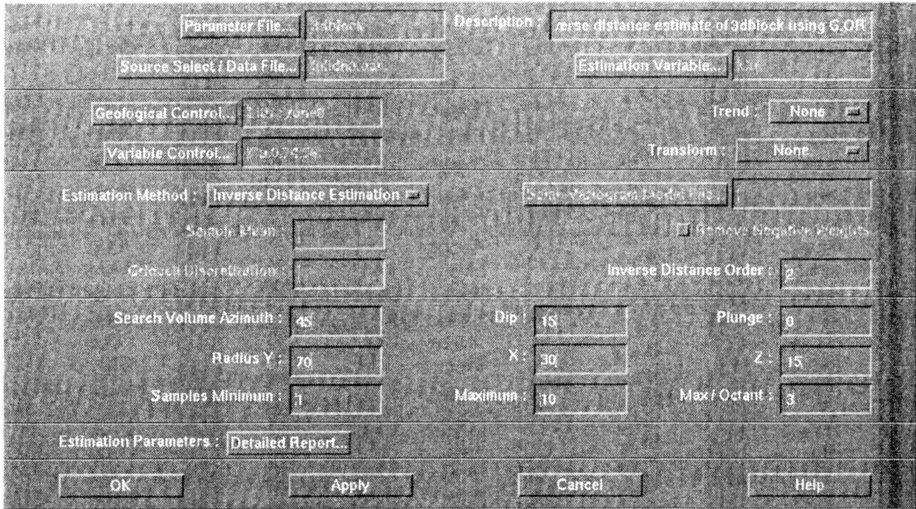

Figure 5.17. 3D Grid Estimation Parameters entry form showing settings for inverse distance estimation of Cu into main ore zone (block model 3dblock).

Control entry form. Here estimation will be limited to only those blocks which intersect the main ore zone: **Characteristic**: <Lith>, **Characteristic Values**: <zone8>. Click on **Variable Control** brining up the Variable Control entry form in which the variable to be estimated is sepcified (**Variable:** <Cu>) as are the **Minimum:** <0> and **Maximum:** <74.56> (defualt dh0 data range for zone8) allowable assay levels used for estimation in the source data. We are estimating block values for Cu in the main ore zone, so the data that is used for estimation should also be limited to drillhole data taken in the same ore zone. This is achieved by setting the drillhole selection field to <Lith> and the selection field value to <zone8>. Now, only Cu assays in drillhole data records for zone8 will be estimated into blocks which intersected with code 2 G.OR components.

Leave the **Trend** and **Transform** push-buttons on the <None> option.

Four kriging **Estimation Method**s are available along with inverse distance. Choose inverse distance with **Inverse Distance Order** set to 2. By convention a value of two is used. To make the estimate more localized increase the power. To make the estimate smoother, decrease the power.

The 'order' p in the weighting scheme $1/d^p$ can range from 0 (equivalent to simple averaging) to 9 (equivalent to polygons of influence). Ordinary, simple, block and universal kriging methods will be discussed later in this chapter. With the selection of inverse distance, four kriging-realted push-buttons become unavailable. The **Semi-Variogram Model File** push-button is used to select an existing variogram model parameter file which would have been stored under the **Data Analysis ⇒ Geostatistics** menu. **Sample Mean** refers to the mean value that is to be used in Simple Kriging which assumes that the average grade is the same throughout the area to be estimated (globally stationary). This mean is then used for the kriging estimate. Ordinary Kriging (which only assumes local stationarity) can generate negative kriging weights when the kriging equations are not based on a contraint for non-negativity of the weights as well as the more common unbiased constraint which required the weights sum to unity. Toggle the **Remove Negative Weights** checkbox to remove from the estimate samples for which a negative weight was calculated. Note that will not produce an estimate that is equivalent to one in which the kriging equations include a non-negativity condition and will potentially increase the estimation error. Block model estimation generally includes block discretization. Unlike Point Kriging which simply estimates a single value per cell, the cell is descritized into a grid of equally spaced points, each of which is estimated. The cell estimate is then based on the average of these point estimates. The default value for **Gridcell Discretiztion** is 4 which results in a $4 \times 4 \times 4$ grid of 64 points to be estimated per block.

Search Volume Azimuth, **Dip**, **Plunge**, and **Radius** *Y*, *X* and *Z* are the parameters defining the size and orientation of the search ellipsoid. The y radius corresponds to model north, which was defined as being approximately along strike (project east/west). Az is defined as being the clockwise rotation of the y axis from block model and Dip is the deviation of the x axis from horizontal. With Az = 0 and Dip = 45, the y axis will parallel strike and the x axis will parallel the dip. Since the z axis is orthogonal to x and y, z will be perpendicular to dip and represent a search aligned with the true thickness of the ore zone. It is likely that the major direction of continuity in grade will be along strike, that the second axis will be perpendicular to dip extending to depth and that the minor axis of continuity will be across the narrowest dimension of the deposit. Since drillholes are spaced approximately 60 m apart on strike, the y axis search distance should be somewhat greater than this distance, say 70 m. Since we are assuming that the x axis continuity is lower than for the y axis, the search distance should also be shorter, say 50 m. The z axis is across the thickness of the deposit, and for a thin deposit need be no greater than the width of the main ore zone (15 m).

Setting **Samples Minimum:** <1>, **Maximum:** <10> (defaults) and **Max/Octant:** <3> controls which samples found within the search ellipse are used and the data requirements for estimation. In this example, if at least one sample is found within the limits of the search ellipse, then estimation of the cell will proceed. Setting a higher minimum can be used to ensure both improved estimation quality and to identify blocks which have insufficient nearby data for estimation. I more samples than the maximum samples setting are located within the search ellipse, the closest (10) samples will be selected. The maximum number of samples used in estimation can be reduced to produce estimates which display more localized variability or increased to smooth the estimate. Estimate smoothing is generally an undesirable side effect of using weighted average estimation even though using a larger number of samples for estimation will, to a point, reduce the estimation error, a factor which cannot be measured directly when using inverse distance estimation (see Section 5.14 for a discussion of jacknifing). The **Max/Octant** parameter setting is used to control the maximum number of samples used per octant of the search ellipse. Inverse distance estimation doesn't account for the spatial orientation of the data beyond the distance (not vector) of separation between each sample and the point to be estimated. Situations can arise in which the majority of data found within the search ellipse are concentrated in one or two adjacent octants. This commonly happens when geologic control is used to limit data to one or more geologic structures. When the data's location is concentrated to one area, the resulting estimate can be biased (see Example 5.8 and Fig. 5.26). A setting of 3 ensures that out of the maximum of 10 samples used for this estimate, only the 3 closest will be taken from any one octant.

Finally, select **Estimation Parameters: Detailed Report ⇒ File ⇒ Save As** <cuid2est> to save a report of the estimation parameters to an ASCII file in the project's ./reports directory for later reference.

Continue by creating additional data selection files and defining the estimation and search parameters for the secondary ore zone (code=3) and for any other assays of interest. The same parameter settings can be used, but it should be noted that the continuity and geometry of different lithologic units can vary widely and should be understood before the parameters are set.

Now that the data selection file has been created and the estimation parameters have been selected estimation can proceed by selecting **3D Grids 3D Grid Estimate** bringing up the 3D Grid Estimation Parameters entry form as shown in Figure 5.18. Note that the active 3D grid is displayed in the entry form's title block. This is the grid in which estimates will be made and is also the grid that was active during geologic intersection. Click on **Parameter File** to select the estimation parameter file <3dblock>. Select the **Estimation Variable:** <Cu> and **Volume Code:** <2>. Lynx refers to the block model indices as *x*, *y* and *z* sections corresponding to 2D slices of the grid having the same *x*, *y* or *z* model orientation. Grid estimation can be limited to a subset of the grid by restricting the **Estimation Limits** to a range of these sections. Accept the defaults which are the limits of the block model. Another option that will restrict the gridcells to be estimated is the **Estimate UDF Grid Values Only** toggle

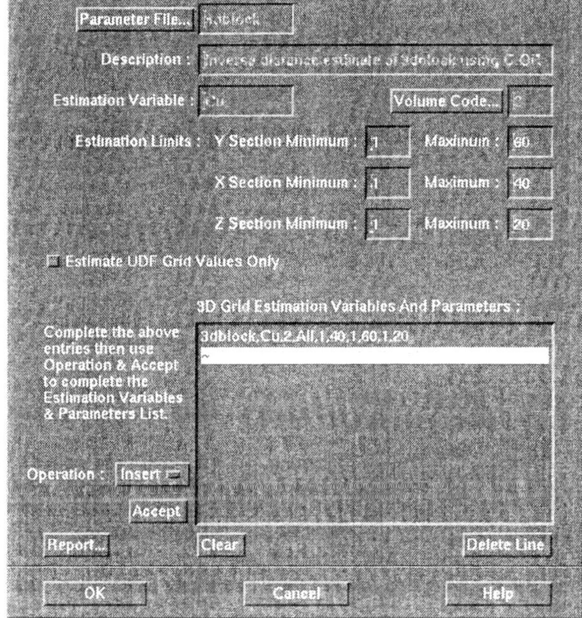

Figure 5.18. 3D Grid Estimation entry form showing values for estimation of Cu into the main ore zone for the block model 3DBLOCK.

which by default is off. When toggles on only those blocks that are currently undefined in value will be estimated. This option is useful in cases where previous estimates left a portion of the grid undefined, either because the sections to be estimated were restricted or when a data seletion file has been updated with additional drillhole data. Click on **Accept** to include these estimation parameters and restrictions in the estimation process. The process can now be repeated for another variable, geologic code or subset of the grid by using **Insert**. As always it's a good idea to use the **Report** facility to output an ASCII report of the estimation parameters. To start estimation click on **OK**. The 3D Grid Estimation Process window will appear to monitor estimation progress. This is followed by a window detailing the results of estimation in terms of cells estimated.

Even when values have been estimated into the block model, the contents of the block model must be reviewed as a check on their validity. A visual approach to this is to display the block intersections and values as part of a sectional background display. The geologic volume mode and drillhole values can also be included in the display to provide a basis for checking the veracity of the estimated block values. To display the grid select **Viewplane** ⇒ **Setup** and enter the desired viewplane orientation (22300,38000,1300,0,0), scale (2500:1) and a global grid (100,100,100). Select **Background** ⇒ **Maps** to intersect the surface topography, **Volumes** to display the intersection of the geologic volume components with the block model and **3D Grids** to display the block model with its values (see Fig. 5.19). Details are given in Step 4

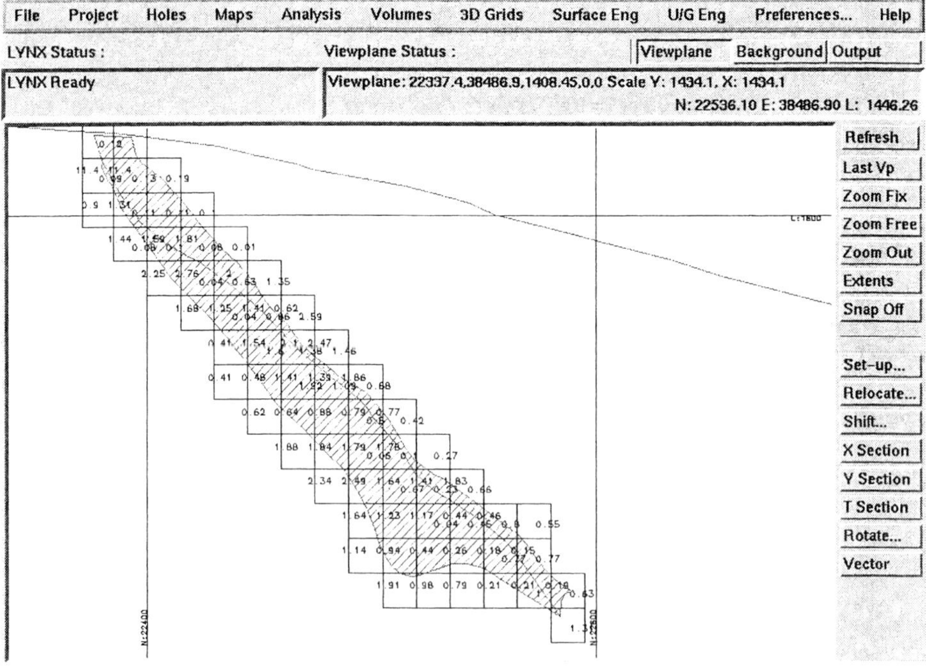

Figure 5.19. Volumes of intersection (cu m) between zones 7 and 8 and the block model with estimated copper grades posted for both zones.

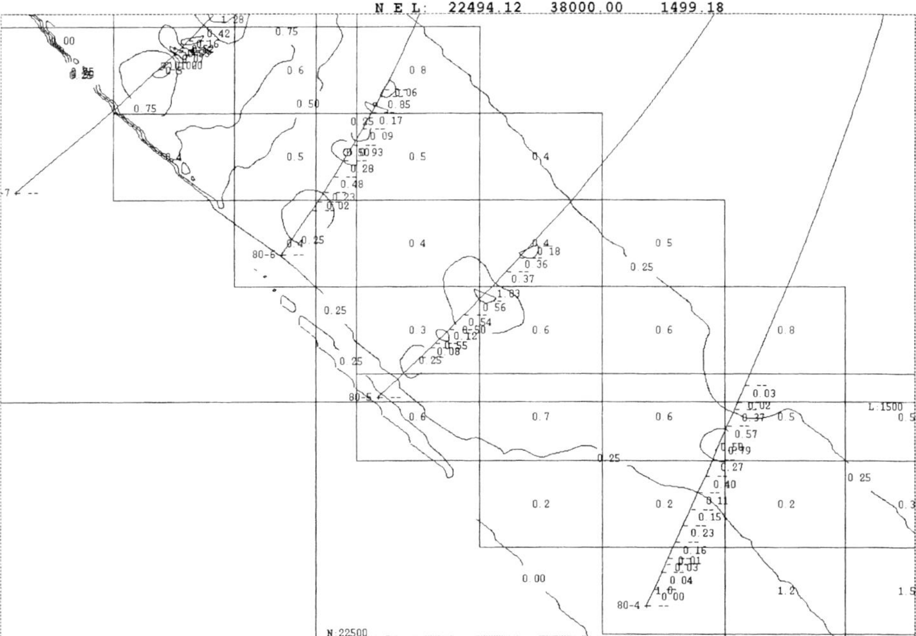

Figure 5.20. Inverse distance estimated Cu grades for the main ore zone displayed in vertical section with drillhole traces, sample values and Cu.

of the following *Summary of procedure*. When the blocks are small relative to the display area, the estimated value can be contoured as in Figure 5.20. First check the block volume against the limits of the volume component to see if the intersection volumes seem to be reasonable. Remember that the blocks and components both have a depth, so that even if there is no intersection on this viewplane and there is a nonzero volume, the intersection probably occurs at a slight offset from the current viewplane. Also display the estimated values against the drillhole data using the Dholes and Dhpost facilities. Due to the small size of the drillhole sampling intervals when compared to the block dimensions, the display will probably have to be zoomed and contoured to make the comparison. Use the display facilities on several sections of the block model to confirm the veracity of the estimation process. The following Summary of Procedure gives the steps to be taken for displaying 3D grids in Version 4.7.

Summary of procedure: Example 5.6
Initial data requirements: For volume integration a volume model and a 3D grid are required as pere the examples of Chapter 4 and Example 5.5. For grade estimation into a grid, 3D grade data is required and must be active, in this casc from drillhole assays.

Procedure:
1. Determining geology intersections for a grid: **3D Grids** ⇒ **3D Grid Geology** (3D Grid Geology entry form)
   ⇒ **Volume Data Select** (Volume Data Selection entry form)

VOLUME: <G.OR> UNIT: <*> COMPONENT: <*> VOLUME CODE: <*>
⇒ **OK**

2. Create a data selection file to be used for Cu estimation in zone8: **Analysis ⇒ Data Analysis** (Data Analysis menu) ⇒ **File ⇒ Data Select ⇒ Select hole and Map Data** (Data Select Hole and Map Data entry form).
   ⇒ **Selection File** ... <fulldh0>
   ⇒ **Create Data File** <on>
   ⇒ **Data File Name** <fulldh0>
   ⇒ **Variable/Characteristic Select** ... <Cu, Lith>
   Minimum/Maximum <default>
   Lith <zone8>
   ⇒ **OK**
3. Initialize 3D grid estimation parameters: **3D Grids ⇒ Estimation Parameters** (3D Grid Estimation Parameters entry form)
   ⇒ **Parameter File** ... <3dblock>
   ⇒ **Source Select/Data File** <fulldh0.dat>
   ⇒ **Geologic Control** (Geologic Control entry form)
   CHARACTERISTIC: <Lith>
   CHARACTERISTIC Values: <zone8>
   ⇒ **Variable Control** (Variable Control entry form)
   VARIABLE : <Cu>
   MINIMUM: <0> MAXIMUM: <74.56> (double click for defaults)
   Estimation Method: <Inverse Distance Estimation>
   Inverse Distance Order: <2>
   Search Volume Azimuth: <45> **Dip**: <15> **Plunge**: <0>
   Radius Y: <70> **X**: <50> **Z**: <15>
   Samples Minimum: <1> **Maximum**: <10> (defaults)
   Max/Octant: <3>
   ⇒ **Estimation Parameters**: <Detailed Report> ⇒ **File** ⇒ **Saves As** <cuidest>
   ⇒ **OK**
4. 3D Grid Estimation: **3D Grids 3D Grid Estimate** (3D Grid Estimation entry form)
   ⇒ **Parameter File** ... <3dblock>
   ⇒ **Estimation Variable** ... <Cu>
   ⇒ **Volume Code** <2>
   ⇒ **Estimation Limits** (accept defaults)
   ⇒ **Accept**
   ⇒ **Report ⇒ File ⇒ Save As**
   ⇒ **OK**
5. 3D grid display: **Background ⇒ 3D Grids** (Background 3D Grid Data Selection entry form, see Fig. 5.21)
   BLACKGROUND DISPLAY FOR: <display 1> (toggle Display 1 <on>)
   3D GRID: <3DBLOCK>
   3D GRID VARIABLE: <cu>
   VOLUME CODE: <2>
   DISPLAY FORMAT: <value>
   DISPLAY GRID CELLS: <on>
   DISPLAY GRID OUTLINE: <on>
   ⇒ **OK**
   ⇒ **Refresh**

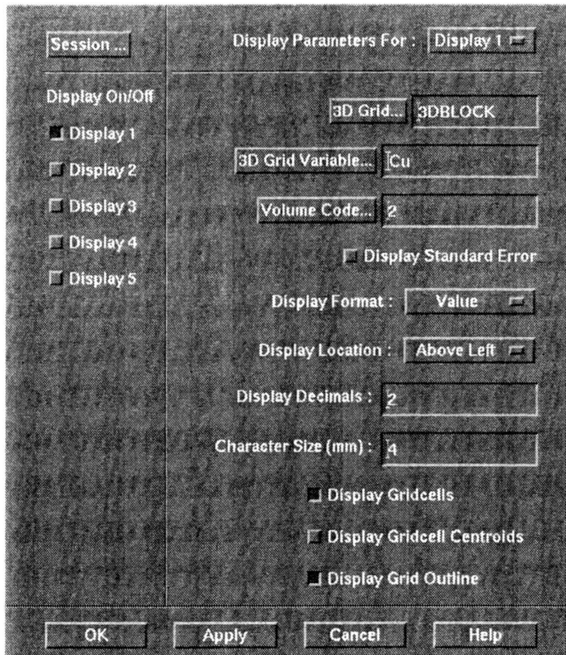

Figure 5.21. Background 3D Grid Data Selection entry form setup for 3DBLOCK.

5.7 GEOLOGIC RESERVES

Preliminary mine design and feasibility analysis requires a model of orebody geometry, depth and an estimate of geologic reserves. These parameters are derived from the geologic volume model, gridded surface topography, and a block model which contains estimates of assays and volumes. Lynx provides facilities to intersect volume and block models as demonstrated above. Once the block mode has been estimated, geological reserves can be written to a report and then used to select an initial mine design, production rate and mine life.

Example 5.7: Reporting geologic reserves in Lynx
From the Lynx main menu select **Analysis** ⇒ **Volumetrics Analysis** ⇒ **Simple Volumetrics** bringing up the Simple Volumetrics entry form shown in Figure 5.22. A simple volumetrics report can be either generated for a single or 'primary' volume or for the intersection of two volumes. In the case of reporting intersection volumes, a secondary volume is selected by which the primary volume will be reported. For instance, consider a case that will arise in Chapter 7 where volume components will be created to represent mine openings such as stopes. The stope volume components can be selected as being the secondary volume while the geologic volume model (say G.OR, codes 2 and 3) is the primary volume. In this way, the tonnage of primary and secondary ore along with undefined material, can be reported for each stope. More on this in Chapter 7. For now, the objective is to check the geologic reserves. Click on **Primary Volume Data** to bring up the Simple Volumetrics Primary Data Selec-

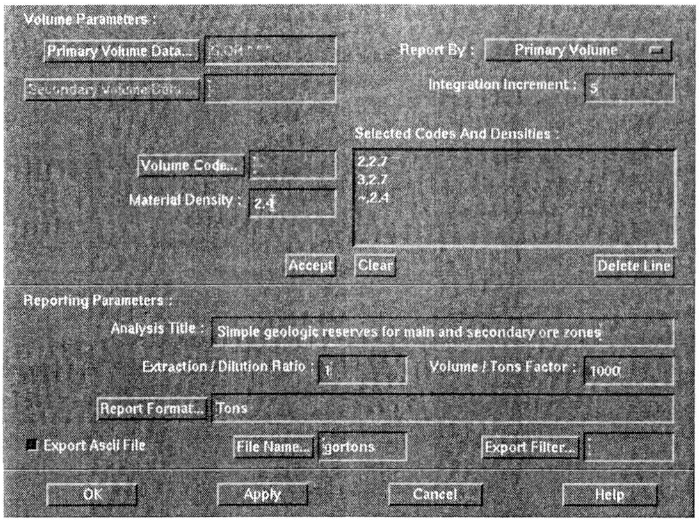

Figure 5.22. Simple Volumetrics entry form for reporting G.OR main, secondary and undefined units.

tion entry form and select <G.OR> as the **Volume** model to be reported while accepting the default value <*> for **Unit**, **Component**, and **Volume Code**. Select **Report By** to indicate how the report will be generated, either by volume, unit or component for either the primary or secondary volume model. In this example, a report will be generated for the primary volume <G.OR>. Note that one of the 'secondary' options must be selected before the Secondary Volume Data push-button can become active. Select the **Integration Increment** to control the slice thickness that will be used during numerical integration between the volume and block model. Note that the intersection will be on the currently active 3D grid (3DBLOCK). An overly thin slice will result in excessive computational time without any real improvement in accuracy, while a slice thickness which approaches the block dimension will result in reduced accuracy. In this example, an increment of 5 m is sufficient. Under Selected Codes and Densities the volume codes to be reported and their respective material densities (g/cc) are input. In this example, codes 2 and 3 will be reported along with any remaining undefined material in the block (indicated by ~ in a code list). Use **Accept** after inputting each code. The Reporting Parameters which must be entered included the **Analysis Title**, an **Extraction/Dilution Ratio** which is not very relevant to a geologic reserve report, a **Volume/Tons Factor**, for instance for listing tonnage by thousands and the **Report Format** in which either tons or volume can be selected. The report can be save to an ASCII file by toggling on **Export ASCII File**. Selecting **Apply** will produce the report as shown in Figure 5.23. Try reporting by unit and code. Also try using a secondary volume such as M.UG.STP, the volume model used in Chapter 7 for stopes.

```
Simple Volumetrics                              Wed Jul  8 12:42:51 1998
Active Project : TUTORIAL                                User : martin

 Analysis : Simple geologic reserves for main and secondary ore zones
 Report By : Primary Volume
      Primary Volume Selection : G.OR,*,*,*
        Integration Increment : 5
Volume Codes And Material Densities : Code Density
                                      ---- -------
                                     :  2   2.7
                                     :  3   2.7
                                     : UDF   2.4
    Extraction / Dilution Ratio : 1
         Volume / Tons Factor : 1000
 Reporting  Reporting Primary
    Unit  Component  Code       Tons

      *        *    2       9008
      *        *    3       6459
---------- ---------- ----- ------------
      *        *    *      15467
```

Figure 5.23. Simple volumetric report of geologic reserves from volume model G.OR.

Summary of procedure: Example 5.7
Initial data requirements: A 3D grid containing volumes of intersection from a geologic volume model and estimated grades (make 3DBLOCK active).

Procedure:
From the main menu **Analysis** ⇒ **Volumetrics Analysis** ⇒ **Simple Volumetrics** (Simple Volumetrics entry form)
⇒ **Primary Volume Data** ... (Simple Volumetrics Primary Data Selection entry form)
Volume: <G.OR>
Unit: <*>
Component: <*>
Volume Code: <*>
OK
⇒ **Report By**: <Primary Volume>
Integration Increment: <5>
Selected Codes and Densities:
Volume Code: <2>
Material Density: <2.7>
⇒ **Accept** (repeat for code 3 <2.7> and UDF <2.4>)
Reporting Parameters:
Extraction/Dilution Ratio: <1>
Volumes/ton Factor: <1000>
Report Format: <tons>
Export ASCII File: <on>
⇒ **Apply**

5.8 DERIVATION OF THE KRIGING WEIGHTS

As in the case of inverse distance, kriging estimates a value, $\hat{u}(\mathbf{x_0})$, for a regionalized variable located at position $\mathbf{x_0}$ which is the vector of position (x_o, y_o, z_o) in R^3 space. This estimate is based on a weighted average of the surrounding data, $u(\mathbf{x_i})$, which fall within a search ellipsoid (Isaaks & Srivastava 1989).

$$\hat{u}(\mathbf{x_0}) = \left(\sum_{i=1}^{n} w_i \times u(x_i) \,\middle|\, \sum_{i=1}^{n} w_i = 1 \right)$$

The constraint specifying that the sum of the weights is unity is the unbiased condition. The unbiased condition used in kriging is absent for non-statistical estimators such as inverse distance. The approach used in selecting values for the weights parallels parameter estimation in regression in which parameter values are determined that will minimize the estimation error.

$$R(\mathbf{x_0}) = \hat{u}(\mathbf{x_0}) - u(\mathbf{x_0}) = \sum_{i=1}^{n} w_i \times u(x_i) - u(\mathbf{x_0})$$

The true value of $u(\mathbf{x_0})$ is unknown. Therefore, a random function must be estimated which describes the variability of the regionalized variable in terms of a statistical distance. The following derivation demonstrates how the estimation error can be minimized.

The expected value of the distribution of errors between the estimated (grid) values and their true, but unknown, values, $E\{R(\mathbf{x_0})\}$, should be zero and unbiased.

$$E\{R(\mathbf{x_0})\} = E\left\{ \sum_{i=1}^{n} w_i \times u(x_i) - u(\mathbf{x_0}) \right\} = \sum_{i=1}^{n} w_i \times E\{u(\mathbf{x_i})\} - E\{u(\mathbf{x_0})\}$$

If the random function is stationary, then the expected value is not a function of location and the notation, for location can be dropped. Stationary data have the same expected distribution regardless of location and there is no reason to discriminate between data and the distribution of data at locations to be estimated.

$$E\{R(\mathbf{x_0})\} = 0 = E(u) \sum_{i=1}^{n} w_i - E(u)$$

$$E(u) = E(u) \times \sum_{i=1}^{n} w_i$$

$$(\therefore) \sum_{i=1}^{n} w_i = 1$$

Therefore, $R(\mathbf{x_0})$ must be minimized subject to the unbiased weights constraint in order to derive weights which will minimize the estimation error and are unbiased. Since the actual errors cannot be calculated, a functional form of the error variance, $R(\mathbf{x_0})$, must be developed and then this function minimized.

5.9 DERIVATION OF AN EXPRESSION FOR $R\,(\mathbf{x}_0)$

The variance of a weighted linear combination of random variables, u_i is found as

$$\mathrm{Var}\left(\sum_{i=1}^{n} w_i \times u_i\right) = \sum_{i=1}^{n}\sum_{j=1}^{n} w_i \times w_j \times \mathrm{Cov}\left(u_i \times u_j\right)$$

For the sum of two random variables, u and v, this becomes

$$\mathrm{Var}\,(u+v) = \mathrm{Cov}\,(u \times u) + \mathrm{Cov}\,(u \times v) + \mathrm{Cov}\,(v \times u) + \mathrm{Cov}\,(v \times v)$$

Therefore, the variance of the estimation error, $R\,(\mathbf{x}_0) = \hat{u}\,(\mathbf{x}_0) - u\,(\mathbf{x}_0)$, is

$$\begin{aligned}\mathrm{Var}\left\{\hat{u}\,(\mathbf{x}_0) - u\,(\mathbf{x}_0)\right\} &= \mathrm{Cov}\left\{\hat{u}\,(\mathbf{x}_0) \times \hat{u}\,(\mathbf{x}_0)\right\} - \mathrm{Cov}\left\{\hat{u}\,(\mathbf{x}_0) \times u\,(\mathbf{x}_0)\right\} \\ &\quad - \mathrm{Cov}\left\{u\,(\mathbf{x}_0) \times \hat{u}\,(\mathbf{x}_0)\right\} + \mathrm{Cov}\left\{u\,(\mathbf{x}_0) \times u\,(\mathbf{x}_0)\right\} \\ &= \mathrm{Cov}\left\{\hat{u}\,(\mathbf{x}_0) \times \hat{u}\,(\mathbf{x}_0)\right\} - 2\,\mathrm{Cov}\left\{\hat{u}\,(\mathbf{x}_0) \times u\,(\mathbf{x}_0)\right\} \\ &\quad + \mathrm{Cov}\left\{u\,(\mathbf{x}_0) \times u\,(\mathbf{x}_0)\right\}\end{aligned}$$

The first term is simply the variance of the estimates or the variance of the weighted average of the surrounding data

$$\mathrm{Var}\left\{\hat{u}\,(\mathbf{x}_0)\right\} = \mathrm{Var}\left(\sum_{i=1}^{n} w_i \times u\,(\mathbf{x}_0)\right) = \sum_{i=1}^{n}\sum_{j=1}^{n} w_i \times w_j \times \mathrm{Cov}\left(u_i \times u_j\right)$$

where u_i and u_j are two of the data values selected for estimating $\hat{u}\,(\mathbf{x}_0)$ which is a function of the spatial covariance of the surrounding data. The covariance between the data points is a function on the statistical distance and is obtained from a model of the spatial covariance, $\tilde{C}_{ij}$.

$$\mathrm{Var}\left\{\hat{u}\,(\mathbf{x}_0)\right\} = \sum_{i=1}^{n}\sum_{j=1}^{n} w_i \times w_j \times \tilde{C}_{ij}$$

Likewise, $\mathrm{Cov}\,\{u\,(\mathbf{x}_0) \times u\,(\mathbf{x}_0)\} = \mathrm{Var}\,\{u\,(\mathbf{x}_0)\}$ which by the assumption of covariance stationarity should equal the variance of the data set, $\mathrm{Var}\,\{u\,(\mathbf{x}_0)\} = \tilde{\sigma}^2$. As for the second term the covariance

$$2\,\mathrm{Cov}\left\{\hat{u}\,(\mathbf{x}_0) \times u\,(\mathbf{x}_0)\right\} = 2\,\mathrm{Cov}\left\{\left(\sum_{i=1}^{n} w_i \times u_i\right) u_o\right\}$$

between two random variables is

$$\mathrm{Cov}\,(u \times v) = E\left[\{u - E\,(u)\}\{v - E\,(v)\}\right] = E\,(u \times v) \quad E(u)\,E\,(v)$$

$$(\therefore)\ 2\ Cov\left\{\left(\sum_{i=1}^{n} w_i \times u_i\right) u_o\right\}$$

$$= 2\ E\left[\left\{\sum_{i=1}^{n} w_i \times u_i - 2\ E\left(\sum_{i=1}^{n} w_i \times u_i\right)\right\} \times \left\{u_o - E\left(u_o\right)\right\}\right]$$

$$= 2\ E\left\{\left(\sum_{i=1}^{n} w_i \times u_i\right) u_o\right\} - 2\ E\left(\sum_{i=1}^{n} w_i \times u_i\right) E\left(u_o\right)$$

$$= 2 \sum_{i=1}^{n} w_i \times E\left(u_o u_i\right) - 2 \sum_{i=1}^{n} w_i \times E\left(u_i\right) E\left(u_o\right)$$

Since Cov $(u \times v) = E\,(u \times v) - E\,(u)\,E\,(v)$ then

$$2 \sum_{i=1}^{n} w_i \times \left\{E\left(u_i \times u_o\right) - E\left(u_i\right) E\left(u_o\right)\right\} = 2 \sum_{i=1}^{n} w_i \times \mathrm{Cov}\left(u_i \times u_o\right)$$

$$= 2 \sum_{i=1}^{n} w_i \times \mathrm{Cov}\left(u_i \times u_o\right) \quad = 2 \sum_{i=1}^{n} w_i \times \tilde{C}_{io}$$

Define the Var$\{R\,(\mathbf{x}_0)\} = \tilde{\sigma}_R^2$ as the error variance,

$$\tilde{\sigma}_R^2 = \tilde{\sigma}^2 + \sum_{i=1}^{n} \sum_{j=1}^{n} w_i \times w_j \times \tilde{C}_{ij} - 2 \sum_{i=1}^{n} w_i \times \tilde{C}_{io}$$

Thus, the error variance can be defined for the kriged estimate, $\hat{u}\,(\mathbf{x}_0)$ given the data values, $u\,(\mathbf{x}_i)$, their weights, w_i, and the model variance, $\tilde{\sigma}^2$ and covariance, $\tilde{C}_{ij}$. The model variance and covariance is obtained from the covariogram, or, following traditional practice, from the variogram. Later in this chapter, definition of the model covariogram/variogram will be covered, but the covariance will be available as a fitted function which can be evaluated for any vector of separation, **h**. Since the variance and covariance are derived from a function which only requires the vector of separation between the location to be estimated and the surrounding data and the separation between the data points, then the only remaining parameters to be determined are the kriging weights.

5.10 UNCONSTRAINED OPTIMIZATION OF A SINGLE FUNCTION AND THE LAGRANGE PARAMETER

Minimization of the estimation error is the criterion that is used to determine a set of optimal kriging weights. Unfortunately, there are some complications. When treating the weights as variables whose values are to be determined, note that the expression for the error variance is nonlinear in its second term due to the product $w_i \times w_j$. Also, remember that the unbiased condition requires that the sum of the weights is unity. Therefore, there is a constraint on the minimization of the error variance. The partial derivatives of the error variance function with respect to the weights can be used to determine weights that minimize the error, but this will not ensure that the weights will sum to one or even be strictly positive in value. The Simplex method could be used to account for the constraint, but the objective function would be nonlinear. As will be shown, a Lagrangian multiplier can be used to overcome both of these problems.

Consider the problem of finding a globally maximum or minimum value for a continuously differentiable nonlinear function, $f(x)$ as shown in Figure 5.24. In clas-

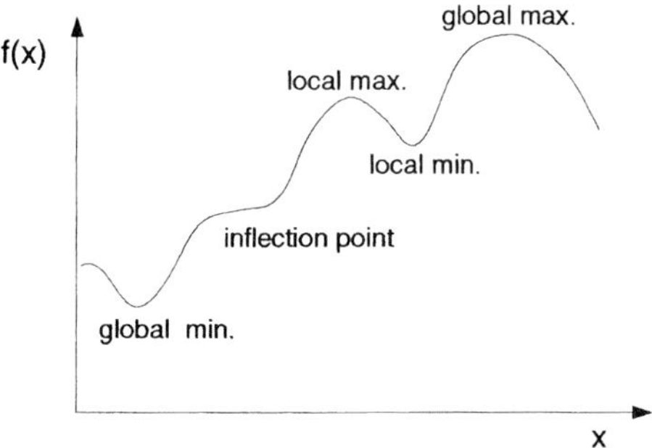

Figure 5.24. Finding a global maximum or minimum of a complex nonlinear function.

sical optimization theory, a necessary but not sufficient condition for a solution $x = x^*$ to be a minimum or maximum is for the derivative to be equal to zero. If the second derivative is strictly greater than zero, then x^* must be at least of local minimum or maximum for all points x in the neighborhood of x^*. If $f(x)$ is convex or concave, then these conditions ensure that either a global minimum or maximum, respectively, has been found at x^*. The same procedure can be used to optimize a nonlinear function of more than one variable. In this case the partial derivative of $f(\mathbf{x})$ is set to zero for each of the variables in the vector $\mathbf{x}$ and the system of equations is solved simultaneously to determine the optimal location $\mathbf{x}^*$.

The method of Lagrange multipliers can be used to extend this method of unconstrained optimization to functions involving constraints by including the constraints as part of the objective function. If the objective function, $f(\mathbf{x})$, is constrained by $i = 1, \ldots, m$ constraints of the form $g_i(\mathbf{x}) = b_i$,then an equivalent unconstrained formulation would be the Lagrangian function.

$$h(\mathbf{x}, \mu) = f(\mathbf{x}) - \sum_{i=1}^{m} \mu_i \{g_i(\mathbf{x}) - b_i\}$$

where the new variables, μ_i, are the Lagrange multipliers. Note that for any feasible values of $\mathbf{x}$, that $g_i(\mathbf{x}) - b_i = 0$. Therefore, the optimal values of $\mathbf{x}^*$ and μ^* can be found be setting the $(n + m)$ partial derivatives to zero.

$$\frac{\partial h}{\partial x_j} = \frac{\partial f}{\partial x_j} - \sum_{i=1}^{m} \mu_i \frac{\partial g_i}{\partial x_j} = 0, \quad \text{for} \quad j = 1, \ldots, n$$

$$\frac{\partial h}{\partial \mu_i} = -g_i(x) + b_i = 0, \quad \text{for} \quad i = 1, \ldots, m$$

The method of Lagrange multipliers requires the objective function and constraints to be continuously differentiable, and depending on the dimension of $(n + m)$ and the complexity of the functions there may be a very large number of solutions to local maxima or minima from which the globally optimal solution must be found. In the case of kriging, there is only a single linear constraint, and the function $\tilde{C}_{ij}$ is constructed from a summation of accepted functions which are continuously differentiable and either linear, convex or concave.

5.11 DERIVING THE KRIGING SYSTEM OF EQUATIONS

The method of Lagrange multipliers can be used to minimize the error variance subject to the unbiased condition, such that

$$\tilde{\sigma}_R^2 = \tilde{\sigma}^2 + \sum_{i=1}^{n}\sum_{j=1}^{n} w_i \times w_j \times \tilde{C}_{ij} - 2\sum_{i=1}^{n} w_i \times \tilde{C}_{io}$$

subject to $\sum_{i=1}^{n} w_i = 1$

$$\Rightarrow \min \tilde{\sigma}_R^2 = \tilde{\sigma}^2 + \sum_{i=1}^{n}\sum_{j=1}^{n} w_i \times w_j \times \tilde{C}_{ij} - 2\sum_{i=1}^{n} w_i \times \tilde{C}_{io} + 2\mu\left(\sum_{i=1}^{n} w_i - 1\right)$$

In which there are $(n + 1)$ parameters to be determined, i.e. the n kriging weights and the Lagrange parameter. Setting the partial derivatives to zero for each of the parameters will yield a system of $(n + 1)$ equations having $(n + 1)$ unknowns which are referred to as the kriging equations.

The first term, $\tilde{\sigma}^2$, is not a function of either μ or of the w_i and drops out of their partial derivatives. Differentiating the second term with respect to w_1 yields

$$\frac{\partial\left(\sum_{i=1}^{n}\sum_{j=1}^{n} w_i w_j \tilde{C}_{ij}\right)}{\partial w_1} = \frac{\partial\left(w_1^2 \tilde{C}_{11} + 2w_1 \sum_{j=2}^{n} w_j \tilde{C}_{1j}\right)}{\partial w_{1n}}$$

$$= 2w_1\tilde{C}_{11} + 2\sum_{j=2}^{n} w_j \tilde{C}_{1j}$$

$$= 2\sum_{j=1}^{n} w_j \tilde{C}_{1j}$$

and the other $(n - 1)$ derivatives for the kriging weights are derived similarly. Differentiating the third term with respect to w_1 yields

$$\frac{\partial\left(\sum_{i=1}^{n} w_i \tilde{C}_{io}\right)}{\partial w_1} = \frac{\partial\left(w_1 \tilde{C}_{1o}\right)}{\partial w_1} = \tilde{C}_{1o}$$

which is the covariogram function value evaluated for the vector of separation $\mathbf{h}_{1o}$ between data point 1 and the location being estimated. The same derivation is carried out for each other data value found in the search ellipse. Deriving the last term with respect to w_1 yields

$$\frac{\partial\left\{\mu\left(\sum_{i=1}^{n} w_i - 1\right)\right\}}{\partial w_1} = \frac{\partial\left(\mu w_1\right)}{\partial w_1} = \mu$$

Putting all three terms together for the minimization of the error variance with respect to w_1 results in the first of the $(n + 1)$ kriging equations.

$$\frac{\partial\left(\tilde{\sigma}_R^2\right)}{\partial w} = 2\sum_{j=1}^{n} w_j \tilde{C}_{1j} - 2\,\tilde{C}_{1o} + 2\,\mu = 0$$

$$\Rightarrow \sum_{j=1}^{n} w_j \tilde{C}_{1j} + \mu = \tilde{C}_{1o}$$

Solving for the full $(n + 1)$ variables yields the system of kriging equations.

$$\sum_{j=1}^{n} w_j \tilde{C}_{ij} + \mu = \tilde{C}_{io} \quad \text{for} \quad i = 1, \ldots, n$$

$$\sum_{i=1}^{n} w_i = 1$$

In matrix notation, the system of kriging equations is

$$\begin{bmatrix} \tilde{C}_{11} & \cdots & \tilde{C}_{1n} & 1 \\ \vdots & \ddots & \vdots & \vdots \\ \tilde{C}_{n1} & \cdots & \tilde{C}_{nn} & 1 \\ 1 & \cdots & 1 & 0 \end{bmatrix} \times \begin{bmatrix} w_1 \\ \vdots \\ w_n \\ \mu \end{bmatrix} = \begin{bmatrix} \tilde{C}_{10} \\ \vdots \\ \tilde{C}_{n0} \\ 1 \end{bmatrix}$$

For every point to be estimated on the grid, the system $\mathbf{Cw} = \mathbf{D}$ must be solved for the optimal kriging weights plus the Lagrange multiplier, $\mathbf{w} = \mathbf{C}^{-1}\mathbf{D}$. Likewise the ordinary kriging variance is found as $\tilde{\sigma}_{OK}^2 = \tilde{\sigma}^2 - \mathbf{wD}$.

Example 5.8: Calculation of kriging weights in comparison with inverse distance
Consider a situation in which four data points are to be used to estimate a cell centered value. First, inverse distance squared will be used followed by several approaches to ordinary kriging.

In Figure 5.25, samples 1-4 are located at 10, 20, 30 and 40 ft and azimuths of 90°, 180°, 270° and 360°, respectively, from the point to be estimated. Using inverse distance squared (ID2), the estimated value (Zest) is 1.46 with 70% of the weight of

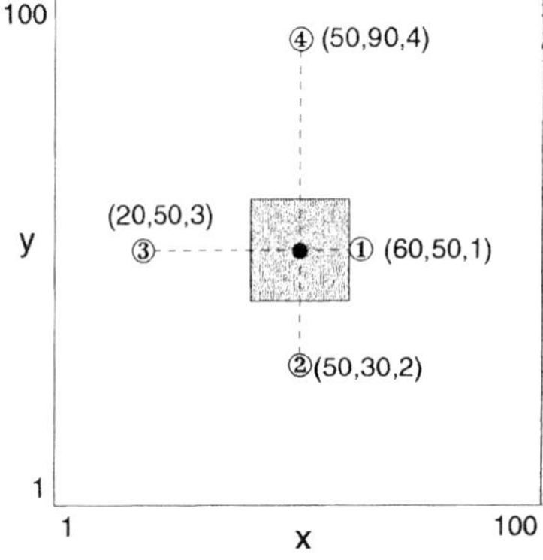

Figure 5.25. Data (x,y,z) evenly surrounding a cell centered point to be estimated for z.

the estimate coming from sample 1. Ordinary kriging was used with the same data set and a global linear variogram, $\gamma(\mathbf{h}) = |\mathbf{h}|$. Linear, exponential and Gaussian variograms approximate inverse distance weighting using different powers. For instance, a nugget effect model would yield the same weights as would inverse distance with $p = 0$, i.e. a simple average. Using a linear variogram, the variogram values will be the separation distances between the data, $\tilde{\gamma}_{ij}$, and the distance from the data to the point to be estimated, $\tilde{\gamma}_{io}$. In matrix notation, the kriging equations for this example are

$$\begin{bmatrix} \gamma_{11} & \cdots & \gamma_{14} & 1 \\ \vdots & \ddots & \vdots & \vdots \\ \gamma_{41} & \cdots & \gamma_{44} & 1 \\ 1 & \cdots & 1 & 0 \end{bmatrix} \begin{bmatrix} w_1 \\ \vdots \\ w_4 \\ \mu \end{bmatrix} = \begin{bmatrix} \gamma_{10} \\ \vdots \\ \gamma_{40} \\ 1 \end{bmatrix} \Rightarrow \begin{bmatrix} 0 & 22.36 & 40 & 41.23 & 1 \\ 22.36 & 0 & 36.06 & 60 & 1 \\ 40 & 36.06 & 0 & 50 & 1 \\ 41.23 & 60 & 50 & 0 & 1 \\ 1 & 1 & 1 & 1 & 0 \end{bmatrix} \begin{bmatrix} w_1 \\ w_2 \\ w_3 \\ w_4 \\ \mu \end{bmatrix} = \begin{bmatrix} 10 \\ 20 \\ 30 \\ 40 \\ 1 \end{bmatrix}$$

Solving simultaneously for $\mathbf{w} = \mathbf{C}^{-1}\,\mathbf{D}$ yields kriging weights of .62, .15, .18, and .04 for samples 1-4, respectively. The resulting kriged estimate is 1.64, which is significantly different from the ID2 estimate, but it should be noted that the estimated value using either inverse distance or kriging is greatly influenced by either the inverse distance power or by the form of the variogram model. These results are summarized in Table 5.1. With this uniform arrangement of data values, a set of parameters could be found using either method that would result in equivalent estimates. It is more interesting to compare the relative weights that were used for the two estimates. Since

Table 5.1. Inverse distance squared and kriging weights for samples evenly distributed around the estimation point.

Sample Z	Distance	ID2 weight	% Zest	OK weight
1	10	0.01	70	0.62
2	20	0.0025	18	0.15
3	30	0.0011	8	0.18
4	40	0.0006	4	0.04

kriging weights are constrained to sum to unity, they can be directly compared with the relative percentage weights from inverse distance which for this example correspond fairly closely to the kriging weights.

Next consider a configuration of data around the point to be estimated in which three of the four samples are quite close to the point to be estimated but are all clustered in the NE quadrant while a fourth more distant sample is in the SW quadrant as shown in Figure 5.26. Inverse distance will still yield weights based on the distance

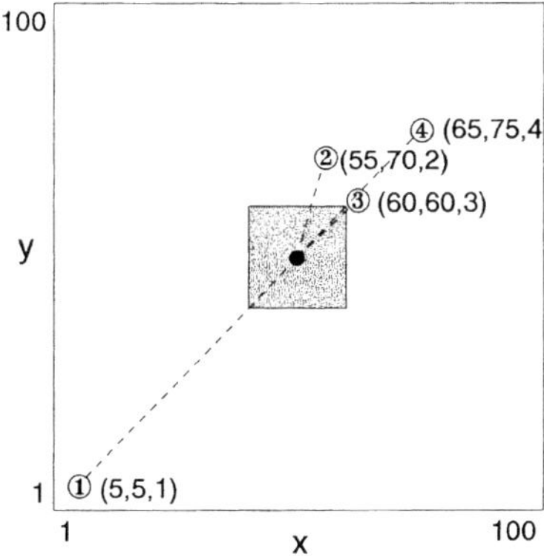

Figure 5.26. Data (x,y,z) unevenly surrounding a cell centered point to be estimated for z.

of separation. Sample 4 will be applied to 11% of the estimate, while sample 1 will account for only 3%. The ID2 estimate is 2.81. Considering this from a practical point of view, a simple application of distance of separation as the criteria for establishing the weights doesn't make much sense. Samples 1-3 are all clustered to one side of the cell and should be highly correlated ($\tilde{C}_{ij}$) since they represent the covariance over the same piece of ground. Even though sample 1 is more than twice the distance from the cell as the other three samples, it alone is available to represent the covariance in the SW quadrant. Moreover, sample 4 is masked from the cell by samples 2 and 3, so except for the possibility of a trend (which ordinary kriging ignores through the assumption of stationarity) sample 4 is largely redundant of samples 2 and 3. The effect of the relative position of the samples with the point to be estimated and the arrangement of the samples with one another is reflected in the kriging equations. Ordinary kriging with a linear variogram assigns a weight of .17 to sample 1, nearly 6 times the ID2 weight, and $\Rightarrow -0.112$ to sample 4! The low negative weight should be viewed as a reflection of the low importance of sample 4 in this estimate due to its relative position. Actually, negative weights really don't make much sense. Remember that the kriging equations only incorporate one constraint, that the weights sum to unity; here is no nonnegativity criteria, although it could easily be included along with a second Lagrange parameter. The reason for the negative weight is probably due to the use of a linear variogram model. The linear model is nontransitive, i.e. it doesn't reach a sill, and is not necessarily positive-definite without the nonnegativity constraint. As a result, certain combinations of sample locations can result in negative weights. The kriging estimate is 2.30, notably lower than the estimate of 2.81 using ID2 and reflecting the much lower weight for sample 4 (at a value of 4) and higher weight assigned to sample 1 (at a value of 1). These results are summarized in Table 5.2.

Table 5.2. Inverse distance squared and kriging weights for samples unevenly distributed around the estimation point.

Sample Z	ID2 weight	% Zest	Pure nugget	Linear $R = 0$
1	0.00025	3	0.25	0.174
2	0.0024	28	0.25	0.703
3	0.005	56	0.25	0.236
4	0001	11	0.25	−0.112

The variogram model used for this example was chosen arbitrarily. What would happen if a different variogram model or different parameters were used? In general, as the range of the variogram increases, the weights assigned to more distant data will be greater. An unfortunately common experience is that experimental variograms are generated which have very short ranges with respect to the separation between drillholes. What happens when the search radius is greater than the range of correlation due to the need to find enough data to make an estimate? By definition, locations which are separated by more than the range of the variogram are equally unrelated regardless of the distance. Thus, a nugget effect model would assign the same weight to any samples found within the search radius. In both of the above examples, all weights would be .25. The same weights will be assigned to any data found outside of the range of correlation. If a variogram with a range of 10 was used in the previous example, then $\tilde{\gamma}_{3o} = |\mathbf{h}| = 10$, while the remaining $\tilde{\gamma}_{ij}$ values would evaluate to the sill of 50. In summary, a data set with widely spaced data with respect to the range of correlation will result in two dominant types of estimates. For locations which are separated from the data by distances greater than the range of the variogram, the estimate will be a simple average of the data found within the search radius, assuming that the search radius is sufficiently great and longer than the variogram range. Where there are data within the range of correlation, probably from adjacent down-hole samples, the estimated value will be based almost entirely on the nearest data.

The choice of variogram parameters has another important aspect, the magnitude of the estimation variance. Almost all variogram models are based on one or more nested monotonically increasing functions. For this reason, the variance will increase with the separation distance (i.e. the covariance decreases). As the separation distance between the data increases, so will the estimation variance. Likewise, as the range of the variogram decreases, the estimation variance will increase. The above example was estimated with variogram ranges of 10, 20, 40 and 90 ft with resulting kriging estimation variances of 62.5, 56.2, 30.9 and 12.41, respectively. The estimation variance can be examined to determine the quality of estimation as a function of location in order to identify areas in need of additional sampling.

Example 5.9: Using kriging estimation variance to evaluate estimation quality in Techbase

The kriging estimation variance is a function of the distribution of the data and of the variogram. The variogram model is fit to the experimental variogram, which should be invariant under the assumption of stationarity. In reality, variogram models are only interpretations and can vary widely, but in effect, the only factor over which there is some control is the distribution of the sample locations. In general, more regular symmetric sampling grids will yield the lowest estimation variance. Otherwise, the estimation variance is a function of the sample spacing.

Data sets are rarely taken on a regular grid. Areas that are apparently anomalous during field investigation or as a result of examination of initial assays will be sampled more densely. Areas which are difficult to access will be poorly sampled. Consider the surface topography of the Boland Banya deposit. The volcanic neck in the NE was densely sampled and a traverse of elevation was run up the center of the valley. Otherwise, samples were taken on a somewhat evenly spaced irregular grid over the rest of the property. Even though there is a strong trend around the valley, we will assume local stationarity and estimate a grid elevation using ordinary kriging. The estimated elevation will be stored in a cell table along with the estimation variance which will be contoured and examined to see if there are areas of poor estimation quality.

Before continuing with this example, the database must include a cell table defined to include the extent of the topographic data which includes fields to hold the kriging estimate and kriged estimation variance. The cell dimensions used in this example are 30 ft^2. The variogram must have also been modeled based on the experimental variogram. A global spherical model without a nugget, a sill of 1680 and range of 340' was found to provide an excellent fit for this data set (see Fig. 5.27). Actually, the experimental variogram displays a strong trend at distances exceeding 400', but since there are plenty of data within the range of correlation of the variogram, and since the variogram shape indicates local stationarity within this range, the ordinary kriging estimate will be assumed to be reasonably valid.

From the Techbase main menu select **Modelling** ⇒ **Krige** ⇒ **Krige** ⇒ **Fields** and enter the field names from the flat table holding the topographic data. The data VALUE will be the elevation field which has only x and y coordinates. The results VALUE will be the field in the cell table that will hold the kriged estimate of the elevation. Since this is a field in a cell table, the default coordinate fields of the cell are not entered. Select **Search** and enter the search parameters. Only enter a search radius of 300' without limiting the number of samples used for the estimate so that the impact of data density will be more apparent in the resulting estimation variance. Select **Variogram**, and under **Variogram** enter the single spherical model with a range of 340' and a sill of 1680. Under **Parameters** enter the KRIGING TYPE as being ordinary and enter the field name that is to contain the estimation variance. Select **Estimate**.

Examining the **Summary** statistics for the estimation variance reveals that the average estimation variance for the 7189 cells is 17.27 with a standard deviation of 6.56 and a maximum of 49.85. This information was used to determine the contour interval. **Gridcont** was used to contour estimation variance intervals into two maps

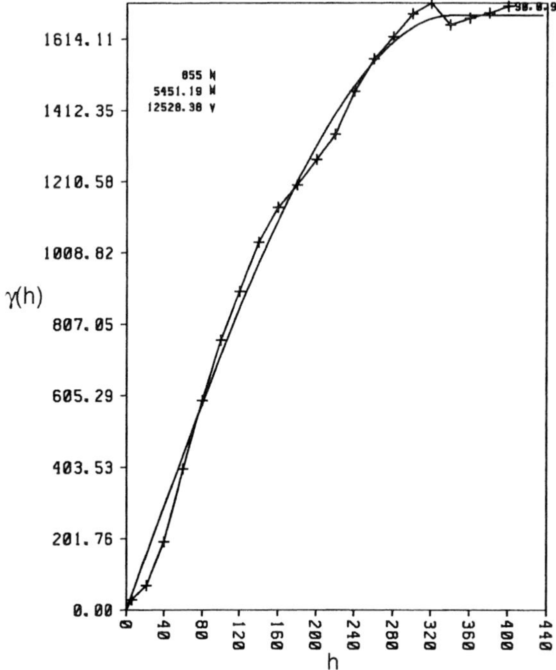

Figure 5.27. Experimental and model variogram {γ (h) = 1680 Sph (30)} for Boland Banya topography.

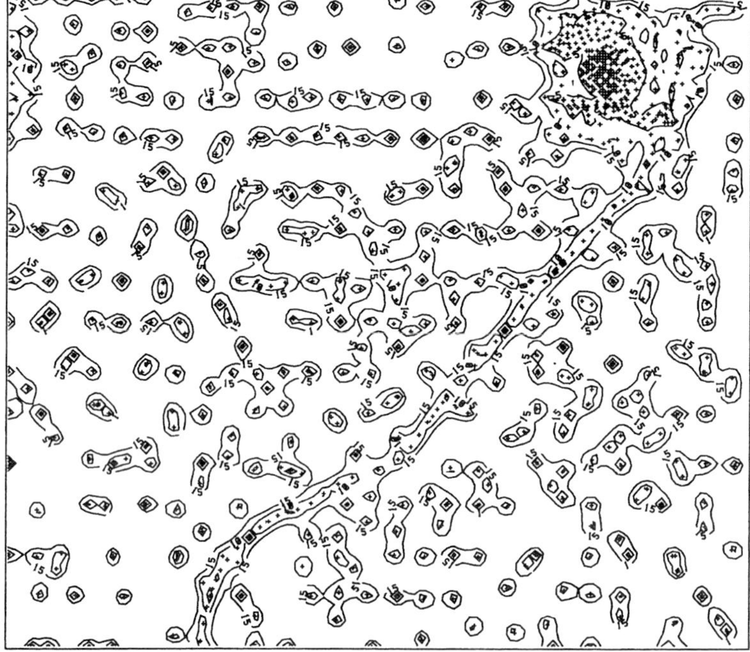

Figure 5.28. Contoured ordinary kriging estimation variance for values from 0 to 15, 'low' variance areas around data locations (+).

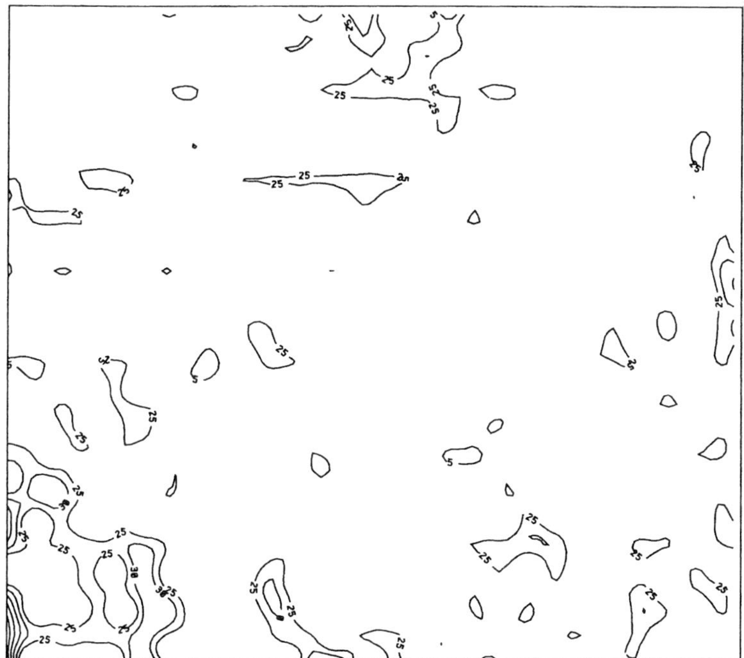

Figure 5.29. Contour of ordinary kriging variance from 25 to 50, 'high' estimation error areas.

in order to highlight areas of low (see Fig. 5.28) and high (see Fig. 5.29) estimation variance. Using an interval of 5, low estimation variance cells were arbitrarily selected as being from 0 to 15, while high estimation variance was contoured for values from 25 to 50. **Poster** was used to post the location of data marking them with a '+' on the map showing the low estimation variance contours. This data set actually has good coverage in terms of estimation quality. There are no areas which are isolated from available data due to the relatively long variogram range. At a scale of 30 × 30 ft the estimation error is essentially nil in the areas of high sample density around the volcanic neck, in the valley bottom and in the immediate neighborhood of other data values. By contouring areas of high estimation error, say from 25 to 50, areas having estimation problems can be identified. In the case of this data set, there are no areas in particular that should be of concern, but the addition of 15 to 30 samples in the centers of the 25 contoured islands would probably eliminate virtually any cells with estimation variances greater than 25.

In summary, the availability of the estimation variance is one of the most useful aspects of using kriging methods in comparison to other estimators such as splines and inverse distance. Depending on the data set and the amount of familiarity which the modeler has with the geology, inverse distance can produce as accurate an estimate as does kriging, but the kriging estimation variance provides the modeler with a measure of certainty which other methods do not provide. The estimation variance provides a degree of confidence in the resulting model which would otherwise be lacking or unfounded. It also provides a basis for locating new samples in order to decrease the probability of error: since the estimation variance is only a function of the variogram and the location of the data, the estimation error function can be used to optimally locate new sampling locations.

Summary of procedure: Example 5.9
Initial data requirements: the topographic data for the Boland deposit and a cell table to hold kriged estimates and the estimation variance.

Procedure:
1. Create two new fields in a cell table <surfaces> (see Examples 4.4 and 5.2) to hold the kriged estimate <topo_k> and its estimation variance <topo_var>.
2. Generate an experimental variogram of topography and model it (see Example 5.10). From the main menu **Modelling** ⇒ **Vario** (name the metafile: <topovario.met>) ⇒ **Setup**
   2.1 ⇒ **Fields**
   Data Points Value: <easting> X: <elevation> Y: <northing>
   2.2 ⇒ **Directions**
   Azimuth: <0> Dip: <0> Tolerance: <90>
   2.3 ⇒ **Parameters**
   Lag Distance: <20> Number: <22>
   ⇒ **eXit**
   ⇒ **Variogram**
3. Model the experimental variogram ⇒ **Model**
   Type: <sph> Sill: <1680> Range: <340>
4. From the main menu **Modelling** ⇒ **Krige** ⇒ **Krige**
   4.1 Define input and output data ⇒ **Fields**
   Data Points Results
   Value: <elevation>Value: <topo_k>
   X: <easting>
   Y: <northing>
   4.2 Define estimation search parameters ⇒ **Search**
   Max Samples: <100> Min Samples: <1> (an extreme example)
   Search Length: <300>
   4.3 Input variogram model parameters determined in Step 3 **Variogram** ⇒ **Variogram**
   Type: <sph> Sill: <1680> Range: <340>
   ⇒ **eXit**
   4.4 Input kriging parameters ⇒ **Parameters** (all others accept defaults)
   Kriging Type: <ordinary> Estimation Variance: <topo_var>
   4.5 ⇒ **Estimate**
5. Post the data locations as per Example 5.3 Step 4.
6. Contour the low and high variance results as per Example 5.3 Step 3 using the same scaling as in Step 5.
7. Overlay the two metafiles as per Example 5.3 Step 5.

5.12 MODELLING THE EXPERIMENTAL VARIOGRAM

There is one remaining topic which needs to be covered more fully before applying kriging: modelling variograms. In Examples 5.8 and 5.9, models of the variogram were used instead of the experimental variograms themselves. Likewise, the discussion of the derivation of the kriging weights and estimation variance was couched entirely in terms of variograms as continuously differentiable functions. Why? To understand this it is necessary to return to the formulation of the system of kriging equations in which the kriging weights are solved in terms of the covariances between the samples, $\tilde{C}_{ij}(\mathbf{h})$, and between the samples and the point(s) to be estimated, $\tilde{C}_{io}(\mathbf{h})$. There are two issues necessitating the use of a specific class of functions: that $\mathbf{h}$ is a vector that can connect an infinite combination of data locations and points to be estimated and the positive definite requirements of the Lagrangian approach to optimization. The experimental variogram is an incomplete guide to the covariance since it only measures the covariance for the discrete set of vectors, $\mathbf{h}_{ij}$, that exist in the data. What is needed is a function that can be evaluated for any possible combination of locations. In the case of $\tilde{C}_{io}(\mathbf{h})$, i is the index of the existing data, and o represents the nodes that are to be used in estimating a grid. The vectors $\mathbf{h}_{io}$ will not cover the same vectors that were used in determining the experimental variogram. Thus, any function which is to be used to model the experimental variogram must be positive-semidefinite and have a shape that can be closely fit to a class of covariance functions which are commonly found in nature.

The procedure for variogram modelling is just an extension of the procedure used for generating experimental variograms in Chapter 3. The modeler attempts to generate stable and interpretable experimental variograms following to the most likely directions of maximum and minimum continuity, as indicated by the geometry, origin or other main structural features of the geology. If there is significant anisotropy, then the experimental variograms along the major, secondary and minor axes of the anisotropy are retained along with their orientation and inclination. Otherwise, a single global experimental variogram is retained. The modeler then visually fits one or more variogram models to the experimental variogram(s). Nonlinear regression is usually not used. Instead, different combinations of variogram models and their parameters (sills and ranges) are tried until a satisfactory combination is found. This approach will be demonstrated in Example 5.10. Note that it is the model of the variogram that is used for kriging and not the experimental variogram which is only used as a guide in selecting the model. The following discussion aims at explaining the application of several popular variogram models and how they can be combined to model more complex covariance functions which can incorporate anisotropy into their formulation.

Most geostatistical programs include options for a very limited number of functions. These generally include the nugget, linear, spherical and exponential models as shown in Figure 5.30. Usually, a linear combination of two or three models, including the nugget effect, are sufficient to fit any variogram function as a nested model. A summary of the functional form and application of these models follows.

5.12.1 *Nugget effect*

The nugget effect is not actually a model, but is included in this discussion since it is almost always included as the first element of a nested model in kriging routines. As explained in Chapter 3, the nugget effect represents that portion of the variance in a variogram which cannot be explained by other models. Usually, the nugget effect accounts for very short-scale variability and sampling and analytic error and must be estimated by interpolating a slope from the first few lags of the experimental vario-

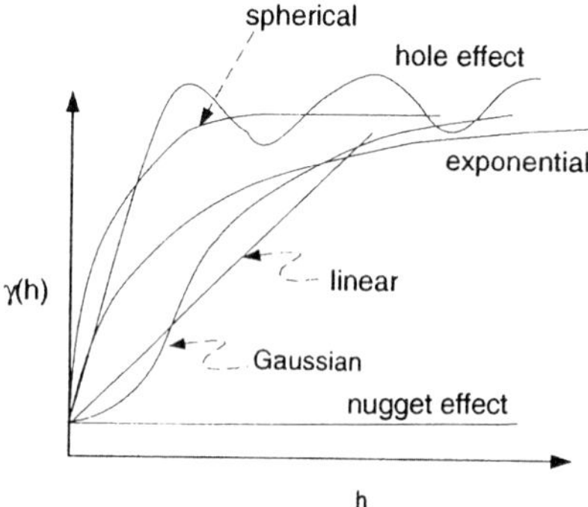

Figure 5.30. Common variogram functions.

gram to $h = 0$. Occasionally, the experimental variogram values fluctuate consistently around a constant value implying that there is no spatial correlation. This is referred to as a pure nugget effect and can be modeled as shown in Example 5.9; using kriging with a pure nugget effect is equivalent to a simple average. Inexperienced modelers working with difficult data and limited variogram programs often mistake erratic experimental variograms for pure nugget effects. It should be noted that pure nugget effects are uncommon, and when they are encountered, they are more likely to be the result of bad data or data taken from mixed populations. A true pure nugget effect will display very little fluctuation between lags from the sill/variance of the data.

5.12.2 *Linear models*

Linear models do not reach a sill, although a sill can be applied to an arbitrary range such that $\gamma(\mathbf{h}) = |\mathbf{h}|$ for $h < r$ and $\gamma(\mathbf{h}) = \text{Sill}$ for $h \geq r$, where r is the arbitrary range of the linear model. When the span of the data used in calculating the experimental variogram is not considerably greater than the range of correlation, the apparent model will be one that does not make a transition to a sill as in Figure 5.31. Even if the true underlying variogram function is transitive in shape, the portion of the function seen within the span of the data can be treated as being linear. A covariance

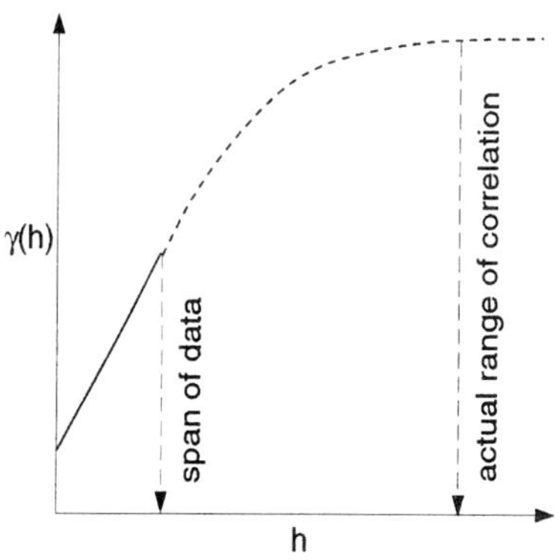

Figure 5.31. An apparent linear variogram where the range of correlation exceeds the data's span.

function with a long range might also be incorrectly interpreted as arising from data which is nonstationary, but this possibility can be resolved by checking for a drift as discussed in Chapter 3. Even if a trend surface is fit to the data, the resulting residuals should still not reach a sill. Remember that a linear model might contribute to assigning negative kriging weights.

5.12.3 *Spherical models*

This is probably the most commonly applied variogram function. Spherical models are characterized by a rapid transition to a sill. The model is linear near the origin and then makes the transition to the sill within a relatively short distance. This model can be used to model phenomena which display rapid linear increases in variance for their initial lag distances and then come to a sill at a clearly identifiable range. The spherical function has the following formulation based on a sill of one and a range r.

$$\gamma(\mathbf{h}) = \begin{pmatrix} \dfrac{3h}{2r} - \dfrac{1}{2}\left(\dfrac{h}{r}\right)^3 & \text{for} \quad h < a \\ 1 \quad \text{otherwise} & \end{pmatrix}$$

5.12.4 *Exponential models*

The exponential model is also linear near the origin, but can rise more steeply near the origin and then flatten out to approach the sill asymptotically. This model can be used for phenomena which have rapidly increasing variance at short ranges but don't quite level out to a constant sill, implying greater continuity at longer lags. Generally, a range is selected as being the point at which the variance has reached 95% or so of the final sill. Whereas a tangent line near the origin for a spherical model will reach the the sill at about two thirds of the range, a tangent line near the origin for an exponential model will hit the sill at about a third of the range. For a sill of one, the exponential function is

$$\gamma(\mathbf{h}) = 1 - \exp\left(-\frac{h}{r}\right)$$

5.12.5 *Gaussian models*

Like the exponential model, the Gaussian model asymptotically reaches a sill. Again, the range is taken at 95% of the sill. The Gaussian model is parabolic near the origin, implying a high degree of continuity in the phenomena as is found, for instance, for thicknesses or coal quality. For a sill of one, the Gaussian function is

$$\gamma(\mathbf{h}) = 1 - \exp\left(-\frac{3h^2}{r^2}\right)$$

5.12.6 *Hole effect*

This model does not reach a sill, but has a sinusoidal fluctuation about the sill such that the variogram goes from high to low states of variability. This can be explained by samples being taken in alternating strata. At short lag lengths, the variogram is being calculated within strata, but as the lag increases beyond the thickness of the strata, the variogram is being calculated from samples which belong to two different populations with significantly different distributions as the variance increases. For instance, there may be alternating strata of shale and limestone that are being sampled in the drillhole. When the lag corresponds to the average distance between either shale or limestone so that most of the variogram pairs are from the same rock type, the variogram value will be lower, but when the lag distance corresponds to shale/limestone pairs, the average variance will be higher. The hole effect model can be used to account for this phenomenon, but the modeler should first consider using increased geologic control of estimation as was described in Chapter 4.

These models can be combined, or 'nested', to fit a wide variety of experimental variograms. Nested models usually include a nugget effect in combination with one or two other models, most commonly the spherical model. Each element of the nested model is associated with a portion of the sill so that the nested model for n variogram components will have the form

$$\gamma(\mathbf{h}) = \sum_{i=1}^{n} |w_i| \, \gamma_i(\mathbf{h})$$

For instance, an experimental variogram with a sill of 100, a nugget of 30 and initially steep increase in variance to a value of 65 at a range of 50 over the shorter lags and more gradual increases in variance that asymptotically reaches about 95% of the sill at 150 might be modeled with the nested model

$$\gamma(\mathbf{h}) = 30 + 35 \, \mathrm{Sph}_{50}(h) + 35 \, \mathrm{Exp}_{150}(h)$$

Note that each component of the nested model contributes to the total sill of 100. In this discussion of models of covariance, nesting is treated as just a means of combining various models to fit more complex single variable variograms, but there is a deeper significance to the existence of nested models which accounts for different proportions of the total variance/covariance over different ranges. The phenomenon being modeled may actually be a manifestation of several other 'coregionalized' variables. For instance, the data used to generate the experimental variogram may be of some measurement used to quantify permeability, such as the rate of compressed gas diffusion through a sedimentary core, but the controlling physical variables for permeability may be the particle size distribution and the percentage of clay. These two unmeasured regionalized variables probably have their own distance covariances, which, if modeled, would have variograms with distinctly different sills, ranges and shapes. The variogram for permeability might be a composite of the short range covariance of clay and the long-range covariance of grain size. To complicate matters still more, these two regionalized variables may also be correlated in space. Such Coregionalization models form the basis of multivariate geostatistical studies, topic not covered herein.

5.13 MODELLING ANISOTROPIC VARIOGRAMS

When the shape or parameters of the variogram are dependent on direction, the covariance function is said to anisotropic. An isotropic variogram is one that is independent of direction and is commonly referred to as a global variogram. Anisotropy is commonly manifested as a change in range or sill with direction. Anisotropy in the range of correlation is referred to as being geometric as in the geometry of the deposit where there may be greater continuity along a deposit's primary axis (see Fig. 5.32a). When only the sill changes, this is referred to as zonal anisotropy as in the sharp increase in variance of grades that can be observed between rock types (see Fig. 5.32b). Variables which are taken from geologic structures which have distinct directional features can be expected to be anisotropic. An example would be a metal-

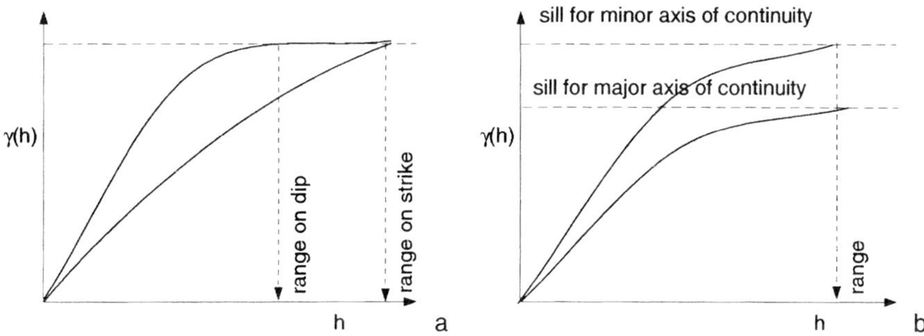

Figure 5.32. Common forms of variogram anisotropy: a) geometric, and b) zonal.

lic vein having an easily defined strike, dip and plunge. The same is true of phenomena in which the deposition had directional controls such as stream deposited placers or lead levels in a tailings pond. In any case, the possible presence of anisotropy should be evaluated by generating directional experimental variograms. Typically, variograms are generated for every 45° of azimuth, or along major structural axes when they are apparent, such as the strike, dip and plunge of an ore zone. Since data are never available on a regular grid, the search angle and lag tolerances are included as part of the experimental variogram search parameters in order to ensure that sufficient variogram data can be found to generate stable variograms in the directions of interest. When the range of correlation is short relative to the spacing of drillholes, it might be necessary to use the vertical variogram of down-hole samples for the lags that are shorter than the between hole distance and the horizontal between-hole variogram for the longer lag distances. Unless there is a strong zonal anisotropy in the horizontal, it may not be possible to establish the nature of anisotropy for short range covariance functions using drillhole data.

Example 5.10: Modelling anisotropy using Techbase

Consider the topographic elevation data that was kriged in Example 5.9. At the time, a single global variogram model was assumed, but was this correct? Select **Modelling** ⇒ **Vario** and assign a new metafile name to hold the directorial variograms. Under **Select** ⇒ **Fields** enter the *xyz* flat table data as in Example 5.9. Define Directions for every 45° of azimuth (0°, 45°, 90°, 135°). The data are two-dimensional, so the DIP is zero. TOLERANCE refers to the search angle tolerance. A tolerance angle of 45° to each side of the direction of search will generate a global variogram in the horizontal (dip = 0). Since variograms are being generated every 45° of azimuth, a tolerance of 22.5° will allow all the available data to fall into one of the directional variograms without any overlap. Under Parameters enter a LAG DISTANCE of 30' and 25 as the NUMBER of lags. The resulting experimental variogram will display the average variogram function value starting at 30' and for every multiple of 30' up to 750'. Select **Variogram** to generate a metafile of the resulting experimental variograms as shown in Figure 5.33.

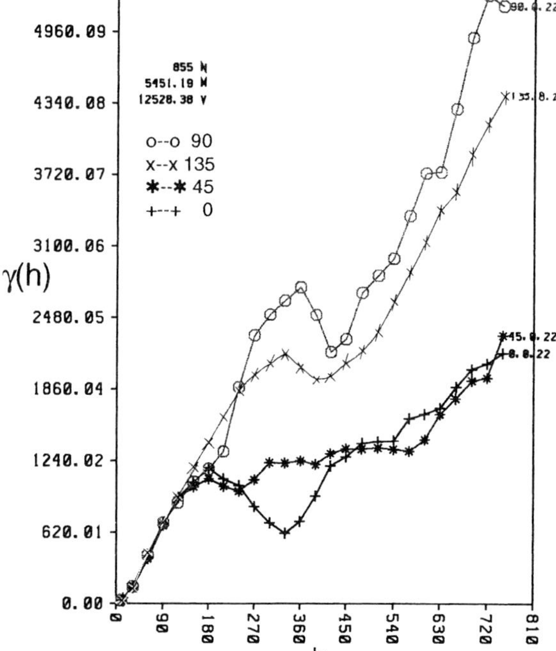

Figure 5.33. Directional experimental variograms of Boland Banya topographic elevation using lags of 30' and 22.5° search angle tolerances.

The variograms still display drift, but we will continue to ignore this problem using as an excuse the relative stability of the variogram up to lag distances in excess of the search distance that was used for kriging in Example 5.9. There is strong anisotropy between azimuths of 0/45° and 90/135°. All of the directional variograms are essentially equivalent up to 120'. Referring back to the random contoured map of topography in Chapter 4, it seems likely that this apparent anisotropy arises from the presence of the valley which runs NE-SW. Experimental variograms running perpendicular to the valley display rapid increases in variance for lag distances in excess of 120'. Variograms running more or less along the bottom of the valley or parallel to it display greater continuity. There are two other interesting features to note in the directional variograms: the hole effect at 360' for an azimuth of 0° and at 450' for 90°; and the much more rapid increase in drift for the 90/135° directions. The variograms suggest that elevation is locally stationary for lags up to 120-150', but since there is a major difference in the elevation between the valley and the surrounding highlands, when the lag distance is in the range of 100-300' at 0/45° or 240-390' at 90/135°, there are a large number of widely different elevations that are being compared. One possible reason is the difference between elevations on the valley, but the valley is so broad that this seems unlikely considering the scale of the apparent hole effect. Remembering that nearly 30% of the data is clustered in the neighborhood of the volcanic neck, it is possible that the sharp change in elevation at the neck is responsible for the hole effect. If this is so, then the clustering of data about the volcanic neck will bias the estimation variance of the kriged estimate since the clustered data results in a higher sill on the directional variograms. This should be checked by filtering out the data around the neck and treating this as an area of local nonstationarity. The much greater drift in the 90/135° variograms is due to the more rapid changes in elevation that are seen transverse to the valley. There is also a trend to increasing elevations to the SW where the head of the valley is, but the trend is much more gradual in this direction.

For now, the hole effect and drift will be ignored and two directional variograms will be fit for azimuths of 45° and 135°. Select **Model ⇒ Variogram** to enter model definitions and have the model drawn on the experimental variogram. To fit the steeply increasing variance up to 175', a spherical model with a sill of 956 can be used, but the variogram increases gradually from there on and never really reaches a sill. To account for the non-transitive nature of the variogram, include an exponential model with a range of 635' and a sill of 600 to the nested model (see Fig. 5.34). Note that the total nested model's sill is approximately 10% greater than the apparent sill of the experimental variogram at 635'. In Techbase, the exponential model only reaches about 90% of the assigned sill. Due to the steep linear portion of the directional variogram at 135°, a single linear model will be used with a sill of 2000 and range of 260' (see Fig. 5.35).

Summary of procedure: Example 5.10
Initial data requirements: Same as per Example 5.3.

Procedure:

1. Generate directional experimental variograms **Modelling ⇒ Vario** (assign a metafile name <topovardir>) ⇒ **Setup**
   1.1 Input field names ⇒ **Fields**
   VALUE: <elevation> X: <easting> Y: <northing>
   1.2 Define directions for generating variograms ⇒ **Variograms**
   AZIMUTH: <0> DIP: <0> TOLERANCE: <22>
   AZIMUTH: <45> DIP: <0> TOLERANCE: <22>
   AZIMUTH: <90> DIP: <0> TOLERANCE: <22>
   AZIMUTH: <135> DIP: <0> TOLERANCE: <22>
   1.3 Input variogram generation parameters ⇒ **Parameters**
   LAG DISTANCE: <30> NUMBER: <25>
   1.4 **Variogram**
2. Model and view the variogram ⇒ **Model ⇒ Variogram**
   TYPE: <sph> SILL: <956> RANGE U: <175>
   TYPE: <exp> SILL: <600> RANGE U: <635>
3. Repeat Step 2 for the directional variogram at 135°.

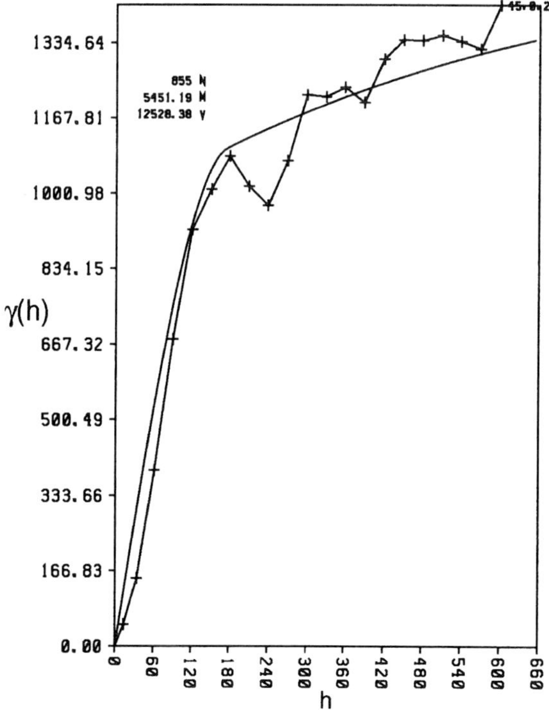

Figure 5.34. Directional experimental variogram of elevation at 45° $\{\gamma(h)_{45} = 956\ \mathrm{Sph}_{175}(h) + 600\ \mathrm{Exp}_{635}(h)\}$.

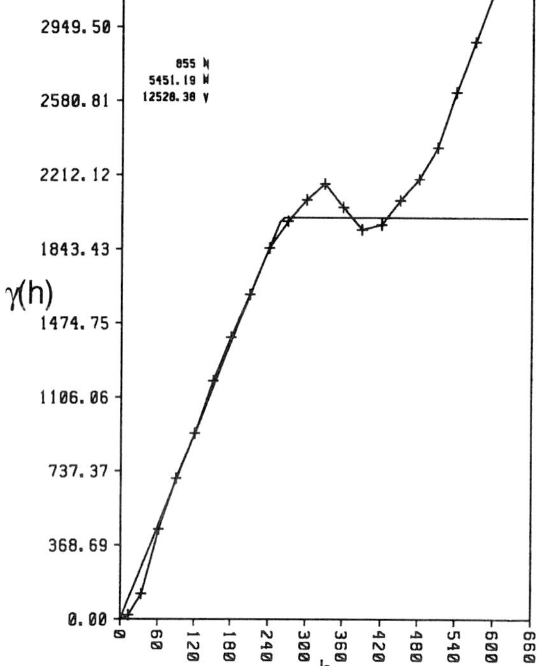

Figure 5.35. Directional experimental variogram of elevation at 135° $\{\gamma(h)_{135} = 2000\ \mathrm{LIN}_{260}(h)\}$.

5.14 USING JACKKNIFING TO EVALUATE ESTIMATION ACCURACY

This chapter has introduced a wide variety of estimation methods and parameters, but to this point no method of confirming the accuracy of an estimate has been presented. Kriging offers the estimation variance as a means of checking the likely magnitude of the error, but it doesn't provide a measurement of the actual error. We have seen how the selection of the inverse distance power can completely change the estimated value, yet what basis is there for selecting a power of one, two, three or more? For all of the estimation routines, there are a host of search and estimation parameters which can have a marked influence on the final estimate. Which parameter settings will minimize estimation error and bias?

In regression analysis and Analysis of Variance, the model's estimate is compared against the actual data that were used to fit the model, and the Sum of Squares error and Mean Squared error is used to evaluate the goodness-of-fit of the model to the data. The same basic approach can be applied to spatial estimation. For example, in regression the model can be used to generate an estimated value, $\hat{z}$, for each of the original data, z, from the equation $\hat{z} = a + bx$, where a and b are first order linear regression parameters that have been estimated by fitting the dependent variable z to the independent variable x. Similarly, a trend surface can be used to remove drift from data and the residuals can then be evaluated to determine the accuracy of the estimate. In both cases, the model is used to estimate a value which corresponds to the original data. In regression, the equation parameters are estimated from the data pairs for x and z, so x in the regression equation represents the position of z. By evaluating the equation for one of the original independent data values, x_i, we get an estimate of the corresponding independent variable, $\hat{z}_i$. The error or residual is simply the difference of the actual and estimated value, $e_i = z_i - \hat{z}_i$. When evaluating the accuracy of a trend surface, the same procedure is followed, but in this case the estimated value is a function of position in space, $\mathbf{x}$, and $Z(\mathbf{x})$ is a regionalized variable. The trend surface polynomial equation is fit to the data $z_i(\mathbf{x})$ at the positions where they were sampled $\{\mathbf{x} = (x_i, y_i, z_i)$ for $R^3\}$ and the residual is calculated, $e_i(\mathbf{x}) = z_i(\mathbf{x}) - \hat{z}_i(\mathbf{x})$.

In both examples (linear regression and trend surfaces), the fit equation can be used to estimate z or $Z(\mathbf{x})$ at any position within the range of $\mathbf{x}$ seen in the data set. This same approach can be used for any of the single equation-based estimation methods including triangulation and splines, but the calculations that have to be made for weighted average routines, such as inverse distance and kriging, are somewhat more involved. These methods are normally used to estimate locations which do not correspond to the original datum location: typically, random point data are used to estimate a grid. Since $Z(\mathbf{x})$ is a function of its location, the estimated value must be at the same location as the datum. Thus, instead of estimating into a grid, the original data locations must be estimated using the surrounding data. This approach to estimation is referred to as Jackknife estimation.

Jackknife estimation is used to evaluate the estimation error for weighted average estimators. Each datum value is estimated based on the surrounding data using the same search and estimation parameters that are being considered for use in grid estimation. This corresponds to estimating random point locations which correspond in number and location to the data set. As each datum location is estimated from the surrounding data, the datum value at that location is cut from the data set. The result is that there is an estimated value for each data record. This procedure can be repeated for two or three parameter values such as inverse distance powers of one, two or three and the residuals can be calculated for each of the estimate values. The three sets of residuals can be compared to see which inverse distance power results in the lowest error variance and bias. The residuals can also be posted or contoured using random contouring to check for spatial bias.

Residual variance is a measure of the magnitude of estimation error. The perfect estimator would have residuals with a mean and variance of zero as shown in Figure 5.36a, i.e. all the estimators match the data exactly. Unless the data values are of a flat surface, this will never happen; the residuals will have a skewed distribution in which the mean is not zero as in Figure 5.36b. The residual distribution which has

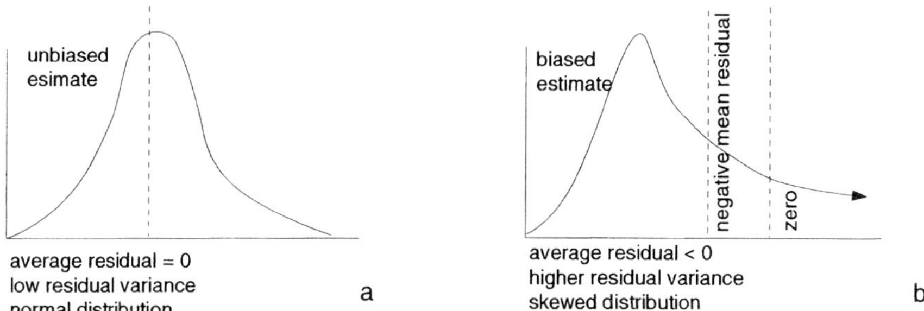

Figure 5.36. Comparing residual distributions for: a) unbiased estimates, and b) biased estimates.

the smallest variance, the mean closest to zero and which is closest to a normal distribution with the smallest skewness is probably the best estimator. One final comparison should be made. While the residuals for the entire data set may be normally distributed with a mean near zero, there might be local bias in which some areas are underestimated and others are overestimated. If the number of over-and under-estimated data locations is approximately equal, the histogram of the residuals will indicate that the estimator is unbiased when in fact it is not. A map of the residuals should be examined to check for spatial bias. One method would be to contour the residuals and look for islands of positive or negative contours. Another approach is to post the residuals as '+' and '-' signs and check for unexpected concentrations of positive and negative residuals.

Example 5.11: Cross-validation, jackknifing and universal kriging using Techbase
We will return to the topography of the Boland Banya project one last time and evaluate the effectiveness of several estimators: trend surfaces, inverse distance and universal (residual) kriging. It was noted in Examples 5.9 and 5.10 that drift is apparent in the experimental variogram and that the volcanic neck is an area of significantly high elevation in which clustered data will bias both the mean elevation and the estimation variance. Each of these estimators will be used to generate sets of re-

siduals that can be compared in order to determine which of the methods is the minimum error estimator and has the least global and spatial bias.

The estimated fields and residuals for each of the methods will be included in the same flat table that holds the *xyz* topographic data. Therefore, two new fields have to be added to the topography table for each estimator. A record in the topography field will now include the original *x*, *y* and *z* fields plus estimated elevations and corresponding residual fields. The residual fields will be calculated fields equal to the difference of the actual *z* and the estimated *z*. Use **Techbase** ⇒ **Define** ⇒ **Fields** ⇒ **Autotable** ⇒ **Create** to create the new actual real fields to hold the topography. Also create the three residual fields by entering 'calculated' for the CLASS. An Equation line will be displayed in which the calculation steps for the new field are entered in Reverse Polish Notation. For instance, if the new residual field for a trend surface of order 1 that was estimated into the field zrt1 is to be called rt1, the corresponding RPN equation would be 'z zrt1 –' for the definition of rt1.

In Example 5.10, we noted that the apparent hole effect in the experimental variogram might be an artifact from the clustering of data about the volcanic neck. This was confirmed by adding four filter groups placing limits on the coordinates of *x* and *y*; specifically individual filters (filter groups have an OR relationship) specifying that $x > 12463$, $x < 11945$, $y > 22340$ and $y < 21810$ will filter out all records within the rectangular area bounding the neck. The resulting variogram (see Fig. 5.37) is

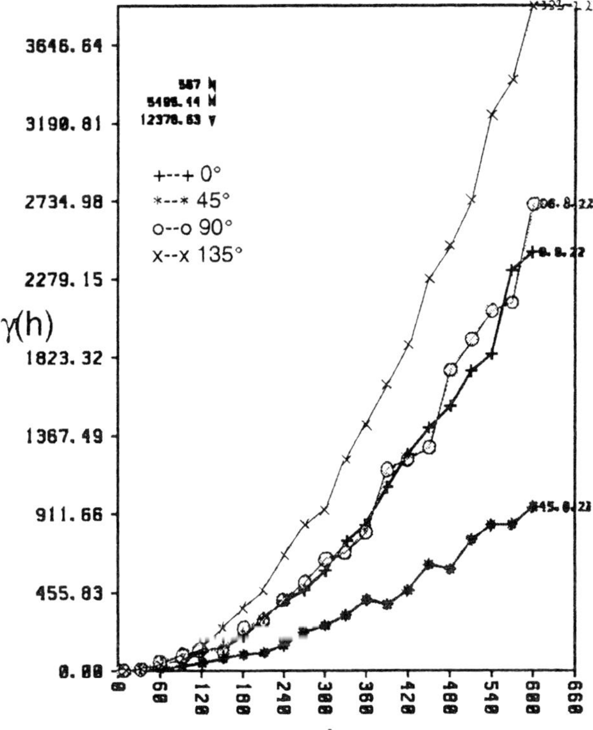

Figure 5.37. Directional variograms of elevation after removal of clustered data.

drastically changed with each directional variogram displaying a parabolic increase in variance. This is now a classic example of drift. The trend so dominates the variogram that the short-scale variability is hidden. If this surface is to be accurately modeled, both long and short-scale variability will have to be accounted for. In order to do this, the trend will have to be first modeled to remove the drift from the experimental variogram. The variogram can then be used to model short-scale variability.

Since it is apparent that the elevation has a dominant trend, the data will have to be de-trended before continuing with ordinary kriging. As was mentioned earlier in the development of the kriging equations, kriging assumes that the mean and covariance function are stationary, i.e. that the average value of the distribution of the regionalized variable and its variogram are not dependent on location. This may seem to be an overly restrictive assumption, but other estimation methods implicitly have the same requirement. The main requirement for estimation is that the data are locally stationary. If either inverse distance or kriging is used for estimation, it is assumed that the data selected within the search radius all come from the same distribution. With broad gradual trends and relatively close data, nonstationarity of the mean is not a problem since the data will be locally stationary, but for the application of kriging, the trend still causes a practical problem in that the trend will cause drift in the variogram, making it difficult, if not impossible, to model. The problem is that the variogram of the raw data captures all of the spatial variability both on the longer scale of the trend surface as well as the short-scale variability found within the search radius. Since the longer-range variability increases with distance (drift) and is often orders of magnitude greater than the short-scale variability, the drift in the variogram obscures the covariance function. Therefore, the data must be de-trended in order to remove the drift and uncover the underlying variogram. Typically, this is accomplished using a kriging method referred to as universal kriging which combines trend surface polynomials and kriging equations. The trend surface is used to fit the large-scale variation and generate a set of residuals which are the difference of the trend surface estimate at the data locations and the data values at those same locations. In theory, the trend surface should remove the large-scale variation which is responsible for the drift in the variogram; the variogram of the residuals can then be modeled and used in kriging of the residuals. A final estimate that combines both long and short-scale variability is obtained by combining the trend surface estimate and the kriging estimate from the residuals. Universal kriging combines both of these steps, but trend surfaces of different orders should still be fit and their residuals examined for bias and drift using histograms, map postings and variograms. The lowest order trend surface that results in disappearance of trends in both the map postings or contours and the residual variogram should be used to model the long-range variability. Note that in either case, that de-trended residuals will have to be used to define the variogram.

After filtering out the data in the immediate vicinity of the volcanic neck, trend surfaces of increasing order were applied to de-trend the data. Both histograms (see Fig. 5.38 and Table 5.3), variograms and postings of the residuals (see Fig. 5.39) were examined to determine if the trend had been removed. The residual histogram from a first order trend surface fit (Fig. 5.38a) is positively skewed, has a high variance, and, even though the mean is zero, both the median and mode are negative.

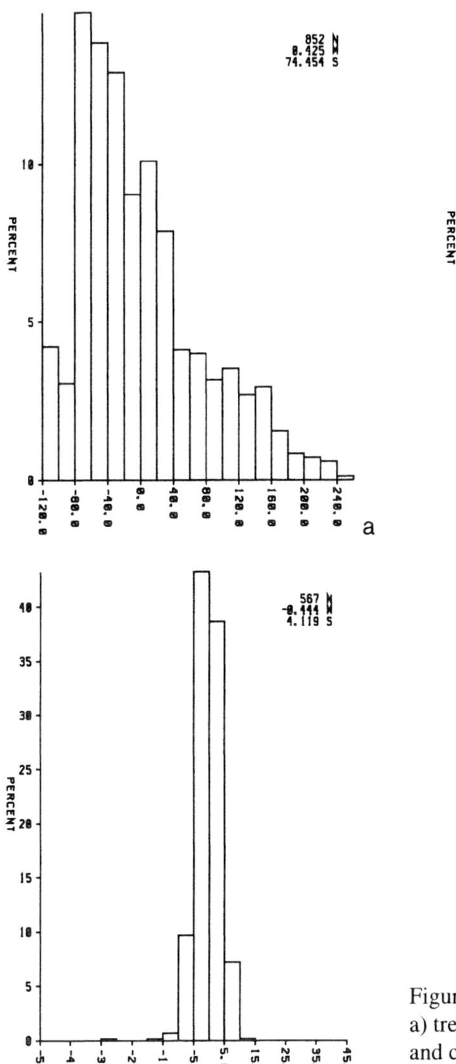

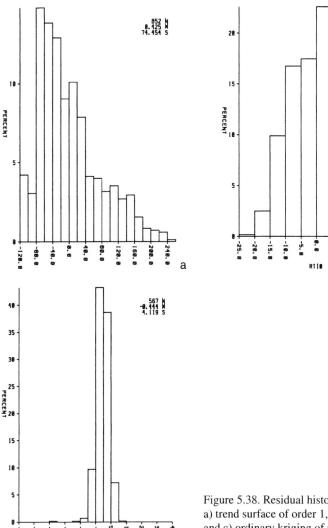

Figure 5.38. Residual histograms resulting from: a) trend surface of order 1, b) trend surface of order 10, and c) ordinary kriging of residuals from Jackknife inverse distance of power 0.5.

Table 5.3. Distribution statistics of residuals from trend surfaces of orders 1 and 10, inverse distance of powers .5 and 1 and universal kriging.

Statistics	Trend 1	Trend 10	Inverse distance .5	Inverse distance 1	UK 1
Mean	0	0	0.96	0.52	–0.44
Median	–18.10	0.65	0.92	0.28	–0.39
Mode	–121.60	–21.10	–54.80	–40.00	–33.80
Variance	6966.71	65.06	50.17	28.00	16.97
Skewness	0.62	–0.14	–.02	–0.14	–1.19
Kurtosis	–0.50	–0.74	10.57	11.04	7.34

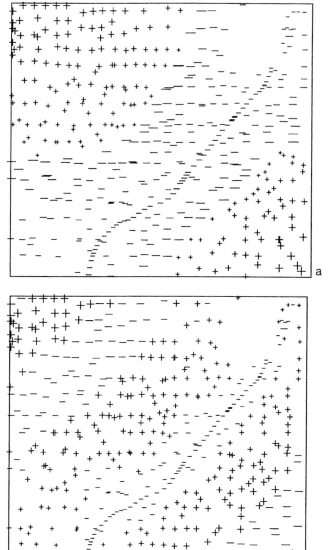

Figure 5.39. Posting of residuals from trend surfaces of: a) order 1, and b) order 10.

This implies a poor fit and bias. Posting of the residuals (see Fig. 5.39a) confirms that the bias is still there in the original trend. It is not in the least surprising that a planar surface failed to de-trend such a complex topography, but increasing the power of the trend surface even up to an order of ten didn't completely remove the trend. Even though the order ten residual histogram (see Fig. 5.39b) and statistics (see Table 5.3) indicate low variance, symmetric and unbiased estimation, the directional variograms still exhibited the same characteristics as were seen in the original data (see Fig. 5.40): higher rates of increase and total variance at an azimuth of 135°.

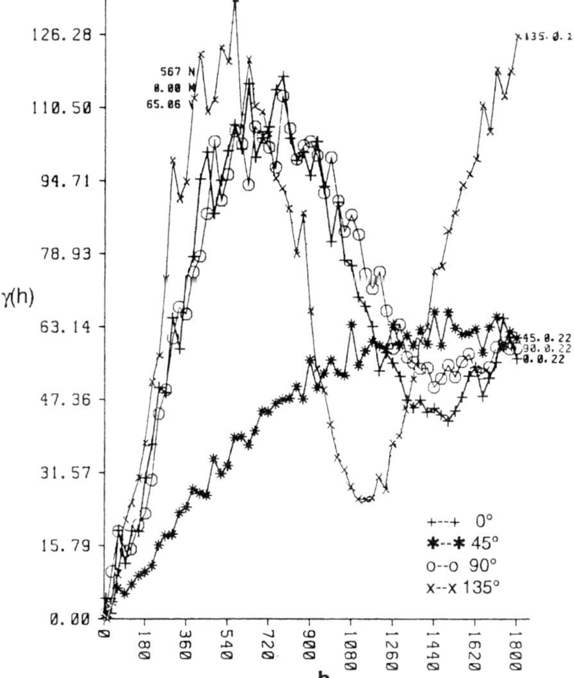

Figure 5.40. Directional variograms of elevations of a 10th order trend surface still displaying anisotropy resulting from directional drift.

Once again, the posted residuals show a pattern of alternating positive and negative residuals trending NE-SW. If the trend had been removed, the pattern of residuals would appear to be random.

Apparently, the topography presents too complex a surface for fitting with a trend surface and some other method must be used. The only requirement for fitting the large-scale variance is that the estimator must be smooth and not remove the short-range variability. 'Sheet' fitting methods like regression and splines are well suited to this, but Techbase's spline routine, MINQ, is based on grid estimation and cannot be applied to jackknife style cross-validation since, by definition, a spline will exactly fit the data. Weighted average methods are not particularly well suited to fitting trends, but by manipulating the search and estimation parameters the estimator can be made to be very smooth yet capable of fitting any type of surface. Inverse distance was used for this purpose with the following parameter settings: a circular search radius of 300', a power of 0.5, 1-100 of the closest samples within the search radius used for estimation and jackknifing to remove the closest sample. This set of parameters will produce a very smooth estimate of the original data. The search radius was selected to approximate the scale of the valley. The low power and large number of points used for estimation should capture elevation variations on the scale of 300' but miss local variability. The jackknifing technique is necessary since the data is being estimated at the same locations where d = 0. The ID.5 residual variogram was plotted (see Fig. 5.41) modeled as $\gamma_G(h) = 12 + 20\ \text{Sph}_{434}(h) + 20\ \text{Exp}_{1200}(h)$.

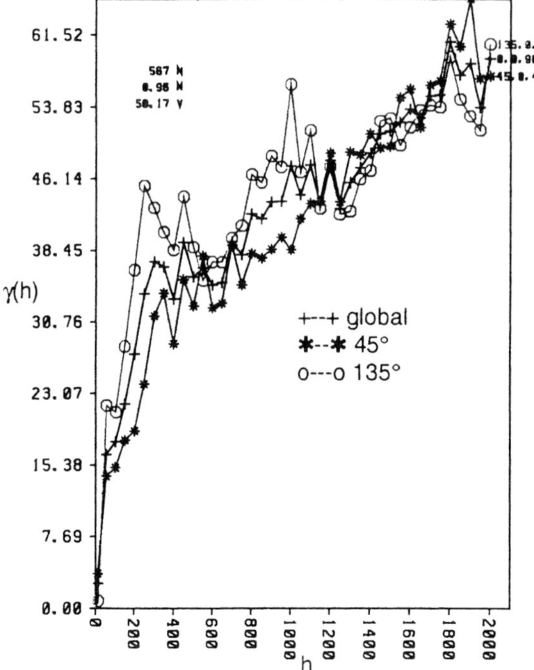

Figure 5.41. Directional variograms of residuals from inverse distance power .5.

The statistics of the ID.5 residuals (see Table 5.3) is somewhat better than what was observed even for the residuals from a tenth order trend surface. Posting of the residuals from ID.5 shows a random pattern except in the bottom of the valley where the elevation is still overestimated (see Fig. 5.42a). The residuals from ID.5 were

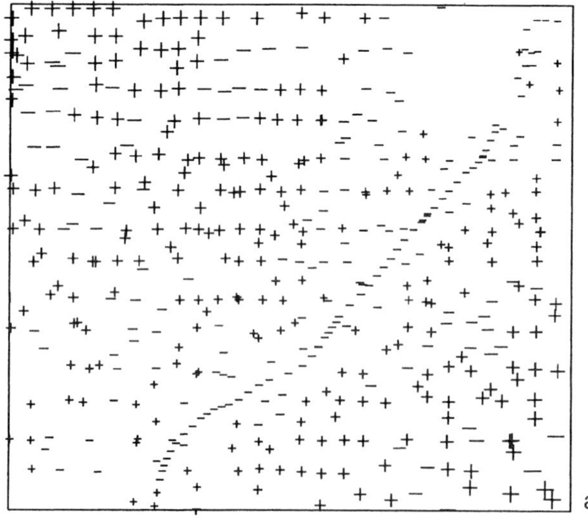

Figure 5.42a. Posting of Jack-knife residuals from: Inverse distance (p = 0.5). Residuals are posted as positive (+) or negative (–) values (symbol size denotes relative magnitude of error).

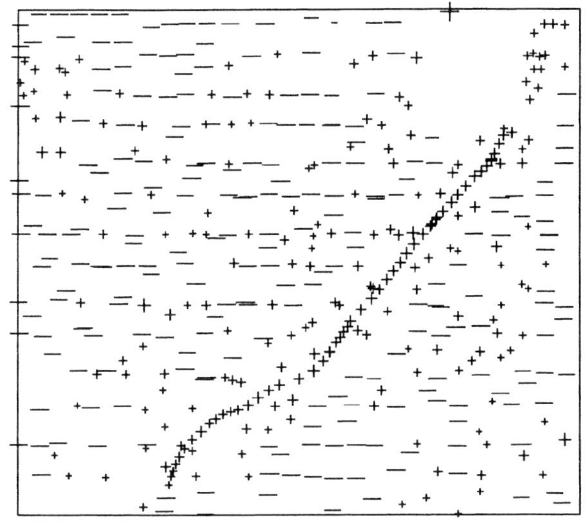

b Figure 5.42b. Posting of Jackknife residuals from: Residuals following inverse distance (p = 0.5) plus residual ordinary kriging. Residuals are posted as positive (+) or negative (–) values (symbol size denotes relative magnitude of error).

jackknife estimated using ordinary kriging, and then the kriged estimate of the residuals from ID.5 were added to the ID.5 estimates to obtain a universal kriging estimate (actually, a residual kriging estimate based on ID.5). An overall residual error was calculated between the original elevation data and the 'universal kriging' estimate and then plotted as a histogram and residual error posting. The histogram (see Fig. 5.38c) shows an improvement over ID.5 in terms of the mean error and variance, but is slightly more skewed. The residual error posting displays spatial randomness with the exception of the transverse data in the valley bottom which are now consistently underestimated (see Fig. 5.42b). This is a normal artifact of smoothing which is common to all estimation methods which are not constrained to strictly honor the data.

Summary of procedure: Example 5.11
Initial data requirements:

Procedure:

1. Generate an experimental variogram filtering out the volcanic neck data **Modelling ⇒ Vario**
   1.1 Filter out the volcanic neck **Database ⇒ Add filter** (filter group1)
   FIELD: <northing> RELATION: <<> FIELDVALUE: <21800>
   (filter group 2) ⇒ **Add filter**
   FIELD: <easting> RELATION: <>> FIELDVALUE: <12400>
   (filter group 3) ⇒ **Add filter**
   FIELD: <easting> RELATION: <<> FIELDVALUE: <12000>
   1.2 Generate directional experimental variogram of elevation as per Example 5.10 Step 1 and note the loss of the apparent hole effect and lack of transition to a sill for all directions.
2. Create two fields for each estimator in the topography flat table, one to hold the estimate and the other to hold the residual error. For a trend surface of order 1 **Techbase ⇒ Define ⇒ Fields ⇒ autoTable**
   TABLE NAME: <topo>
   2.1 Create the estimation field for the trend surface of order 1

FIELD NAME: <z_tl> TYPE: <real> CLASS: <actual>

2.2 Create the estimation error field for the trend surface of order 1

FIELD NAME: <z_tlerr> TYPE: <real> CLASS: <actual>

Repeat Steps 2.1 and 2.2 for a trend surface of order 10 (<z_t10>, <z_t10err>), inverse distance of p = .5 (<z_id5>), the kriged residual estimator (<res_k) and inverse distance of p =1 (<z_id1>).

For creating the residual fields for inverse distance and universal kriging, Step 2.2 is replaced by Step 2.3.

2.3 **Create** the estimation error field for inverse distance of p = .5

FIELD NAME: <z_id5err> TYPE: <real> CLASS: <calculated>

EQUATION: <elevation z_id5 - >

Repeat Step 2.3 creating the residual error field for p =1 (<z_id1err>).

2.4 **Create** the estimation error field for universal kriging

FIELD NAME: <res_kerr> TYPE: <real> CLASS: <calculated>

EQUATION: <elevation z_id5 - res_k ->

3. Estimate values back into the topograhy table using the data in the topography table **Modelling ⇒ Trend ⇒ Trend**

3.1 **Fields**

DATA POINTS	RESULTS
VALUE: <elevation>	VALUE: <z_tl>
X: <easting>	X: <easting>
Y: <northing>	Y: <northing>

3.2 **Parameters**

TREND ORDER: <1> RESIDUAL: <z_tlerr>

3.3 **Trend**

4. Repeat Step 3 for the trend surface of order 10.
5. Estimate inverse distance values back into the topography table **Modelling ⇒ Inverse ⇒ Inverse**

5.1 **Fields**

DATA POINTS	RESULTS
VALUE: <elevation>	VALUE: <z_id5>
X: <easting>	X: <easting>
Y: <northing>	Y: <northing>

5.2 **Search**

MAX SAMPLES: <100> MIN SAMPLES: <1> SEARCH LENGTH: <300>

5.3 **Parameters**

INVERSE DISTANCE POWER: <.5> JACKKNIFE: <yes>

5.4 **Estimate**

6. Repeat Step 5 for inverse distance p = 1 (<z_id1>)
7. Generate and model (see Step 8.3) directional experimental variograms of <z_id5err> as per Step 1.2.
8. Krige the residuals z_id5 **Modelling ⇒ Krige ⇒ Krige**

8.1 **Fields**

DATA POINTS	RESULTS
VALUE: <z_id1err>	VALUE: <res_k>
X: <easting>	X: <easting>
Y: <northing>	Y: <northing>

8.2 **Search**

MAX SAMPLES: <100> MIN SAMPLES: <1> SEARCH LENGTH: <300>

8.3 **Variogram**

TYPE: <nugget>	SILL: <12>	
TYPE: <spherical>	SILL: <20>	RANGE U: <434>
TYPE: <exponential>	SILL: <20>	RANGE U: <1200>

⇒ **eXit**

8.4 **Parameters**

KRIGING TYPE: <ordinary> JACKKNIFE: <yes>

8.5 **Estimate**

9. Post the universal kriging residual <res_kerr> **Graphics** ⇒ **Poster** ⇒ (metafile name: <ukresid.met>) ⇒ **Scaling** (see Example 5.9)

9.1 Post positive residuals **Database** ⇒ **Filter** ⇒ **Add filter** (delete filter of Step 1 as it is no longer needed)

FIELD: <res_kerr> RELATION: <>> FIELD/VALUE: <0>

9.2 **Point** ⇒ **Field**

X-COORD: <easting> Y-COORD: <northing> VALUE 1: <topo_nul>

9.3 **Marker**

TYPE: <2> PROPORTIONAL TO: <res_kerr>

9.4 **Draw** ⇒ **Graphics** ⇒ **Review**

Repeat Steps 9.1-9.4 with the filter relations set to <res_kerr <0> the marker type set to <126>.

5.15 ESTIMATING AREAS AND VOLUMES OF INTERSECTION

Estimation of the value of a regionalized variable into a grid has been discussed in the preceding sections, but average grade is only part of the estimate; the volume or tonnage must also be estimated. In Chapter 4, matrix operations between grids and numeric integration of volume models were examined as a means of estimating total volumes, but projects are concerned with specific increments of the total volume of ore and waste. Ore and waste is mined in incremental amounts which correspond to a production schedule. In the case of surface mining, an ultimate pit can be defined as a grid. This grid can then be intersected with the surface topology and other grids and volume models of ore and waste zones to determine volumes of intersection. For underground mining, volume models of mine openings can be intersected with grid or volume models to determine tons of ore and waste.

In open pit projects, production scheduling is built around benches. Polygons are defined on a bench that represent an area to be mined over a specific period of time. The polygon is intersected with the mining blocks that correspond with that level and the intersection of the polygon with the block model is used to calculate tonnages and average grades as shown in Figure 5.43. In a similar fashion, ownership bounda-

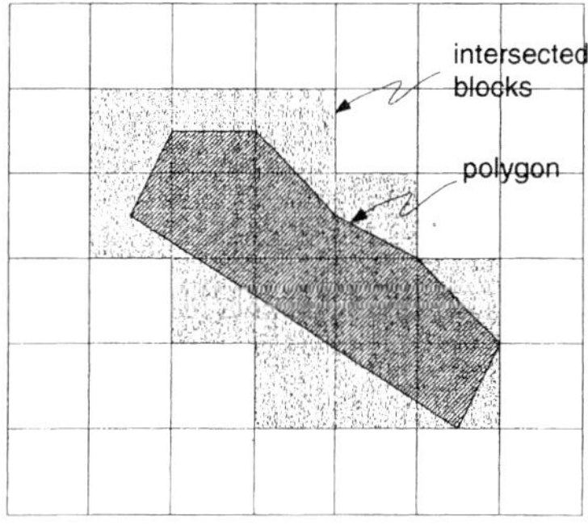

Figure 5.43. Intersection of polygon representing a mining cut with a block model layer representing a bench.

ries can be intersected with grids of coal seam thickness and topography to determine royalty payments as part of a reserve report. Polygons of intersection can be used to limit estimation. Most estimation methods extrapolate at least some distance beyond the available data. Grid nodes that are identified (located) as being within the polygon can be estimated while those outside of the polygon are left unestimated.

This approach has important consequences for estimating grids. It was noted in Chapter 3 that estimation should be limited to only those values which belong to the same statistical population. When working with geologic data, sample populations are typically associated with rock types, or, in the case of regionally nonstatioanary data, with localities. The rock type or location of the data is used to filter sample data so that only the samples associated with one population are used for estimation. For instance, if permeability is to be estimated for layers of limestone and shale, then only samples from the limestone should be used in estimating limestone permeability. But when grids are estimated with methods that extrapolate beyond the limits of the data, then nodes which are outside of the limits of the rock type will be estimated. In order to avoid this, the block model can be intersected with grids or volume models that represent the limits of the geologic unit to be estimated. The blocks or cells that are located as being at least partially within the limits of a polygon, a hanging-wall and footwall grid, or a volume model of the zone to be estimated can be flagged. The flag can then be used to limit estimation.

Volumes are estimated in Lynx by numeric integration (Holding 1994). Consider the intersection of a volume with a plane. Since the volume model midplane is based on a digitized polygon, the resulting volume model will also be a polygon with a faceted surface. Therefore, the area of intersection of the plane with the volume will be a polygon (see Fig. 5.44). If this area of intersection were assigned a constant thickness and the process of slicing the volume by this incremental thickness were extended to the limits of the volume, the sum of the areas of the polygons of intersection times the slice thickness would be an approximation of the true component volume. This is the approach used in Lynx where the slice orientation is taken from the

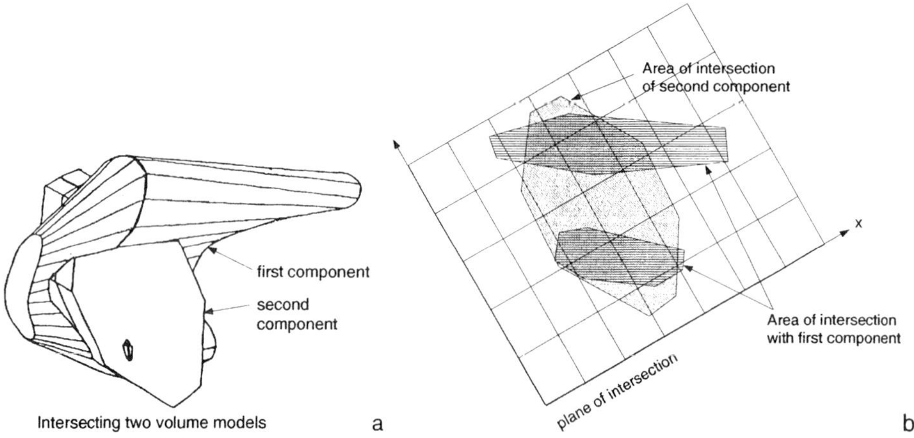

Figure 5.44. Intersecting volumes (a) and areas (b).

current viewplane and the thickness is a user defined parameter. When two volume components are being intersected, each component will have a polygon of intersection on the viewplane for which the area can be calculated. Where the two components intersect each other, the area of intersection will also be a polygon. Again, these areas of intersection can be summed and multiplied by a constant slice thickness. Examples of intersections of volume models with mining block models, ultimate pit limit grids and underground mine opening volume components will be presented in Chapters 6 and 7 as part of surface and underground mine design.

Example 5.12: Using Locate to limit estimation of blocks in Techbase
In previous examples, the top and bottom structure of the OA main ore zone of the Boland Banya deposit have been gridded and block table values for gold have been estimated. A filter on the lithology was used to restrict estimation to samples that were from the ore zone, but there were no controls on the block model so that only those blocks that were in the ore zone were estimated. The **Locate** program will now be used to identify those blocks that are at least partially between the top and bottom structure of OA by assigning the fraction of the block volume that is within OA to a new field in the block model. This fraction will then be used in the **Edit** program to assign null values to all assay estimates in the block model that are not within the limits of OA.

Before continuing you will need gridded estimates of the OA top and bottom structure, a block model containing an estimated value for gold assay, and a real actual field to contain the fraction to two decimal places in the block table.

Select **Modelling** ⇒ **Locate** ⇒ **Setup** ⇒ **Fields** and enter the fields that identify the location of the point to be estimated. These fields could represent random point values, but in this example we are locating blocks. For blocks, the centroid location is stored in the default fields: name_xc, name_yc and name_zc where 'name' is the name of the block table. Select **Surface** and enter the fields containing the gridded estimates for the TOP ELEVATION and BOTTOM ELEVATION. Under **Assignment** all field value assignments are made that are dependent on whether they are in the ore zone. Enter the fraction field beside the heading POLYGON/VERTICAL FRACTION. **Locate** will assign a fractional value to all blocks that lie within the two surfaces. Locate can also be used with a polygon in order to specify horizontal limits. In this example only vertical limits are given in the form of the two surfaces. Upon selection of **Locate**, Techbase will respond with, 'No polygon was given. Location will be with respect to Z surfaces only'. Each block coordinate will be compared against the corresponding z values of the two grids and the total number of blocks that were assigned a positive fraction will be reported. Note the number of positive fractions assigned to see if the results look reasonable. When using polygons to locate areas, there may be more than one polygon. For example, polygons may be available that define lease boundaries or ownership. A polygon record consists of the x and y coordinates followed by a polygon identifier. Under **Assignment**, this polygon identifier can be assigned to a field just as the fraction was beside the heading POLYGON ID MARKER. This polygon identifier can then be used to calculate royalty payments and lease fees during the scheduling of mine production. When using multiple polygons or surfaces to locate blocks, some of the blocks might be within the limits of more than one polygon or set of grids. If **Locate** is used with all polygons simultaneously, then there can be only one fraction field. Any block which falls within more than one polygon will be assigned a value which is the sum of the fractions from each polygon. Therefore, it is possible to end up with a fraction greater than one. This can be avoided by having a separate fraction field for each polygon and executing **Locate** multiple times, once for each polygon.

Now the fraction field can be used to limit gold estimation to only those blocks with a positive fraction. Execute **tbEdit**, and enter the **Fields** for the block fraction and gold assay. Select **Browse** followed by **Assign**. **Assign** allows the value of a field to be changed according to its own value or the value of another field. In this example, for all block fractions which are less than, say, .01, the gold assay will be changed to a null value.

FOR RECORDS MATCHING
FIELD: <fraction> RELATION: < >= > VALUE: <.01>
ASSIGN
FIELD: <gold> RELATION: < = > VALUE: < >

Note that nothing is entered under ASSIGN VALUE, i.e. a null value. A common mistake is to enter a value of zero when there actually shouldn't be any value at all.

Once the fraction has been assign to all blocks, the total volume of the deposit can be estimated and used in the calculation of mine life and production rates as a first step towards mine planning. Before continuing, new fields have to be defined in the block table to hold volume, tonnage and total metal. These will all be calculated fields based on the fraction of the block in the ore zone and the dimensions of the block. If the block dimensions are fixed, then the volume of ore in the block is simply defined as the fixed volume times the fraction of the block in the ore zone. For variable blocks, the default fields for row, column and level size can be used. In RPN the equation for ore volume would be: name_rsz name_csz * name_lsz * fraction. Multiplying the volume field by a fixed tonnage factor or a estimated specific gravity yields the tons of ore, and if the assay is in oz/st, the tonnage field times the estimate for the assay yields the ounces in the block. To get the total volume, tons and ounces execute **Report** ⇒ **Setup** ⇒ **Page Layout** and for 'FIELD VALUES DISPLAYED?' enter 'no' so that only the totals will be written to the report file. Enter the volume, tonnage and ounces fields followed by a '(t)' to get the totals.

Summary of procedure: Example 5.12
Initial data requirements: a block model with records (cell and block tables have no records prior to estimation or use of **Define** ⇒ **tbEdit** ⇒ **Tables** ⇒ **Insert records**) and a coincident cell table with estimated top and bottom structures that bound the ore blocks that are to be located.

Procedure:
1. **Create** a field in the block table <oablk> to contain the fraction of the block between the bounding surfaces <oafrac> that is a real number displaying to two decimal places.
2. Locate blocks within ore zone OA **Modelling** ⇒ **Locate** ⇒ **Setup**
   2.1 Enter the coordinate fields for the block model ⇒ **Fields**
   X: <oablk_xc> Y: <oablk_yc> Z: <oablk_zc>
   2.2 Specify limiting surfaces ⇒ **Surface**

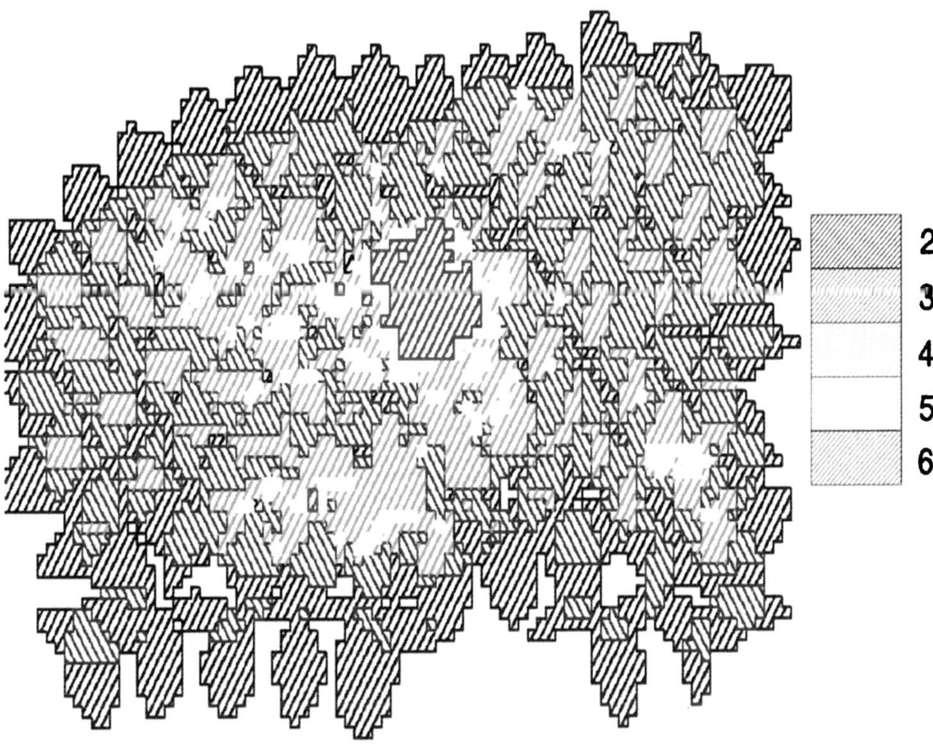

Figure 5.12. Color cell posting of number of data points used for estimation of the OA top structure using a search radius of 200'.

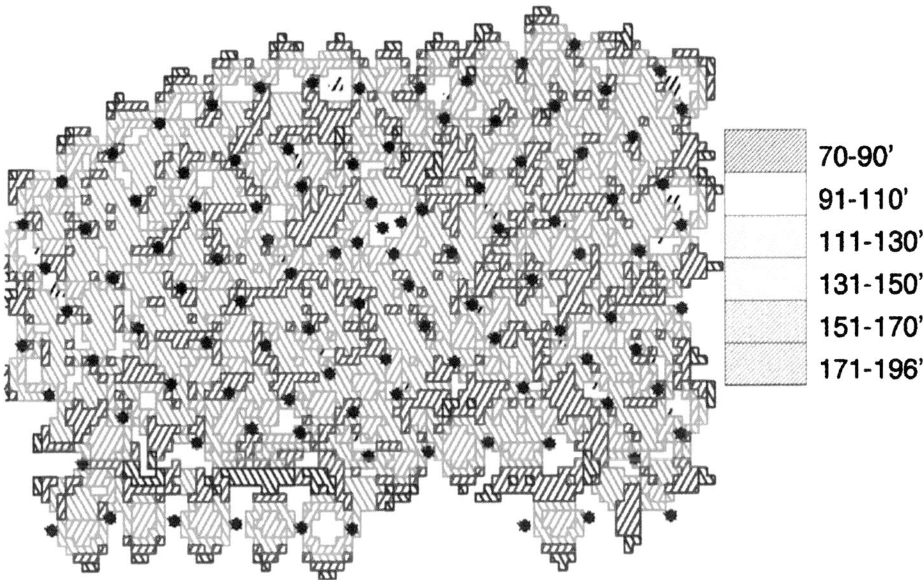

Figure 5.13. Colored cell posting of the average distance of estimation using inverse distance squared with a search radius of 200' and the sample data.

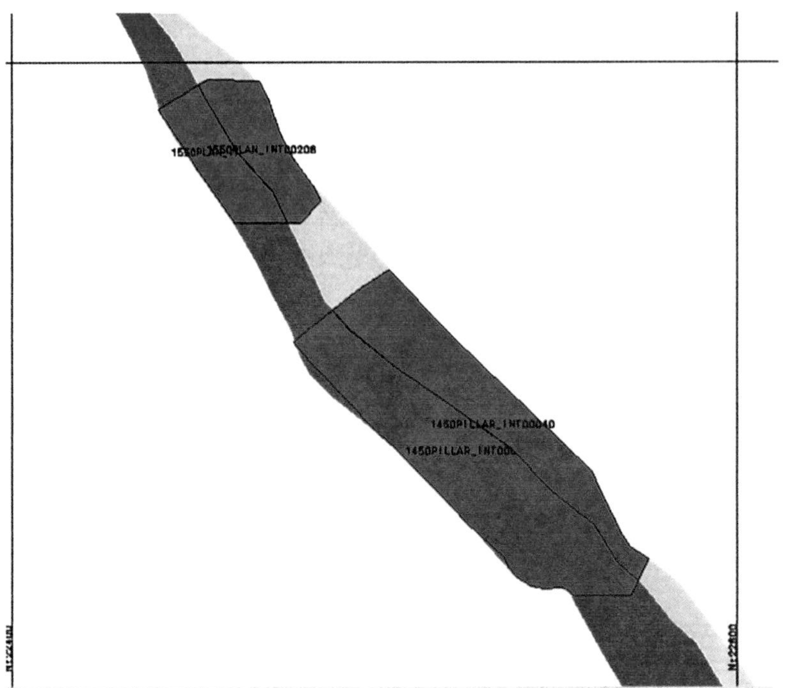

Figure 5.47. Side view (normal to strike) of volumes of intersection including ore zones.

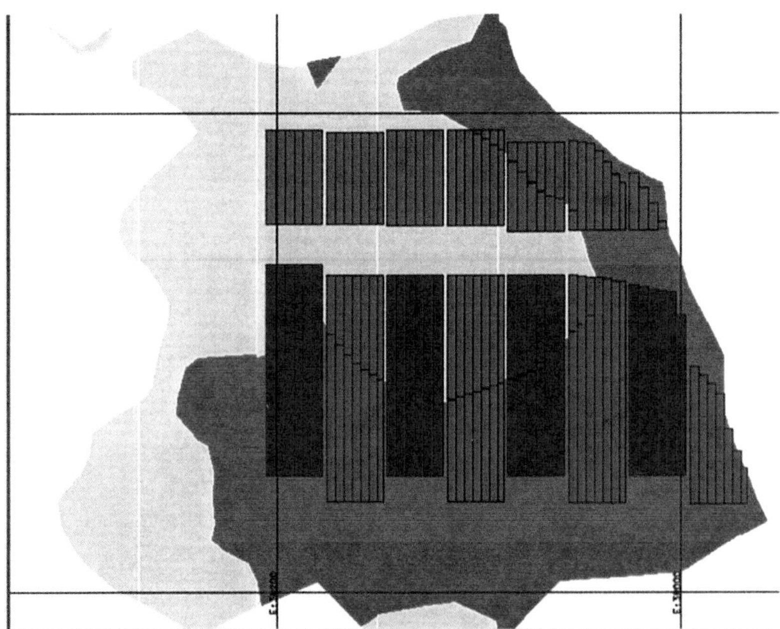
Figure 5.48. Front view (in the plane of the deposit looking down) of the stope-based volumes of intersection including ore zones.

TOP ELEVATION: <oat_c> BOTTOM ELEVATION: <oab_c>
2.3 **Assignment** of values to blocks located in the polygon/surfaces
Field: <oafrac> SET TO: = POLYGON/VERTICAL FRACTION
⇒ **eXit**
⇒ **Locate**
3. Eliminate estimated values for blocks not located within ore zone OA **Techbase** ⇒ **tbEdit**
3.1 Identify **Fields**
FIELD AND TABLE LIST: <auoablk oafrac>
3.2 Assign null values to unlocated blocks **Browse** ⇒ **Assign**
FOR RECORDS MATCHING FIELD: <oafrac> RELATION: <<=> VALUE: <.01>
ASSIGN FIELD: <auoablk> RELATION: <=> VALUE: <>
Repeat for all estimated block table fields used for calculating reserves.

Example 5.13: Calculating volumes of intersection using Lynx
As of Version 2.52 of Techbase, 'volumes' of intersection are actually accomplished by intersection of a 3D grid, or block model, with 2D polygons and surfaces. Lynx uses a potentially more accurate numeric volumes of integration approach in which volume components representing the intersection of two volumes are generated and added to the active volume model. Primary and secondary volume components are selected for intersection in which the primary components might be a series of stopes and the secondary components are the ore zone. These primary and secondary volume models might be defined on differing midplane orientations, so the orientation of the primary components is used for the definition of the components representing the volume of intersection.

The numeric approximation of the intersection volume parallels the algorithm of integration by slices that is used for estimating component volumes and geologic reserves. The intersection volume component's thicknesses are based on an intersection increment specified by the user. A thinner increment results in more components being generated and potentially greater accuracy. The resulting components will be approximately the thickness of the increment value (since the total length of the volumes' intersection normal to the primary components' midplane is unlikely to be exactly divisible by the increment) and will be of constant cross-section, i.e. the fore, back and midplane of each component will be identical. Following their generation, the resulting intersection components are available for further manipulation as per any other volume component. They can be displayed in background, modified by linear interpolation for improved realism, and used for calculating volumes and mining reserves.

From the main menu use **File** ⇒ **Volume Data** ⇒ **Volume Select** to make an existing volume model active. This volume will contain the volumes of intersection and thus could be a new volume model used strictly for this purpose, for instance a volume model that will be used only for stope and geology model intersections.

Once a volume model is active, select **Volumes** ⇒ **Intersection Volumes** bringing up the Volume Intersection entry form (see Fig. 5.45). Select **Primary Volume Data** and enter the volume model that is to be used both for intersection and as the basis for the definition of the orientation of the volumes of intersection. Then select the **Secondary Volume**. The INTERSECTION INCREMENT is the preferred spacing

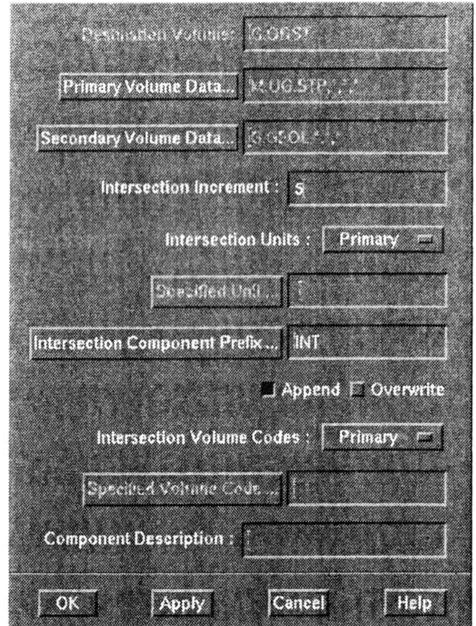

Figure 5.45. Volume Intersection entry form.

of the intersection components and has a minimum spacing of .1 m or ft depending on the project definition.

The naming of components is based on the unit names of the volume model selected under INTERSECTION UNITS, in this case <primary> and an INTERSECTION COMPONENT PREFIX. In this example the primary volume is M.UG.STP which contains the stopes. The prefix is INT and the code is arbitarily selected as 7. Individual components will be generated parallel to the midplane of the stopes and numbered 1 to however many approximately 5 m thick components are needed to cover the depth of the stope. In the case of M.UG.STP's component 1450PILLAR, the components' names would start with 1450PILLAR, INT00001, 7 as shown in Figure 5.46. Later, when reporting mining reserves and production schedules, wise selection

File Search Help

LYNX : Volume Data Report Mon Feb 10 19:16:36 1997
Active Project : LYNXMINE User : martin

VOLUME: G.ORST *,*,*
DESCRIPTION: intersection of G.OR & stopes

UNIT	COMPONENT	CODE	FORE-THICKNESS	BACK-THICKNESS	NORTH	EAST	ELEVATION	AZIMUTH	INCLINATI
1450PILLAR	INT00001	7	1.25	1.25	22300.00	37908.75	1350.00	0.0	0
1450PILLAR	INT00002	7	2.12	2.12	22300.00	37933.38	1350.00	0.0	0
1450PILLAR	INT00003	7	2.12	2.12	22300.00	37929.12	1350.00	0.0	0
1450PILLAR	INT00004	7	2.12	2.12	22300.00	37924.88	1350.00	0.0	0
1450PILLAR	INT00005	7	2.12	2.12	22300.00	37920.62	1350.00	0.0	0
1450PILLAR	INT00006	7	2.12	2.12	22300.00	37916.38	1350.00	0.0	0
1450PILLAR	INT00007	7	2.12	2.12	22300.00	37912.12	1350.00	0.0	0
1450PILLAR	INT00008	7	1.25	1.25	22300.00	37908.75	1350.00	0.0	0
1450PILLAR	INT00009	7	2.12	2.12	22300.00	37933.38	1350.00	0.0	0
1450PILLAR	INT00010	7	2.12	2.12	22300.00	37929.12	1350.00	0.0	0
1450PILLAR	INT00011	7	2.12	2.12	22300.00	37924.88	1350.00	0.0	0
1450PILLAR	INT00012	7	2.12	2.12	22300.00	37920.62	1350.00	0.0	0
1450PILLAR	INT00013	7	2.12	2.12	22300.00	37916.38	1350.00	0.0	0

Figure 5.46. Volume Report of volumes of intersection using stope definitions as the primary volume model and 5 m increments.

of the primary volume and naming conventions can be used to sort out pillar from stope volumes.

The volumes of intersection can also be displayed in background. The intersection between the stopes and ore zones is shown in side view in Figure 5.47 and in the plane of the deposit in Figure 5.48. Note that when viewing the volumes that the intersection components will not necessarily correspond exactly to the mine workings. In this example, reporting of volumes and geologic reserves for the stope intersection volume components can only be used for ore tonnages, not for total development tonnage or dilution. To get these figures the volumes of the mine workings must be compared against the intersection volumes.

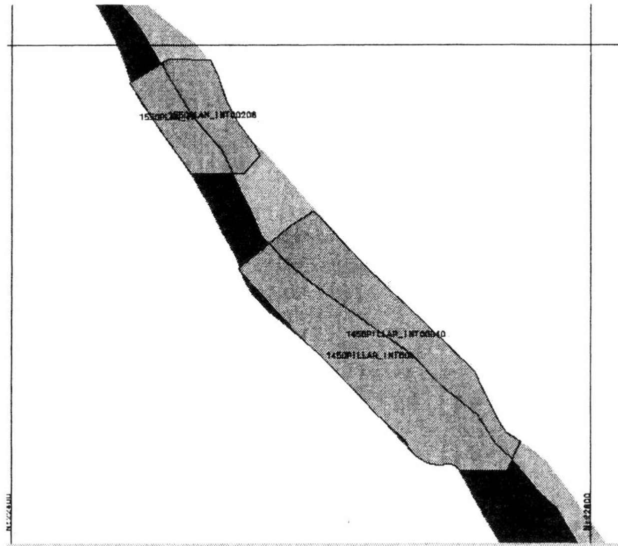

Figure 5.47. Side view (normal to strike) of volumes of intersection including ore zones. (Figure in colour, see opposite page 215.)

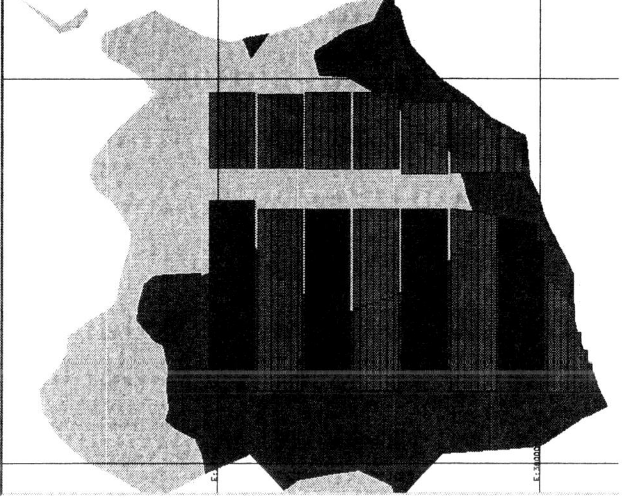

Figure 5.48. Front view (in the plane of the deposit looking down) of the stope-based volumes of intersection including ore zones. (Figure in colour, see opposite page 215.)

Summary of procedure: Example 5.13

Initial data requirements: intersecting volume models, in this case stopes and pillars versus the ore zone (note that components from the same model cannot be intersected).

Procedure:

1. Repeat Step 1 of Example 5.12 to make a volume active, but this time use **File ⇒ Volume Data ⇒ Volume Create** to create a new volume <G.ORST> to only hole intersection volumes.
2. Intersect the volumes **Volumes ⇒ Intersection Volumes** (Volume Intersect entry form)
   ⇒ **Primary Volume Data** (Volume Intersect Primary Data selection list)
   VOLUME: <M.UG.ST> (stopes & pillars)
   UNIT: <*> COMPONENT: <*> CODE: <*>
   ⇒ **Secondary Volume Data** (Volume Intersect Secondary Data selection list)
   VOLUME: <G.OR> (ore zone)
   UNIT: <*> COMPONENT: <*> CODE: <*>
   INTERSECTION INCREMENT: <5>
   INTERSECTION UNITS: <primary>
   INTERSECTION COMPONENT PREFIX: <int>
   INTERSECTION VOLUME CODES: <7>
   ⇒ **OK**
3. Use **Background ⇒ Volumes** to display the intersection components with the primary and secondary source volumes. Either use hatch fill or place the intersection components last on the list to avoid hiding of the volume of intersection behind the source volumes.

CHAPTER 6

Open pit mine design

Open pit configurations are based on mining blocks which are to be removed and on the addition of haul roads that provide access to the mining blocks. At the feasibility stage of mine design, the pit limits are roughly based on algorithms which determine which blocks should be mined in order to maximize the value of the project subject to constraints on slope angles, bench widths and minimum equipment operating radius in the pit bottom. The general procedure for pit design is as follows:

Pit generation algorithms work directly on the three-dimensional matrix of mining blocks by determining which blocks are to be mined. The block model is necessarily rectangular and is not well suited to representing the surface topography which is best modeled as a two-dimensional grid. The pit which is expanded through the block model must be intersected with the surface topography in order to locate the pit crest line and the tonnage of ore and waste in those blocks which are not completely covered. Therefore, before proceeding with a mining block-based pit-generating algorithm, a grid of topography must be available.

Chapter 5 covered the estimation of assay values into block models but not the calculation of the net mining value of the blocks. The decision to mine a block is based upon the net value of the block and of all other blocks which would have to be removed to make the block accessible for mining. Before the net value can be assigned, a cost analysis has to be conducted to determine a mining block's gross value as a function of all recovered minerals, total mining and milling costs as a function of position, and the cutoff grade or ore classification.

Once a block model has been generated that includes an estimate of the net value of each block, one of several approaches to pit optimization can be used. Pit expansion most closely approximates the traditional approach to surface mine design. A pit bottom is defined along with slope angles and bench widths and then the pit bottom is expanded to the surface topography. This pit can then be modified by hand to include pushbacks for haul roads and other necessary modifications. The pit limit is then intersected on the mining block model to determine the net value of the pit. Finding a better valued pit is a matter of trial-and-error. The most common approach to pit design is to first use an algorithm which selects the blocks to be mined in order to produce an optimal or near optimal pit in terms of maximizing net value while conforming to geometric mining constraints. Since these algorithms do not accommodate the full complexity of mine operations and are based on a coarse grid of mining blocks, the resulting pit limits have to be displayed and modified to include haul roads and other pit expansions. Typically, the final modified pit will have a substantially higher stripping ratio than the 'optimal' pit.

Once a final pit has been produced, it is saved in the form of either a grid that represents the ultimate pit 'skin' or as a volume model in which each bench may be treated as a component. This final pit can then be intersected with the block model to determine metal recovered and tonnages of ore and waste. Pits can be generated based on annual production requirements with each year's pit using the previous year's pit limits as a starting point. In this manner, annual mine plans can be used to determine cashflow to establish a NPV or ROR for the project.

Many of the parameters used in determining the final pit should be subjected to a sensitivity analysis, including factors such as forecasts of metal prices, international rates of exchange, slope angles and mining costs. The ultimate pit should be generated for the more uncertain parameters using best case/worst case combinations of values. Parameters for which the project is sensitive will result in changes in pit size, production, life and profitability. A project is of relatively low risk when the ultimate pit remains unchanged during a rigorous sensitivity analysis.

A comprehensive prefeasibility analysis would conclude with a cashflow analysis based on the original net value calculations and resulting annual pits, but it must be recognized that pit optimization based on net value mining block assignments which don't account for the sequence or time in which the blocks are mined is far from perfect. One of the most glaring flaws of the current state of the art in open pit mine design is the manner in which the net value of mining blocks is assigned, i.e. the cost of mining a block has to be assigned before a mine plan or production schedule is available. Thus, the resulting pit limits are of questionable reliability. For this reason, a full feasibility analysis should include detailed scheduling of production so that the order of accessing mining blocks can be determined and used to recalculate mining costs for each block. In this case, the annual pits from the initial pit optimization are used only as a template and more specific net block values can be determined. These schedule-based net values can then be used as input to the pit optimizer to generate a new optimal pit which can be compared with the original. If the discrepancy between the two pits is significant, the new pit can be used as the basis of a second cycle of optimization and this cyclic process of pit optimization can be continued until convergence.

In summary, the general procedure for the prefeasibility stage of open pit design is: (1) gridding of pre-mining topography; (2) assigning net block values; (3) generating annual 'optimal' pits based on the block model values; (4) modifying the optimal pits to account for operational limitations including haul roads; (5) converting the pit contours into a grid or volume model; and (6) intersecting the pit model with the mining block model to determine volumes and revenues over time. More detailed feasibility studies should also include: (7) a sensitivity analysis; and (8) top-down scheduling of production as part of cyclic cost analysis and true pit optimization. The remainder of this chapter will cover each of these topics in greater detail and present examples of their application.

6.1 GEOLOGICAL VERSUS MINING BLOCK MODELS

Mining block models are based on a rectangular array of mining blocks which are individually defined by their dimensions and location. The definition of the mining block model doesn't necessarily coincide with the geologic block model. The geologic blocks are sized to suit the spacing of the data, while the overall extent of the model needs to encompass the dimensions of the orebody. Since the purpose of the geologic block model is to hold estimated values such as assays, the dimensions of the blocks are selected to achieve the best resolution without producing an unnecessarily large model. The mining block model must be based on the geologic block model's estimated values, but the extent of the model and the dimensions of individual blocks are selected with entirely different goals. In the case of the mining block model, the blocks represent mining units. The volume of the mining block represents a quantity that is suitable for scheduling purposes and each level in the mining block model corresponds to a bench in the final mine. Obviously, mining blocks based on production-sized units may not be suitable for estimation purposes in which the block dimensions are based solely on the density of the data. Geologic block height is usually set to a fixed bench height, and then estimation is carried out using drillhole composites in which the composite interval is also based on the bench interval. The *xy* dimensions of the geologic block are then based on a reasonable estimation resolution which is then carried over into the mining block's *xy* dimensions. Drillhole data density is at its greatest along *z*. In terms of resolution alone, it would be reasonable to use very shallow geologic blocks which are much narrower than the bench height. On the other hand, there is no reason to retain the geologic block's dimensions in the mining block. Unlike the geologic block model, the mining block's height should vary with the bench height and depth should be some multiple of the berm width. The mining block's length can then be selected based on blasting limitations, or more generally, on a manageable volume of material for short-term scheduling. The extent of the geologic and mining block models may also not coincide. The geologic block model often needs to be no larger than the extent of the orebody. In the case of multiple orebodies wherein the coordinates do not overlap, it makes sense to use a separate block model for each deposit, especially if the drillhole density is substantially different between the deposits. It always is advisable to work with the smallest possible block model without sacrificing statistical resolution. In the case of open pit mining, the *xy* extent of the pit will probably be much greater than for the orebody due to the waste that has to be removed to maintain stable pit slopes, berms and haul roads, but for deep plunging orebodies, the practical depth or the mining block model may be much less than the depth of the geologic model. This argument extends itself to underground projects in which the shaft and access openings will typically be exterior to the ore zones. In general, geologic block models will have smaller block dimensions and more limited extent than mining block models. The smaller extent of the geologic block model is used to limit the computational requirements of working with an excessively large array. Even though the extent of the mining block model may be much greater than for the geologic block model, the array size can be kept within reasonable bounds by the use of larger block dimensions.

Some software packages recognize this divergence between mining and geologic block models by providing a means of converting a geologic block model into a mining block model. Note that the matrix manipulations to convert between models having different extents and block dimensions are relatively simple, as long as the block dimension in one model is a multiple of the other and the origin and extent of the smaller model places it within the extent of the larger, with each set of smaller blocks matching a larger block. When geologic blocks are not coincident with mining blocks, more complex calculations are necessary to determine the intersection of a geologic block on multiple mining blocks. The values that were estimated into the geologic blocks can then be averaged in the mining blocks.

If all else fails, for example if conversion software is not available, or if the blocks of the two models are not coincident, then the values and locations in the geologic block model can be treated as data and used for estimation of the mining block model. Note that estimation using estimated rather that actual values has statistical consequences. Data values are assumed to be known with certainty. If the initial grid is estimated using kriging, then the kriged estimates have an associated estimation variance. For that matter, any estimated value has an associated estimation error, but only kriging provides an estimate of this uncertainty. If an estimated grid is used to estimate another grid, then how is the estimation variance to be defined? A more reasonable approach is to define the mining block model such that the locations of the geologic blocks are coincident with the mining blocks and have dimensions which result in an integer number of geologic blocks fitting within each mining block without overlap. For instance, a $4 \times 4 \times 4$ pattern of geologic blocks having dimensions of $10 \times 10 \times 10$ might be merged into a mining block model having block dimensions of $40 \times 40 \times 40$ such that there are 16 geologic blocks coincident (no overlap) on each mining block. The estimated values in the geologic block model can then be averaged by groups of 16 and assigned to the mining blocks.

Large pits having a high stripping ratio do not have to be based on huge mining block models capable of including waste blocks all the way to the crest of the pit. Some pit generation and gridding software allows the mining block model to be smaller than the pit extent. In this case, the 'missing' blocks which are within the pit limits but exterior to the block model can be treated as generic waste blocks and assigned a fixed value and volume. These missing blocks are then mined based on maintaining slopes.

6.2 INTERSECTING PITS, BLOCK MODELS AND SURFACE TOPOGRAPHY

Since the mining block model (henceforth referred to as the block model) is necessarily rectangular, many of the upper blocks will be 'air' blocks. If the range of elevation in the overlying topography is essentially nil, then this won't be the case, but even if mining commences on a flat plane, the process of mining itself will produce air blocks as the pit progresses downwards. The process of mine planning based on block models must include a gridded surface topography which can be intersected with the block model in order to determine the location of the pit crest and identify the volume of waste, ore or air in each block. The method used for intersecting the block model with topography follows that discussed in the previous section for converting geologic block models to mining blocks when the block dimensions are different. If the cells of the gridded topography are coincident with the blocks in the *xy* dimensions, then the elevations of matching cells can be compared against the under-

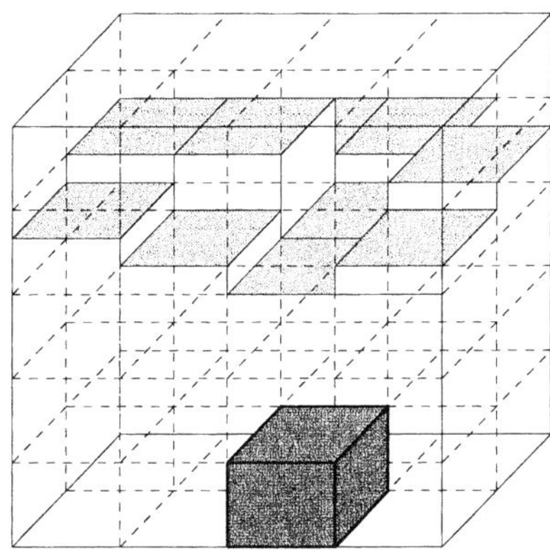

Figure 6.1 Gridded topography intersecting a coincident block model.

lying blocks to determine how much, if any, of the block is air (Fig. 6.1). Also, the *xy* extent of the topography grid must be at least as great as for the block model before this comparison can be made for every column of blocks. Note that it is possible to have a pit whose extent is greater than its supporting block model by using the missing block method. In this case, the gridded topography must be at least as extensive as the projected pit crest. Gridded topography is usually estimated from a spatially dense data set such as digitized contours. Thus, the grid cell dimensions used for topography are commonly smaller than those used at any other stage of estimation. While this small cell size provides contour maps which approximate the detail found in the original map, it is not suitable for intersection with a block model whose *xy* block dimensions are much larger. As a result, topographic maps should be based on either triangulation or fine grid sizes which take advantage of the higher data density which is typical of topography, while another, coarser topographic grid is used for intersecting with block models. Again, the grid used for intersection should match the block model's *xy* definition in extent and block dimensions.

Example 6.1: Topographic grid/mining block model definition in Techbase
Consider the Boland Banya project in which topography was gridded based on digitized data.

The average sample spacing can be determined using any of the weighted averaging estimation methods by using the jackknife method and setting the maximum number of samples used for estimation to one. Jackknifing will then determine the distance from the data point to the nearest data value which can be saved in a field in the topography flat table for which the average value can be determined using **Summary**. Using all 855 topographic data values, the average distance was found to be 54 ft, but this includes the densely sampled area of the volcanic neck that has been noted in earlier examples. Excluding this area, the average distance between samples is 72 ft. In selecting a reasonable cell dimension, the one third the average sample spacing rule yields a cell size of about 25 ft on a side. Note that the average distance for the area about the neck is less than 24 ft, which would lead to a cell size in this area of 8 ft on a side. In the interest of computational efficiency, a cell size of 30 ft on a side was used as part of the following grid definition:

LOWER LEFT X-COORD: 10000 COLUMN SIZE: 30 NUMBER: 91
Y-COORD: 20000 ROW SIZE: 30 NUMBER: 79
BASELINE AZIMUTH: 90

The mining block model was sized to be 50 ft on a side and to have the same origin:

LOWER LEFT X-COORD: 10000 COLUMN SIZE: 50 NUMBER: 52
Y-COORD: 20000 ROW SIZE: 50 NUMBER: 46
TOP Z-COORD: 5650 LEVEL SIZE: 50 NUMBER: 9
BASELINE AZIMUTH: 90

These two models do not coincide, although both are appropriately defined for their tasks. Another grid model needs to be defined to hold estimates of topography whose definition matches the *x* and *y* definition of the block model.

Summary of procedure: Example 6.1
Initial data requirements: The data that is to be estimated into the grids so that the cell and block dimensions can be determined from the data spacing.

Procedure:
1. Find the average distance for topographic data.
   1.1 Create the distance field **Techbase ⇒ Define ⇒ Fields ⇒ autoTable**
   TABLE: <topo>
   ⇒ **Create**
   FIELD NAME: <dave> TYPE: <real> CLASS: <actual>
   1.2 Use inverse distance to calculate the nearest neighbor data distance as per Exmaple 5.4 Step 2 **Modelling ⇒ Inverse ⇒ Inverse ⇒ Fields**
   DATA POINTS RESULTS
   VALUE: <elevation> VALUE: <z_jID>
   X: <easting> X: <easting>
   Y: <northing> Y: <northing>
   ⇒ **Search**
   MAX SAMPLES: <1> MIN SAMPLES: <1> LENGTH: <1000>
   ⇒ **Parameters**
   INVERSE DISTANCE POWER: <1> JACKKNIFE: <y> DISTANCE: <dave>
   1.3 Find average distance **Statistics ⇒ Summary ⇒ Field name**
   FIELD: <dave>
   ⇒ **eXit**
   ⇒ **Summary stats** (mean = 54, std. dev. = 37, max. = 190)
   1.4 Find average distance excluding volcanic neck as per Example 5.11 Step 1.1.
   ⇒ **Summary stats** (mean = 70, std. dev. = 32, max. = 190: this suggests cell dimensons of 20 to 35 ft)
2. Repeat Steps 1.1-1.3 using the collars table to determine the average horizontal hole data spacing (mean 138, std. dev. = 36, min. = 72, max. = 240: this suggest *xy* block dimensions of 30 to 50 ft).
3. Based on the results of Step 1.4 define the block table **Techbase ⇒ Define ⇒ Table ⇒ Create** (note that bock dimensions in this example are based on computational efficiency and drillhole spacing rather than topographic density)
   TABLE: <pit> TYPE: <block>
   BASELINE AZIMUTH: <90> (default)
   LOWER LEFT X-COORD: <9900> COLUMN SIZE: <50> NUMBER: <53>
   Y-COORD: <20410> ROW SIZE: <50> NUMBER: <33>
   TOP Z-COORD: <5630> LEVEL SIZE: <30> NUMBER: <13>

4. Create a cell table having coincident cells as per Example 4.4 Step 1, but with the same coordinate origin, row and column size and extent as in Step 3 above.

6.3 CALCULATING MINING BLOCK VALUES

Mining revenues are based on the calculation of net block values. When the mining block model is intersected with the grid or volume model of the mine openings, the volume and tonnage of ore and waste, mineral recovery and net block value within the opening can be calculated and used to determine the economic feasibility of the project. Ore and waste volume and tonnage is determined from the block dimension and the fraction of the block intersected by the pit. When a variable block model is used, the fields containing the row, column and level dimension are used to calculate volume. Otherwise, the block volume is fixed and a constant can be used for the volume in subsequent calculations. Except in the case of disseminated deposits, orebodies are delineated either by surfaces or volume models. Thus, the calculation of volumes and tonnages is based on the intersection, for instance, of the volume model and the pit surface grid with the block model. Different methods can be used to determine the volume of intersection. A common approach would be to determine the fraction of each block that is within the orebody, calculate the volume of ore from the block dimensions and the fraction of the block that is within the orebody, assign or estimate a tonnage factor, and then calculate the tonnage of ore. The waste tonnage is based on the remaining volume in the block and its own tonnage factor. Once the volume and tonnage of ore is determined, the mineral contained in the block can be calculated based on the estimated assay. Gross block value is then simply the mineral content of the block adjusted for mining and processing recovery.

Up to this point, determination of the net block value is relatively straightforward. The only source of uncertainty in calculating gross block value is the uncertainty associated with the estimate of the mineral content and the boundary of the orebody. Unfortunately, the procedure used to assign mining and processing costs to blocks is somewhat flawed in that costs must be assigned to blocks prior to having a mine design or production schedule. This limits the validity of the cost assignments since the procedure and time of mining is unknown. Typically, a fixed mining cost is assigned to all blocks regardless of position. A cutoff to distinguish between ore and waste is used to include the cost of processing the ore blocks. There is not much room for flexibility in assigning costs in this format. Haulage costs might be varied by adding an incremental cost of haulage out of the pit which is a function of the block level so that deeper blocks are assigned a slightly higher mining cost, but there is no way of estimating the actual transportation costs prior to detailed scheduling. Finally, mining and processing costs are subtracted from the gross block value to determine the net value of the block. This net block value is then used as the basis for a pit optimization algorithm and cashflow analysis.

Net value calculations vary depending on commodity type. Coal, industrial minerals and other bulk commodities often only require net value calculations based on tonnage of ore. Cutoffs that discriminate between various ore/waste classes and differing mining/processing costs are often not applicable to bulk commodities. When the ore must be processed in order to recover mineral values and when cutoff grades are used as a criterion for the level of processing involved (delivery to waste dump, heap leach pad, or mill), net value computations can be much more complex, requiring IF THEN type statements as part of the net value calculation in order to determine which set of costs is applicable to a mining block. Note that the use of a gross value or grade cutoff criterion for selection of mining and processing costs is itself a fallacious approach which by strength of tradition continues to be accepted. Using a fixed cutoff grade as an ore/waste selection criterion in circumstances where the cost of mining a block is a function of the sequence in which covering blocks are removed ignores the interaction of positively valued mining blocks: a high valued mining block may by itself justify the removal of waste that at least partially uncovers a lower valued block, which in itself would not be worth mining, but once overburden has been removed to mine a neighboring block, the original cutoff criterion should no longer apply since the cost of mining the lower valued block is now lower than originally assumed. This argument will be explored more fully later in this chapter as part of the discussion of cone mining and optimal pit generation algorithms.

To illustrate the previous discussion, two examples of net value assignments will be used: one based on a steeply dipping coal property and another for the Boland Banya deposit discussed previously.

Example 6.2: Assigning net block value as a calculated field for a coal deposit using Techbase

Consider a coal deposit consisting of two steeply dipping wings of a multi-seam anticline, two seams of which will be used in this example. This type of steep folded structure is very awkward to model using 2D grids and is best represented as a volume model. Still, Techbase's **Locate** and **polyEdit** routines are capable of providing a reasonable estimate of the intersection of these seams with a block model. The procedure used for modelling this deposit will be briefly reviewed before focussing on the calculation of net block values.

The data for this project consists of a set of planview hand drawn sections at 10-m intervals and a topographic map. Drillhole data were also available, but were sparse and in any case unsuitable for gridding due to the steep folded nature of the deposit. The topography and the sections were digitized to provide a basis for structural modelling. This is not an uncommon approach for projects wherein the geology was delineated prior to the use of geological modelling software. Hand-drawn sections often provide an excellent geologic interpretation on which to base computer-generated volume models and volumetric calculations. The digitized ASCII files of the coal seams were cleaned up and converted into polygon files using **polyEdit**. Next a variable block model was defined to cover the extent of the deposit. Smaller blocks were used where the seams were the steepest, and larger blocks where there were lower angle areas and areas of waste. **Locate** (see Example 5.12) using the polygons and the gridded surface topography (also defined to have the same variable *xy* dimensions as the block model), was used to identify the fraction of the volume of each block that was within the limits of the polygons. Eighteen fraction fields were defined, one for each of the coal seams and each of the nine levels of the block model.

Each polygon and associated fraction field corresponded to the intersection area of the seam on the block model at a level in the block model, so a filter on the block level was used so that only blocks on the elevation corresponding to the polygon would be assigned fraction values. For each level in the block model, these fraction fields were used to calculate the volume of coal and waste. For instance, if the name of the variable block model was 'vblock' and the fraction fields for level one were 'frac11' and 'frac12' for seams one and two, respectively, then the volume of coal on level one of the block model (corresponding to the uppermost bench) could be placed into a calculated field 'vcoal1' with the equation:

frac11 frac12 + vblock_*xcz* vblock_*ycz* * vblock_*lcz* * *

Note that **tbCalc** and calculated fields use Reverse Polish Notation (RPN) in which values (constants or fields) are placed into memory registers and then followed with an operator or function call. In this example, frac11 is placed into register 1 and frac12 into register 2. This is following by a '+' operator that sums the two registers. The total fraction of coal in the block is now in register 1. vblock_*xcz* and vblock_*ycz* are next placed in registers 2 and 3 and then multiplied together; the product going into register 2 (note that these operators act on two values, or registers, at a time, clear those two values and place the result in the next free register, which in this case is 2). vblock_*lcz* (block height) is then entered into register 3 and multiplied with register 2 which is the block area. The resulting total block volume is then saved into register 2. Finally, the product of registers 1 (total fraction of coal) and 2 is formed and placed into the calculated field vcoal1. Similarly, the volume of waste can be calculated by replacing the equation entries 'frac11 frac12 +' used for coal volume with '1 frac11 frac12 + –'. The results of these block volume calculations are shown in Figure 6.2. To get the tonnage, multiply by the appropriate tonnage factor. Since the fraction fields were generated in **Locate** with a filter on for the block level, the volume and tonnage values will be positive only for blocks which intersected the coal seam at that level. Elsewhere on the level these values will be zero, and for other levels they will be null. Thus, the total tonnages can be calculated into a new field which is the sum of all of the block level tonnage fields.

To get the net coal value in the block, multiply the tonnage by a dollar per ton constant and subtract from that the block's mining cost which can also include any cleaning costs and royalties. Costs in a block model can be varied by location and any other estimated variable and need not be constant. In this example, mining cost will be varied by an incremental haulage cost which increases with depth and by a royalty charge which is a function of location. To include the royalty charge, **Locate** will have to have been used to set fraction fields which identify how much of a block is within the limits of a polygon of a lease's boundary. If there are two leases wherein the fraction fields in the block model are flease1 and flease2, then the mining cost could be calculated as:

tonscoal coalcost *	# fixed coal mining/hauling/cleaning cost/block
vblock_lev hcoalcost * +	# incremental coal haulage cost flease1
flease2 + tonscoal * royalty * +	# royalty charge on tonnage in block

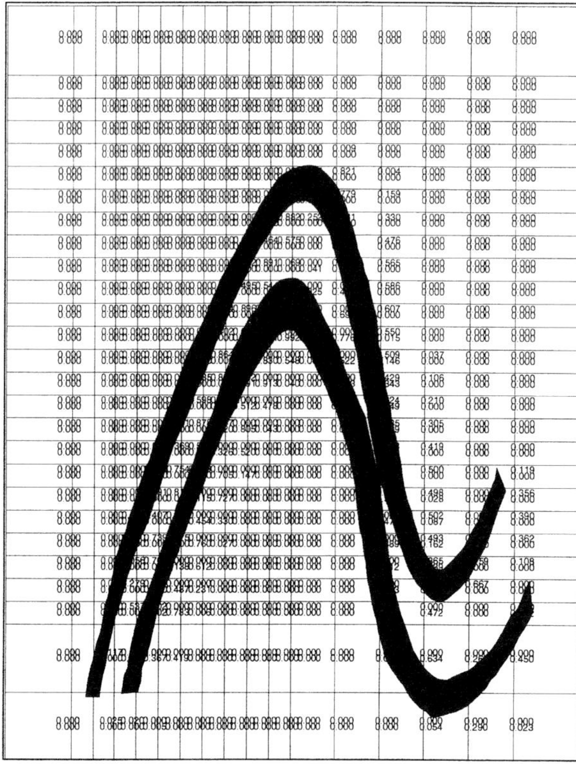

Figure 6.2. Posting of polygons of intersection of a variable block model with polygons representing the limits of two inclined coal seams. Volume fractions of intersection of the blocks with the seams are displayed.

```
volwaste wastcost *           # fixed stripping/hauling cost/block
vblock_lev hwastcost * +      # incremental waste haulage cost
```

The net value of the block would be obtained by subtracting this from the gross coal value, i.e. the tonscoal field times a constant dollars per ton. If the value is a function of the coal quality, then the quality characteristics can be estimated into the block table and an adjusted coal value calculated field can be formulated.

Example 6.3: Using logical operators to assign net values to actual fields in Techbase

In previous examples, calculated fields have been defined as simple functions of other variables. For instance, seam thickness was defined as the difference in elevations from a top and bottom structure in the Boland Banya deposit (dip=0). When the calculation process is more complex than what can be accomplished within a single equation, such as for a inclined deposit or a net block value, the field value has to be defined using **tbCalc**, which incorporates a variety of functions that can be used to make complex field assignments. Unlike the CALCULATED fields whose equations are entered in **Define**, **tbCalc** works with previously defined ACTUAL fields which have not been assigned values. Also, the number of lines available for entering code in **tbCalc** is limited. Therefore, the lengthy computations that are typical of net value assignments can be entered in **tbCalc** via an ASCII file.

The Boland Banya project has two orebodies wherein the gold grades have been separately estimated; OA is the lithologic code for the larger outcropping deposit, OB is deeper. The block estimated gold assay fields are OAgold and OBgold. These fields have only been estimated within the limits of their top and bottom structure grids. Outside of the orebody limits, these assay fields have null values. The fixed mining cost has been determined to be $1.55/ton for either ore or waste, but there are additional costs for ore processing. We will start with a simple example of net value calculations that discriminate between ore and waste.

Before continuing, **Define** an ACTUAL field to be included in the block model, 'netvalue', that will be assigned a net block value in tbCalc. Next, use a text editor to create an ASCII file that contains the RPN calculation steps for netvalue as follows:

```
OAgold OBgold +& 0.00 max  # determine the block's grade (oz./ton)
.90 *                      # metallurgical recovery
.85 *                      # mining recovery
360 *                      # gross block value
1.55 –                     # less mining cost ($/ton)
9.89 –                     # less processing cost
0.25 –                     # less G & A
–1.55 max                  # the block is waste if the $/ton value is less than
                           # the waste removal cost
1284.4 * 2 *               # calculate total block tonnage and calculate net
                             block value
= netvalue                 # store results in block table field netvalue
```

The function 'max' is used to select the block grade and to assign costs based on the block being either ore or waste. The +& operator sums the assays treating any null value as zero. Otherwise, any operation involving a null value equates to a null value. No block is in both OA and OB, and most blocks are in neither. The calculation step OAgold OBgold +& 0.00 max will place a positive gold assay in register 1 if the block has one or will set the assay stored in register 1 to zero (zero is greater than null). If the block grade was zero or too low to produce a positive value after adjusting for mining and milling recovery and subtracting mining, milling and G & A costs from the gross recovered metal value, then the material value will be less than the fixed waste removal cost of $1.55/ton, in which case the block should be treated as waste and assigned that cost.

From the Techbase main menu select **Techbase ⇒ tbCalc ⇒ Setup ⇒ Equation** and under CALCULATION STEPS enter the name of the file containing the netvalue assignment. If the ASCII file was stored as netval_pit.calc, then enter (f,netval_pit.calc). Select **Calculate** to place net block values into the block table. Note that calculated fields wherein the CLASS is <actual> differ from fields wherein the CLASS is <calculated> in that they are assigned values using **tbCalc**. Conversely, a field wherein the CLASS is <calculated> has its equation defined at the time it is created. Both classes of fields are calculated based on other variables, but a field wherein the CLASS is <calculated> will have its value updated automatically when any of the fields it is a function of changes value whereas an actual field whose value is calculated must be updated by repeating its assignment in **tbCalc**.

Summary of procedure: Example 6.3
Initial data requirements: mining and metallurgical processing costs and recoveries, ore and waste densities and estimated grades in a mining block model.

Procedure:
1. Create a new field <netval_pit.calc> in the block model <pit> of TYPE: <real> and CLASS: <actual>.
2. Use a text editor to create an ASCII file <net_val.pit> containing RPN calculations steps for the net mining value for the mining blocks as given above.
3. From the main menu **Techbase** ⇒ **tbCalc** ⇒ **Setup** ⇒ **Equations**
   CALCULATION STEPS: <f,netval_pit.calc>
   ⇒ **eXit**
   ⇒ **Calculate**

Example 6.4: Calculating net block values using equivalent metal grades in Lynx
The previous example assumes that only gold will be recovered and sold. It is much more common for metallic deposits to consist of a number of minerals which can be recovered at a profit. Thus, the net block value has be based on an equivalent metal grade which combines the recovered grades. The Lynx project TUTORIAL has three recovered metals whose values have been estimated into 3D grid GEORES: copper, zinc and silver in order of descending importance. These three assays are reported in either percent or oz/tonne, so using the material density each assay has to be converted to \$/cu. m, referred to as the \$Factor in Lynx's net values calculations. This \$Factor is defined generally as the product of the assay in unit weight (%Cu/100, oz/tonne, etc), the value/unit weight (\$/lb, \$/oz, etc.), the density (tons/cu. m) and the recovery (percent). For example, in this example copper assay has been estimated into the block model as %Cu. The copper price is taken at \$1.00/lb. There are 2000 lb/ton. With a material density of 3.17 tons/cu. m and a recovery of 85% the \$Factor for copper is calculated as

$$\$\text{Factor Cu} = (\%\text{Cu}/100)(2000\ \text{lb/tonne})(\$1.00/\text{lb})(3.17\ \text{tonnes/cu. m})(.85) = \$65.60/\text{cu. m}$$

Note again that all units must cancel to yield \$/cu. m. The \$Factors for zinc is found as:

$$\$\text{Factor Zn} = (\%\text{Zn}/100)(2000\ \text{lb/tonne})(\$0.54/\text{lb})(3.17\ \text{tonnes/cu. m})(.85) = \$35.42/\text{cu. m}$$

For silver:

$$\$\text{Factor Ag} = (\text{oz/tonne})(\$5.00/\text{oz})(3.17\ \text{tonnes/cu. m})(.85) = \$13.47/\text{cu. m}$$

The average values in the main ore zone (zone7) for Cu, Zn and Ag are 1.3%, 0.71% and 18 oz/tonne, respectively. Thus, the average recovered gross value per cu. m in the main ore zone (code 2) is

$$1.3(65.6) + .71(35.42) + 19(13.47) = \$366{,}41/\text{cu. m}$$

The equivalent metal variable ($Equivalent) is then based on the relative values of the assays. In the TUTORIAL project's database, Cu and Zn are reported as percent and Ag in gm/ton. Ag copper equivalent grade can be used by converting Ag to %/cu. m and then multiplying percent Zn and Ag by the ratio of their market values to Cu. Using equivalent grades substantially simplifies calculating and tracking net values.

From the Lynx main menu select **Surface Eng ⇒ Net Value Generation** bringing up the 3D Grid Net Value entry form as shown in Figure 6.3. Note on the entry form's title the active grid. This should be GEORES or any other 3D Grid which includes estimates for all three (or four) metal grades in both ore zones (use grid reporting for confirmation). In **$Equivalent** enter the grid variable that will be used to store the equivalent metal grade <Cueq> which will be used to calculate the **Net Value Variable** <Net>. Select **Topographic Map** and enter a pre-mining topographic map that has been triangulated (includes a TIN) <TOP>. Toggle **Generate Pit Optimization Files** on. This will cause files to be generated that hold information needed later for pit optimization, specifically a 3D grid of netvalues and a 2D grid of pre-mining elevation which corresponds to the location and extent of the block model.

Under Mineral Value Parameters are entered the $Factor and cut-off grade for each metal to be recovered. This information is given for each ore zone (codes 2 and 3). In all other zones the netvalue calculation will be based on waste. The cutoff value should be selected based on the break-even value of the block. The primary

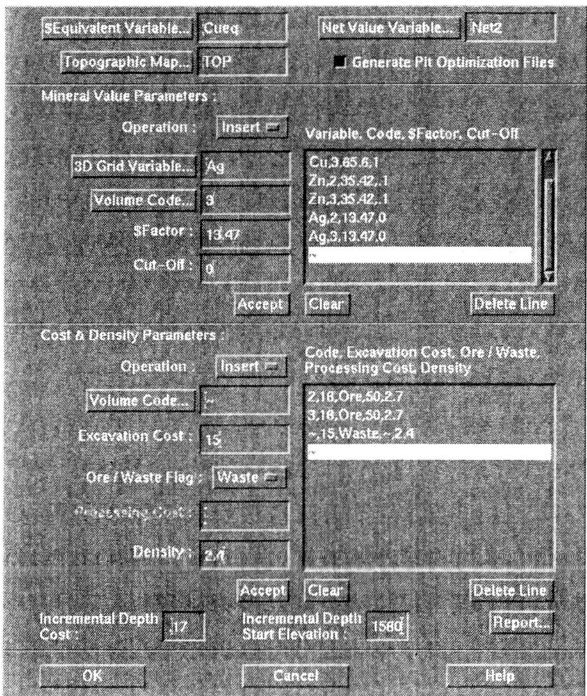

Figure 6.3. 3D Grid (GEORES) net value calculation.

metal value recovered is most important in this calculation and a value of 1% Cu has been used for this example. The cutoff can also be based on copper equivalent grade. Unfortunately, a single cutoff is somewhat meaningless due to the interaction that occurs between block values during the pit optimization process. The important point for now is that selection of too low a cutoff will result in the calculation block net-values based on ore for block's that will in reality be treated as waste. Since ore blocks have significant additional processing costs, these low grade blocks will be assigned much higher negative net values with the result that the optimized pit will be smaller and shallower than would otherwise be the case.

Cost and Density Parameters are entered for each ore zone code and for undefined material (~). The **Excavation Cost** is base on mining only and not haulage or processing. No **Processing Cost** is entered for waste of undefined material. The densities are based on the material specific gravity (g/cc).

The **Incremental Depth Cost** is optional. For an open pit operation, this would be the additional cost per vertical meter of haulage which would have to be approximated based on the truck size, road grade/width and cycle time. For this example a cost of $.17/meter was used. For an underground project, the incremental cost may be irrelevant as a separate item if ore handling is by orepass to a single ore bin at shaft bottom. The **Incremental Depth Starting Elevation** corresponds to the elevation at which the haul road will cross the pit crest. Referring to Figure 5.15, the lowest likely elevation for this is 1580 m. A full set of parameter values is given in Figure 6.3. If the reader is disturbed by the approximate values that are used in netvalue calculation, values which are subsequently used for defining an 'optimal' pit, he should be. All of these values are approximations. Some, such as 3D grid grade estimates have at least a statistical justification, but others, such as haulage cost, are based on parameters that aren't even known at this point, such as haul road design and truck size. This example demonstrates the critical need for cyclic design where net values are dependent on pit design, which is in turn based on pit optimization which uses netvalue calculations. Once the initial cycle of mine design is completed, the procedure should be repeated with cost parameters and cutoffs now based on a preliminary mine design which at least approximates the final form.

Selection of **Report** generates a ASCII file reporting netvalue calculation parameters. Click on **OK** to start net values assignment. In the main menu, **Background 3D Grids** should be used on several sections to confirm net value assignment to GEORES. An example 3D grid display of GEORES is given in Figure 6.4.

Summary of procedure: Example 6.4

Initial data requirements: a mining block model with an equivalent metal grade variable in the project definition <GEORES> and its component estimates for Cu, Zn and Ag and volumes of intersection with the volume model of the ore zones. Additionally, a pre-mining topographic map including a TIN (TOP).

Procedure:

1. Define net value calculation parameters **Surface Eng** ⇒ **Net Value Generation** (3D Grid Net Value entry form).
   a) ⇒ **$Equivalent Variable**... <Cueq>
   ⇒ **Net Value Variable**... <Net>
   ⇒ **Topographic Map**... <TOP>

Generate Pit Optimization Files <on>

b) Mineral Value Parameters:

Operation: <insert>

⇒ **3D Grid Variable**... <Cu>

⇒ **Volume Code**... <2>

$Factor: <65.6>

Cut-off: <1>

Repeat Step 1b for each recovered metal and ore zone (codes 2 and 3) as per Figure 6.3.

c) Cost and Density Parameters:

Operation: <insert>

⇒ **Volume Code**... <2>

Excavation Cost: <18>

Ore/Waste Flag: <Ore>

Processing Cost: <50>

Density: <2.7>

Repeat Step 1c for each ore zone and for undefined material (waste, ~) as per Figure 6.3.

Incremental Depth Cost: <.17>

Incremental Depth Start Elevation: <1580>

Report ⇒ **OK**

2. Use 3D grid display facilities to visually confirm the calculation of net and eq. From the main Version 4 menu **Background** ⇒ **3D Grids** (Background 3D Grid Data Selection entry form). For displays 1 and 2 <on> use the following settings:

3D GRID: <GEORES>

3D GRID VARIABLE: <net> in display 1 and <eq> in display 2

VOLUME CODE: <0>

DISPLAY FORMAT: <value>

DISPLAY LOCATION: <below right> in 1 and <above left>

1000599	1534729	527879	−85813	−85813	−85813	−85813	−85813	−85813	−85813	−85813
75941	1531394	1228069	−72390	−72188	−72188	−72188	−72188	−72188	−72188	−72188
−78583	524458	1663827	983825	−78583	−78583	−78583	−78583	−78583	−78583	−78583
−84938	−85017	1507429	2094723	757228	−84938	−84938	−84938	−84938	−84938	−84938
−91313	−91313	813322	2201239	2145785	825735	−91313	−91313	−91313	−91313	−91313
−97688	−97688	−97788	1151222	2016033	1052959	247461	−97688	−97688	−97688	−97688
−104083	−104083	−104083	−104118	394332	735030	1846283	−104308	−104083	−104083	−104083
−110438	−110438	−110438	−110438	−110511	473843	4423241	888087	−110438	−110438	−110438
−116813	−116813	−116813	−116813	−116813	311498	4590348	3323778	179942	−116813	−116813
−123188	−123188	−123188	−123188	−123188	−37918	3591399	4945265	1973783	−38884	−123188
−129563	−129563	−129563	−129563	−129563	−129563	2821487	4758122	4488263	849046	−12089

Figure 6.4. 3D grid net mining values (code 0) in neighborhood of ore zones.

6.4 PIT EXPANSION METHODS

There are a number of approaches to generating pits. These methods generally fall into one of three categories: conical pit expansion routines, moving cone heuristics and pit optimization. All three methods have their unique advantages and can be used together in pit design. Pit expansion routines most closely resemble the pre-computer practices of open pit mine design and productions scheduling in which a pit bottom would be defined on a series of parallel geologic sections. Overall slopes, possibly including haul roads and benches, would then be used to expand the pit bottom up to the surface. Using a planimeter, the area on the section of ore and waste would be calculated. The section would be assigned a thickness to the midpoint of the adjacent sections to estimate the volume of material, and the sum of all section volumes through the proposed pit would form the basis of the economic analysis of the design's feasibility. The economics of several alternative pits at different depths would be used to select the approximate 'best' pit. This 2D approach to a 3D problem is very limited. There are an infinite number of possible pit bottoms having different widths, depths and lengths, not to mention that the pit bottom can be split among several benches or even between several pits.

Pit expansion routines work on the same the trial-and-error approach to mining, but are based on a 3D expansion of the pit bottom to the surface topography. A typical approach would be to digitize a polygon that represents the expected shape of the pit at a fixed depth that corresponds to the toe of some bench. Pit slopes are then defined along with the surface topography. The pit is then expanded by benches until a crest line is defined by the intersection of the cone of the expanded pit and the topography. With many types of mining operations, a relatively simple expansion routine is all that is required for design and production scheduling. This would be the case for most quarries where production rates are defined by local demand and where the starting point for mining is at the base of the outcrop. Since quarries often mine into a slope, mine design proceeds from the bottom up in a series of pushbacks that progress across the face of the hillside. Pit optimizers and cone miners are entirely unsuited to this mode of production and would yield unpredictable pit bottoms. Even with more geologically complex deposits where pit optimization can be taken advantage of, pit expansion routines are essential for shorter-term production planning and for modifying pits to include features such as haul roads. These expansion routines can be used to modify the pit face with pushbacks that represent the volumes of ore and waste removed during production or the inclusion of haul road segments. Pushbacks and the inclusion of haul roads in the pit will be discussed at the end of this chapter. Production scheduling will be addressed in Chapter 8. For now, pit expansion will be discussed in terms of mine design and of making volumetric calculations.

Example 6.5: Using CAD tools to generate an expanded pit in Techbase
CAD drafting packages form the basis for many operators' and consultants' mine designs and production schedules. In fact, for mining purposes CAD software only lacks the ability to work with 3D grids. Mining software commonly includes CAD software which is enhanced for specialized mining problems such as spiral ramp de-

sign or pit expansion. Mining software packages integrate CAD and grids by providing routines which will intersect mine designs with grids enabling the calculation of volumes of intersection. Still, mining CAD packages cannot compete in terms of performance with commercial CAD software due to the much greater market size and competition that exists in general engineering design. Most engineering firms which use comprehensive mining software packages such as Lynx or Techbase also use a CAD package. Designs, 2D grids or volume models that are generated using the mining software are exported to the CAD package (for instance, as a DXF file) where detailed engineering specifications can be more easily included on the drawings. The inclusion of C programming and 3D graphical ability in advanced CAD software has enabled some mining firms to do all of their mine planning and design without the use of specialized mining software. This option is especially attractive to companies which are heavily committed to a CAD package or feel that none of the available geological modelling software meets their needs. Needless to say, this route involves a heavy investment in programming time and geostatistical expertise, yet it will continue to be an attractive alternative, especially if the technological gap between 3D CAD software and mining software continues to grow.

Techbase's **polyEdit** can be used for creating and editing polygons using familiar CAD polyline commands. **polyEdit** will be used in conjunction with other Techbase facilities to illustrate the basic methods and utilities needed for pit expansion. It should be noted that other Techbase routines are available which automate this example, specifically the routine **opSched** which will be used in later examples.

Consider the steep coal deposit that was introduced in Example 6.2. A pit is to be designed for the mining of both seams on ten meter benches. The toe of the lower bench will be at the outer edge of the outer seam which encompasses three sides of the pit. There is no coal to the south, so the limits of the pit bottom in that direction will be selected to minimize waste removal while maintaining a pit floor which is wide enough to operate trucks and shovels. The surface topography gradually slopes from 80 m in the north to a 40 m saddle near the southern extent of the deposit (Fig. 6.5). The logical choice for a low production pit would be to mine both seams down to an elevation of 40 m with coal and overburden haulage along that contour.

The pit expansion approach will be to digitize a proposed pit bottom along the 40 m bench to mine both seams, and then expand the 40 m toe by 10 m increments to replicate a 45° slope with 10 m bench heights and widths. This will be repeated up to 70 m, which is the highest elevation that this pit will intercept. The result will be a 40 m high irregularly shaped 45° frustum which will have to be intercepted on the surrounding topography. The expanded pit's contours will then be modified to match the surface topography along the pit crest, leaving a level haulage route out of the pit at 40 m. The resulting polygon file represents the post-mining topography and can be treated as a data set for gridding. There will then be two gridded surfaces which can be used for volumetric calculations: the pre-mining and post-mining topography. The two surfaces can then be intercepted on the mining block model to determine volumes of coal and waste and the net value of the pit. A polygon that defines the extent of coal at 40 m will be required before continuing.

Generate a contour map of the topography and then export the contours into a polygon file. Using **Gridcont**, create contour lines wherein the values correspond

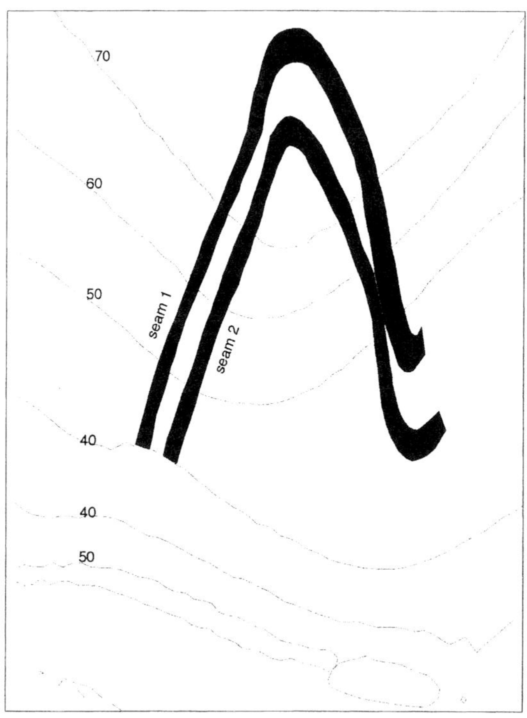

Figure 6.5. Contours of topography exported as polygons and polygons of seam limits at 40 m.

with the elevations that are to be used in the pit. Since the pit is to start at 40 m and have 10 m benches, contour from 40 to 80 m using 10 m intervals. **Write polygons** into a file that will be loaded into **polyEdit** later.

From the main Techbase menu select **Graphics ⇒ polyEdit ⇒ Load ⇒ polygon File** and enter the name of the polygon file that defines the limit of coal at 40 m. Note that this project was based on data digitized from hand drawn horizontal sections. A more common approach would be to generate a horizontal section (using **Poster**) that displayed the intersection of the top and bottom structural grids of the seams and then to display this metafile as background under **background Metafile**. This background metafile could then be used as a template for digitizing in a polygon representing the 40 m pit bottom. For this example, an existing 40 m coal limit polygon will be loaded. Next, modify the polygon so that the 40 m toe follows the outer contact of the exterior seam and then continues across the southern edge of the pit to form an inverted heart with each lobe being wide enough to accommodate a digging face for small trucks and hydraulic shovels. To modify the existing polygon using Techbase's polygon facilities, select **Edit**. Within the Edit menu are a variety of common CAD utilities for manipulating polygons. This example is not intended as a guide to using these utilities. Instead, the general procedure will be outlined. For further details, refer to the *Techbase Users Guide*. There are two options for producing the pit bottom. One is to select **Line ⇒ Modify line ⇒ Create** and enter a polygon ID, which would be the elevation of the pit bottom (40), and use **Add** to trace out the pit using the coal seam limits as a template. While tracing out the pit bottom will certainly work, it is difficult to exactly match an existing line. A better approach would be to use the **Follow** utility so that the pit bottom exactly follows the outer edge of the seam 1 polygon. Once the pit bottom is completed, use **Close line** to form a closed polygon and **eXit** back to the main **Edit** menu and select **Operations** where the pit bottom will be **Expand**ed by 10 m increments. As the pit is expanded, enter the corresponding contour elevation as the polygon ID. Note that expansion routines have trouble with inside corners such as the ^ formed in the base of this pit. As these corners are expanded, loops or other oddities might be formed. Use **Modify line** ⇒ **Delete** to clean up any such problems before continuing with pit expansion to the next higher bench. The final expanded pit will be similar to that shown in Figure 6.6.

Finally, the bench polygons must be merged with the pre-mining topography. In the **Modify line** menu, use **Delete** to delete all vertices on the 70 m bench polygons that extend beyond the topographic contour and to delete any 70 m contour vertices that are within the bench polygons. Switch to the **Operations** menu and **Split** the modified 70 m contour polygon at the gap left by deleting the vertices. Retain the same polygon ID for the two resulting polygons (70). This results in three 70 m polylines. Use **Join** to connect all three. The resulting polygon represents the post-mining 70 m contour. Repeat this procedure for every intersecting contour line and bench polygon which are at the same elevation. This procedure is illustrated in Figure 6.7 and the completed contours are given in Figure 6.8.

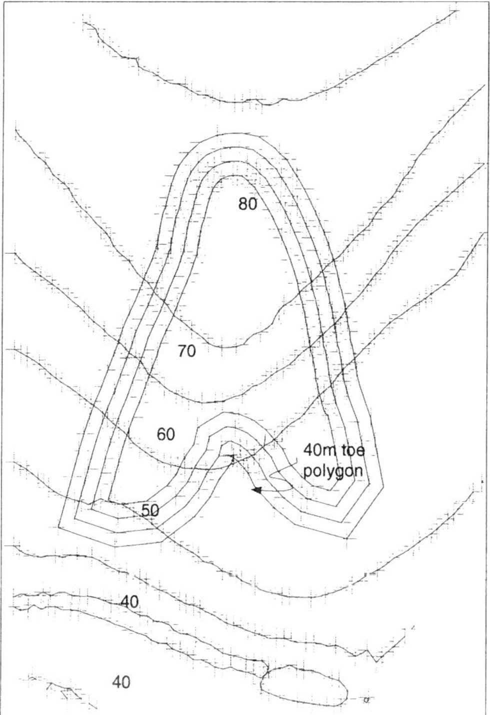

Figure 6.6. Polygons of topography and the expanded pit prior to intersection of benches and topographic contours.

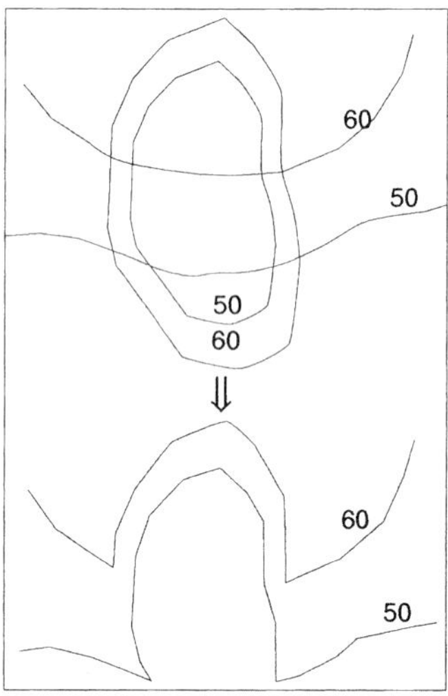

Figure 6.7. Editing of two contour and two expanded pit bench polygons to form the post-mining topography.

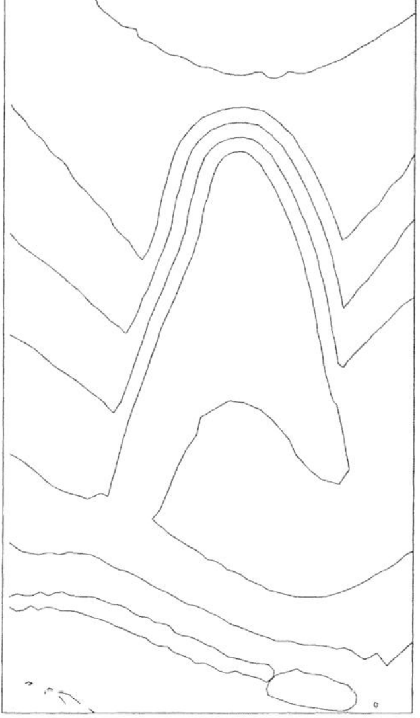

Figure 6.8. Final post-mining contours for a 40 m pit bottom.

The post-mining topography polygons can now be loaded into a Flat table and treated as *xyz* data for gridding. **Define** a new field that will hold the gridded post-mining surface in the cell table that holds the pre-mining topography. The cell table that holds these surfaces should be defined to match the *xy* definition of the block table. Additionally, **Define** a fraction field in the block table that will identify the fraction of each block that is within the pit and a set of calculated fields that are the product of the fraction of the block within the pit and any other block field of interest such as the net value, tons of coal or volume of waste. To hold the final pit contours as point data, **Define** a new flat table and create three new REAL fields to hold the polygon *x*, *y* and ID values. ID will now be treated as the *z* field. **Load** the polygon file into the new flat table's *xyz* fields.

Since the 'data' represents the contours of the pit, a grid estimation routine that exactly honors the data without smoothing or interpolation, such as **triGrid**, should be used. Once the pit *xyz* data has been gridded, use **Locate** to set the fraction of each block that is between the pre-mining and post-mining grids. **Report** can then be used to get total tonnages of ore and waste and the net value of the pit.

Summary of procedure: Example 6.5
Initial data requirements: A topographic grid for generating contours.

Procedure:
1. Generate and export contours of topography **Graphics** ⇒ **Gridcont**
   1.1 Setup scaling ⇒ **Scaling**
   ⇒ **Scale**
   X-SCALE: <2000> Y-SCALE: <2000>
   ⇒ **Range**
   LEFT X: <0> RIGHT X: <300>
   BOTTOM Y: <10> TOP Y: <500>
   ⇒ **Sheets**
   SHEET SIZE: <A>
   ⇒ eXit
   1.2 Generate and export contours ⇒ Contour
   ⇒ **Field**
   FIELD: <surf>
   ⇒ **Interval**
   CONTOUR INETERVAL: <10> LOW CONTOUR: <30> HIGH CONTOUR: <70>
   ⇒ **Contour** ⇒ **Write Polygons**
   FILE NAME: <topo.pol>
2. Create an expanded pit **Graphics** ⇒ **polyEdit**
   2.1 Retrieve the polygon file and background showing the seams ⇒ **Load**
   ⇒ **Polygon file**
   FILE NAME: <topo.pol>
   ⇒ **background Metafile** (shows seams at 40 m elevation)
   FILE NAME: <seampol.met>
   ⇒ **eXit**
   2.2 Use the Line facilities to digitize a bench **Edit** (Polygon Edit menu level 1)
   ⇒ **Line** ⇒ **Create** (level 2)
   NEW POLYGON ID: <40bench> (level 3)
   (Place cursor at outcrop of outer edge of seam 2 starting at 40 m contour and digitize 40bench polygon limits using **Add** – place cursor at location of vertex and hit A – using **Insert**, **Move point** as needed and **Close** to complete the bench limit. Remember to take advantage of the Zoom and Pan facilities)
   ⇒ **eXit** (level 2)
   ⇒ **eXit** (level 1)
   2.3 Expand the 40 m bench to create the basis for the 50 m bench ⇒ **Operations** ⇒ **Expand**
   NEW POLYGON ID. <50bench>
   EXPANSION DISTANCE: <10>
   2.4 Modify the bench polygon to match topography: select a vertex on 50bench immediately above the 50 m contour and enter **Cut** cutting the polygon in two from the first vertex to the selected point. **Delete** the new cut polygon which is above the surrounding topogra-

phy. **eXit** (Operation menu)⇒ **Line** ⇒ **Delete** (select the remaining 50bench) ⇒ **Modify line**.

To create a new southern boundary of 50bench use **Insert** to input new vertices. Or to create closed mid-bench polygons (see Example 6.15) use **Follow** to select a line segment following the 50 m contour polyline. If there is difficulty in selecting points, either zoom in and/or use ^T to increase the tolerance from .05% of the screen distance to 1 or 2%. Use **Delete** as needed to bring the left wing of 50bench back to the 50 m contour.

To modify both the 50 m contour's polyline and 50bench to create a set of polylines representing the post-mining surface (see Example 6.16) **Cut** the 50bench polygon on a vertex next to the 50 m contour and **Delete** the hanging (southern) polygon as above. Select the 50 m contour's polyline and under the Operations menu use **spliT** to split the 50 m polyline into two components by selecting a vertex on the contour just beyond 50bench polygon as the first point along the split and the vertex on the other side of 50bench as the second point. **Delete** the hanging contour as above. In the Operations menu use **Cut** to open the 50 m contour's polyline with one of the vertices used to split the contour. The two resulting polylines are now exterior to the limit of 50bench. Use **Delete** under the Modify line menu to trim the contour and 50bench polylines until the ends of the contours and bench nearly match. The two 50 m contour polylines can be joined to 50bench under the Operations menu using the following procedure. **Join** will join two segments into one by connecting the last vertex of the first segment to the first vertex of the second segment. Since the order of the vertices on the segments may not result in the first vertex of one segment being adjacent to the last vertex of the second, **Reverse** can be used to reverse the ordering of vertices in a polyline. Therefore, use **Reverse** as needed so that the last vertices are adjacent to the first vertices on the ends of the three polylines that are to be joined. Select the last vertex on one segment and then select **Join** which will prompt for the second polygon. Selecting the second polygon will result in the two segments being joined. Repeat this for the other 50 m contour polyline so that there is only the one continuous 50 m post-mining contour.

Repeat Step 2.4 for each bench that intersects the surface topography.

2.5 Save the polygons of Step 2.3 **Save** ⇒ **polygon File**

FILE NAME: <pit.pol>

The remainder of this example is covered under Summary of procedure for Examples 6.15 and 6.16.

6.5 CONE MINING HEURISTICS

Positive moving cone routines are a heuristic approach to mine planning in that they attempt to determine the limits of a reasonable pit without the guarantee that the pit will be, in some sense, optimal. Cone routines determine a pit limit using break-even analysis. Every block of ore that can be mined will be mined as long as the total cost of mining the block and its associated overburden is not negative. A profit margin could be included in the net value of the block, but a more common approach is to assign a net value or grade cutoff that restricts mining to only those blocks that are considered to be profitable to mine.

The positive moving cone algorithm has been in use since the early sixties and is still commonly used in spite of the ready availability of pit optimization routines. One possible reason for the continued use of the moving cone approach is the ease with which it can be understood, applied and interpreted. To mine any subsurface block requires that a cone of overlying blocks must also be removed (Fig. 6.9). Sim-

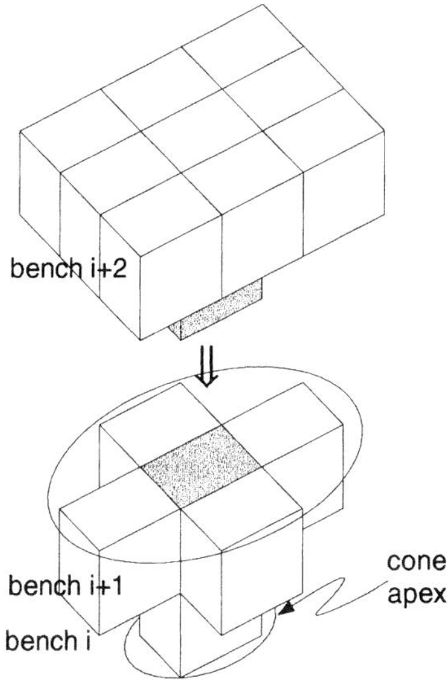

Figure 6.9. Cone mining based on slopes for a 1:5:9 stack of mining blocks.

ply put, any ore block that is mined must have a net value that meets or exceeds the sum of the negative net values of the block that would have to be mined to access it. The blocks that would have to be mined to get at the apex block are in a form of a cone wherein the side slopes conform to the pit's slope angle. Following 'normal' mining procedure, cones are mined from the top down. Starting at the first bench, any positive blocks will be mined. When blocks on any lower benches are evaluated, a cone is formed and the net cone value is calculated. When a positive cone is found, it is necessary to check to see if any of the previously rejected negative cones intersected the mined cone. Since some of the previously evaluated negative cones' waste blocks might have been removed to mine the positively valued cone, the net value of the cone might now be positive. Therefore, when a cone is mined, the net values of its component blocks are zeroed out and cone mining returns to the top and starts over. Past the first few upper benches, most positive cones are the result of this interaction between cones in which stripping cost is shared between a number of positively valued blocks. This simple algorithm works best for massive low variability deposits. More geologically complex and variable deposits require more complex algorithms to ensure that the interaction between cones is thoroughly evaluated.

Strictly speaking, it is not a cone, but rather a predetermined set of blocks which stack to approximate a cone with the desired pit slope. For instance, to mine a positive apex block on bench i, five covering blocks must also be removed on $i + 1$, but for each block on $i + 1$ there might be nine covering blocks on bench $i + 2$. It is apparent that the accuracy of the final pit slopes is dependent on the size and number of blocks with respect to the final pit. A pit which is composed of relatively few blocks will be capable of accommodating a wide variety of slopes.

Example 6.6: Generating a pit with the moving cone algorithm in Techbase

Before continuing with pit generation the following must be available: a block model containing at least the net value of each block, gridded topography of sufficiently great extent to intercept the crest of the expanded pit, and knowledge of the allowable pit slopes.

Techbase's moving cone routine involves a sequence of three programs: **opTopo** in which the the pre-mining topography is defined and where the post-mining topography is saved to a grid, **opSlope** where the pit slopes are defined, and **opCone** where the pit tonnages are calculated. The sequence of operations is to first enter the definition of the block model and the surface against which it is to be intercepted in **opTopo**. This information is saved in a *.seq file. Next, the pit slopes are defined which will control the shape of the cone, i.e. the stacking of blocks, as a function of the area of the pit, bench and azimuth. This slope template is saved in a *.tem file. Since the slopes are defined for the entire pit, there is no reason to return to **opSlope** to modify *.tem unless a new pit is being defined. **opCone** is then executed to determine the shape of a pit that would result from mining a given tonnage of ore and/or waste, such as would correspond to a year's production. The shape of the resulting pit and associated mining period is recorded the *.seq file. To retain a record of this pit, the *.seq file results are saved to a grid in **opTopo**. This sequence of operations is repeated for each mining period until no more positive cones can be found, i.e. **opCone** is executed for a subsequent number of mining periods and the resulting pit is then saved to a grid in **opTopo**.

The block model and surface topography of the Boland Banya deposit will be used to illustrate this procedure, but before continuing, a set of cell table fields must be defined that will hold the post-mining topography. There should be a field available in the topography's cell table for each of the mining periods. In order to make the pit crest extend into the original topography, each of the post-mining fields should be set equal to the pre-mining topography. When the results from **opCone** are saved from the *.seq file into these fields, those cells that are within the crest of the pit will be reset to the elevation of the remaining unmined blocks. Thus, the resulting grids will display the contours of the surrounding topography and of the pit. **Define** REAL ACTUAL fields in the topography cell table, one for each mining period to be saved. Next, set these new fields equal to the pre-mining topography by selecting **tbCalc ⇒ Setup ⇒ Equations**. Enter an equation that equates all of the post-mining fields to the pre-mining topography. For instance, if the pre-mining topography field were 'topo' and there were ten mining periods named 'per_1', 'per_2', ... , 'per_10', then the equation would be:

topo = per_1 = per_2 = ... = per_10

From the main menu select **Open-pit ⇒ opTopo ⇒ design File ⇒ Initialize** and enter the block model definition that will be stored in the *.seq file. This must be the same definition that was used in the original block model. Under **Topography ⇒ Topography** the surface topographic grid field name is entered as FIELD|VALUE. The 'half block rule' refers to how the elevation is to be set for blocks that are intersected by the surface. Elevations in the *.seq file are retained in terms of the bench level which is an integer value. When the half block rule is active, the elevation for any intersected block will be set to the top of the next lower block. Since the pit topography is based on on intersected blocks, the true pit elevation will be on average half a bench higher than the cone generator's results. A **polygon File** and **Bottom grid** can also be used to define the limit of cone mining. For large block models that have surface minable ore in a relatively small areal extent of the block model, use of a polygon to limit the search for positive cones can greatly speed processing time. Setting a bottom grid can be used to ensure a operationally feasible pit bottom while in ULTIMATE mode, or when subsequently in SCHEDULING mode a grid of the ultimate pit limits can be used to ensure that the annual pits do not exceed ultimate pit limits.

eXit from **opTopo** and execute **opSlope** ⇒ **Template file** ⇒ **Initialize** and provide a new file name for the *.seq file that will hold the pit slope definitions. To define the pit slope select **Slope**. Pit slopes are defined within **Zones** which delimit the submatrix in the block model to which the subsequent slope definitions belong. A **polygon File** can be used to define the area assigned to the zone for more complex slope zone boundaries. **Zones** are composed of **Sectors** which are in turn defined by degrees of azimuth and bench intervals. Under **Add Sector** enter the ENDING AZIMUTH, SLOPE and the bench through which that slope is to be applied. THROUGH BENCH will default to all benches. **Calculate** intersects the slope zones on the block model as defined in the *.tem file and determines the allowable slopes.

Once the *.tem and *.seq files are defined, **opCone** can be used to generate a pit. From the **opCone** main menu select **Apex** ⇒ **apex Cutoffs** and enter a set of logical relationships that must be satisfied before a block will be examined as a possible apex of a cone. Typical cutoffs would be associated with the block's net value and grade fields. The logic is the same as for database filters; all relationships have an AND relationship so that all must evaluate to true before a block will be considered as a cone apex. Under the **Values** menu select **Fields**. At this point the output file, which defaults to opcone.out, is created. Be aware that this file will contain the report of the cone miner's results. If the cone miner is going to be run for more than one mining period, the default file name should be changed (in **Database** ⇒ **output File**). Otherwise, the mining period's production results will be lost when the output file is overwritten in the next run of **opCone**. Six fields or constants can be entered in the Fields menu that are used in the calculation of cone values. NET VALUE is usually set to the block net value field, but can be a constant value as might be the case for a quarry or gravel pit. The NET TONS of the block may be a constant if the blocks are of constant dimension and density or can be a calculated field if a variable block model is used or if density varies by material type. ORE TONS and ORE VALUE are optional fields, which can be included in the output file by entering a relevant field or constant. ORE TONS would typically be field based on the fraction of the block in ore, the block dimensions and a density conversion factor. ORE VALUE is redundant of NET VALUE unless NET VALUE was entered as a constant. Block models can be smaller in extent than the pit. This makes sense when the characteristics of the waste material beyond the limits of the block model will be assumed to have a constant mining cost and density. Constants can be entered for MISSING VALUE and MISSING TONS that can be applied to any blocks which are eliminated by a database filter or that are external to the block model but within the stripping limit of the pit. Under **ore Cutoffs**, enter the logical relationships that will be used by the cone miner to discriminate between ore and waste blocks. The format used in this menu is the same as for **apex Cutoffs**. This shouldn't be very different from the apex cutoff that was used before, such as 'netvalue >= 0.' In **Goals**, three production goals can be entered that will determine the extent of mining that is to be performed by the cone miner. The goal that is satisfied first will conclude program execution. Note that these are constant values that represent, for instance, a year's production which can be based on NET TONS if the production equipment is the limiting factor, ORE TONS if a contract based on tonnage, such as for coal, has to be fulfilled, or WASTE TONS as might be true for initial mining periods when all mining activity may be for pre-stripping of overburden.

Parameters controls the sequencing of the cone mining cycles. The default method of RESTARTing the cone miner is 'cone' which will cause the cone miner to return to the top bench each time a positive cone is found. If a quick and dirty pit is desired, then RESTART can be set to 'period', which will eliminate any restarting, or the number of CYCLES that the cone miner goes through in restarting can be limited. Enter the starting and ending mining periods. The cone miner will attempt to produce the tonnages entered in **Goals** for each of these periods. If it is of interest to observe the progression of the pit, then the cone miner should only be run for a few periods so that the resulting pit can be saved from the *.tem file into a grid. **Search** controls the direction of cone mining on each bench. Thoughtful selection of the cone search direction can improve processing time and final results. For instance, if the ore out-

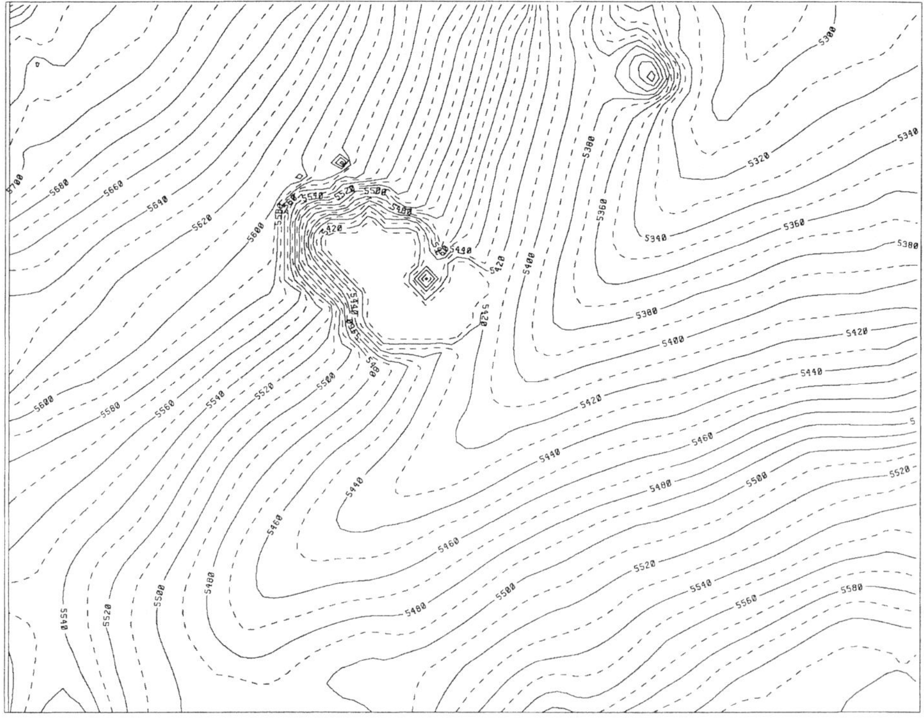

Figure 6.10. Pit following two mining periods (2 × 10 ^ 6 tons ore and waste).

crops on the northeast and the topography slopes to the southeast, then the cone miner should search for positive cones starting at the outcrop. OUTPUT defines the level of detail which is reported to the output file. The default is to report only the total mining period's production for each period; 'verbose' and 'extra' can be specified to report details on the cones themselves. This level of detailed output can be used to determine the actual value of marginal blocks. Selecting **Cone** starts the cone miner.

Once **opCone** has been run for the desired number of periods, the resulting pit limit can be saved into a grid, specifically, one of the grids that were previously set to the pre-mining topography. Return to **opTopo** ⇒ **Topography** ⇒ **Save** and enter the TOPO FIELD name corresponding to the grid for the mining period. This grid can now be contoured in **Gridcont** using intervals which correspond to the bench elevations. The resulting period 1 pit is shown in Figure 6.10.

Summary of procedure: Example 6.6
Initial data requirements: a mining block model containing a net mining value field and gridded topography.

Procedure:

1. Create a field in the cell table that contains the pre-mining topography to hold the post-mining topography **Techbase** ⇒ **Define Fields** ⇒ **autoTable**
   TABLE NAME: <surfaces>
   ⇒ **Create**
   FIELD NAME: <ult>   TYPE: <real>   CLASS: <actual>
2. Set the initial cell values for <ult> to the pre-mining topography **Techbase** ⇒ **tbCalc** ⇒ **Setup** ⇒ **Equations**
   CALCULATION STEPS: <ult = topo_c> (note that an actual field definition and calculation of cell values using tbCalc is necessary since a calculated field would always retain the value of topo_c)
   ⇒ **Calculate**
3. Define a pit sequence file **Open pit** ⇒ **opTopo** ⇒ **design File**
   3.1 Initialize the sequence file ⇒ **Initialize**
   FILE NAME: <ultpit.seq>
   X-COORD: <9900>   COLUMN SIZE: <50>   NUMBER: <53>
   Y-COORD: <20410>   ROW SIZE: <50>   NUMBER: <33>
   Z-COORD: <5630>   LEVEL SIZE:<30>   NUMBER: <13>
   ⇒ **eXit**
   (an output file containing the block table definition can be created and used for reference using **Define** ⇒ **Options** ⇒ **Message output**: <yes>⇒ **eXit** ⇒ **Tables** ⇒ **Show** TABLE NAME: <pit>)
   3.2 Set the upper and lower limits of mining in the sequence file ⇒ **Topography** ⇒ **Topography**
   FIELD VALUE: <topo_c>
   ⇒ **Bottom**
   FIELD VALUE: <oab_c>
4. Define a template file **Open pit** ⇒ **opSlope** ⇒ **design File** ⇒ **Show** (check to see that ultpit.seq is active, otherwise **Open** ultpuit.seq)
   4.1 Initialize the template file **Template file** ⇒ **Initialize**
   FILE NAME: <ultpit.tem>
   ⇒ **eXit**
   4.2 Define slopes ⇒ **Slope** ⇒ **Zone**

ZONE: <1> (defaults to all blocks in the sequence file)
⇒ **Add sector**
ENDING AZIMUTH: <360> SLOPE: <45> (defaults to a 45° pit for all benches)
⇒ **Calculate**

5. Generate an ultimate pit **Open pit** ⇒ **opCone** (use **Show** under both **Template file** and **design File** to confirm active template and sequence files)
   5.1 Select blocks to be examined as possible cone apexes **Apex** ⇒ **apex Cutoffs**
   FIELD: <netval_pit> RELATION: <>> FIELD VALUE: <0>
   ⇒ **eXit**
   5.2 Block field values and ore/waste discrimination **Values** ⇒ **Fields**
   NET VALUE: <netval_pit> MISSING VALUE: <> (only required if the cones extend into
   NET TONS: <pittons> MISSING TONS: <> areas missing an assignment)
   ORE VALUE: <pitauoz> (report mined ounces of gold)
   ORE TONS: <pit_waste> (can report any field under these headings)
   ⇒ **ore Cutoffs** (same as per step 5.1 or different, but while step 5.1 controls cone selection, ore Cutoffs separates ore from waste in the cone)
   ⇒ **Goals** (Default is to mine maximum size pit. Enter values to control scheduling)
   ⇒ **Parameters** (accept defaults unless scheduling, increasing detail of output file or attempting to increase processing speed)
   ⇒ **Cone** (starts moving cone algorithm for final pit generation, results are in opcone.out)
6. Save ultimate pit results to a grid **Open pit** ⇒ **opTopo** ⇒ **Topography** ⇒ **Save**
   TOPO FIELD: <ult> INCREMENTAL: <yes> (topography will only be reset to post-mining topography)
7. Use **Gridcont** to contour the final pit.

6.6 ULTIMATE PIT LIMIT OPTIMIZATION

The moving cone heuristic should be used with the understanding that the resulting pit will be the pit which recovers the most ore on a break-even basis by comparing the value of positive mining blocks against the material that must be removed in order to mine them. There is no insurance that the final pit is either optimal or unique. In fact, different runs with different search parameters might easily result in different final pits. Optimization methods seek to find pit limits which maximize the net value of the pit. It should be noted that none of the methods discussed in this chapter optimize the pit in terms of the time-value of money, such as a ROR. Instead, a pit limit is determined either on a break-even basis, as in the case of the cone mining heuristic, or as a maximum one-time net value as would be the case if the pit were mined instantaneously. This approach to design is far from realistic since it ignores the interaction of the mining schedule with the depth of overburden and spatial variability of the net values of the mining blocks, but for now these non-temporal pit optimization algorithms will be accepted for what they are worth and the topic of production scheduling will be left for Chapter 8.

Example 6.7: Lerchs-Grossman pit optimization using Lynx
Before generating an optimal pit limit for the TUTORIAL project, the following will have had to be completed:

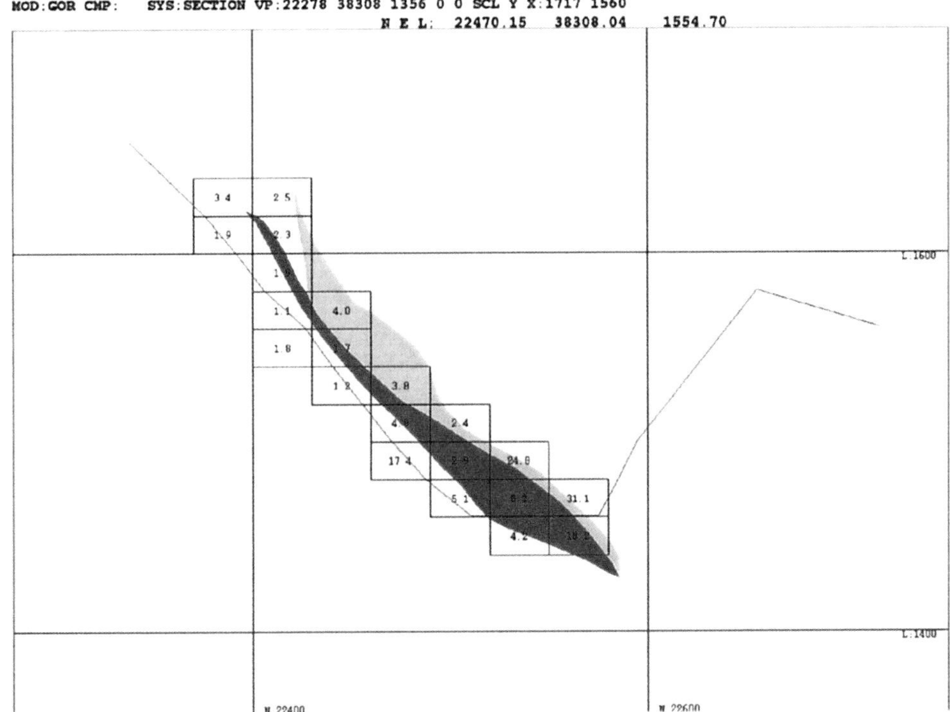

Figure 6.15. Section through optimal pit transverse to strike.

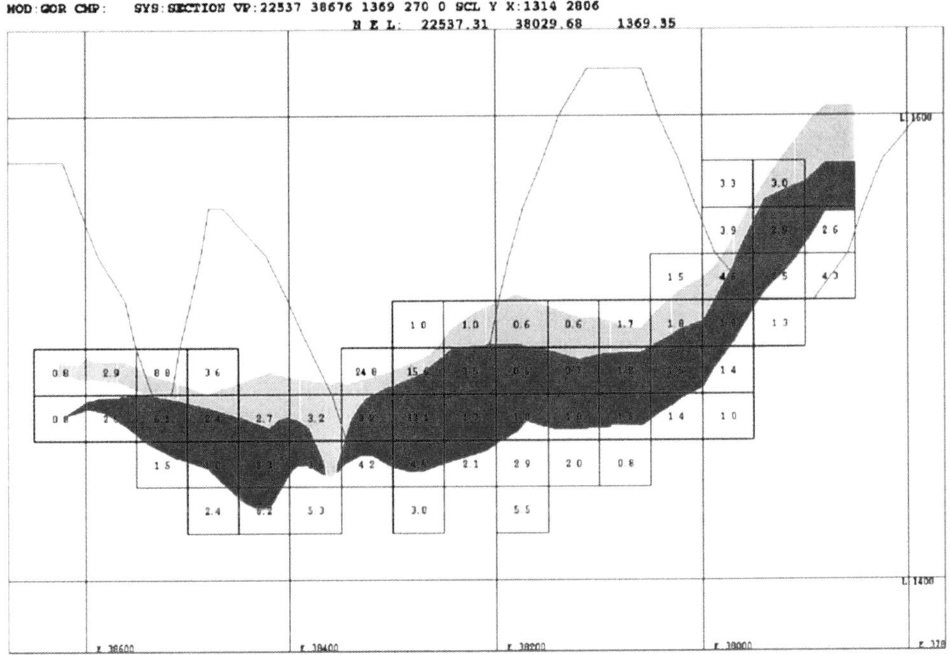

Figure 6.16. Optimal pit limits on section along strike.

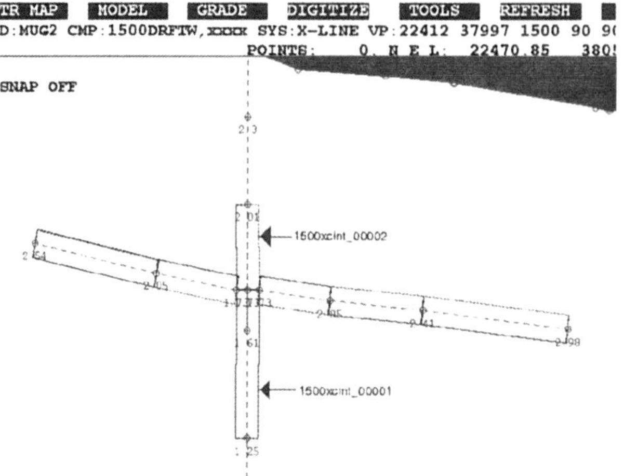

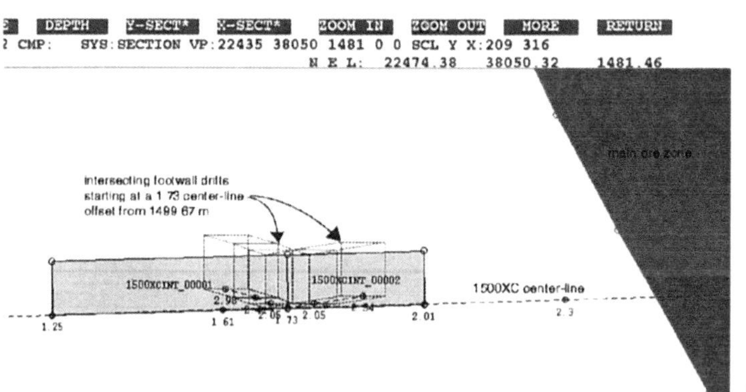

Figure 7.15. Intersection of shaft cross-cut and footwall drifts based on center-line alignment shown in: a) Plan view and b) Side view.

1. A mining block model must be defined which is sufficiently large in extent to contain the pit (Example 5.5);
2. Volumes of intersection, assays and total metal equivalents must be estimated for the block model (Example 5.6); and
3. A net mining value must be calculated for each mining block (Example 6.4).

From the main menu, select **Surface Eng ⇒ Pit Optimization ⇒ Optimization Parameters** bringing up the Optimization Parameters entry form. The upper portion of the entry form will default to an active parameter list with the active grid selected as the **Source 3D Grid** <GEORES>, the recent **Net Value Variable** <Net> and **Topography Map** <TOP>. Remember, that in Example 6.4 these values were stored in a set of Optimization Files. The remainder of the form is occupied with defining pit wall slopes across sectors which are defined by a set of azimuths. More specific examples of using azimuths to define pit slopes will be given later in this chapter under the topic of pit expansion. The wall slop pattern is defined by a sequence of azimuth (0 to 360°) and wall slope pairs (20 to 80°). Each pair defines the overall slope at a specified azimuth. Intermediate slopes are obtained by linear interpolation between adjacent pairs. Enter azimuth/wall slope pairs corresponding to the allowable overall slopes. Note that this includes both the bench width and highwall angle (295° and 115° at 60° slope, 115° and 265° at 50° slope). Figure 6.11 provide a complete listing. **OK** processes the block model prior to optimization to limit the number of

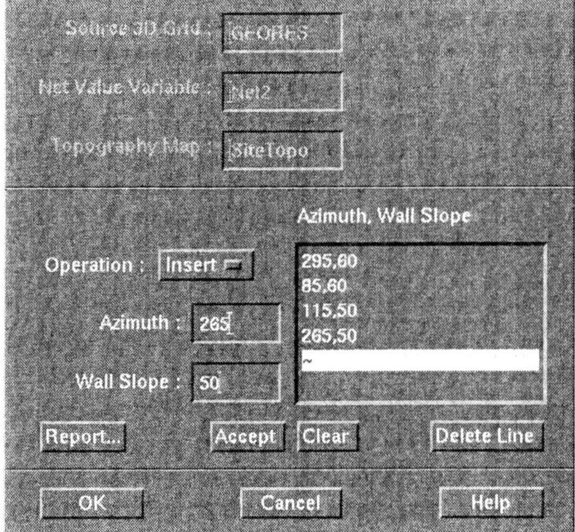

Figure 6.11. Optimization Parameters entry form.

blocks that will be evaluated as root nodes in the network. The lowest positive block in each column is used as the base of a cone which includes all other blocks that fall within the cone's slope limits. The set of all intersecting cones defines the maximum possible pit limit; any blocks falling outside of this maximum allowable pit are dropped from further consideration, which limits the size of the network and speeds solution. Maximum pit tree formation extends to the deepest positively valued blocks in the grid and extends, in accordance with the slope angles, up to the TIN of the pre-mining topographic map. This can easily result in a projected pit crest which extends beyond the limit of the block model. As a result, blocks near the limit of the grid may be pruned from the tree. It is up to the user to decide if the limits and extent of the block model are sufficient to contain any reasonably sized economic pit.

To optimize the pit, click on **Surface Eng Pit Optimization Optimize Pit** bringing up the Pit Optimize entry form (see Fig. 6.12). Lynx's Lerchs-Grossman pit optimizer can be used to optimize a pit from any **Grid Z Section Limit for Optimization** (as represented by the grids z-section number) to a given **Start Surface** as

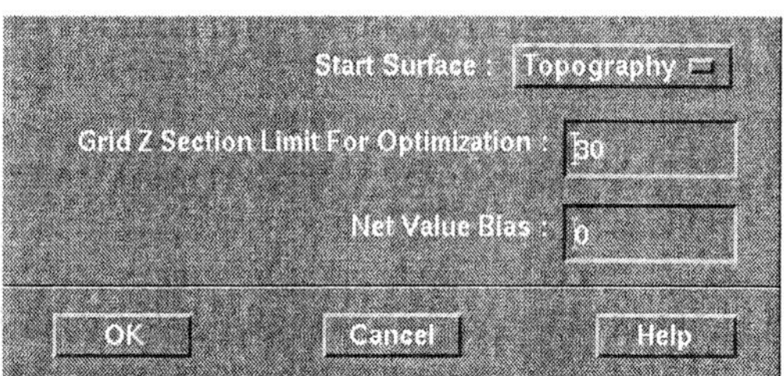

Figure 6.12. Pit Optimization entry form setup for ultimate pit optimization.

represented by a topographic TIN. The TIN can be either a representation of pre-mining topography, existing mine contours or a previous optimization. A **Net Value Bias** can also be applied. This bias is simply added to the block's net value. This in effect adds a profitability cost to the decision to mine a block and will result in a reduced pit tonnage. This is handy for cases where a pit of a given tonnage is to be determined while maximizing profit. One of the major drawbacks of ultimate pit optimization is that the Lerchs-Grossman algorithm will simultaneously optimize the entire deposit for a given set of net values. as a result, it cannot be used directly to determine annual pit limits for long-term schedules. Lynx offers two approaches for long-term scheduling. One is to limit the optimizer to considering only those blocks above a given *z*-section, effectively limiting the pit depth/size. The other method is by including the Net Value Bias and using trail-and-error to find the bias that will produce a pit with the desired tonnage. This procedure can be repeated for subsequent planning periods using the previous pit contours as a Start Surface and by reducing the magnitude of the bias allowing more blocks to be selected. Note that during each period, that the pit is only optimized for that period and is dependent on the results of previous periods. As a result, the final sequence of optimal pits are unlikely to be optimal overall in terms of either cash-flow or in terms of equipment usage (balancing stripping and waste production).

Summary of procedure: Example 6.7

Initial data requirements: a mining block model containing net mining values and volumes of intersection and a map with a TIN of pre-mining surface topography.

Procedure:

1. From the main menu select **Surface Eng** ⇒ **Pit Optimization** ⇒ **Optimization Parameters** (Optimization Parameters entry form).

⇒ **Source 3D Grid** <GEORES>
⇒ **Net Value Variable** <Net>
⇒ **Topography Map** <TOP>
All three of the above parameters should default to values provided by the pit optimization files.

2. Define pit slopes:
   AZIMUTH: <295> WALL ANGLE: <60> ⇒ **Accept**
   AZIMUTH: <085> WALL ANGLE: <60> ⇒ **Accept**
   AZIMUTH: <115> WALL ANGLE: <50> ⇒ **Accept**
   AZIMUTH: <265> WALL ANGLE: <50> ⇒ **Accept**
   ⇒ **OK**
3. Optimize the pit **Surface Eng** ⇒ **Pit Optimization** ⇒ **Pit Optimize** (Pit Optimize entry form)
   Start Surface: <topography>
   Grid Z Section Limit for Optimization: <30>
   Net Value Bias: <0>
   ⇒ **OK**

Example 6.8: Exporting the pit to a map and map contouring

Before blindly continuing with reserve reporting, the ultimate pit grid generated in Example 6.7 should be exported to a map and then contoured for viewing. From the main Lynx menu select **Surface eng Pit Optimization** ⇒ **Optimized Pit Map** bringing up the Optimized Pit Map entry form as shown in Figure 6.13. Select Current Pit as the **Surface** to be converted into a map. Under **Map Select** enter the name

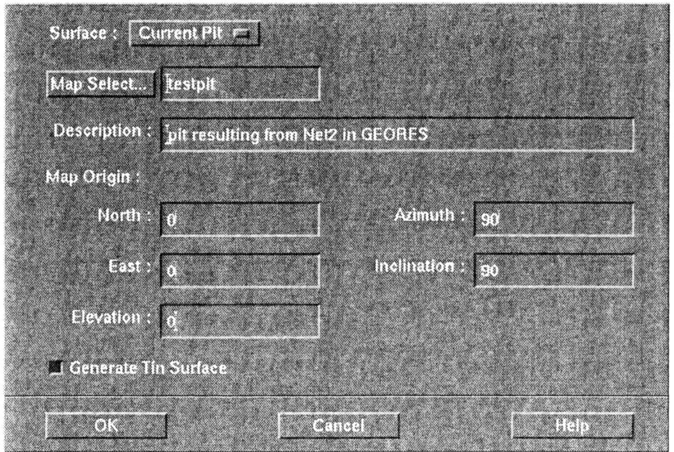

Figure 6.13. Optimized Pit Map entry form.

of the map that will be used to store the grid of pit elevations <testpit>. The pit optimizer created a grid of Z pit (Current) and pre-mining (Topography) elevations. Optimize Pit Map stores these *xyz* values (referred to as a pos feature in Lynx) in the selected map. A word of caution, the default Map origins are based on the origin of the source grid. Using these defaults will result in a map with negative elevations in which the top of the grid is the origin at zero elevation. the resulting contours will be negative values relative to zero. Therefore, be sure to convert the default North, East, Elevation, Azimuth and Inclination to: 0, 0, 0, 90, 90. This will yield pit contour that correspond to project Z coordinates. The **Generate Tin Surface** checkbox should be <on>. A TIN for the output map must be generated before contouring.

Repeat the grid to map conversion for the gridded topography. This will be used later in defining a volume model of the pit and a crest line.

The ultimate pit surface is now available as map data. As such, it can be contoured, triangulated (if not done already), modeled as a surface or used in creating a volume model of the pit. First the ultimate pit map will be contoured to provide a guide for conical pit expansion and haul road design. In Lynx map contouring is based on triangle sets. Therefore, the map must be made active in memory, triangulated and then contoured. If the TIN hasn't already been triangulated, from the Version 4 main menu select **File ⇒ Map Data ⇒ Map Select** and make <testpit> the active map. Next, select **Maps ⇒ Map TIN Define** to bring up the Map TIN Define menu and Parameter form. To see the grid and TIN toggle FEATURES and TRIANGLES on. Select **Generate** to triangulate the map. If inverse distance map to grid conversion was comprehensive there will be two triangles generated for each cell in the grid. For a large pit/3D Grid the resulting TIN might be quite large and take some time to generate. **Save & Exit** from the Map TIN Define menu and select **Maps ⇒ Map TIN Contouring** bringing up the Map Contouring entry form and enter the source map that was created above <testpit>, a destination map <testopc>

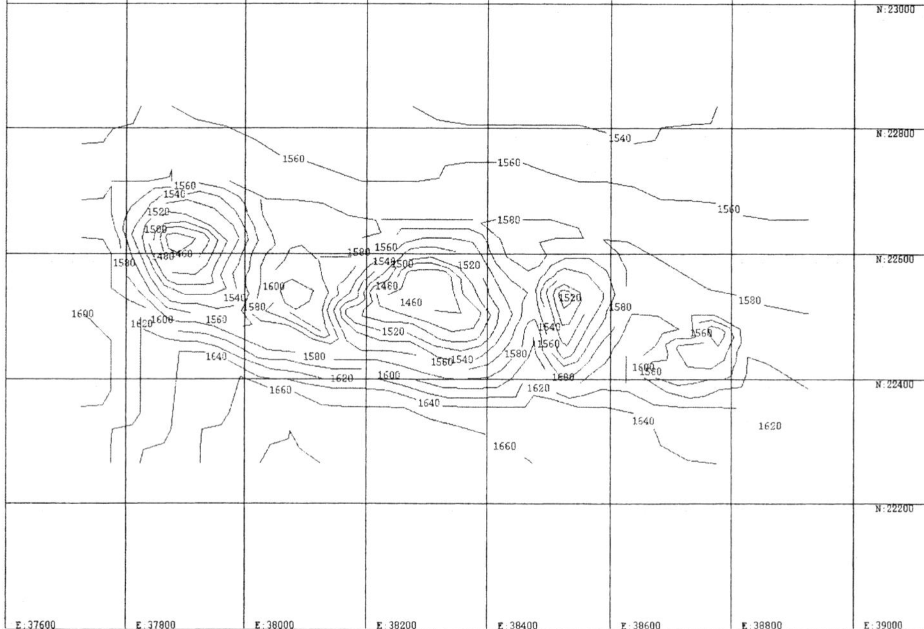

Figure 6.14. Contoured map of the optimal pit.

and map description. Contour the z offset from the lowest to highest benches using an increment that corresponds to the bench interval <10>. Follow the normal contouring procedure.

In order to confirm the generation of the ultimate pit, the ultimate pit contours will be viewed in plan and section along with the ore body model and net values in the block model. Return to the main menu and select **Background ⇒ Volumes**, enter the geologic model that will be viewed <G.OR> and set up a view plan centered over the pit <N,E,L,AZ,INC: 22000,37600,1460,90,90> and a scale of 5000:1. Also, use **Background ⇒ Maps** and display the contoured pit in background (Fig. 6.14). Place the 3D grid into background displaying the net block value <Net>. In a narrow deposit like this, examination of the block model can be used to better understand the nature of the ultimate pit.

Examine a plan view from several elevations, especially from the bottom pit elevation, to confirm that the pit is aligned with the orebody. Also run vertical section transverse (Fig. 6.15) and parallel (Fig. 6.16) to the long axis of the pit. Note that the pit bottom is irregular and does not conform to allowable mining practice. Also, note that the pit slopes are all highwalls without benches, berms or haul roads. It is obvious that this 'optimal' pit will have to be substantially modified to establish a workable ultimate pit limit.

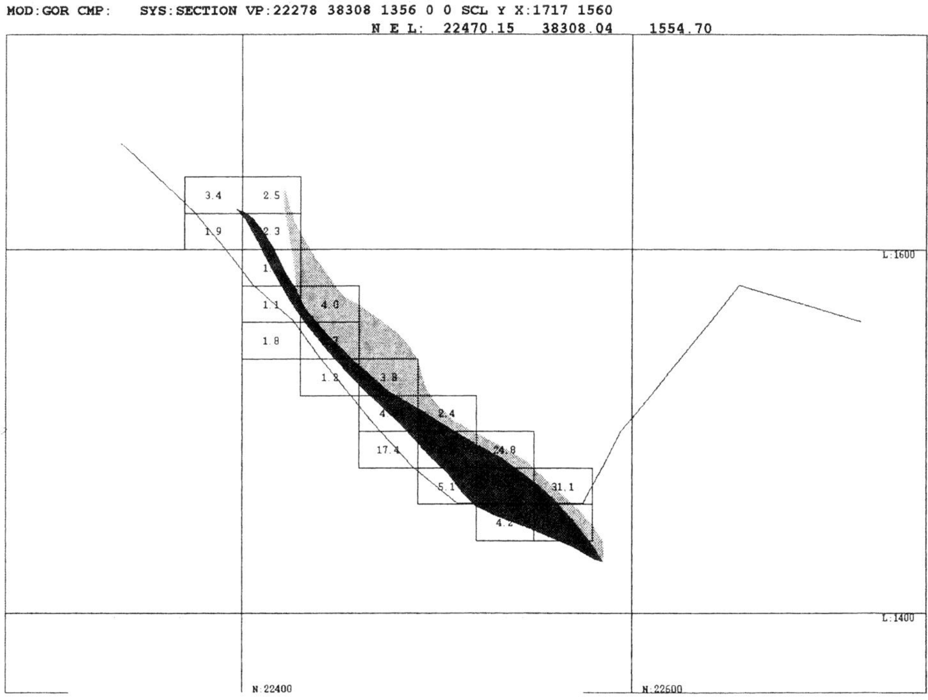

Figure 6.15. Section through optimal pit transverse to strike. (Figure in colour, see opposite page 246.)

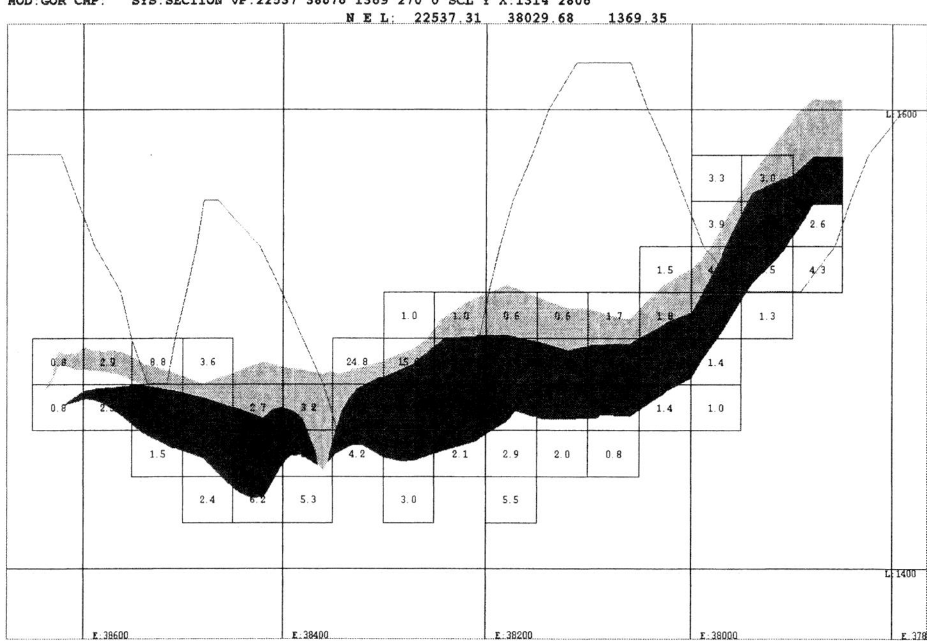

Figure 6.16. Optimal pit limits on section along strike. (Figure in colour, see opposite page 246.)

Summary of procedure: Example 6.8
Initial data requirements: a grid of the optimal pit as generated in Example 6.8 and (optionally) a grid of the pre-mining surface topography (note that the grids' extent will be limited to that of the block model).

Procedure:
1. Output the optimal pit grid to a map **Surface Eng** ⇒ **Pit Optimization** ⇒ **Optimized Pit Map** (Optimized Pit Map entry form)
   SURFACE: <Current Pit>
   ⇒ **Map Select** <testpit>
   Map Origin: <0,0,0,90,90>
   GENERATE TIN SURFACE <on>
   ⇒ **OK**
2. Contour the map (only if not done in Step 1). From the main Version 4 menu:
   2.1 Examine the map data **Maps** ⇒ **Map Data Report**: <testpit>
   2.2 Make the map active in memory **File** ⇒ Map Data ⇒ **Map Select**: <testpit>
   2.3 Triangulate the map **Maps** ⇒ **Map TIN Define** (Map TIN Define menu) ⇒ **Generate** ⇒ **Save & Exit**
   2.4 Contour the map TIN **Maps** ⇒ **Map TIN Contouring** (Map Contouring entry form)
   SOURCE MAP: <testpit>
   DESTINATION MAP: <testopc> (the destination can be the same as the source to overwrite)
   CONTOUR VARIABLE: <Z>
   INTERVAL TYPE: <regular>
   ⇒ **Query Min/Max** (report range of data values in source map)
   MINIMUM: <> MAXIMUM: <>

MINOR INTERVAL: <> MAJOR INTERVAL: <>
⇒ **OK**
3. Use **Maps** ⇒ **Map Data Report** to check resulting contours and **Background** ⇒ **Maps** to view the results.

6.7 PIT EXPANSION

The Lerchs-Grossman ultimate pit limits displayed in Figures 6.15 and 6.16 may be optimal in the sense of the instantaneous total net mining block value, but they are not acceptable in terms of mining practice. At best, the optimal pit serves only as a guide to pit design. In actual design practice, the ultimate pit bench contours are used as a template for a pit expansion similar to Example 6.6. Pit optimization does not incorporate haulage beyond an incremental cost for increased depth, $.01-.005/t/bench in the net value calculation. The pit that was generated in Example 6.10 was composed of three isolated pits, one of which seems to have a pit bottom that is only one block in area. Obviously, haul road construction from such a configuration would not be plausible. It is much more likely that the truly optimal pit will have only one haul road rising from a contiguous pit bottom along the north side of the pit from a pit bottom of 1460 m to a 1560 m crest line. Example 6.10 illustrates the circular nature of open pit mine design: optimization provides an initial design template which must be modified to suit operating practice, such as minimal pit bottoms and haul roads. Conical pit expansion routines can then be used to generate a pit based on the optimal pit contours but modified to suit mining practice. Often, pit optimization packages include options that allow the pit bottom to be fixed. For instance, a grid of the expanded pit can be used as a limit on the blocks that are considered during subsequent optimization, and the slopes can be adjusted to match the pushbacks required by the haul road. If the resulting optimal pit is a reasonably close match to the expanded pit, then the expanded pit is used as the final design. Otherwise, the cycle of pit expansion and optimization is repeated until closure. Note that an exact match between the optimal and designed pit is not terribly important: it is highly unlikely that mining costs, market forces and reserves will remain in balance for the life of the pit.

Example 6.9: Pit expansion using Lynx
Continuing from Example 6.8, the contoured map of the optimal pit will be used as a template for an expanded pit. From the main menu select **Surface Eng** ⇒ **V3 Open Pit Eng** ⇒ **Open Pit Mine Design** ⇒ **Open Pit - Design and Optimization** and enter the PIT DESIGN TO BE SELECTED. Since this is a new pit, rather than a modification of an existing design, enter a new engineering design file name. From the Open Pit – Design and Optimization menu select **Conical Expansion Design** and enter the 3D grid and then select **Conical Expansion Mine Design**. Enter the new PIT ID and description. At this point, a source pit can be entered if the new design is to be based on an existing pit. In this example, the pit is to be generated from scratch. The conical expansion design utility can be used either for pit or dump design. Typically, a pit will be expanded from a bottom bench up to intersect the surface, while dumps are expanded down to the topography. Accept the defaults to expand the pit

up to the topography. Next enter the elevations of the bottom and top benches in the pit and the bench interval. Examine the pit bench contours of Figure 6.14. The bottom of the deepest possible pit is 1460 and the highest point on the crest will be 1660. The bench interval is 10 m. Entering <1460>1660:10> will create design benches from 1460 to 1660 at intervals of 10 m. Next enter the overall average bench slope and width. In the optimal pit, slopes of 50 to 60° were used. This was an overall slope which accounted for the toe to crest angle (90° = vertical), the bench width, and the haul road once its location, width and profile are known. For this example, a bench slope of 75° and bench width of 10 m will be used for better graphic visibility even though this will result in a much lower overall slope than was used for optimization. The expanded pit design parameters are given in Figure 6.17. Finally,

```
Continue (Y/N) <Y>

Pit Id                            :exppit2
Pit Description                   :pit expansion
Source Pit Id                     :
Pit Or Dump Design                :Pit
Top Down Or Bottom Up Design      :Bottom Up
Default Bench Slope,Bench Width:75,10

1460        1590
1470        1600
1480        1610
1490        1620
1500        1630
1510        1640
1520        1650
1530        1660
1540
1550
1560
1570
1580
```

Figure 6.17. Expanded pit design parameters.

enter the TOPOGRAPHIC SURFACE TO EXPAND AGAINST. At this point, the viewplane can be oriented and scaled to cover the area of the pit. Use the same scale and orientation as the map of the optimal pit contours (N, E, L, Az, Inc: 22200,39000,1500,0,90; SCALE: 7500,7500). Place the optimal pit contours in background display.

To define the bottom of the expanded pit, return from the background menu and select **Define** ⇒ **Bndry** ⇒ **Enter** to enter the toe line of the 1460 crest. For this example, the pit bottom will be based on the centrally located 1460 m bench centered on an easting of 38300. Once points have been entered, the **Insert** and **Edit** keys can be used to modify the toe. Note that only one pit bottom can be entered. After digitizing the toe, select **Return** and the 1460 m toe will be expanded to intersect with the surface topography (see Fig. 6.18).

Select **Define** again and use **Slope** to change the pit slopes. This is accomplished by entering the base toe or crest elevation of the benches that are to be modified, digitizing an AZIMUTH CONTROL CENTER roughly in the middle of the bottom of the pit, entering the slope that is to be applied to slopes falling along the vector

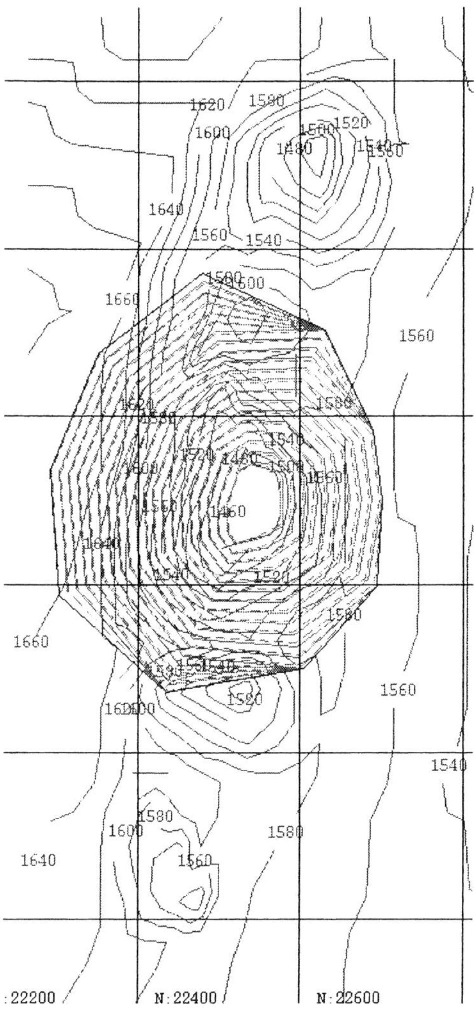

Figure 6.18. Expanded pit based on 1460 toe.

defined by the azimuth control center line and a series of AZIMUTH CONTROL DIRECTIONS. The slopes between azimuth control vectors of the same value will have the slope defined by the bounding vectors. Slopes between vectors of different values will have a linear transition between the two slopes. Benches below the selected elevation will be unaffected and successively higher benches can have their slopes modified. Thus, even though the initial pit expansion is based on a fixed default slope angle, the final slopes can vary both by bench and sector. Figure 6.19 shows the pit of Figure 6.18 modified to have slopes of 70° on the shallower northern side of the pit and 80° on the southern benches. **Bench** can be used in exactly the same manner as **Slope** to modify bench widths. The pit of Figure 6.19 has also been modified to reduce the 10 m benches on the southern face of the pit to 7 m resulting in a steeper highwall.

The pit of Figure 6.19 must still be modified using pushbacks to expand it into the

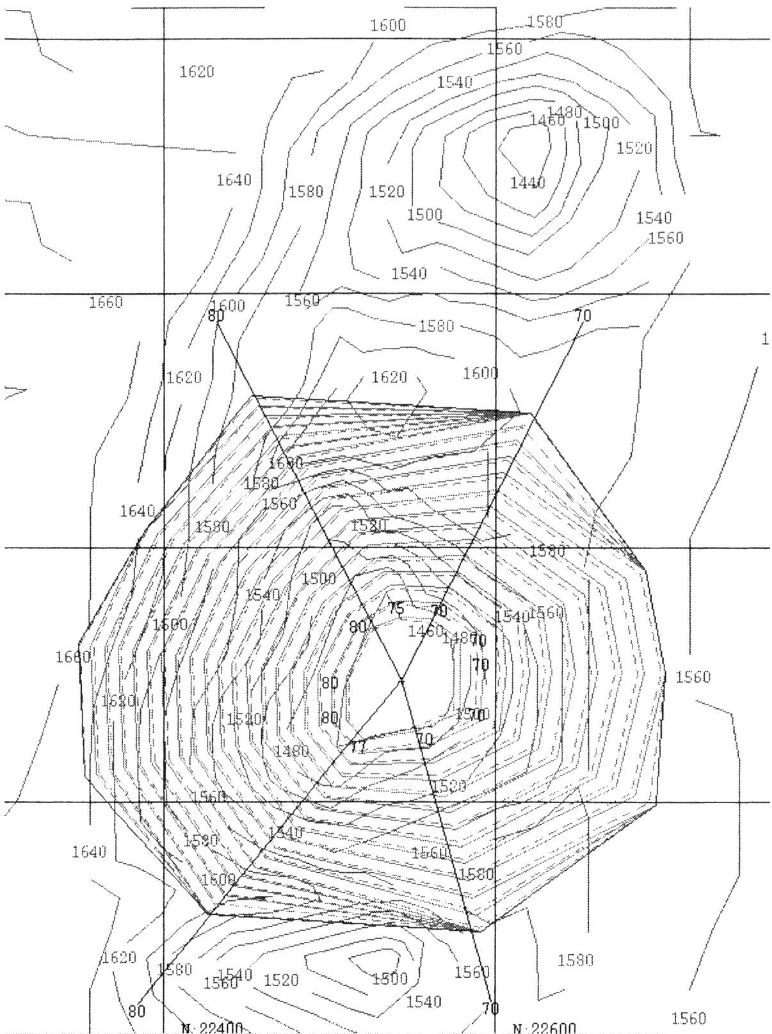

Figure 6.19. 1460 toe based pit, codified to have variable pit slopes.

eastern and western pit bottoms of the optimal pit limits. Since it is not possible at this stage to have multiple toe or crest polygons at the same elevation, the pit will be pushed across to the limits of the optimal pit at higher elevations. Since the 1560 bench looks like it should be pushed across the limits of the ultimate pit, select **Bndry**, enter the toe elevation of the bench polygon to be modified (1560) and use **Enter**, **Insert**, **Edit** to push the 1560 bench back to its optimal limits (see Fig. 6.20). Note that this pit will be suboptimal in comparison to the Lerchs-Grossman pit limits.

Multiple pit bottoms can be included by expanding separate pits up to a common pit bottom. The overlying pit could be expanded above these two additional pit bottoms at the 1560 bench. Two additional pits could be added to this design expanded

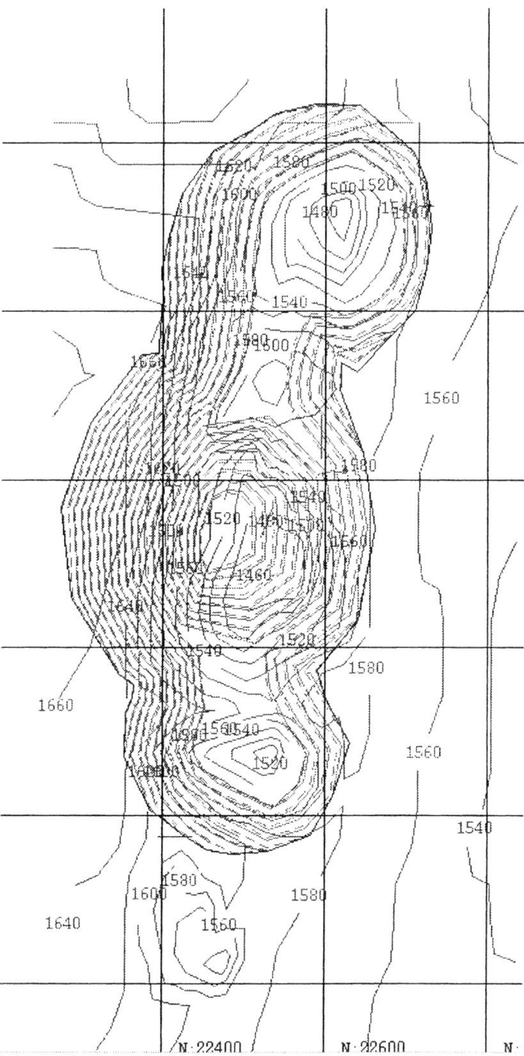

Figure 6.20. 1560 benches to the side of the pit bottom.

to a fixed elevation of 1560 m. The three pits in the design could then be merged into one map. The map data consisting of the three pits can then be triangulated and used to create a volume model of the pit for reserve calculations.

Summary of procedure: Example 6.9
Initial data requirements: a mining block model, a topographic surface to expand against and (optionally) a contoured map of the ultimate pit.

Procedure:
1. Initial pit expansion with fixed slope and bench width settings **Surface Eng ⇒ V3 Open Pit Engineering** (Open Pit Engineering menu) ⇒ **Open Pit Design and Optimization**
   1.1 Define initial pit parameters

Design To Be Selected: <NEW>
Is This A New Design: <y>
⇒ **Conical Expansion Design**
3D Grid Model To Be Selected: <GEORES>
⇒ **Conical Expansion Mine Design**
Pit Id: <new1>
Cannot Read <new1> – Create New Pit: <y>
Pit Description: <>
Source Pit Id: <>
Pit Or Dump Design: <P>
Top Down Or Bottom Up Design: <T>
Enter [Seam], [Bench], [Min>Max: Interval]: <1460>1660:10>
Enter [Seam], [Bench], [Min>Max: Interval]: <>
Default Bench Slope, Bench Width: <75,10>
Continue: <y>
Topographical Surface To Expand Against: <TOP>
(sys: Conical Idle menu)

1.2 Define viewplane orientation and background display
⇒ **Vplane** ⇒ **Orient** (sys: Vplane menu)
New Viewplane N,E,L,AZI,INC: <22200,39000,1500,0,90>
⇒ **Scale**
Viewplane Scales SY, SX: <7500,7500>
⇒ **Grid**
Global Grid Displayed: <y>
Global Grid Spacings DN, DE, DL: <200,200,200>
⇒ **Refresh**
⇒ **Return** (sys: Conical Idle menu)
⇒ **B'grnd** (sys: Bgrnd menu)
⇒ **Volumes**
Initialize Background Memory: <n>
Background Source Model Type, Model Id: <G,GEOL>
Select Unit Id, Comp Id, Code: <all>
Background Source Model Type, Model Id: <>
Volume Display Thickness; Fore, Back: <0,0>
⇒ **Maps**
Overlays: <OPT> (or any ultimate pit map – see Example 6.10)
Map Display Thickness; Fore, Back: <>
⇒ **Return** (sys: Conical Idle)

1.3 Digitize the pit bottom of the 1460 toe ⇒ **Define** (sys: Conical Define) ⇒ **Bndry** ⇒ **Enter**
(Use **Del All**, **Del Pnt**, **Undo**, **Shift**, **Insert** and **Edit** to modify toe as needed. **Zoom** in to the bottom of the pit for editing.)
⇒ **Return** (Busy...Expanding pit) (sys: Conical Define)

2. Modify the pit (continuing from the sys: Conical Define menu)

2.1 Modify the slopes ⇒ **Slope**
Surface Id, Bench Elevation, Or Pit Crest (Topo): <1460> (only a toe is at 1460 m)
Digitize Azimuth Control Center (select a point in the bottom of the pit)
Enter Bench Slope Angle: <70>
Digitize Azimuth Control Direction: (Digitize another point exterior to the pit creating a ray origination from the control center. Repeat the digitization of control directions for the various slope angles required. Note that the slope will change linearly between the control directions only for those benches that are within the limit of the arc

described by two adjacent slope control rays. Enter <> in response to ENTER BENCH SLOPE ANGLE to return to sys: Conical Define)

2.2 Modify bench widths ⇒ **Bench**
SURFACE ID, BENCH ELEVATION, OR PIT CREST (TOPO): <1470>
REGION WIDTH: <10>
DIGITIZE AZIMUTH CONTROL DIRECTION: (The AZIMUTH CONTROL CENTER has already been established in step 2.1. Follow step 2.1 to modify bench widths by sector.)

2.3 Create a pushback ⇒ **Bndry**
SURFACE ID, BENCH ELEVATION, OR PIT CREST (TOPO): <1560>
BENCH TOE OR CREST: <T>
(Use **Snap on**, **Edit** and **Insert** to shift the 1560 toe line to the 1560 optimal pit limit contour.)
⇒ **Return**
⇒ **Return**
STORE PIT: <y>

6.8 INCLUDING HAUL ROADS IN AN EXPANDED PIT

Placement of the haul road requires that additional modifications be made to the expanded version of the optimal pit, which can be accomplished by either burying reserves by pushing the road into the pit or pushing the pit out into waste and increasing the stripping ratio. In the case of a switchback, a combination of pushing the pit both in and out can be used.

The layout of the haul road has to balance several objectives while remaining within the road design's criteria: minimizing both the distance and tonnage hauled, minimizing the loss in reserves or increase in stripping ratio and providing a profile which limits wear and tear on trucks.

The pit should be exited at a point closest to the final destination. Since there are multiple destinations, including at least one for ore and another for waste, the point on the crest where the haul road clears the pit should be selected to minimize the overall distance-tonnage hauled over the road's life. When major destinations are at opposite ends of the pit, it may be desirable to split the haul road as it spirals through the upper benches. Any additions to the haul road system should be left for the upper benches to reduce the consequent increase in the stripping ratio.

Not only should the point at which the road exits the pit be close to the final destination, it should be chosen to limit the vertical distance covered in the pit. Since the road rises from the bottom of the pit and continues at a largely fixed maximum grade, the road should exit the pit at the lowest pit crest elevation. It is much more economical to construct and maintain roads external to the pit since they remain in place for the planned life of the project. In-pit road length must be minimized: in-pit haul roads increase the amount of waste stripping and must be frequently relocated as the pit expands. Also, avoid placing haul roads in the highwall side of a pit, since this will increase the stripping ratio.

Location of the haul road must also take into account the inclination of the deposit with respect to pit slopes. When the dip on the footwall of the orebody approaches the pit slope on the footwall side, it would generally be best to place the road on the hangingwall side of the pit if the road is to be pushed into the pit, subsequently burying reserves on the lower benches. This approach will minimize the loss of ore as shown in Figure 6.21, illustrating the need to place the road to minimize the loss in reserves or increase in waste removal.

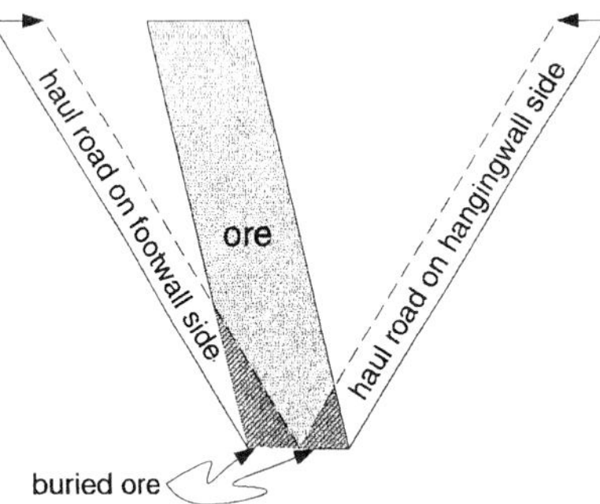

Figure.6.21. Comparison of ore loss for road push-ins to pit on the footwall and hangingwall side of steeply inclined deposits.

Consider the case of designing a road profile by starting at the toe of the lowermost bench and working upwards towards the pit crest. The pushback procedure can be viewed as driving a road ramp into the bench with the pit-inwards edge of the road starting at the expanded pit's toe and continuing to the crest of the next bench as shown in Figure 6.22. In this case, a road cut is being created in the bench which results in additional stripping of waste, which will commonly result in the entire pit above the road being pushed back into waste since the road will commonly be wider than the bench on which it's being constructed. The road will continue at the same grade through the remaining benches with the pit-inwards edge of the road extending from bench crest to bench crest. On each bench, fill material is being pushed down through the bench while the bench is being widened to accommodate the width of the road. Once started, the pushback of the pit above the road cannot be avoided unless the pit slope on the benches is sufficiently low to allow for bench widths as wide as

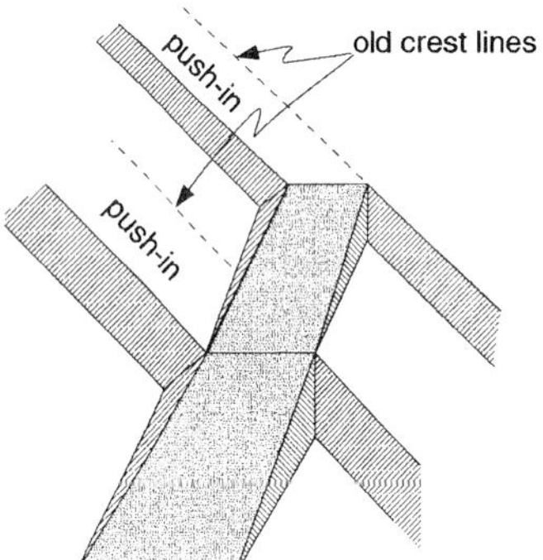

Figure 6.22. Push-in road construction.

or wider than the road. Generally, this will not be the case, and the only way in which the road can be accommodated without pushing the overlying benches back would be by moving the road into the pit. This requires the placement of fill material under the push-in to maintain the maximum allowable pit slope.

At this point, it should be remembered that the procedure described herein is a conceptual approach to modifying a final pit limit to include a haul road. Actual road construction will not be from the bottom to the top, nor will it involve placing fill for a road over existing benches or cutting a haul road into the pit slope. Rather, the haul road system grows organically with the pit from the surface down and middle of the pit outwards as the pit deepens and the benches expand. Thus, in the context of this discussion, the inclusion of the haul road in the pit is necessary only to produce a final pit design which is operationally feasible. Therefore, the push-in/pushback procedure should be viewed as a means of modifying the expanded pit to account for a haul road prior to determining the tons of waste and ore in the final pit. In this context, a haul road can include both push-ins and pushbacks. Consider the oval pit bottom and deposit limits shown in plan view in Figure 6.23. The haul road will be

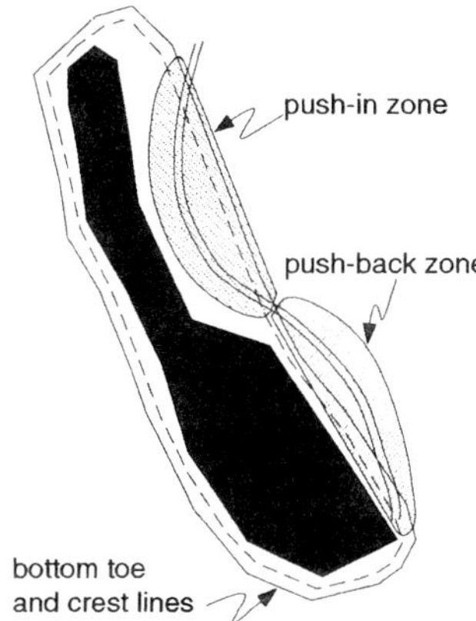

Figure 6.23. Using a combination of push-in and pushback road construction to avoid loss of ore and an increased stripping ratio.

placed to exit from the pit floor in the SE corner of the pit. Since the toe of the bench is up against ore, a pushback will be necessary to avoid burial of reserves, but as modification of the pit continues to the north, a gap opens between the toe of the lowest bench and the ore hangingwall. In this area, inclusion of a haul road can move from a pushback to a push-in of the pit since no ore will be lost and the stripping ratio can be reduced.

Minimizing wear and tear on the haulage trucks along with safety are the primary criteria for road construction. From the point of view of modification of the expanded final pit, the primary effect is maximum road grade, road width and the design of switchbacks. Road grade is primarily a function of pit depth. If a large vertical haulage component is necessary, then the grade should be kept to a maximum to minimize stripping. A widely accepted maximum grade is 10%. An 8% grade may be preferred, especially at a transition from lower slopes such as at the pit bottom, crest and sharp curves. Switchbacks are kept flat along the turn and will be discussed in more detail later in this section. Road width is a function of the number of lanes, the berm and inner ditch width and the width of the largest trucks. Each lane of traffic should have clearance to both the left and right equal to half the truck width. For two-lane haul roads, the rule of thumb is that the road width should be four times the truck width. For 85, 120, and 170 st trucks this works out to road widths of 23, 25 and 30 m, respectively.

6.9 HAUL ROAD CONSTRUCTION IN PLAN VIEW

The methodology used to modify an expanded pit to include haul roads focuses on using road segment templates which represent the plan view geometry of a haul road crossing a bench. In the case of a straight road segment, the length of the segment depends on the grade. For example, a 10% road crossing a 10 m wide bench that is 10 m in height will be 100 m in length. The width of the segment is simply the road width (for this example say 25 m). In plan view, the segment will appear as a trapezoid since the horizontal distance covered in moving from the toe to the crest must be included. The angle that the edge of road (EOR) makes with the crest lines is found as $\theta = \sin^{-1}(25/100) = 14.4°$. The line perpendicular to the crest will be wider than the haul road since it crosses the center line of the segment at 14.4° off the center line. Thus, the apparent width of the road is $W_a = 25/\cos 14.4° = 25.82 \approx 26$ m. If the bench face slope angle is assumed to be 75°, then the horizontal distance from toe to crest is $10/\tan 75° = 2.7 \approx 3$ m (Hustralid & Kuchta 1995).

In plan view, the toe and benches for this example would appear as shown in Figure 6.24a with about 3 m between the toe and crest, a 10 m bench width and 13 m from crest to crest. Figure 6.24b illustrates the road segment template whose geometry was determined above: a trapezoid 100 × 26 m at a 14.4° angle from the toe. The template includes both toe, crest and dashed EOR (Edge Of Road) lines. In Figure 6.24c the pit is being modified by starting with the lowermost bench by pushing the 1480 toe and 1490 crest into the pit by the width of the haul road template so that both lines match and follow the template toe and crest lines. The template is then placed on the 1490 crest at the point where the haul road enters the 1490 bench to provide for a continuous 10% grade and the process of pushing the 1490 toe and 1500 crest lines is repeated. The templates have been removed in Figure 6.24d so that only the modified toe and crest lines are visible, as would be the case for the final expanded pit.

The procedure used for switchbacks is largely the same as for straight road segments: a template for a switchback is used to modify the toe and crest lines. The plan view geometry of the switchback template must reflect some additional road design criteria beyond those used for straight road segments. The curve itself must be level. This means that the road segments leading into and out of the curve must cover the vertical distance of a bench. In the previous example of a straight road segment 100 m in length, a 10% grade was used. For this example, the switchback template will include a 50 m straight segment at 10% on both ends of the curve, although a lower-grade, say 8%, should be used for safety when making the transition on and off the level curve to provide sufficient depth of vision for the driver. The inner curve radius

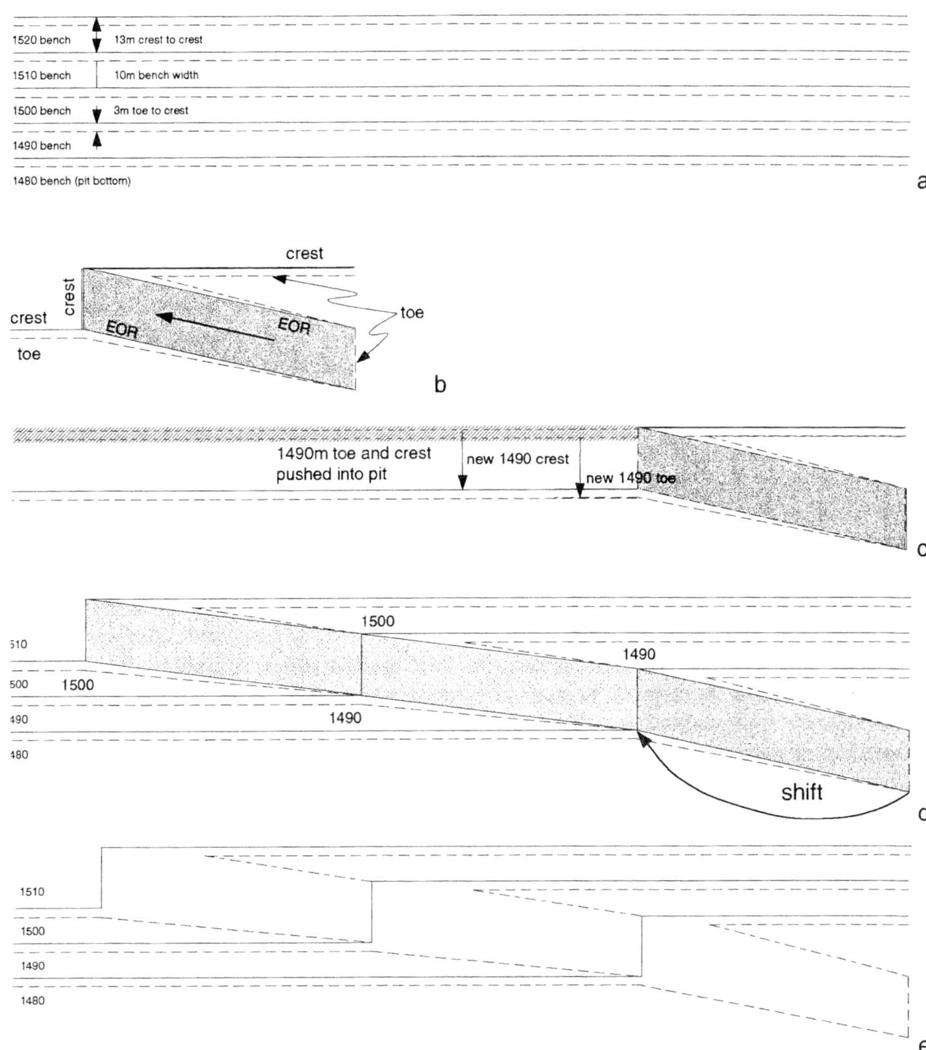

Figure 6.24. Spiral haul road construction using straight road segments. a) Plan view bench geometry without haul road, b) Template for a road segment connecting two benches from toe to crest, c) Lowermost bench pushed into pit by haul road depth with toe and crest lines modified to match road template, d) Push forward of haul road completed for benches 1490 to 1510, and e) Toe and crest lines following removal of road segment templates.

is governed by the turning radius of largest truck that will be used. For this example, the inner radius is 25 m. The curve's width will be greater than for a straight segment. As trucks round a curve, there will be increased vehicle overhang at the nose and tail while the truck is turning. The tighter the turn and larger the truck, the greater the overhang. For this example, the road width on the curve has been increased to 30 m.

As shown in Figure 6.25a, straight road segments are added to the pit up to the bench in which the switchback is to be placed following the same methodology as for Figure 6.24. The switchback and clockwise spiraling road segment templates are shown in Figure 6.25b. Note the path of the toe along the base of the switchback: unless a great deal of space is already available on the portion of the bench in which the switchback is to be constructed, a deep pushback into the pit wall will be required. This pushback is shown in Figure 6.25c. Space needs to be provided to the left of the curve for its construction and a smooth transition made from the 10 m bench to the width required by the switchback. From this point, the haul road continues to spiral up the pit by pushing the benches back using the clockwise straight road segment as shown in Figure 6.25d.

Example 6.10: Generating road templates as map figures with Lynx
The road templates shown in Figures 6.24b and 6.25b could easily be constructed using CAD software. In theory, the drawing could then be imported into either Techbase or Lynx as a DXF file, but in reality importing graphics from one product to another is, at best, problematic. For this reason, generation of a switchback as a map figure will be demonstrated in Lynx. Its plan view geometry will be the same as discussed above.

First create a new map to hold the switchback haulroad segment. From the main Version 4 menu select **File ⇒ Map Data ⇒ Map Create** and enter a new source map name <exswtch>. The switchback will become map data which will be available as a figure for pit modification. Accept the default map origin of <0,0,0,90,90> and give the map a description. When a map is retrieved as a figure into a viewplane, it will appear in the center of the screen regardless of its origin's coordinates. Since the figure will be digitized directly into the map, none of the default Lynx variables need be included in the map's definition. Select **Maps ⇒ Map Data Define** bringing up the Map Data Define menu. At this point, the viewplane is visible and can be adjusted in scale, position and both global and local grids to provide a format convenient to generating the figure. The scale should be set so that the entire figure can be easily viewed on the screen. Additionally, working from a starting position of 0,0,0 makes it easier to calculate the relative positions of successive points on the figure. Lynx does not include a snap-to-grid option, but both the global and local grids can be used as guidelines. Select **Viewplane ⇒ Setup** and shift the viewplane by <–100,–100> and set the scale to 1000:1 so that 0,0,0 is about midscreen and use a 50 m global grid. Set cursor tracking to global coordinates.

Different types of features are available to represent the road. Since the figure will be composed of a polyline with vertices in x, y and z, a convenient feature type to work with is a traverse. The default traverse feature's attributes can be modified to meet the desired attributes for a road segment and then saved as a new feature. To create this new feature, select **Map ⇒ Map Feature Attributes**. to bring up the Map Features Attributes entry form and select <trav> as the FEATURE TYPE. The attributes of the traverse feature are displayed in the entry form. Any of these can be changed. For instance, if the toe and crest lines of the expanded pit are displayed in cyan (pen 5), then the same default color that will be used for the road. Change the pen color to red (2), the line type to dotted (3), and the fill type to single hatch at 2

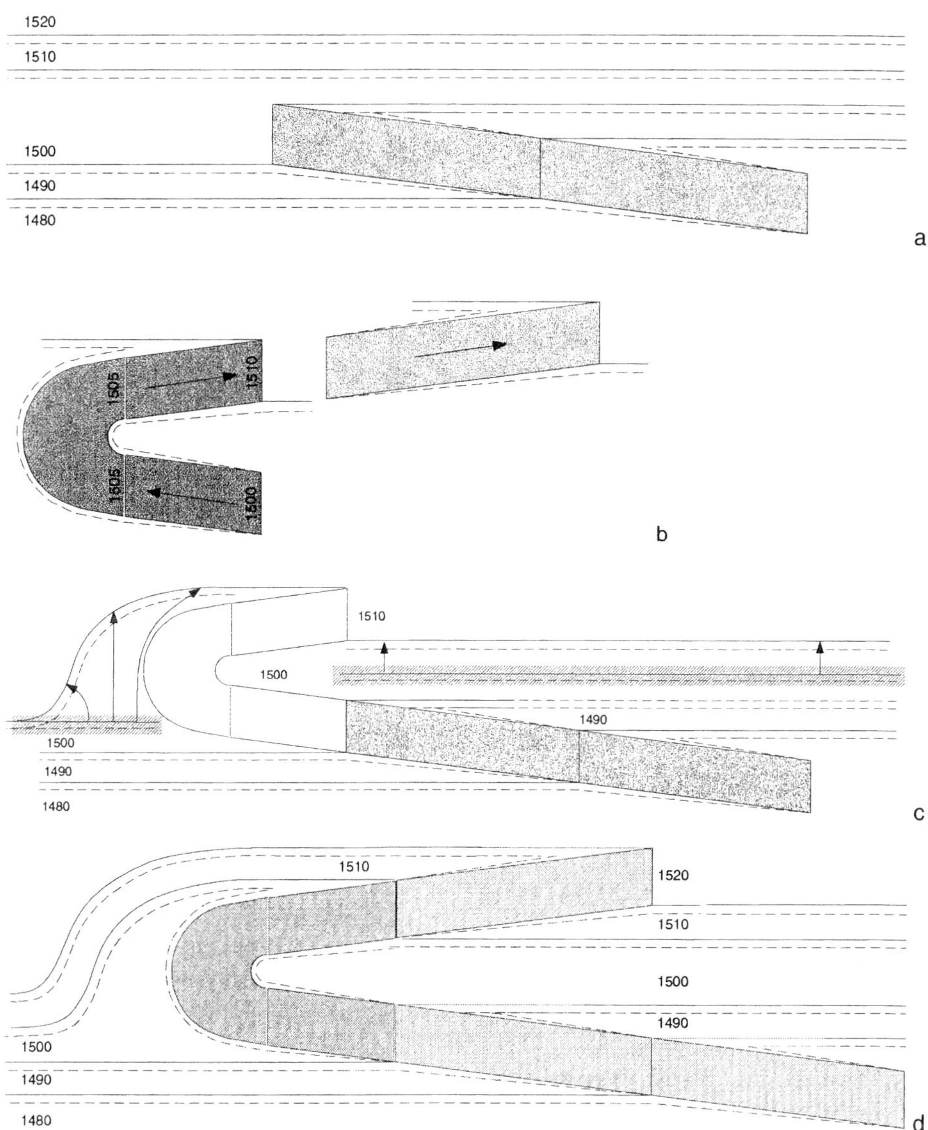

Figure 6.25. Including a switchback in haul road construction. a) Push forward of haul road into pit for benches prior to switchback, b) Templates for a switchback and right facing haul road segment, c) Switchback added and 1500 m toe and crest lines pushed back to make room for level section of the turn and two ramps, and d) 1510-1520 m ramp added by pushing back overlying benches. 1500 m bench pushed back sufficiently far from toe of switchback curve for construction.

mm spacings (2) so that the road template will be more distinctly visible against the expanded pit contours. To save these traverse feature attributes as a new road feature, select **Feature Type Save** and enter the NEW FEATURE TYPE NAME: <road>. Example attributes of a road are given in Figure 6.26.

Group number: 18
Description : in pit haul road (required)

Group name (5 char)	Pen colour (1-32)	Line type (0-4)	Feature label (0-6)	Fill type (0-9)	Symbol type (0-255)	Points label (0,1 or 2)
road	5	1	0	0	0	1

Label direction (0-360)	Char size (1-640)	Points dimension (2 or 3)	Feature dimension (0 or 1)	Feature value flag (Y or N)	Feature Format flag (Y or N)
0	20	3	1	N	N

Figure 6.26. Group attribute table for a road.

The template can now be digitized using the map entry/edit facilities by selecting <road> as the active feature type in the Parameters form. Start the figure by entering the first point on the inner radius of the curve approximately at the 0,0 grid coordinate. The figure will be started by digitizing the two vertices that establish the line that represents the inner edge ramp entrance (see Fig. 6.25b). Select **Define ⇒ Enter** and digitize the point. Note that Lynx does not have a snap to grid option, making it difficult to get exact point coordinates when using the mouse. Instead, point coordinates can be entered from the keyboard into the Parameters form which can be used both for entering new vertices in the feature or for changing the coordinates of existing vertices. Toggle TRACKING off in the Parameters form and key in 0,0,0 for x, y, and z. To determine the location of the second point, the horizontal change in distance and angle must be calculated for the given road grade, bench height, bench width and toe to crest slope angle. For this figure, these factors result in a 50 m ramp up to the switchback's curve which will be at a constant 5 m (the elevation change from the start of the ramp on the lower bench's toe at 0 m). Going from the start of the ramp to the start of the inner curve requires a distance of 50 m and angle change of 13.4°. Therefore, the Bearing facility is the most convenient approach to digitizing the figure. Select **Bearing** and pick the first point as the origin of the vector and enter DISTANCE: <50> and ANGLE: <283.4>. The Bearing facility only defines the x and y coordinates. The second point must be at 5 m relative elevation. Select **XYZ** and use the Parameters form to change z to 5.

Note that for this and the following examples that will use the template, the z value is not important: only the plan view geometry will be used to modify the toe and crest lines of the expanded pit. However, the figure which is being created could be merged into another contoured map of the pit. In this case, the z values would need to match the toe and crest elevations that are being modified by the switchback, and the final merged map would show the actual EOR position as the haul road spiraled through the pit topography. In addition, this final map could then be triangulated and matched against triangulated pre-mining surface topography to obtain a more accurate estimate of volumes.

Before continuing to enter points for the inner road radius, another figure of a circle can be included in the display to use as a template for digitizing the inner and outer curves of the switchback. **Save & Clear** to remove the road figure from active memory. This enables the figures facilities. Select **Figures ⇒ Fetch Figures** bringing up the Fetch Figures selection list and select the circle figure <F.circle>. F.circle is a previously digitized map figure that was created using a digitizing tablet. Use **Fetch** to select the circle, **Size** to reduce its radius to 25 m and **Shift** to position its base at 0,0. **Exit** from the figures menu and use **Enter** with **Snap on** to digitize the inner curve.

To digitize the upper ramp segment, use **Bearing**, remembering to toggle **Snap off**. Use **XYZ** to set the elevation which will increase to 10 m at the top of the segment of ramp exiting the curve. Now select the last point on the inner curve and enter from the keyboard the azimuth and distance to the end of the switchback <76.6,50>. This point is now on the crest of the next bench, so remember to enter a z value of 10. Continue to use **Bearing** until the upper ramp is completed to the point where the outer radius of the curve begins as shown in Figure 6.27. You may have to shift the viewplane at this point to keep the figure within the window. Here the circle figure is required again to digitize the outer curve radius. **Save & Clear** and select **Figures** and use **Fetch** and **Size** to create a 50 m radius circle, using **Shift** to move the left-most edge of the circle 54 m left of the center of curvature of the inner curve and then continue to digitize the outer curve at a z value of 5. This provides for a road width of 30 m on the curve. Once the outer curve is completed, use **Bearing** to add the lower ramp down to a z value of 0. The same procedure can be continued to

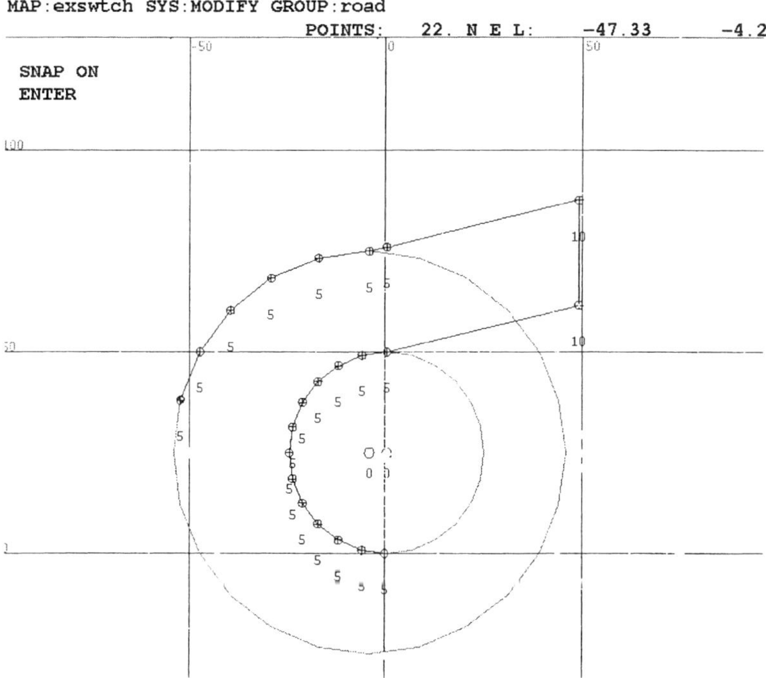

Figure 6.27. Using a circle as a template to construct a curve.

add a toe and crest line on the template. Exit the Map Data Define menu using **File ⇒ Exit & Save**. The final curve template without toe and crest lines will be similar to Figure 6.28. The same procedure can be followed to create clockwise and counter-clockwise spiraling straight road segments as well as curves for corners of the pit. Each figure should be stored as its own map to avoid clutter.

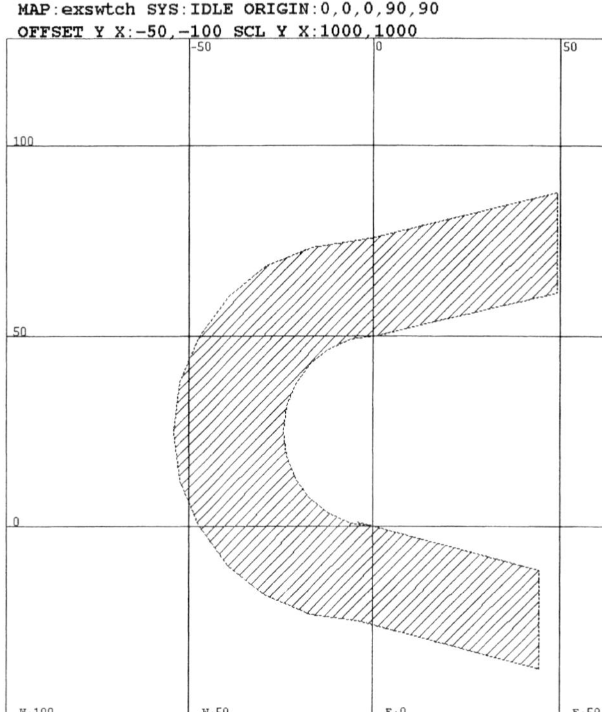

Figure 6.28. Switchback map figure.

Summary of procedure: Example 6.10
Initial data requirements: none.

Procedure:
1. Create a map to hold the figure **File ⇒ Map Data ⇒ Map Create** (Map create entry form)
   MAP NAME: <exswtch> (no data needed ⇒ no map variables needed)
2. Define the road template **Maps ⇒ Map Data Define** (Map Data Define menu)
   2.1 Set up a viewplane conducive to figure definition ⇒ **Viewplane ⇒ Setup** (Viewplane Set-up entry form)
   OFFSET Y: <-100> OFFSET X: <-100>
   SCALE Y: <1000> SCALE X: <1000>
   GLOBAL GRID: <on> N: <50> E: <50> L: <50>
   TRACKING: <global>
   ⇒ **OK**
   2.2 Define a new map feature for the road template. From the Map Data Define menu **Map ⇒ Map Feature Attributes** (Map Feature Attributes entry form)

FEATURE TYPE SELECT: <trav>
FILL TYPE: <hatch>
Z VALUE DISPLAY: <on>
⇒ **Feature Type Save**
NEW FEATURE TYPE NAME: <road>

2.3 In the Parameters form, select <road> as being the active feature type

2.4 Define the inner EOR of the ramp entering the curve
Define ⇒ **Enter** (Digitize the first figure point at 0,0,0. To be precise, set TRACKING: <off> in the Parameters form and key in 0,0,0 for *x*, *y*, *z*.)
⇒ **Bearing** (Selecting the previously digitized point will bring up the Bearing entry form. With TRACKING: <off> the exact distance and angle to the start of the inner curve can keyed into the form)
DISTANCE: <50> ANGLE: <283.4>
(The *z* value is not defined by Bearing. So use ⇒ **XYZ** to select the second point and enter a *z* value of <5> in the Parameters form.)
⇒ **Save & Clear** (Clear the figure from the active memory so that the Figures menu is available.)

2.5 Fetch a circle figure to use as a template ⇒ **Figures** ⇒ **Fetch Figures** (Fetch Figures selection list)
AVAILABLE MAPS: <F.Circle>
⇒ **Pick Figure** (pick circle)
⇒ **Size** (pick size point as circle center)
SIZE FACTOR: <.25>
⇒ **Move** (pick the move point to be a vertex just left of center on the circle's base)
⇒ **Snap On** (Click on the second point of the road figure. F.Circle will now serve as the inner curve.)

2.6 Continue with digitizing the road
⇒ **Select** (Click on the road segment to make it active
⇒ **Define** (**Snap On**) ⇒ **Enter** (Select points around the circumference of the circle to complete the inner radius.)
or ⇒ **Pick** (select the road figure) ⇒ **Segment** (Pick first and last points needed from the circle's radius, but note that the 5 m *z* value will not be automatically assigned using **Segment**. Use **XYZ** and the Parameters form to assign *z* to these points.)
⇒ **Snap Off** ⇒ **Bearing** (to complete the inner EOR select the last point of the curve and enter the last point on the top of the ramp.)
DISTANCE: <50> ANGLE: <76.6>
(Use **XYZ** to enter Z: <10> in the Parameters form)
⇒ **Bearing**
DISTANCE: <25> ANGLE: <0>
(Use **XYZ** to enter Z: <10> in the Parameters form)
⇒ **Bearing**
DISTANCE: <50> ANGLE: <256.6>
(Use **XYZ** to enter Z: <5> in the Parameters form)
⇒ **Save & Clear** (ready to commence outer EOR curve using circle template)

2.7 Expand circle to 50 m radius ⇒ **Figure** ⇒ **Pick Figure** (pick circle) ⇒ **Size** (pick center as size point)
SIZE FACTOR: <.5>

2.8 Complete figure definition
⇒ **Select** (road figure again active)
⇒ **Define** (Snap On) ⇒ **Enter** (or use **Segment** and **XYZ** and complete outer curve of road)
(**Snap off** and use Tools facilities to determine approximate distance and angle to final point on road figure.)

⇒ **Tools** ⇒ **Vector** (place cursor at (-25,0))
⇒ **Define** ⇒ **Bearing**
DISTANCE: <57>   ANGLE: <103>
(Or use **XYZ** for precision.)
⇒ **XYZ** (and in Parameters menu place the last point at the exact coordinate)
Y: <-25>   X: <0>   Z: <0>
⇒ **Save & Clear** ⇒ **Refresh**
⇒ **File** ⇒ **Exit & Save**

Example 6.11: Including a spiral haul road using Lynx
A simplified version of the optimal pit from Example 6.10 will be modified to include a spiraling haul road. The pit design will remain the same as in the previous discussion: bench heights and widths of 10 m and toe-to-crest bench face slopes of 75°. The road will be based on a straight road template as shown in Figure 6.24b with an apparent width of 26 m, length of 100 m, and angle from the toe and crest lines of 14.4°.

From the main Version 4.7 menu select **Surface Eng** ⇒ **V3 Open Pit Engineering** bringing up the Version 3.12 Lynx-Open Pit Design menu. Enter the 3D Grid to be selected GEORES> and enter the engineering design name <EXPPIT>. Note that the design name relates to a family of related designs. In this case, EXPPIT will eventually contain a series of pit designs related to an interpretation of the optimal pit contours generated in Examples 6.9 and 6.10. Files holding the alternative designs for these pits reside in the path /apps/lynx/projects/TUTORIAL/designs/EXPPIT. The expanded pit that was generated in Example 6.11 was named 'exppit'. For this example, a simplified oval pit was generated as per Example 6.9 and named 'EXPPIT'. Both of these designs can be found in the EXPPIT subdirectory as the files 'CPEXPPIT' and 'CPexppit'.

Select **Conical Expansion Mine Design** ⇒ **Conical Expansion Mine Design** and enter the PIT ID as <spiral> and the topographic grid that was created in Example 6 <TOP>. Since the new pit called 'spiral' is to be based on the expanded pit 'EXPPIT', enter EXPPIT as the source pit. An alternative would be to enter EXPPIT as the PIT ID, but the changes would be stored back into the original expanded pit file 'CPEXPPIT'.

Set the viewplane orientation (N, E, L, Azi, Inc: 22100,38600,1500,0,90), scale (6000:1) and grid (G: 200:1) and **Refresh** the screen. **Return** from the vplane menu and select **Zoom In** on the area in the bottom of the bench in which the haul road is to be placed, and select **B'grd** ⇒ **Figures** and enter the map in which the straight road segment was stored. A map called 'rdtmplt' containing both the straight road segment and another switchback is available for this purpose. Use **Fetch** and **Shift** to position the road segment at an appropriate starting point in the bottom of the pit and **Rotate** the road template into position as shown in Figure 6.29 using either the mouse to define the rotation or **Kbd** for keyboard entry.

Commence pushing the pit inwards so that the 1480 toe follows the template by selecting **Define** ⇒ **Bndry** and entering the lowest toe elevation <1480>. After completing the push-in, selecting **Return** will result in the pit being expanded. Now the 1490 crest must be modified so that the crest line crosses the top of the template.

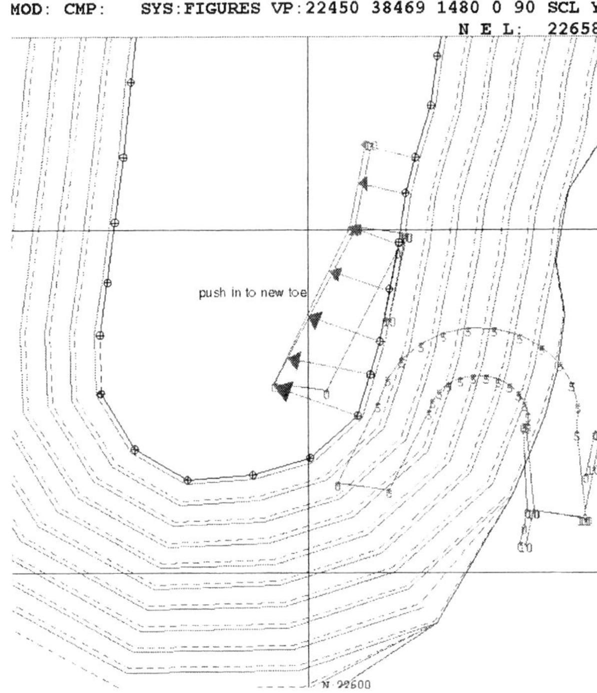

Figure 6.29. Using a map figure as a template to modify a pit to include a spriraling haul road using a pit push-in.

Again, select **Bndry**, enter <1490> and <crest> and modify the crest line. From the tools menu, use **Shift** and **Rotate** to position the base of the road template on the point where the 1490 crest line met the ramp, then **Return** to expand from the 1490 crest.

As this process is repeated, note that the road will continue to spiral into the left side of the pit and that push-ins turn into pushbacks due to the width of the road. As a result of the pit being expanded into deeper cover, it is apparent that a switchback should be considered as a means of reducing the stripping ratio. The final spiral pit is shown in Figure 6.30.

Example 6.12: Switchback design using Lynx

It was apparent in Example 6.11 that a spiral haul road would involve pushbacks into the highwall side of the pit. No starting location for the road in the pit bottom could have avoided this since the road requires nearly 360° to exit at the crest. Generally, a switchback should be avoided due to the much greater pushback that is required for its construction, but by using a switchback the road can be kept on the shallower side of the pit.

The switchback design for this example follows the discussion earlier in this section and will be based on the template created in Example 6.11, i.e. a 25 m road which widens to 30 m on a flat curve of 25 m inner radius. The curve will be accessed and exited by a 50 m straight segment of 10% grade.

Follow the same procedure as given in Example 6.11. Return to conical pit expan-

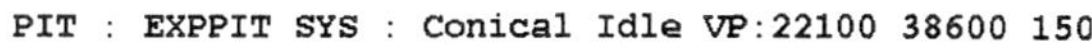

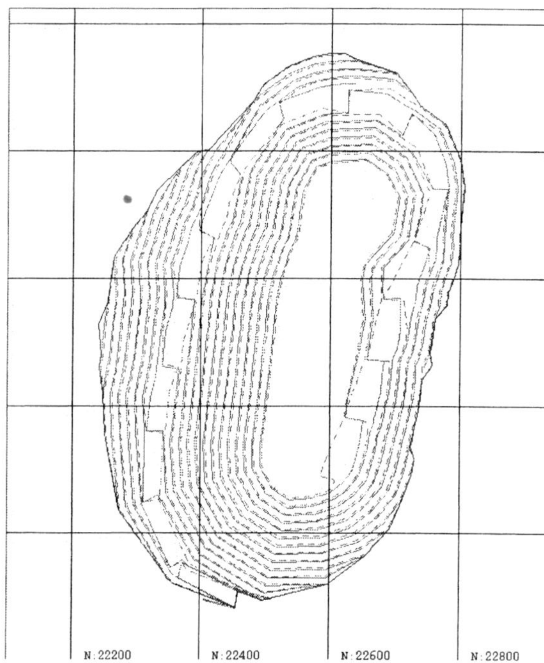

Figure 6.30. Spiral haul road based on bottom to top bench modification.

sion entering a new pit name to be stored under the design subdirectory ./EXPPIT. Enter a new pit name <switchbk> to create a file in the EXPPIT design subdirectory named 'CPswitchbk'. Enter the spiral pit design <spiral> of Example 6.11 as the source design. Set the viewplane orientation as per Example 6.11 and use the Tools menu to fetch the map 'rdtmplt' as a figure.

In order to exit the pit near the lowest pit crest elevation, the switchback will have to be placed on the 1510 bench. Thus, the spiral pit modifications for benches 1480, 1490 and 1500 can be retained. Start the pit modification with the 1510 toe. Note that once this boundary has been modified, all higher features will be reset to fit an expansion of this toe following the basic slope and bench width criteria for this pit design. Use **Fetch**, **Shift** and **Rotate** in the Tools menu to position the switchback template into position on the 1510 crest and modify the toe as shown in Figure 6.31. Provide sufficient room on the 1510 bench on the inside of the curve for subsequent straight road segment pushbacks. The modification of the pit continues with the road spiraling clockwise out of the pit. The final design is shown in Figure 6.32.

6.10 PIT VOLUME MODELLING AND MINING RESERVE ANALYSIS

Once a final pit limit has been generated, it must be converted into a data structure which can be used for mine reserve reporting. This is accomplished by intersecting the pit model on the geologic volume and mining block models to determine total re-

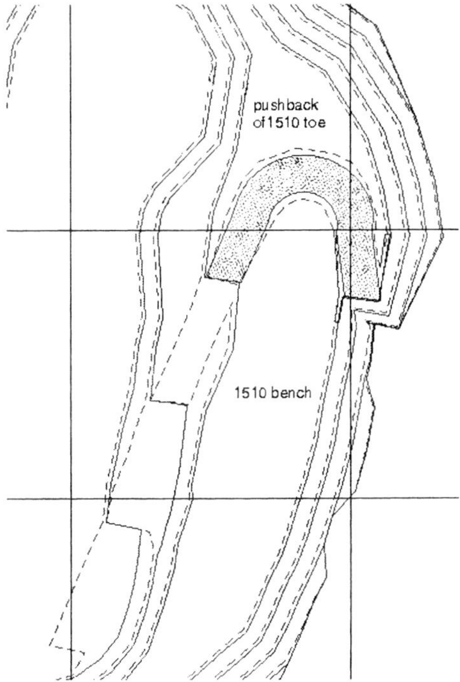

Figure 6.31. Using a switchback map figure to modify the toe and crest lines by pit pushback.

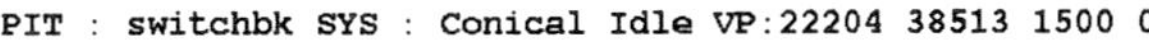

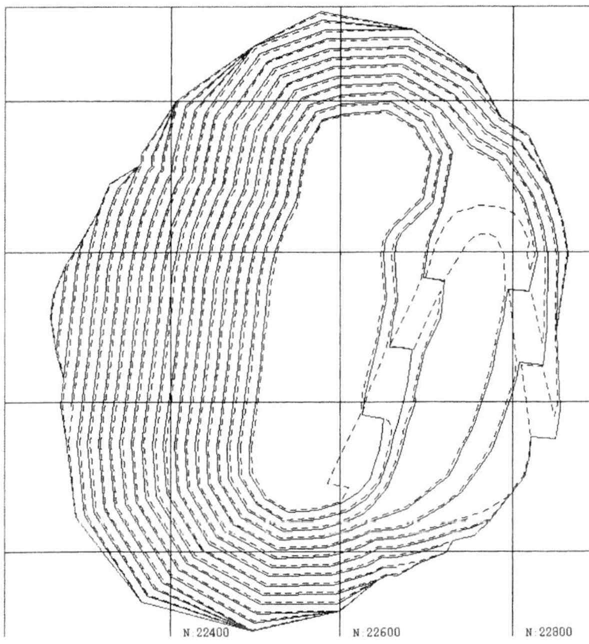

Figure 6.32. Expanded pit with a switchback haul road.

covered values along with the tonnages of ore and waste that will have to be moved during the projected mine life. The pit surface can be represented in several ways that are suitable for intersection with the block and geologic models, either as bench polygons, a gridded surface or a volume model. Note that the geologic model will have already been intersected on the block model using one of these methods. Thus, the block model will already contain values for ore and waste tonnages so that intersection of the pit limit with the block model will provide mining reserves by each geologic unit in the geologic volume model.

Representing a pit limit as a set of polygons is a handy way of determining mining reserves. In this case, there will be one polygon representing the limit of each bench. This is the most direct approach to reserve estimation when the pit has been generated by polygon expansion methods as illustrated in Example 6.5. Typically, the polygon will be located at the midbench elevation. The set of polygons is then used to determine the fractional volume of each block that is lying within the pit limits. This method proceeds bench by bench, which expedites reserve reporting by bench. On the upper benches that outcrop on the surface, the bench polygons will not be closed and contours of the pre-mining surface topography must be used to complete the polygon.

Another option is to generate 2D grids of the pit limits and pre-mining topography and then intersect these on the 3D block model grid. In this case, all three grids have to be defined on the same grid interval in plan view as discussed in Chapter 4. This method requires somewhat more effort than using bench polygons, since the grids must be estimated, but it has the potential of being more accurate since the pit limits and topography can be represented in greater detail. The estimation method used for gridding the pit limits must be carefully selected and implemented to avoid the smoothing effect: toe and crest lines must be honored. A weighted average gridding algorithm can smooth away toe, crest and haul road survey data, leaving a featureless surface. On highwalls, this effect might not matter, but where the pit slope changes rapidly, as in the transition from a highwall with narrow catch benches to a bench with a road, the grid will underestimate the contained volume. In general, weighted average and extrapolation algorithms should be avoided entirely for pit limit gridding. In contrast to these methods, triangulation can be used to strictly honor toe, crest and EOR data.

A third approach to determining mining reserves is to represent the pit as a volume model using the same methods as for geologic volume models as discussed in Chapter 4. In Lynx, either triangulated surfaces or interactive modelling can be used to generate the volume model. The final expanded pit can be stored as map data. In fact, Lynx does this automatically, saving the pit design as a design file in the ./designs/EXPPIT subdirectory and as a map in ./overlays in the project directory. The map data can be triangulated and map features honored. Since crest and toe lines are map features, the triangles will be formed between adjacent toe and crest lines. A map containing the surface topography is also triangulated, and prism-shaped volume components are defined by mapping the pit's triangle set on the pre-mining topography.

An alternative volume modelling method is to interactively define volume components. This parallels geologic volume modelling, but in this case the midplane orientation will be in plan view and the map of the expanded pit is shown in background. The midplane interpretation of the bench limit can be based on either the bench's toe, crest or an interpretation of the midbench elevation, in which case the toe and crest would be used to interpret the foreplane and backplane of the volume component.

Example 6.13: Pit volume modelling and mining reserve estimation using Lynx
Surface handling will be used to generate a volume model based on the spiral pit of Example 6.12. First the pit volume model will be generated using the surface handling facilities. Once the volume model is available, volumes and reserves can be reported using several facilities that will report either the volumes of the model components, the volumes of intersection of two components, the intersection of a volume model and a 3D grid, or the intersection of two volume models and a 3D grid.

To create the files for a new volume model, select **Files ⇒ Volume Data ⇒ Volume Create** and enter the model's name and description. The components for this volume model will be based on TINS of the post- and pre-mining topography as generated in Examples 6.11 and 4.7, respectively. To generate the TINS make the map active and select **Maps ⇒ Map TIN Define** to bring up the Map TIN Define menu. Note that the default tolerance level for identifying duplicate vertices may result in numerous points being removed from the pit map before triangulation. This is because many of the toe to crest vertices are very close when viewed in the horizontal, especially if the pit slope is steep. Also note that a large and complex pit consisting of many contours may require map contour thinning in order to reduce memory requirements. Select **Generate ⇒ Save & Exit** to create the TIN. Repeat TIN generation for the pre-mining topographic map if necessary.

With TINS of the pre-mining and post-mining topography available, surface handling can be used to generate a volume model. Select **Volumes ⇒ Surface Handling** to bring up the Surface Handling menu. Select **Primary** and enter the map that will be used for the primary surface. Since the triangle sets are based on the primary surface, defining the pit TIN as being the primary surface will result in greater accuracy. At this point the toe and crest line vertices will be displayed if FEATURES is toggled on in the Parameters menu. The triangulated pit will appear as in Figure 6.33. Select **Secondary** and enter the pre-mining TIN.

Next, thicknesses have to be calculated between the triangle set of Figure 6.33 and the topography. Select **Manipulate**, bringing up the Surface Manipulate entry form. The same procedure is used to generate the pit volume model as was used for geologic volume modeling between the hangingwall of zone 7 and the footwall of zone 8 in Example 4.6. Define register A as being the primary surface and register B to be the secondary surface. Select **Function** to bring up the Manipulate function entry form, placing into register C the function A – B. Upon selecting OK, the Manipulate Primary Values form will prompt for a Z OFFSET REGISTER: <A> and a THICKNESS REGISTER: <C>.

Thickness data will be based on the mapping of the triangle vertices of the primary surface from post-mining map onto the secondary triangle set. The thickness is the difference between the B and A registers: this difference will then be stored in the C register. Unfortunately, the pit crest line will not be in exact agreement with the

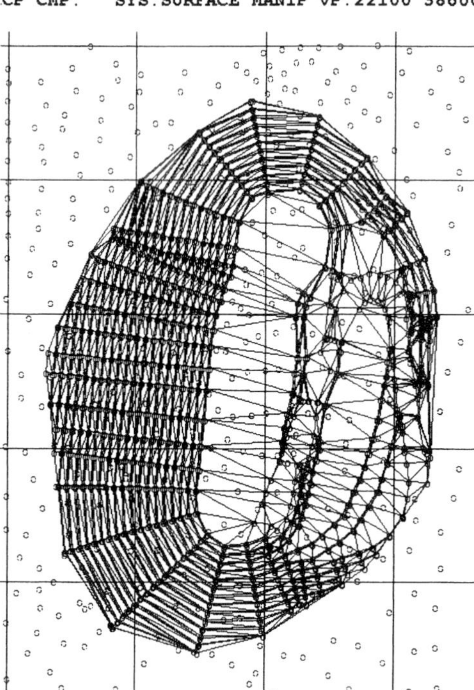

Figure 6.33. Pit map used as the primary triangle set mapped onto TOP.

surface topography because the pit crest was based on the intersection of the pit with a gridded surface which was estimated using inverse distance and the map data contained in 'TOP'. Due to the smoothing that results from using inverse distance and the coarseness of the grid, the gridded crest will not quite honor the original topography and some negative thickness might result. Therefore, negative thicknesses will have to be identified and replaced with zero thickness values in the following calculations.

To generate volume components from the primary triangle set and its associated z and thickness values, select **Generate** from the Surface Handling menu and enter a UNIT ID for the pit along with a PREFIX, a UNIT CODE and a description. At this point, the volume components will be generated and displayed as green fill on the primary triangle set. Look for missing components which may occur if the topography map doesn't completely encompass the pit limits or where prisms have zero thickness. Figure 6.34 shows the PIT volume components in **Y-sect**.

Lynx has several options for generating reports of mining reserves. One method uses volumes of intersection as demonstrated in Example 5.7 in which the volumes of the geologic units were reported using numeric integration. This same procedure can be used with the pit volume model (M.PIT) to report the gross pit volume or tonnage, but since the pit was modelled as a set of prismatic components between a premining and post-mining TIN, the individual components have no relation to either the geologic model or to design volume components. In order to get a meaningful re-

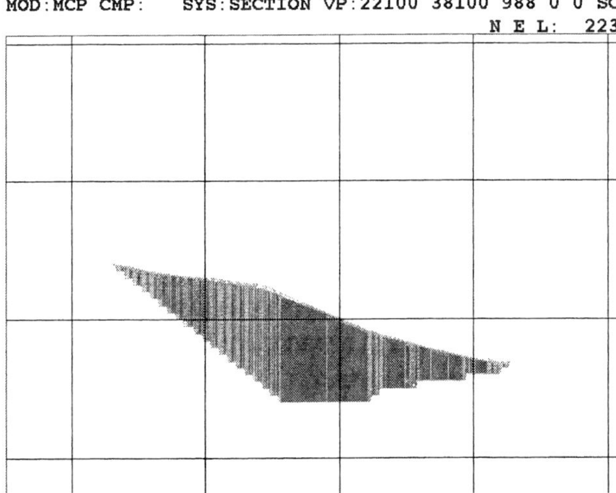

Figure 6.34. Pit volume model in sectional view transverse to strike.

port of mining reserves the geologic model (say, G.OR) must be intersected with the mining model (M.PIT). In this example the reserves should be reported by the geologic units, since the TIN-based mining model's components are individually of no interest. Note that the mining volume model might have been based on using component codes that corresponded to, say, annual pit limits. In this case the TIN's of succeeding years would have been used to generate components and a report of volumes of intersection would yield annual tons for each geologic code. Another interesting approach to open pit volume modelling is to use volume interpretation as in Example 4.10. In this approach the midplane of the volume model would be oriented in the horizontal and each mining volume component would be based on the limits of a bench's toe and crest. With this type of mine volume model, mining reserves can be reported by bench and geologic code. The toe and crest outlines can be taken from the contours of a pit map. While this approach is a very attractive alternative to the TIN-based approach of this example, using interactive volumetrics to match the bench components with a topographic surface is highly problematic. In Example 4.10, the geologic components defined in a vertical section view that included a TIN of the pre-mining topography. A tight correspondence between the limit of the components and the topography was easily accomplished by matching the uppermost vertices of the volume components with the intersection of the map's TIN. In the case of having horizontally defined bench volume components, the intersection of the topographic TIN will be displayed as a polyline of intersection with constant elevation (the elevation of the current viewplane). This TIN intersection must then be used to define a closed polygon for the mid, fore and backplane of each bench above the lowest elevation of the pit's crest. The resulting volumes for these upper benches are likely to be overestimated. In any case, the work involved in obtaining accurate volume component's using interactive volume modeling will be greater than when using pairs of TINs.

To generate a mining reserve report using volumes of intersection select **Analysis ⇒ Volumetrics Analysis ⇒ Simple Volumetrics** (or **Estimation Volumetrics**) bringing up the Simple Volumetrics entry form. **Simple Volumetrics** generates reserve estimate from the direct intersection of two volume models and is only capable of reporting volumes and tons. The entry form for simple volumetrics is shown in Figure 6.35. Of primary interest is generating a report of total mine reserves by geologic unit. Therefore, the **Primary Volume Data** is selected to be the geologic model <G.OR> while the **Secondary Volume Data** <M.PIT> is the mining model. The Secondary Volume Data pushbutton is made available by selecting **Report By**: <Secondary Volume>. Under **Volume Code** the codes to be included in the report for units in the primary volume model are selected along with an associated material density. The remainder of the parameter values are given in Figure 6.35 and explained in Example 5.7.

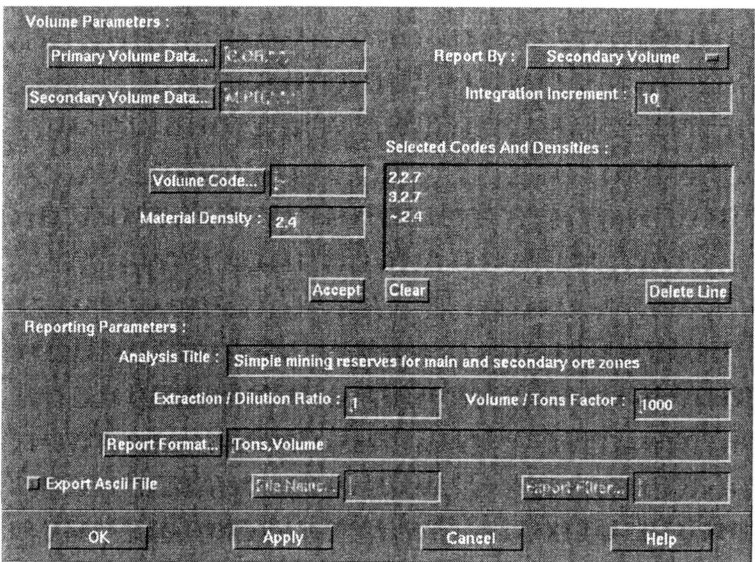

Figure 6.35. Simple Volumetrics entry form to report volumes of intersection between mining and geologic volume models.

Estimation Volumetrics intersects the volume model(s) with a 3D Grid so that values estimated into the grid can also be reported. A completed Estimation Volumetrics entry form for this example is shown in Figure 3.36 with the resulting reserves report in Figure 3.37.

Note that the processing time required for calculating volumes intersection when one of the volume models is TIN-based can be very great due to the large number of components needed for modelling a pit.

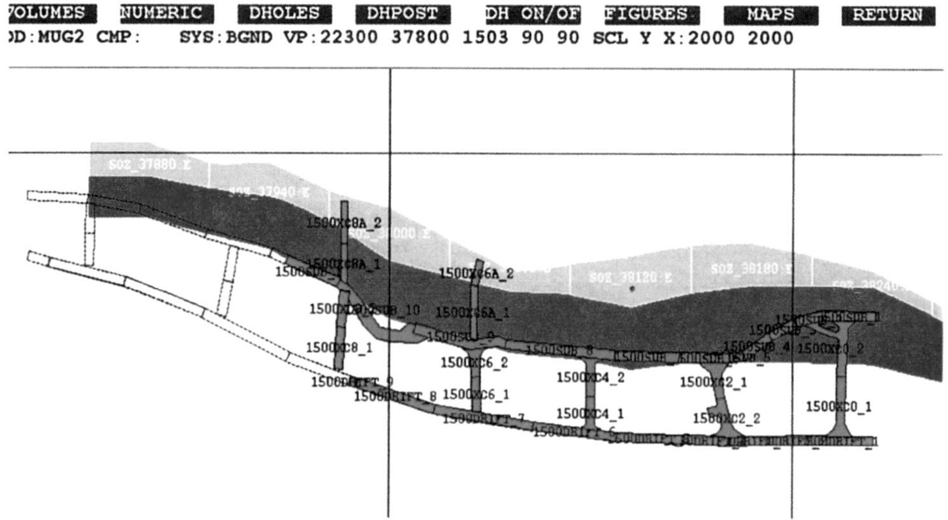

Figure 7.19. An example of development drifts centered on the 1500 m sublevel.

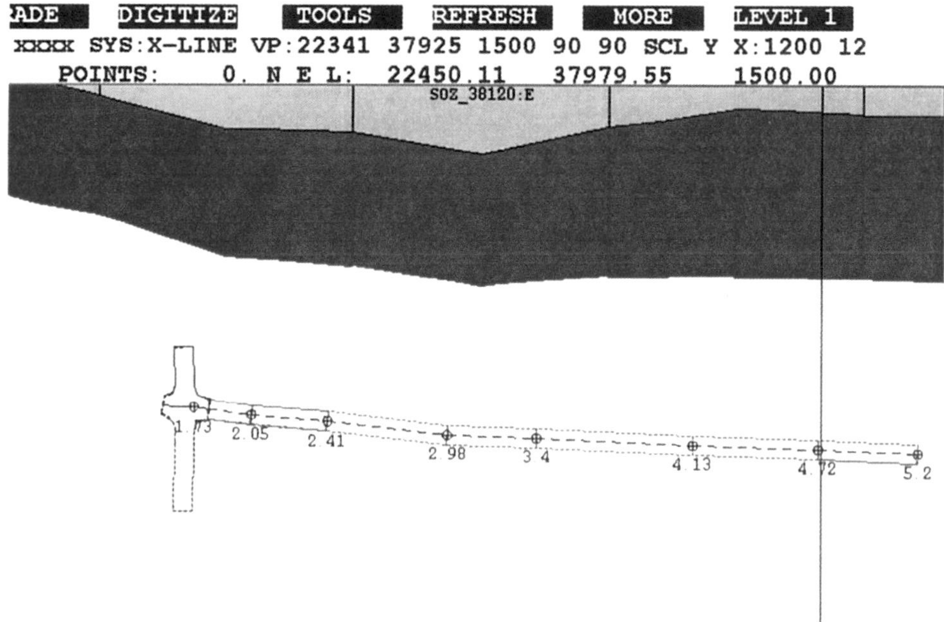

Figure 7.20. Completed extension of 1500 drft E.

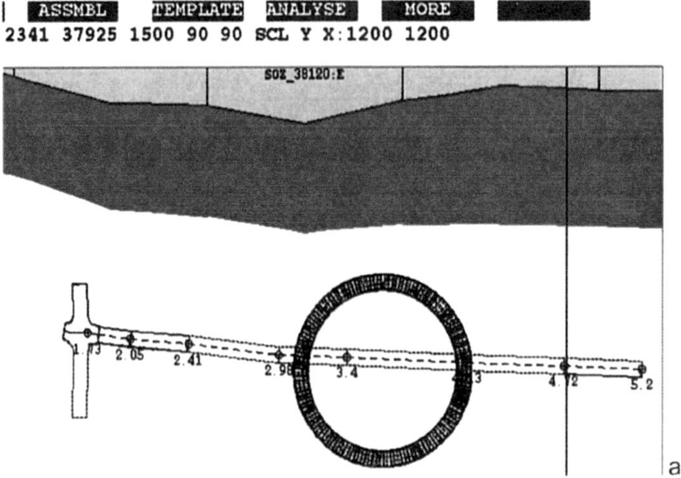

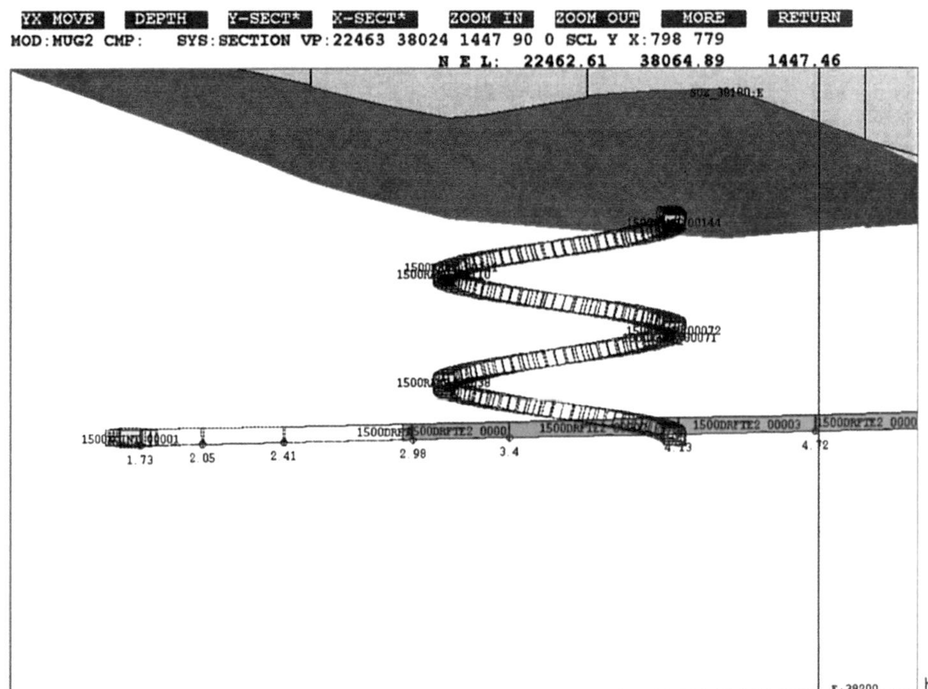

Figure 7.22. 1500 level spiral ramp in: a) Plan view and b) Side view.

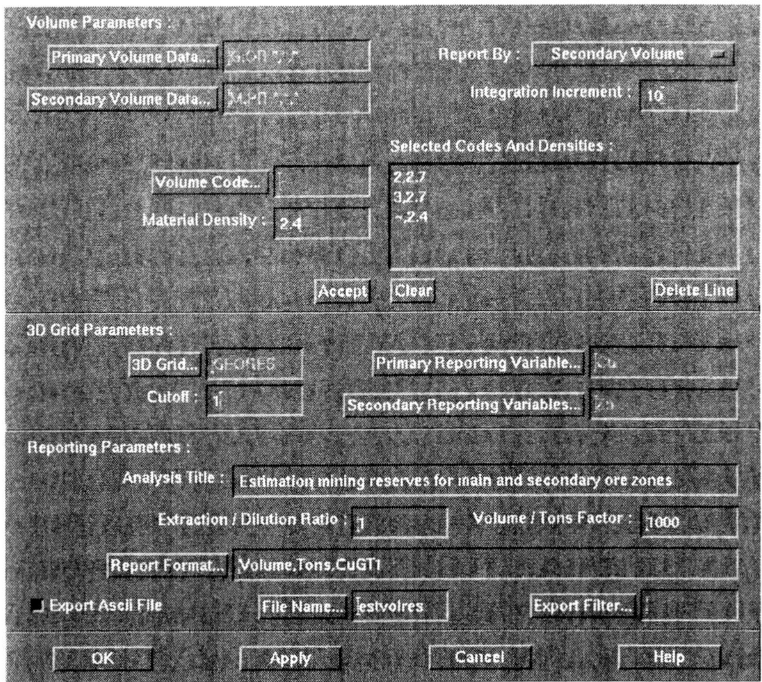

Figure 6.36. Estimation Volumetrics entry form for open pit mining reserve calculations on a 3D grid.

Analysis : Estimation mining reserves for main and secondary ore zones
Report By : Secondary Volume
Primary Volume Selection : G.OR,*,*,*
Secondary Volume Selection : M.PIT,*,*,*
Integration Increment : 10
Volume Codes And Material Densities : Code Density

	Code	Density
:	2	2.7
:	3	2.7
:	UDF	2.4

3D Grid : GEORES
Cutoff : 1
Primary Variable : Cu, Secondary Variables : Zn, Extraction / Dilution Ratio : 1
Volume / Tons Factor : 1000

Reporting Unit	Reporting Component	Primary Code	Volume	Tons	CuGT1
*	*	2	1547	4176	3.33
*	*	3	1289	3481	4.64
*	*	*	18006	44064	3.77

Total Undefined Volume : 15169, Undefined Tons : 36407, Average Concentration 0.273

Figure 6.37. Estimation Volumterics mining reserves report from intersection of M.PIT, G.OR and 3D grid GEORES.

Summary of procedure: Example 6.13
Initial data requirements: contoured maps of pre-mining and post-mining topography.

Procedure:
1 Create a new volume model **File ⇒ Volume Data ⇒ Volume Create**
VOLUME: <M.PIT>
2. Triangulate the pit contours
2.1 Make the map active **File ⇒ Map Data ⇒ Map Select**: <new1>
2.2 Generate the TIN **Maps ⇒ Map TIN Define** (Map TIN Define menu) ⇒ **Generate ⇒ File ⇒ Save & Exit**
(Repeat Step 2 for the pre-mining topography map <TOP>)
3. Use Surface Handling the create the volume components **Volumes ⇒ Surface Handling** (Surface Handling menu)
⇒ **Primary**: <new1>
⇒ **Secondary**: <TOP>
⇒ **Manipulate** (Surface Manipulate entry form)
⇒ **Define** (Manipulate Define entry form)
REGISTER: <A> SURFACE: <primary>
VARIABLE: <z offset>
⇒ **OK**
⇒ **Define** (Manipulate Define entry form)
REGISTER: <B> SURFACE: <secondary>
VARIABLE: <z offset>
⇒ **OK**
⇒ **Function** (Manipulate Function entry form)
REGISTER: <C> SUBJECT: <A>
OPERATOR: <-> OBJECT: <B>
⇒ **OK**
⇒ **OK** (Manipulate Save Confirm)
⇒ **Generate** (Volume Generate entry form)
VOLUME: <M.PIT>
UNIT: <P> PREFIX: <SUR>
CODE: <4> APPEND: <y>
⇒ **OK**
⇒ **File ⇒ Save & Exit**
4. View the resulting volume model **Background ⇒ Volumes** (Background Volume Data Selection entry form)
VOLUME: <M.PIT> (accept)
FILL TYPE: <intensity
⇒ **OK**
⇒ **Viewplane ⇒ Y Section ⇒ Setup** (change elevation as needed)
5. Generating a Simple Volumetrics report Analysis **Analysis ⇒ Volumetrics Analysis ⇒ Simple Volumetrics** (Simple Volumetrics entry form)
⇒ **Primary Volume Data**... <G.OR> (accept defaults to *)
Report By: <Secondary Volume>
⇒ **Secondary Volume Data**... <M.PIT (accept defaults to *)
Integration Increment: <5>
⇒ **Volume Code**... <2> **Material Density**: <2.7>
⇒ **Accept** (repeat for code 3 and undefined material '~')
Reporting Parameters: (same as in Example 5.7 and Fig. 6.35)
6. Generating a Estimation Volumetrics report **Analysis ⇒ Volumetrics Analysis ⇒ Estimation Volumetrics** (Estimation Volumetrics entry form). Enter same parameter values as in Step 5 except for 3D Grid Parameters:

⇒ **3D Grid**... <GEORES>
⇒ **Primary Reporting Variable**... <Cu>
Cutoff: <1>
⇒ **Secondary Reporting Variable**... <Zn>

Example 6.14: Calculating reserves from bench polygons using Techbase
The Boland Banya project will be used to illustrate calculation of reserves from an expanded pit generated as in Example 6.5. In an expanded pit, polygons are used to delineate bench limits. Near the pit bottom, the polygon limits are based on either the contours of the base of the deposit, or if the deposit continues to great depth, the pit bottom is based on an economic stripping limit and reasonable working space for the production equipment. The bench limits are expanded from the bottom of the pit following the contours of the deposit to a point where the deposit pinches out or is otherwise uneconomical to mine. Succeeding benches are simple expansions of the previous bench with the plan view expansion distance being based on the required bench face angle and bench width. Bench expansion continues until the bench polygon intersects a topographic contour. From this elevation to the highest point on the pit crest, the expanded polygons must be cut and spliced to match the topography. An example of such an expanded pit for the Boland Banya deposit is given in Figure 6.38.

Figure 6.38 includes contours for the bottom structure of the OA deposit, contours

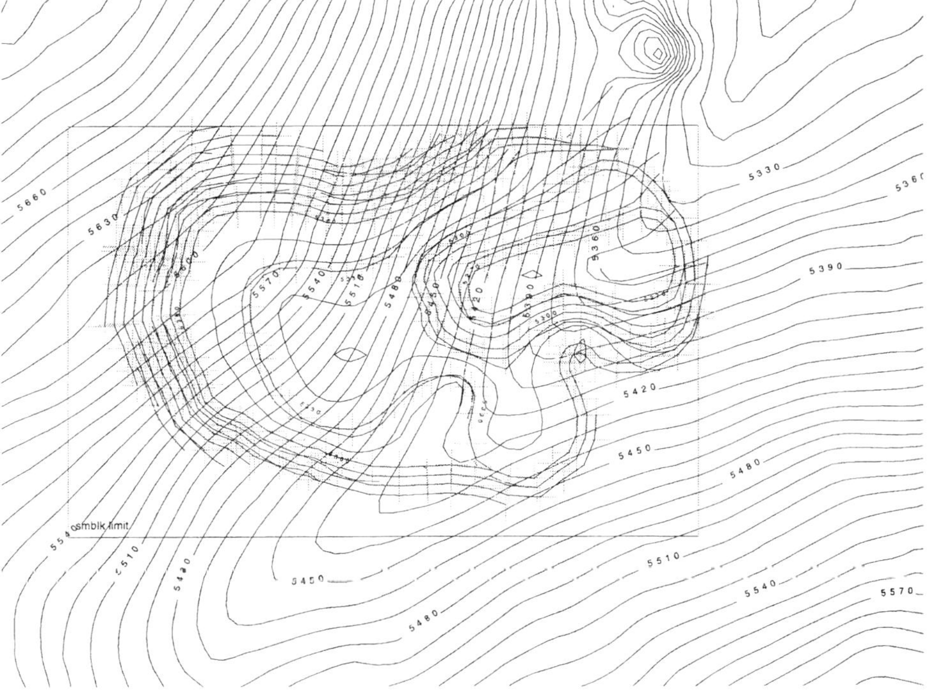

Figure 6.38. Elements needed for estimation of mining reserves using midbench polygons.

of the surface topography, the bench polygons with their vertices and boundary of the block model that will be used for calculating mining reserves. It's a good idea to graphically check the block model limits against the pit crest and topography to ensure that both are large enough to cover the full extent of the pit. This example will start with the set of midbench polygons contained in the file 'exppitmid.pol' and a metafile of topographic contours based on the same elevation base and a contour interval matching the bench height. Both the midbench polygons and the the topographic contours will then be imported into **Polyedit** so that the upper benches can be closed using their corresponding contours. These closed midbench polygons will then be used in **Locate** to determine the fraction of each block which is inside the pit limit and **Report** will then be used to summarize the mining reserves.

To generate an expanded pit based on the midbench elevation, use **Gridcont** to contour the bottom elevation of AO <oab_c> from an elevation of 5270 using a contour interval of 30 ft. The contour elevation must match the midplane elevation of the block model since the block height is 30' with the lowest block level corresponding to an elevation range of 5255 to 5285 ft. More specifically, the midbench elevations must match the values of the 'smblk_zc' default field in the block model table 'smblk'. Under **Contour ⇒ Write polygons**, export the bottom structure contours to a polygon file (oab_midb.pol). Repeat the procedure for the surface topography <topo_c> using the same parameter settings for the contour interval and exporting the topographic contours to 'topo_midb.pol'.

Select **polyEdit ⇒ Load ⇒ polygon file** and input the bottom structure polygons (oab_midb.pol) and repeat to load the topography polygons (topo_midb.pol). A different color and line style can be used for the two sets of polygons to make them more visually distinct.

Select **Edit** and zoom into the lowest contours for the bottom of OA. Select the 5285 polygon and use **Line ⇒ Modify line ⇒ Add** to complete the eastern side of the 5285 polygon forming a closed bench. **Close line** closes the polygon. Use the 'S' hot key to select the 5315 polygon from OA's bottom structure, but note that OA at this elevation is nearly outcropping on the 5315 topography polygon. The 5315 polygon can be opened on the 5300 contour line to provide haul road access, but the objective of this exercise is to determine the bench limits only. Watch for broken polylines in the polygons when closing them. For instance, there may be multiple 5315 polygons along the OA contour. These broken segments will be noticeable when the polygon is closed since they will be excluded from the polygon's circumference. **Operations ⇒ Join** polylines can be used to correct this. In closing the eastern side of the 5315 polygon, the polygon should be held as close to the 5285 polygon as the maximum pit slope will allow. If an overall slope of 45° is to be maintained, then the horizontal distance of separation between the two should be 30'. Cursor tracking in x and y coordinates can be used to locate an approximate pit limit, but for a more exact limit, the 5285 polygon can be **spliT** or **Cut** at locations corresponding to the eastern edge of the pit and then the eastern segment can be **Expand**ed by 30' as shown in Figure 6.39. A less problematic approach than using **SpliT** is to **Expand** the completed bench and assign it a name such as 'tmp', then **Create** a new polygon and use **Follow** to snap the new polygon to the 'tmp' polygon for areas beneath the topography and to the topographic contour where the bench

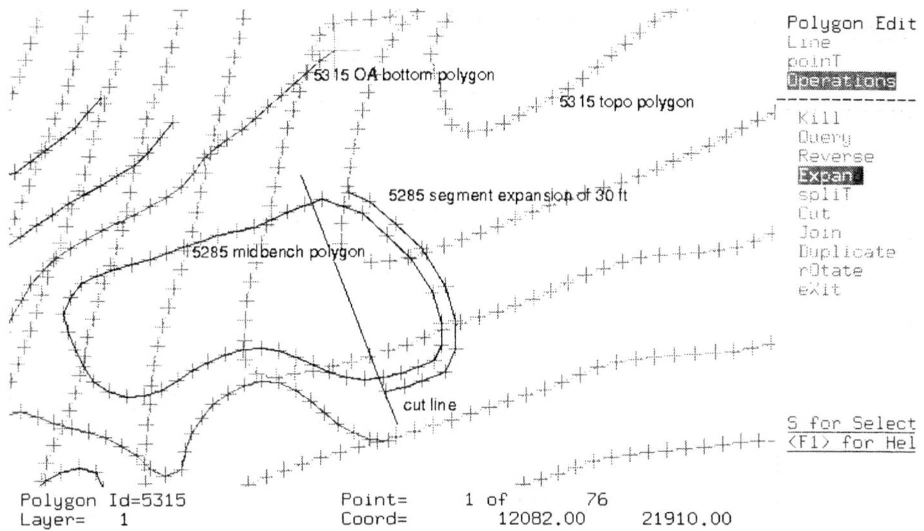

Figure 6.39. Using polyEdit to create a mid-bench expanded pit.

outcrops. Either **Join** under **Operations** or **Line** ⇒ **Modify** ⇒ **Follow** can be used to move the 5315 polygon to an expansion distance of 30' from the 5285 polygon. Using **Follow**, vertices can be added to the 5315 polyline until it is close to the point at which the expanded 5285 segment is to be adhered to, then select **Follow** and select the initial vertex and final vertex along the segment that the 5315 polygon is to follow. The 5315 midbench polygon will now exactly match a 30' expansion of the 5285 midbench polygon. **Delete** the 5285 expanded 'tmp' segment when finished.

The same procedure can be repeated for the 5345 midbench polygon, except that at this elevation the bench daylights on the 5345 topographic contour in the middle of the eastern bench face. In this case, both a pit temporary expansion of the 5315 midbench polygon and the 5345 topography will be used to define the 5330 midbench polygon.

Continue creating the midbench polygons until the highest pit crest is above the matching topographic contour. This should happen at the 5645' midbench polygon. Finally, select and **Delete** the topographic contour polygons until only the bench polygons are left and then **Save** them in a polygon file 'exppit_midb.pol'. A full set of midbench polygons is shown in Figure 6.40.

Before continuing, use **Define** to **Create** a REAL field in the block model (smblk) to hold the fraction of each block that is within the limits of the pit (**Techbase** ⇒ **Define** ⇒ **Fields** ⇒ **Autotable** <smblk> ⇒ **Create** <midb_f>). Also, create tonnage <midb_tons> and ounces <midb_au and midb_ag> calculated fields which are tied to the fraction field. The total tonnage of the block can be calculated in RPN as:

smblk_rsz smblk_csz * smblk_lsz * minb_f * 174 * 2000 /

or as <6525 midb_f *> since the blocks are of constant size and density. The ounces per ton of gold and silver have already been estimated in the fields <ausmblk> and

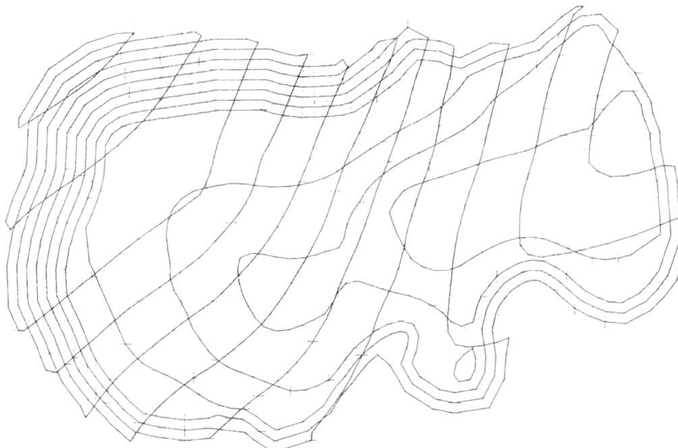

Figure 6.40. Complete set of mid-bench polygons for the Boland Banya expanded pit.

<agsmblk>. The ounces per block can be found as: <ausmblk midb_tons *> and <agsmblk midb_tons*>.

Next, the fraction <midb_f> field will be assigned a value by locating the portion of each block lying within the midbench polygon associated with that level. When using polygons, **Locate** will assign a fractional value to every block lying within the limits of the polygons read from the polygon file. Since the polygon file 'exppit_midb.pol' was saved with all of the midbench polygons, many of the blocks will be located within multiple polygons resulting in fractional values which exceed one. Therefore, before continuing, the file 'exppit_midb.pol' must be broken up into thirteen separate files. Use a text editor to create new single bench polygons files 'midb1.pol' to 'midb13.pol'.

Once the thirteen polygons have been created and checked for errors, continue with the **Locate** program. From the main menu select **Modelling ⇒ Locate ⇒ Database ⇒ Filter ⇒ Add filter** and add the filter <smblk_zc = 5270>. Select **Setup ⇒ Fields** and enter the default *xy* coordinate fields for the block model (<smblk_xc> and <smblk_yc>). Since **Locate** is to be based solely on a 2D polygon, the _zc field shouldn't be entered. Under **Polygon**, enter the polygon file name <5285>. Under **Assignment**, enter <midb_f> as the polygon fraction field. **eXit ⇒ Locate** will assign values to <midb_f>. Repeat this procedure for each bench by changing the filter setting and polygon file name.

Any operation that involves as many repeated tasks as in this example has a high risk of error. A good way to confirm that the blocks on each bench have been correctly assigned a 'midb_f' value is to use **Poster** to display the located cells on each bench along with their associated midbench polygon. Start with the uppermost bench by setting a filter to <smblk_lev = 1> and <midb_f > .01>. Display the blocks on each level by also using <smblk_lev> as the block color. Visually check each of the thirteen levels to confirm that the blocks are located within the polygons as expected.

Finally, the block model total tonnage and ounces can be sent to an ASCII report file of mining reserves. From the main menu, select **Techbase ⇒ Report ⇒ Report file** and enter the name of the reserve report file <mineres.rpt>. In **Page layout**, provide a report title, but set FIELD VALUES DISPLAYED to <NO> since only totals for tonnage and ounces are to be reported. Under **Fields**, enter <midb_tons (t) midb_au (t) midb_ag (t)>. Select **Report** to generate the totals and **View file** to see the results. Using the 2D polygon approach, the reserves are reported as:

mid_b polygon reserve estimates for smblk

	midb_tons	midb_au	midb_ag
TOTALS:			
	19,867,791.09	2,031,583.46	39,389,986.84

Summary of procedure: Example 6.14

Initial data requirements: gridded surfaces of topography and the bottom structure of the OA deposit for the Boland project.

Procedure:

1. Export midbench contours from the bottom of OA and the pre-mining topography **Graphics** ⇒ **Gridcont** ⇒ **Scaling** (set **Scale** and **Range**) ⇒ **Contour** ⇒ **Field**
   CONTOUR VALUE: <oab_c>
   ⇒ **Intervals**
   CONTOUR INETRVAL: <30> LOW CONTOUR: <5285>
   ⇒ **Write polygons**
   FILE NAME: <oab_midb>
   (Repeat Step 1 for <topo_c> and **Write polygons** to <topo_midb>)
2. Generate the midbench polygons **Graphics** ⇒ **polyEdit** ⇒ **Load**
   2.1 Load the polygon and any background metafiles ⇒ **Polygon file**
   FILE NAME: <oab_midb> LAYER: <1>
   (repeat for topo_midb.pol assigning it to LAYER: <2>)
   ⇒ **eXit**
   2.2 Editing single polygons to form closed midbench polygons ⇒ **Edit**
   (Place the the cursor on 5285 from oab_midb.pol and hit S to select it.)
   ⇒ **Line** ⇒ **Modify line** ⇒ **Add**
   (Note that in the level 3 menu for **polyEdit** the hot key for the function must be used while the cursor is positioned on the location of the action. Thus to **Add**, **Enter**, **Insert**, and any other function used to manipulate vertices, the capitalized letter in the function name must be used while the cursor is positioned, for instance, to **Add** to add vertices to the polygon. Add vertices to 5285 as shown in Figure 6.39 to complete the 5285 bench and close the polygon with **Close line**)
   2.3 Expanding a polygon to represent the limit on the next higher bench (5285 still active). **eXit** to level 1 of **Edit Operations** ⇒ **Expand**
   NEW POLYGON ID: <5315t> EXPANSION DISTANCE: <30>
   2.4 Editing two polygons (5315 from topo_midb.pol and 5315t) to form a single joined polygon. From level 1 of **Edit** select Display
   SHOW DIRECTION: <y> SHOW FIRST: <y>
   (The symbols '<⊗' indicate the starting point and direction on the polygon to be used as an aid in using the Operations functions.)
   Cut 5315t into two polygons, one that will be kept as the 5315 highwall and a remainder to be discarded → **Operations** → **Cut** (Placing the cursor on the vertex at which 5315t is to be Cut will result in two closed polygons cut along a line running from the first point on 5315t to the position of the cursor)
   Delete the remainder of 5315t. **eXit** to **Line** ⇒ **Modify line** (select the polygon to be deleted) ⇒ **Delete** (confirm)

Join 5315 to 5315t. **Open line** (level 3 Modify line menu opens 5315t) ⇒ **eXit** ⇒ **eXit** ⇒ **Operations** ⇒ **Join**

(Before continuing, identify the two polygons to be joined and note where their first points are and what direction, CW or CCW their vertices are ranked. Both polylines (open polygons) must have the same direction and the last point of one polyline must be joined to the first point of the second. If both directions aren't the same, then select one of the two polylines and **Reverse** its direction before using **Join**. Select as the first polyline the one with its last vertex next to the first vertex on the second polyline. Select **Join** and then use **S** to select the second polyline)

Close the 5315 midbench polygon and insert vertices as needed **eXit** ⇒ **Line** ⇒ **Modify line** ⇒ **Close line** ⇒ **Insert**

(Repeat Step 2 until the expanded pit completely reaches the pre-mining topography as shown in Figure 6.40)

2.5 Save the edited polygons to file. From the Interactive Polygon Editor main menu select **Save** ⇒ **polygon File**
FILE NAME: <exppit_midb> APPEND: <n>

3. Create the real and calculated fields needed to hold the fraction of the block inside the midbench polygons, the total tonnage of the block, and the ounces of gold and silver in each block. **Techbase** ⇒ **Define** ⇒ **Fields Autotable**
TABLE NAME: <smblk>
⇒ **Create**
FIELD NAME: <midb_f> TYPE: <real> CLASS: <actual> DECIMALS: <2>
Create the following fields in the block model as CLASS: <calculated>
FIELD: <midb_tons> EQUATION: <6525 midb_f *>
FIELD: <midb_au> EQUATION: <ausmblk midb_tons *>
FIELD: <midb_ag> EQUATION: <agsmblk midb_tons *>
4. Use a file editor to break up the midbench polygon file <exppit_midb.pol> into a series of separate files, each only containing the polygons related to one bench (e.g., midb1.pol ... midb13.pol)
5. Use the Locate facilities to assign the fraction of each block that is within the limits of its respective midbench polygon **Modelling** ⇒ **Locate**
5.1 Define a filter limiting the bench to be located **Database** ⇒ **Filter** ⇒ **Add filter** (e.g., for the lowest bench)
FIELD: <smblk_zc> RELATION: <=> FIELD|VALUE: <5285>
5.2 Provide block coordinates **Setup** ⇒ **Fields**
X: <smblk_xc> Y: <smblk_yc> Z: <>
5.3 Provide the midbench polygon ⇒ **Polygon**
FILE NAME: <midb13>
5.4 Assign the fraction field ⇒ **Assignment**
FIELD: <midb_f> = POLYGON/VERTICAL FRACTION
⇒ **eXit**
⇒ **Locate**
(Repeat Step 5 for each of the midbench polygons)
6. Visually confirm the results of the polygon Locate procedure. From the main Techbase menu **Graphics** ⇒ **Poster**
6.1 Define a filter limiting the blocks to be displayed ⇒ **Database** ⇒ **Filter** ⇒ **Add Filter** (e.g., for first bench)
FIELD: <smblk_lev> RELATION: <=> FIELD|VALUE: <1>
6.2 Set **Scaling**
6.3 Define polygon display ⇒ **Cell/poly** ⇒ **Fields**
VALUE 1: <smblk_nul>
⇒ **Style**
CELL COLOR: <smblk_lev> FILL STYLE: <18> FILL SCALE: <.5>

⇒ **Draw** ⇒ **Outline** ⇒ **eXit** ⇒ **Graphics** ⇒ **Review** (verify results)
(Repeat Step 6 for each bench that has been located using **Graphics** ⇒ **Undraw** to clear display if necessary)

7. Report mining reserves. From the main menu **Techbase** ⇒ **Report** ⇒ **Setup** ⇒ **Report file**
FILE NAME: <mineres>
⇒ **Page layout**
TITLES ON EACH PAGE: <n> FORM FEED ALLOWED: <n>
EACH PAGE NUMBERED: <n> FIELD VALUES DISPLAYED: <n>
FIELD NAMES DISPLAYED: <y> COLUMN WIDTH: <12> GUTTER WIDTH: <2>
⇒ **Fields**
FIELD AND FORMAT LIST: <midb_tons (t) midb_au (t) midb_ag (t)>
⇒ **Report** ⇒ **View file**

Example 6.15: Basing mining reserves on gridded surfaces using Techbase
A more common approach to reserve estimation using Techbase is to locate the blocks lying within the pit limits using 2D grids of the pre-mining surface topography and pit limits. Examples 6.6 and 6.9 illustrated the basis for this approach in that either the moving cone or Lerches-Grossman algorithm was used to determine the pit limits by identifying the lowest mined block in each column of the block model. This 2D matrix of the pit bottom was then contoured to provide an ultimate pit limits map. In Techbase, the gridded ultimate pit limit could be used to directly determine mining reserves by using **Locate** with top and bottom bounding surfaces only rather than a series of midbench polygons as in Example 6.14. Instead of using the grids generated in Example 6.6, a new pit limit grid will be estimated using the expanded bench polygons of Example 6.14 so that the mining reserves using the two different methods can be compared; the ultimate pit that resulted from the moving cone algorithm is based on a different set of criteria and cannot be compared to the pit of Figure 6.14.

The midbench polygon file 'exppit_midb.pol' first must be loaded into a flat table in the Boland Banya geologic database so that is can be used for estimation of an ultimate pit limit grid. Use **Define** to create the flat table 'exppit_midb' and place *x*, *y* and *z* coordinate fields in it (midb_x, midb_y, and midb_z). Also create a new field <z_midb> in the cell table that currently holds the gridded surface topography ('surfaces') to hold the estimated grid value for the pit limits. **Load** the polygon table *x*, *y* and *z* values into the corresponding fields in 'exppit_midb'.

The ultimate pit limits contained in 'exppit_midb' should be gridded using triangulation. From the main menu select **Modelling** ⇒ **triGrid** ⇒ **Grid** ⇒ **Fields** and enter 'midb_x', 'midb_y' and 'midb_z' as the data points and 'z_midb' as the resulting estimated value. Under **Parameters** enter the minimum possible number of derivative points <4> and a gradient of zero. This will constrain estimation to planar patches. Note that Techbase does not use map data or have the ability to honor toe and crest lines as does Lynx and that the resulting estimate can smooth away the flat bench intervals and toe to crest bench faces. In fact, the midbench polygons that have been used in this example have nothing to do with toe and crest lines. A triangle set and contour map of the resulting pit grid is shown in Figure 6.41. Note how chaotic both the triangle set and the corresponding contours are. This is due to the fact that

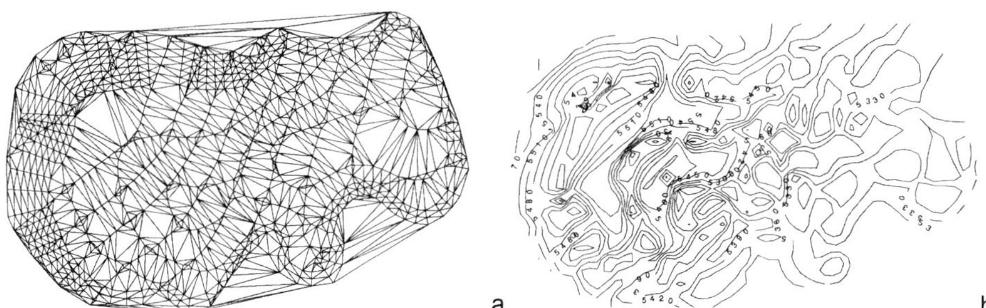

Figure 6.41. Triangulation based gridding using mid-bench elevations. a) Mid-bench triangle set, and b) Resulting mid-bench contours.

there aren't any toe and crest line elevations in the polygon file 'exppit_midb.pol' and the resulting pit surface is simply a linear interpolation across the mid-bench elevations.

Instead of using only mid-bench polygons, gridding can be based on toe and crest elevations. The mid-bench polygons were once again edited using **polyEdit** to remove the portion of the polygon that outcrops with topography and expand the mid-bench polygon by 4 ft inwards and outwards of the mid-bench line to represent a toe and crest horizontal separation distance of 8 ft for a 75° bench face. The resulting toe and crest polygons were concatenated into the file 'exppittc.pol' and then loaded into the 'exppit_midb' flat table as before. **triGrid** was then used to estimate the cell table 'z_midb' field. The triangle set and contour map for the estimation of 'z_midb' based on toe and crest elevation is shown in Figure 6.42. Note the drastic improvement over Figure 6.42b. Even though Techbase doesn't have any provisions for honoring toe and crest lines, as in the case of Lynx, the near proximity of the toe and crest elevations and the use of Delanay triangles ensures that the triangles will usually honor the toe and crest lines.

Now the fraction of each block that is between the gridded ultimate pit limits and surface topography can be estimated using **Locate**. The same fields will be used to hold the fractional value, tonnage and ounces as in Example 6.14. Select **Modelling** ⇒ **Locate** ⇒ **Setup** ⇒ **Fields** and provide the default block coordinate fields in-

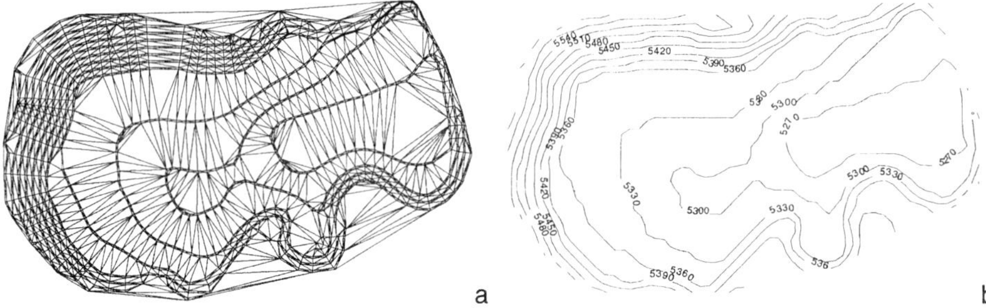

Figure 6.42. Triangulation based gridding using toe and crest elevations. a) Toe and crest triangle set, and b) Resulting contours.

cluding the *z* field for 'smblk': 'smblk_xc', 'smblk_yc', 'smblk_zc'. Select **Surfaces** and enter the top and bottom grids: 'topo_c' and 'z_midb'. Under **Assignment** assign the vertical fraction to 'midb_f' and select **Locate**.

Finally, use **Report** to find the mining reserves as in Example 6.14, appending the results to the ASCII file 'mineres.rpt'. Using gridded surfaces the mining reserves are reported as:

toe and crest 2D grid reserve estimates for smblk

	midb_tons	midb_au	midb_ag
TOTALS:			
	23,013,588.87	2,071,814.67	40,151,675.66

Summary of procedure: Example 6.15

Initial data requirements: a series of bench toe and crest line polylines <exppittc.pol> and a grid of the pre-mining topography <topo_c>.

Procedure:

1. Generate toe and crest line polygons as in Example 6.14 and save them in <exppittc.pol>.
2. Load the polygon vertices into flat table in preparation for gridding.
   2.1 Create new table and fields **Techbase** ⇒ **Define** ⇒ **Tables** ⇒ **Create**
   TABLE NAME: <exppit_midb> TYPE: <flat>
   ⇒ **eXit** ⇒ **Fields** ⇒ **Autotable**
   TABLE NAME: <exppit_midb>
   ⇒ **Create**
   FIELD NAME: <midb_x> TYPE: <real> CLASS: <actual>
   (Repeat for <midb_y> and <midb_z>. Also, **Create** the field <z_midb> in the cell table <surfaces>)
   2.2 Load the polygons **Techbase** ⇒ **Load** ⇒ **Setup** ⇒ **Data file**
   FILE NAME: <exppittc>
   ⇒ **Fields**
   FIELD AND FORMAT LIST: <midb_x midb_y midb_z>
   ⇒ **Load**
3. Grid the polygon vertices **Modelling** ⇒ **triGrid** ⇒ **Grid**
   3.1 Define **Fields**
   DATA POINTS VALUE: <midb_x> RESULTS VALUE: <z_midb>
   X: <midb_y> Y: <midb_z>
   3.2 Gridding **Parameters**
   NUMBER OF DERIVATIVE POINTS: <4>
   MAXIMUM GRADIENT MAGNITUDE: <1>
   3.3 **Grid**
4. Determine the percentage of each block that's within the post and pre-mining topography **Modelling** ⇒ **Locate**
   4.1 Provide block coordinates **Setup** ⇒ **Fields**
   X: <smblk_xc> Y: <smblk_yc> Z: <smblk_zc>
   4.2 Provide the bounding surfaces ⇒ **Surfaces**
   TOP ELEVATION: <topo_c> BOTTOM ELEVATION: <z_midb>
   4.3 Assign the fraction field ⇒ **Assignment**
   FIELD: <midb_f> = POLYGON/VERTICAL FRACTION
   ⇒ **eXit**
   ⇒ **Locate**
5. Report mining reserves as per Step 7 in Example 6.14.

Example 6.16: Reporting mining reserves using Techbase's opRept

In Example 6.6, Techbase's open pit mine planning tools were used to generate an ultimate pit using the moving cone algorithm. The pit limits were saved into a grid file based on the original surface topography, which could be used as in Example 6.15 to obtain mining reserves using **Locate** and **Report**. Rather than using this somewhat involved procedure, Techbase has an automated program for reporting mining reserves, **opRept**.

Prior to continuing with this example, an ultimate pit must be generated as in Example 6.6. The same block model will be used as in the previous two examples (smblk) which includes fields for the block net value, ounces of gold and silver (auoz and agoz) and block tonnage (tons) which is set to a constant 6525 tons/block. For this example, **opCone** has been used to generate a single ultimate pit using the mining sequence file 'mvc.seq' and template file 'mvc.tem'. Pit slopes were maintained at a constant 47° to match the bench width height and bench face slope of Example 6.6.

From the main menu select **Open-pit ⇒ opRept ⇒ design File ⇒ Open file** and enter the sequence file 'mvc.seq'. Select **Values ⇒ Fields** and enter NET VALUE <netval_sm>, NET TONS <6525>, MISSING VALUE <-10115> and MISSING TONS <6525>. These parameters can either be entered as fields or as constant values and have the same meaning as in **opCone** in Example 6.6. Material classifications for generating the reserves report are specified in **Add class**. Typically, this facility is used to specify ore grade ranges that are relevant to the ore/waste and ore processing categories, but in this example only the total tonnage and ounces within the pit limits are of interest. Therefore, enter <smblk_lev >= 1>. Since all blocks meet this constraint, all tonnage within the ultimate pit will be reported.

Select **Setup ⇒ Report fields** to define the contents and format of the mining re-

Table 6.1. Mining reserves reported by bench using opRept.

BENCH	NET_VALUE	NET_TONS	Netval_sm	Tons	Auoz	Agoz
1	0	0	0.00	0	0.00	0.00
2	-80910	52200	-80910.00	52200	0.00	0.00
3	-465232	300150	-465232.50	300150	0.00	0.00
4	-1304674	841725	-1304673.75	841725	0.00	0.00
5	-2285708	1474650	-2285707.50	1474650	0.00	0.00
6	-3175718	2048850	-3175717.50	2048850	0.00	0.00
7	-4075841	2629575	-4075841.25	2629575	0.00	0.00
8	-5117558	3301650	-5117557.50	3301650	0.00	0.00
9	-5946922	4156425	-5946921.62	4156425	1686.47	33117.31
10	50652949	5304825	50652948.89	5304825	201071.54	3938656.29
11	167385762	5761575	167385761.93	5761575	593450.45	11505350.85
12	154963577	5376600	154963576.61	5376600	552039.73	10649343.97
13	99518925	4378275	99518925.20	4378275	361286.26	6938989.41
14	39432040	3223350	39432040.27	3223350	151010.54	2961922.88
15	13318837	2642625	13318837.00	2642625	59300.08	1221044.67
16	0	0	0.00	0	0.00	0.00
Total	502819528	41492475	502819528.28	41492475	1919845.08	37248425.38

serves report that will be sent to the output file (default name, 'oprept.out') and enter <netval_sm (12) (t) tons (12) (t) auoz (12) (t) agoz (12) (t)>. The report file will print totals for each field in columns of twelve. Under **Bench**, specifying a bench label type of 'number' will cause the report to include totals of each of the fields by bench number (smblk_lev). **Report** generates the report file. Using **opRept** the mining reserves are reported as shown in Table 6.1.

Summary of procedure: Example 6.16
Initial data requirements: an ultimate pit sequence file <mvc.seq> generated as per Example 6.7 and a corresponding block model with fields for tonnage, ounces of metal, etc.

Procedure:
1. Open the sequence file **Open-pit** ⇒ **opRept** ⇒ **design File** ⇒ **Open file**
   FILE NAME: <mvc>
2. Set the fields to be used for block values and define ore and waste classes as per Example 6.7 Step 5.2 ⇒ **Values**
   2.1 Provide block fields ⇒ **Fields**
       NET VALUE: <netval_sm>   NET TONS: <6525>
   2.2 Provide ore classes ⇒ **Add class**
       FIELD: <smblk_lev>   RELATION: <>>   FIELD|VALUE: <1>
       ⇒ **eXit**
3. Set up fields and constraints on reserve reporting ⇒ **Setup** ⇒ **Report fields**
   FIELD AND FORMAT LIST: <netval_sm (12) (t) tons (12) (t) auoz (12) (t) agoz (12) (t)>
   ⇒ **Bench type**
   BENCH LABEL TYPE: <number>
   ⇒ **eXit** ⇒ **Report**

CHAPTER 7

Underground design

Numerous interrelated and often conflicting goals must be satisfied as part of an underground mine design. Mine drainage, materials handling, ventilation and ground support requirements must all be met while extracting the maximum mineral value with the least waste rock. While all of these factors must be accounted for in the feasibility study, some of these systems have less influence on the design than others.

For instance, mine drainage will not be a major consideration at the design stage. For one, the prediction of inflow is very difficult to make, but even though drainage may be a major operating expense, the effect on design is minor, being limited to the sizing and location of sumps and pump stations and providing for low gradients towards the shaft or other sump locations. Design determines the rate of inflow into the mine workings: the deeper and greater the extent of the openings, the greater the inflow will be, but the extent and depth of the workings are a consequence of the depth and breadth of the deposit.

In contrast to drainage, ventilation requirements are entirely a consequence of design and can be accounted for during the feasibility stage of a project. Shafts, raises and all mine workings must be designed with ventilation in mind. Production shafts usually serve as intakes for fresh air and their ventilation compartment must be sized accordingly. Additionally, at least one exhaust shaft will have to be located at the other extreme of the mine workings opposite to the intake/production shaft. Level development between stopes must be planned to provide a circuit through stopes. Still, as important as ventilation is, it is surprising how minor a role it often plays at the feasibility level, where it is largely assumed that ventilation can be worked out at a later date and is largely a matter of providing large enough fans.

Materials handling, along with ventilation, determines the layout of the mine within the constraints imposed by the configuration of the deposit and topography. Shaft diameter is a function of skip capacity. In the case of a decline, the slope and dimension are determined by the maximum angle of inclination and width of the conveyor belt. Ramps, orepasses, haulage levels, crosscuts, orepockets and chutes are all designed around the efficient movement of material, equipment and men between the stopes and development headings and the surface. Materials handling plays a key role in design at the feasibility stage not only because it is critical, but also because the materials handling parameters can be quantified during design, unlike drainage, which will have to wait for resolution until mining commences.

The most pressing issue at the feasibility stage is the influence of ground control on the dimension of mine openings and the need for artificial support. Ground support limitations determine both the stoping method and the layout of the mine. In the case of vein mining, competent ground with low stress may be minable using open stopes of large dimension with relatively few pillars. In weak ground, overhand cut and fill might be required, and then only if the ore is competent. The difference in extraction ratios and mining cost between the two methods is considerable and may well be the deciding factor in the decision to develop the deposit. In room and pillar operations, the deciding factor is the allowable unsupported roof span. The larger the opening, the lower the production cost, the greater the extraction ratio and the less spent on development, ventilation and maintenance of mains and submains. Unfortunately, the level of detailed information on rock strengths and ground stress is very limited at the feasibility stage, since the only available information is from drill cores. Still, mapping of RQD and faults can be used to provide guidance on the type of stoping method, the sizing and location of pillars and the need to isolate the shaft and other major access entries such as the haulage level. Surprisingly, spatial modelling of rock mass quality as described for assays in Chapter 5 is rarely practiced.

In summary, the emphasis, or lack thereof, which is placed on the different aspects of underground design is largely a function of the availability and implementation of data and the complexity of the interactions between critical systems. Even though ground control may be the ultimate arbiter of design, there is typically insufficient information to do more than take a qualitative approach to the design of openings at the feasibility stage. In contrast, the materials handling system can undergo detailed design since the capacities, haulage profiles and equipment specifications are well known. Thus, some issues, such as pillar dimensions and rates of groundwater inflow, must be generalized during the design stage, while other systems, such as the sizing of the conveyor and the ventilation network, can be designed in detail. Some of this uncertainty can be quantified during underground mine costing using sensitivity analysis to determine the potential impact a change in, say, pillar size will have on project feasibility, but within the limitations of a trial-and-error approach the number of interacting variables in underground design are daunting.

A gap has grown between the practice of design for surface and underground mines which is a direct result of the application of operations research methods to pit design. In the previous chapter, several algorithms were presented that could be used to determine a final pit limit. This pit limit was then used as a starting point in the design process. No such single algorithm has been developed and accepted for underground design which has had so profound an influence on the design process. The process of underground design is little changed from the days prior to computers. Even though CAD design is the norm for underground projects, software which integrates design with ventilation circuit calculation, simulation of materials handling or other optimal design methodologies are not integrated with commercial mine design software. As such, the primary purpose of CAD at the feasibility stage is for layout and volumetric calculations. It is hoped that the day will soon come when optimization is an accepted aspect of underground design. Certainly, there is no fundamental reason why mathematical programming cannot be applied to underground design.

This chapter is not aimed at a comprehensive discussion of mining methods or of the crucial related mining systems. That would require several texts ranging from hydrology to rock mechanics and is better treated elsewhere. Instead, this discussion will provide an overview of the layout and design of the principal openings which constitute an underground mine with the objective of providing a basis for the design and layout of the mine openings required during mine development. The following examples will included stoping methods, but a detailed discussion of mining methods is outside the focus of this text. Many excellent texts are available on underground mining methods to which the reader is referred to. Note that the difference between, say, cut and fill, shrinkage stoping and open stoping are not relevant to the level of engineering detail possible with mine design software, which is only oriented towards representing the geometry of a stope and volumetric calculations. Other software is readily available for ventilation, finite element analysis of opening stability and materials handling simulation. Several examples of underground mine design will present both the potential and limitations of underground mine design software.

7.1 MINE DEVELOPMENT OPENINGS

The design criteria for underground mine layout and opening design is based on the safe flow of men, equipment, supplies, water, backfill, air and ore between the surface and ore body. The objective of design is to maximize the value of the extracted ore and minimize cost, but safety is the overriding constraint on design. Potentially, there are as many mine designs as there are ore bodies. This makes it difficult to generalize mine design and layout, but most mines are variants on fundamental classes of mining methods. Within these classes, there can be substantial differences in the configuration of the mine workings. For example, compare the development of a high angle vein, as illustrated in Figure 7.1, with the haulage level of a block caving

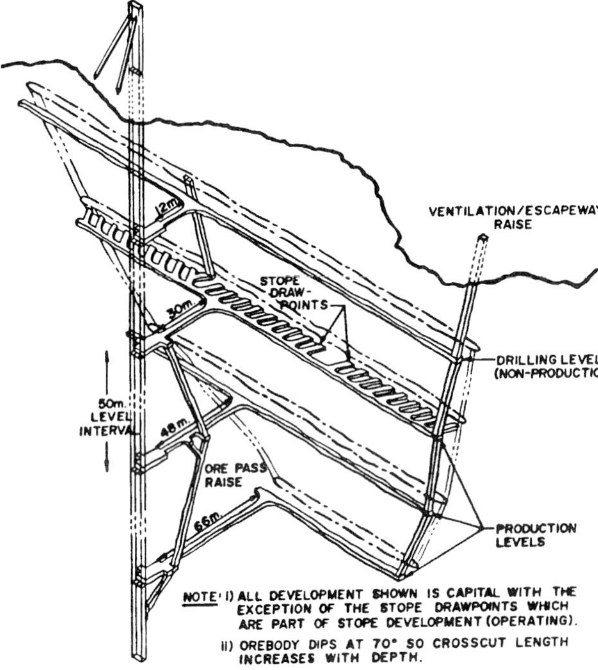

Figure 7.1. Typical layout of mine development openings for a small high angle vein (CANMET 1986).

operation in Figure 7.2. The small metal mine of Figure 7.1 is accessed by a single production shaft in the footwall of the deposit. Crosscuts at 50 m intervals provide access to the footwall production drifts with their crosscuts/drawpoints driven into the ore deposit. Not shown are the three levels of stopes and their development and the drawpoints for the third and fourth levels. Ore is shuttled back to an orepass which terminates at a loading pocket above the shaft bottom, leaving enough room for skip loading and a sump. Due to the small size of the mine, only a single ventilation raise is needed for exhaust air.

A completely different layout is required for the large scale caving operation partially displayed in Figure 7.2. Whereas the small scale mine of Figure 7.1 would use diesel LHDs, rail haulage was used at the Inspiration Mine, an option that would still be used today considering the project's long life, high production and the long haulage distances involved in this type of operation. Both operations relied on gravity flow to bring ore to the draw points, but in the case of the caving operation the scram drifts, ore passes and haulage level must be placed under the blocks due to the massive nature of the deposit. The great extent of the mine workings, and the greater air loss and resistance encountered in an undercut and cave mining method requires numerous ventilation shafts. Air requirements for the high production rate requires multiple production shafts placed well beyond the limit of disturbance of the caving blocks.

In spite of the differences in the mine layout used for these two very different classes of deposits, both mines share the same components: shafts; crosscuts from the shafts to haulage and production levels; drifts for haulage and stope development; raises for access, ore passes and ventilation; and other auxiliary openings used for level stations, shops, drys, pump rooms and ore storage. The design and configuration of these openings is a function of the specifics of the deposit and will not be

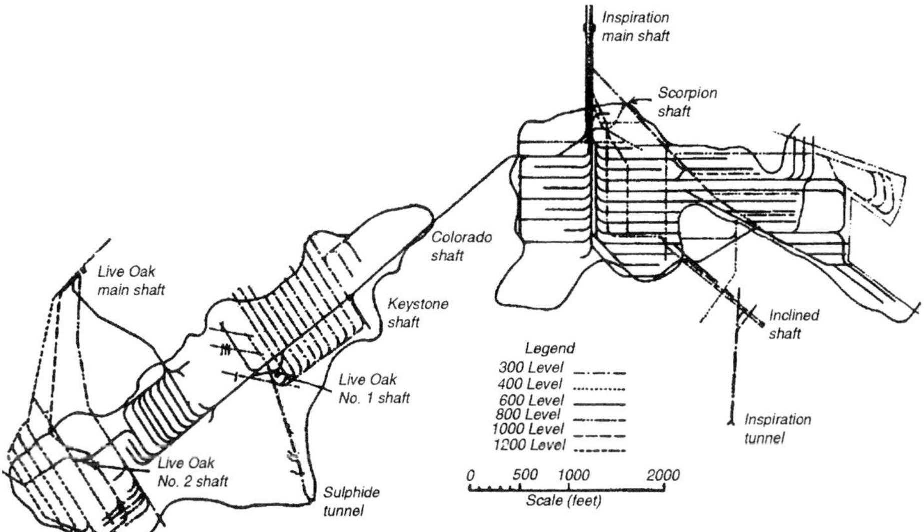

Figure 7.2. Layout of haulage level and shaft locations for a block caving operation (Jackson & Hedges 1939).

treated herein, but the basic design of the components is discussed in the following sections in order to provide the basic background required before applying mine design software.

7.2 MINE ACCESS AND HAULAGE METHOD

Underground mine access can be by shaft, incline or adit. In the case of a vertical shaft, ore transport will include a skip/cage system. There are two types of shafts: ventilation shafts which serve for mine air exhaust and which have limited facilities for access, and production shafts, which often serve a dual role of hoisting and air intake and will contain a manway, pipes and cables. For inclines and adits, ore transport from the mining face can be continuous and commonly involves some combination of wheeled transport and conveyors with the choice between the two being a function of the mining method, production rate and distance to the portal. Thus, the first decision to be made is the type of mine access and from this the material handling system can be designed. The mine access decision involves a number of criteria: topography, overburden depth and type, extent of deposit to depth, production rate, capital and operating costs, ventilation, and indicated ore at depth.

In rugged terrain, the location of the deposit relative to the surface topography might make it possible to drive an adit or decline from a mountain side or valley bottom to the lower levels of the deposit. An adit should be considered whenever the terrain is suitable, since the capital cost for drifting is far lower than for sinking either inclines or shafts. Also, operating costs with an adit will be far lower for mine drainage and ore transport. Adit entries can be used in wet ground to drain the overlying rock. Access to the different levels of the mine depends on the distances involved. Multiple adits opening to different elevations on the hillside can be used when the distance from the adit portals to the mine galleries is not excessive. Otherwise, a single haulage level combined with ramps and orepasses can be used. When the adits can be driven parallel to strike, they can serve for ventilation, haulage and stope access, but when the adits are driven in waste, the economics of adits versus a shaft entry rapidly diminishes as the heading distance from the portal to the production face increases.

When the overburden is composed of poorly consolidated fill or includes weaker porous strata, the choice between access methods may be influenced by the need to limit the distance that the opening passes through the overburden. Even though the depth of the deposit or the topography may make an incline or adit an attractive option, the increased distance required in weak and water bearing material may rule out low angle access. A vertical shaft is the most direct method of access through overburden. Additionally, the use of concrete lining and freezing technology makes it possible to sink shafts through very difficult ground. Inclines, being rectangular, are subject to both horizontal and vertical pressure and are inherently less stable than concrete lined circular shafts and require more maintenance over time.

Even when a deposit can be accessed by adit or incline, the total depth of the deposit below the lowest feasible level of an incline must be considered. If a significant portion of the reserves will be below the incline, then it must be recognized that the incline will be used to develop the upper level of the deposit and that it will be necessary to sink a shaft in the future to develop the lower levels. In the past, inclined shafts running from lower levels up to a haulage adit or level were not uncommon, but this was for non-mechanized mines of modest capacity. An example of an inclined underground shaft is shown in Figure 7.3.

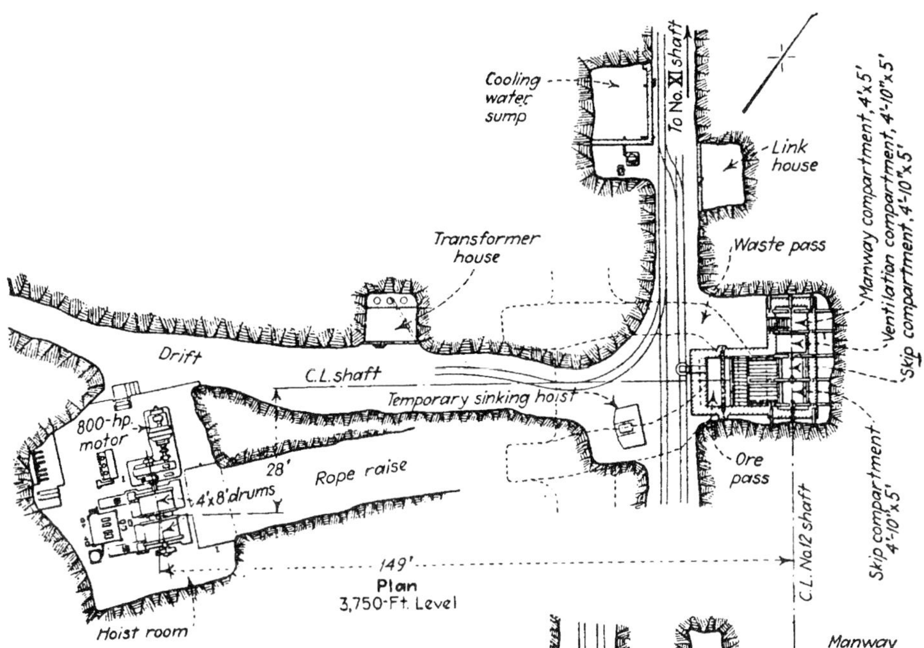

Figure 7.3. Vertical shaft extension by means of an underground incline (Tillson 1938).

Shafts limit the maximum production rate by eliminating the advantages of continuous materials handling. When a shaft is included in the haulage circuit, ore is hauled from the face and/or dumped down an ore pass in a multilevel mine, stored in an ore pocket, measured into a skip, hoisted from the loading station in the shaft bottom and dumped from the skip into another storage bin before loading onto surface transport. Every time the ore is transferred, it must be rehandled and additional time is added to the haulage cycle. For many types of deposits, ore transport can be by conveyor directly from the face. In the case of coal, trona, and potash, longwalls or continuous miners can feed ore directly onto a flexible, extendable conveyor which transfers ore to a system of stationary conveyors. When an adit or incline is used, ore can be moved by conveyor continuously through a production shift from the face to its final destination on the surface. With such an arrangement, enormous production capacities are possible. While there is little practical limit to the width, length and speed of a conveyor system, shaft production is severely constrained by the maximum allowable acceleration rates and skip sizes due to the resulting dynamic stress placed on wire hoisting ropes.

Even though inclines have great potential advantage in terms of materials handling, their capital cost increases very rapidly with increased depth. According to one source (Dravo 1974), the cost of sinking a shaft becomes less expensive than an incline at around 600 ft, but the lower cost and high production rates possible with conveyors can lower the economic limit of inclines to as much as twice that depth. The maximum allowable grade of an incline cannot exceed 20% and is commonly limited to 15% if it is to include a conveyor, less for wheeled haulage. Thus, the length of an incline must be at least five times that of a vertical shaft to obtain the same depth. Although the per foot cost of shaft sinking is greater than for an incline, the increased length of development that is required by an incline makes a shaft a more attractive option in terms of capital cost for all but very shallow deposits, typically where terrain favors hillside access as in the case of an adit.

Ventilation requirements favor a method of access which results in the shortest path to the mining face with the least resistant opening. Following the preceding discussion on capital costs, shaft access will generally provide the least air resistance.

With steeply dipping or plunging deposits, the ultimate depth of the ore may be unknown beyond the fact that it continues beyond the reach of exploratory drilling. If an adit or incline is used for access, a shaft may have to be sunk later to reach deeper reserves. When the choice between an incline or shaft is uncertain, the presence of indicated ore at depth would favor a shaft. It should be noted that shafts and declines are not mutually exclusive methods of access. The dominance of wheeled haulage and transport has resulted in the use of both inclines and shafts with ramps serving for development and access with a shaft being used for production.

7.3 SHAFT LOCATION AND NUMBER

Depending on the depth, sinking a shaft may account for 60% of the development time and a corresponding proportion of initial capital expenditure on development openings. Of all underground development expenditures, the shaft is the earliest and therefore has the greatest risk of not being recovered in the case of an unsuccessful project. As a result, the siting, number and design of shafts is the first and possibly the most crucial design decision that must be made and requires the most attention to detail.

The production shaft will be located near the center of gravity of the deposit both in the sense of ore tonnage and value. The shaft location should be central to the bulk of the reserves so as to minimize the haulage distance and provide the earliest start-up time for production, but the distribution of ore values should also be taken into account. To optimize the project, the richest ore should be recovered early in the mine's life. Thus, the shaft should be located to minimize ore transportation costs and maximize the NPV of the project by allowing stope production to start in the richer ore. Often a mine will exploit several separate deposits. If there are multiple deposits, a comparison of the cost of driving the haulage level and the increased operating cost for hauling back to a single shaft should be compared to the cost of using multiple shafts of lesser capacity. The initial shaft will be located in the deposit of the greatest value and the value of the secondary deposits must be sufficient to pay for additional shafts. In the case of low angle deposits, the center of gravity of the deposit is simply the center of gravity of the deposit. For horizontal deposits, haulage is a large component of cost. Placing the shaft in the low point of a horizontal deposit will take advantage of gravity for mine drainage to the main sump. If the low point in the deposit is not centrally located, then the sump can still be located at the low point and water pumped directly from the sump to the surface via a small diameter bored raise. The following discussion relates primarily to high angle or irregularly shaped metalliferous deposits.

The next issue is at what side of an inclined deposit the production shaft be should located and at what distance from the stopes. The shaft can be located either on the hangingwall or footwall side of the deposit or even in the deposit itself. In the past, ore haulage costs and production capacity was largely a function of distance, since small ore cars had to be hand trammed to the shaft. To reduce operating cost, an inclined shaft would be sunk in the ore following the dip of the deposit, and additional shafts would be added along the strike as the mine workings extended to distances that limited the productivity of the trammers. With the introduction of rail haulage, wheeled transport and especially where conveyor transport can be profitably used, the relevance of the distance to the shaft has greatly decreased. The main issue is the development cost and the pre-production development time. To maximize the value of the project, the development cost and time must be minimized. Placing the shaft directly in the deposit can accomplish this goal by allowing very early ore production. Since the ore is often softer than the surrounding host rock, driving the opening in the ore can reduce drilling and blasting costs. Driving openings in the ore also has exploratory value. Inclined shafts in the ore are rarely selected since either the shaft will have to sacrificed later in the mine's life or very large barrier pillars will have to be left in the ore to protect the shaft from ground movement. Another option is to sink an inclined shaft along dip in the footwall to maintain the minimum crosscut distance, but vertical hoisting allows for faster hoisting speeds, smaller clearances, larger capacity skips and lower maintenance costs.

Placing the shaft in the hangingwall side of the deposit has the advantage of decreasing the crosscut distance to the ore, but ground control problems are generally more severe on the hangingwall side of the deposit due to subsidence or caving around abandoned stopes. This is less of a concern as the deposit flattens out in inclination and shape or approaches vertical since the subsidence angle will be more nearly the same as on the footwall side. Another consideration is the differences in geologic conditions that can be encountered between the rock mass on the footwall versus hangingwall side. Heavily faulted, unconsolidated and water-bearing strata should be avoided even at the expense of longer crosscuts. A steeply dipping deposit may act either as an aquitard or as a more permeable zone that diverts groundwater flow away from the footwall rock mass, making a footwall shaft more attractive.

Generally, a footwall location is the most attractive location even though the crosscut distance will increase with depth. The shaft must be placed far enough back from the stopes so that upper reaches of the shaft will be beyond any subsidence cracks. Surface facilities and the shaft collar should be a minimum of 100 ft outside of the subsidence limit.

The number of shafts needed depends on the production capacity and the geometry of the deposit and extent of mine workings. Typically, there will be two shafts: a larger diameter production shaft with compartments for ore hoisting, man and material transport and air intake, and an exhaust ventilation shaft. The ventilation shaft will be located at the shallower end of the deposit. When the deposit is elongated, two ventilation shafts can be located at either extreme of the deposit. For higher mine production systems, four shafts can be used with two centrally located intake shafts, one for hoisting and the other for man and material transport, and two exhaust ventilation shafts. For deposits which elongate along the dip, one of the ventilation shafts can be located towards the deeper end of the deposit along with the two production shafts, and the remaining ventilation shaft can be located up dip at the shallower end of the deposit.

7.4 SHAFT DESIGN

Shaft design parameters include: depth, shape, liner, cross sectional area, number and type of compartments, collar, shaft bottom, shaft stations. Additionally, ancillary mine systems such as the hoisting system, loading pocket, ventilation and sump must be considered during shaft design. From the point of view of computer-aided mine design, only a few of these parameters are relevant, principally those related to the geometry of the shaft: its depth, area and point and method of connection with other mine openings. In order to define the geometry of the shaft, a certain amount of attention must be paid to the engineering aspects of its design and functionality.

Shaft depth is often defined by the depth of the deposit when exploratory drilling has discovered its limit, but when the deposit continues to great or unknown depth, the shaft depth must be based on an economic limit. Excessive capital costs cannot be incurred during the development stage of a project before ground stresses and the value and extent of the deposit are better defined. At a later stage of mine production, the shaft can be extended if warranted. Thus, the issue of shaft depth is either trivial, in the case of deposits of limited depth, or must be resolved by cost analysis and cash flow maximization. In either case, the shaft depth must exceed the lowermost level to allow room for the loading pocket and sump, say 90 ft (Fig. 7.4).

The cross sectional area of the shaft must be sufficient to contain compartments for hoisting, man and material transport and pipes and cables, while leaving enough room for air intake without excessive pressure increase. Skips are commonly used for ore. In the case of a single skip with a counter weight (Fig. 7.5), the one large skip and counterweight will occupy less space than two skips in balance, but the cycle time for hoisting will be doubled. A more common arrangement is two identical skips in balance in separate compartments (Fig 7.6). If one production shaft is being used, a cage will also be needed for men and material. For shallower mines, a single cage and counterweight would be reasonable. Cables and pipe and a manway will also be required, probably in an adjoining or in the same compartment for ease of access.

Vertical shafts are normally circular and lined with a foot or more of concrete. Where ground control or water requires a thicker lining, as much as 18 to 24 inches may be used. In addition to the area occupied by skips and cages, the compartment

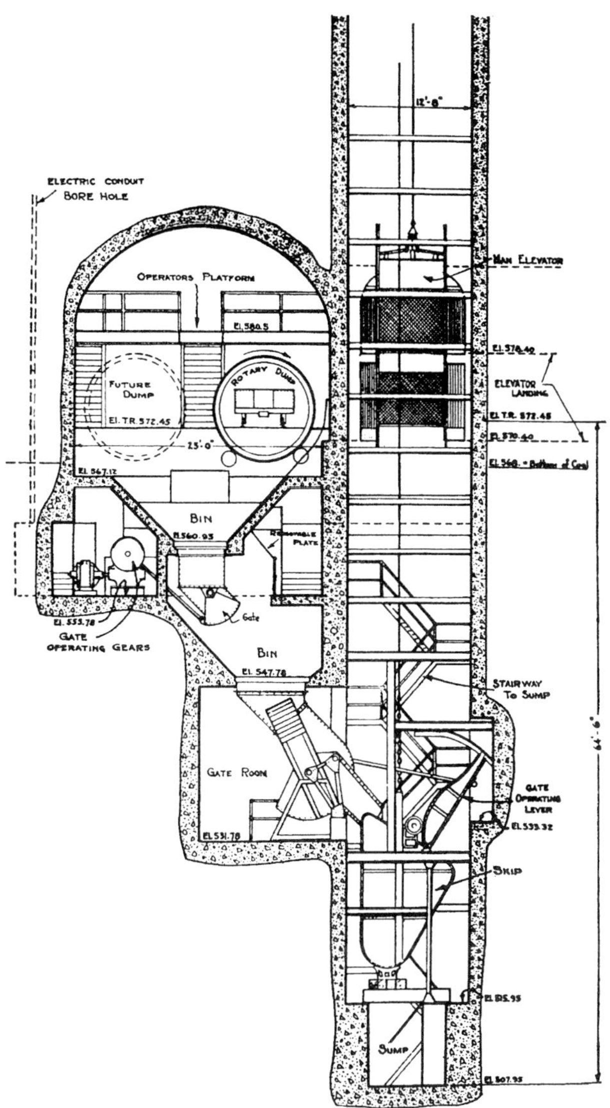

Figure 7.4. Shaft bottom and ore pocket for rotary ore car loading of skips (Tillson 1938).

walls and conveyance guides occupy additional space. When timbers are used for the compartment, the superstructure will occupy 8-9 inches plus 1.5 inches lagging. Fir guides will be approximately 5.5 inches deep while rail guides will be 7 inches deep. Therefore, the compartment walls will add as much as 3 ft to the compartment dimension in addition to the width and depth of the skip or cage. When guide ropes are used, 2.5 inches is maintained between the skip and rope plus another 10' between the guide ropes of adjacent compartments.

Skip capacity is determined by the production rate, hoist cycle time and material characteristics. Due to limits on acceleration rates and rope strength, there are limits to the number of hoisting cycles that can be achieved in a shift for a given depth of

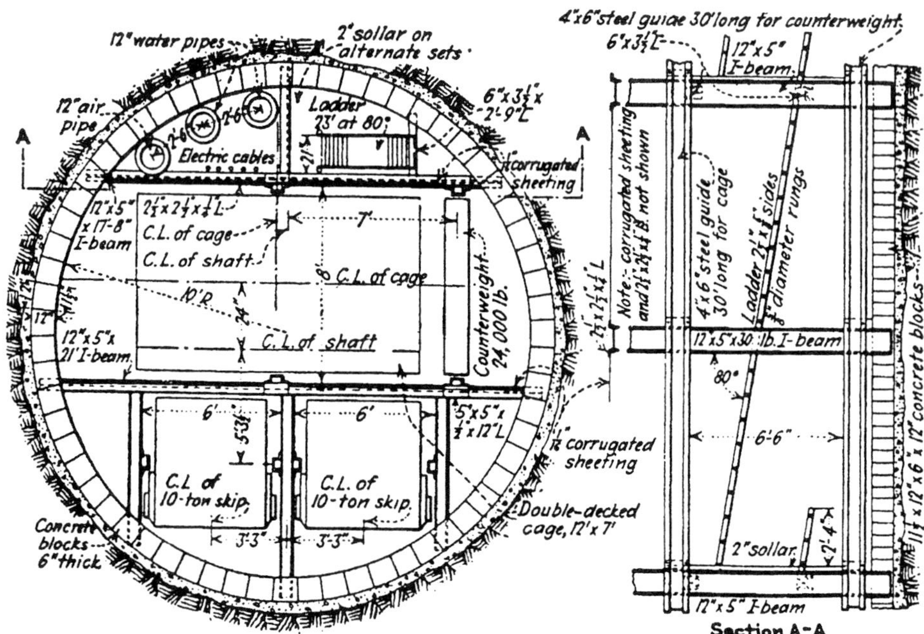

Figure 7.5. Circular shaft design (Eaton 1934).

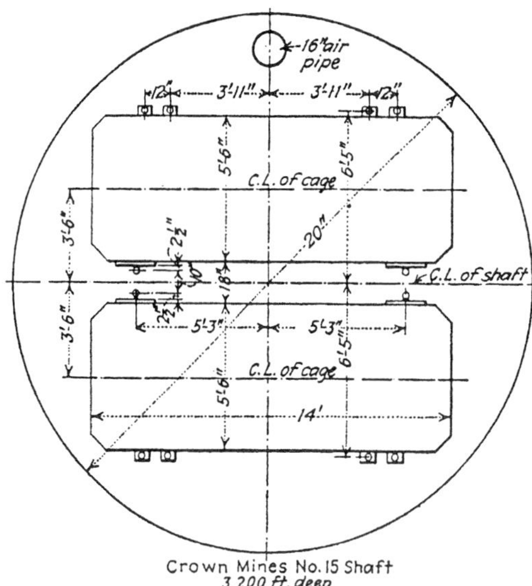

Figure 7.6. Shaft layout for cages in balance using rope guides (Eaton 1934).

shaft. A skip capacity must be selected that will maintain the desired production rate given a hoisting cycle and acceptable acceleration. The skip volume is determined from the ore density and swell factor and should have 10% more capacity than the planned load. To prevent blockage or spillage during loading, the final dimensions of the skip should not be too narrow in comparison to its depth.

The area occupied by the pipe and ladder compartment depends on the size and number of pipes and cables. Even though these can be scattered around the unoccupied perimeter of a circular shaft, they are best kept with the manway for maintenance and to reduce dead space. This might include two or more water pipes, two compressed air pipes and various signal and power cables. The ladder should be at least 3 ft wide and slightly inclined (≤ 80°) with a distance of 20-25 ft and staggered between landings.

Since the conveyances in a shaft are rectangular in section, much of a circular shaft's area is dead space in comparison to a rectangular shaft. Circular shafts require as much as 50% greater excavation than a rectangular shaft of equal capacity, but can withstand heavier ground pressure. This is acceptable since the production shaft must also serve for intake air. Air velocity in production shafts will be around 800 fpm, but velocities up to 1500 fpm are common. Velocities up to 2000 fpm can be used in the ventilation shaft. Thus, there must be sufficient dead space in the production shaft to allow this rate of flow without excessive air velocities and shock losses. In general, intake shaft diameters will be from 23 to 28 ft and the ventilation shafts will be from 18 to 20 ft.

7.5 SHAFT STATION CHAMBERS

Level stations can be as simple as an intersecting drift and shaft inset as shown in Figure 7.7 or consist of a complex of openings that include skip loading pockets, ore and waste storage bins with conveyors and possibly a primary crusher, pump chambers and sumps, mechanic shops, supply rooms, electrical power stations, explosive

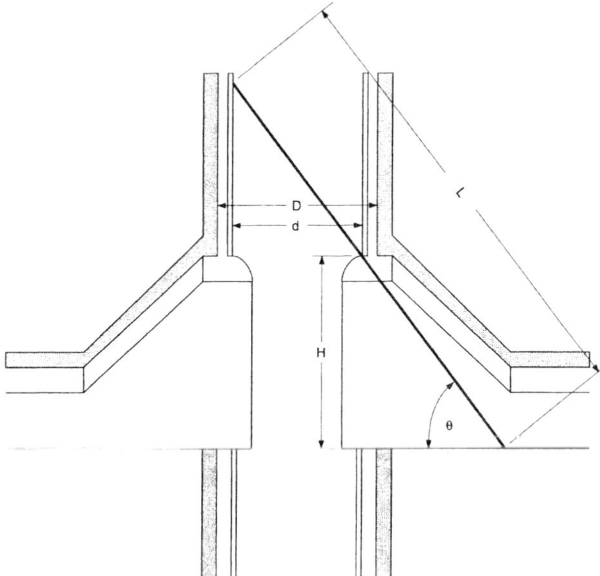

Figure 7.7. Shaft inset height calculation (adapted from Unrug 1992).

chamber, and miner's dry. When rail haulage is used, space must be provided for storage of empty cars and uninterrupted circuits for trains. The number, size and types of openings required will depend on: the materials handling system, i.e. the number, width and type of skip loading system; the production rate; and whether or not the station is on the production/skip loading level.

Shaft insets must be of sufficient height and depth to allow the passage of longer items of supply and equipment components to be moved underground and to limit shock losses to air flow. Figure 7.7 illustrates inset height calculation for transporting material of maximum length L at an angle θ from the horizontal in a shaft of diameter D. For $\theta = 45°$, the free space between the shaft beams $d = .71D$ and the shaft inset height, $H = .71\ (L - D)$.

Ore loading pockets act as temporary storage bins between the skip and mine production. In operations with a single production and haulage level, such as block caving or room and pillar, ore may be dumped directly into a bin or into a storage pocket as shown in Figure 7.4. For multi-level operations, such as shown in Figure 7.1, an orepass transfers ore and waste rock to the pocket. Complex arrangements of ore and waste bins connected by conveyor belts with transfer points can be used as illustrated in Figure 7.8. Ore storage capacity can vary from 3-191% of hoisting capacity, being largely a function of the expected down time for the materials handling system from the skip loading pockets on up the shaft and the importance of uninterrupted flow.

Shops and storehouses need to be larger when the equipment cannot easily be brought to the surface. Thus, the size of shops depends on the production rate, mine life, mining method and degree of mechanization. i.e. the number and size of the production units and supporting equipment. A room and pillar operation using boring machines and continuous miners will certainly need much larger shops than a cut and fill operation using jacklegs and stopers. Base the shop dimensions on the expected availability of the equipment, the dimensions of the larger units and the number of units needed to maintain production. Make the chambers large enough to accommodate an overhead crane. Shop area can range from 5000 to 15,000 ft^2 depending on the amount of inventory that has to be kept on hand. A supply shop can also be kept on the surface, but inventory which is consumed in large quantities (such as roof bolts and timbers) must be kept underground.

Pump station size will depend on the rate of inflow, depth and the amount and type of suspended fines in the mine water. Inflow rates in the US vary from 0 to 35,000 gpm. Inflow rate can be predicted if permeability tests have been conducted on drill cores and if there have been drawdown tests in groundwater monitoring wells. The permeability of the various rock units and the inflow rates observed in the monitoring wells can be used to estimate inflow rates as a function of depth and excavation volumes in the different rock units. The drawdown test may also indicate a reduction of inflow over time unless there is surface recharge of ground water. It may be possible to identify surface stream inflows by using ultraviolet light-sensitive dye.

If a paste or slurry backfill is used, then this can add considerably more water and suspended sand and fines. The number of pumps needed is a function of the head and capacity required. For the centrifugal and plunger pumps used for mine service, the capacity is inversely proportional to the head. Since head is directly proportional to

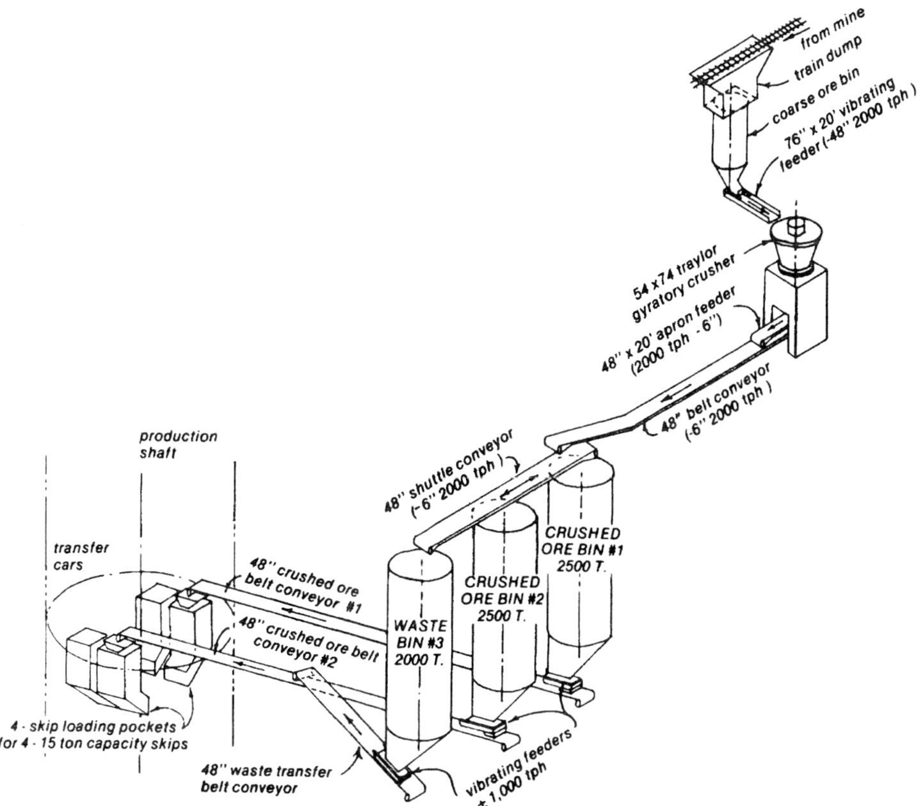

Figure 7.8. Underground ore and waste crushing and storage system for loading four 15 st skips in balance (McArthur 1982).

the vertical lift required of the pump, then as the depth of pump station increases, more pumps operating in series will be need to provide sufficient lift. An additional complication arises when the rate of inflow is seasonal. A pump's best efficiency operating point is within a fairly narrow capacity range. Therefore, enough pumps are needed to handle the maximum flow. Pumps can be arranged in parallel with some pumps down for maintenance or during periods of low inflow. The number of pumps required and the dimensions of their foundation pads can be used to estimate the area needed to house the pump room (Fig. 7.9).

More room will probably be needed for the sumps than for the pumps. Sump size depends on the inflow rate, the reliability of mine power and, perhaps most importantly, the settling rate of solids in the mine water and the type of pump used. Suspended sand will rapidly eat up a pump's impeller blades or scour the cylinder of a piston displacement pump. When suspended sand cannot be avoided, a plunger pump can be used, but generally, the sump must be large enough to provide sufficient residence time for the coarser and more abrasive sand to settle out. In addition, more than one sump might be needed to allow for the periodic removal of accumulated sand.

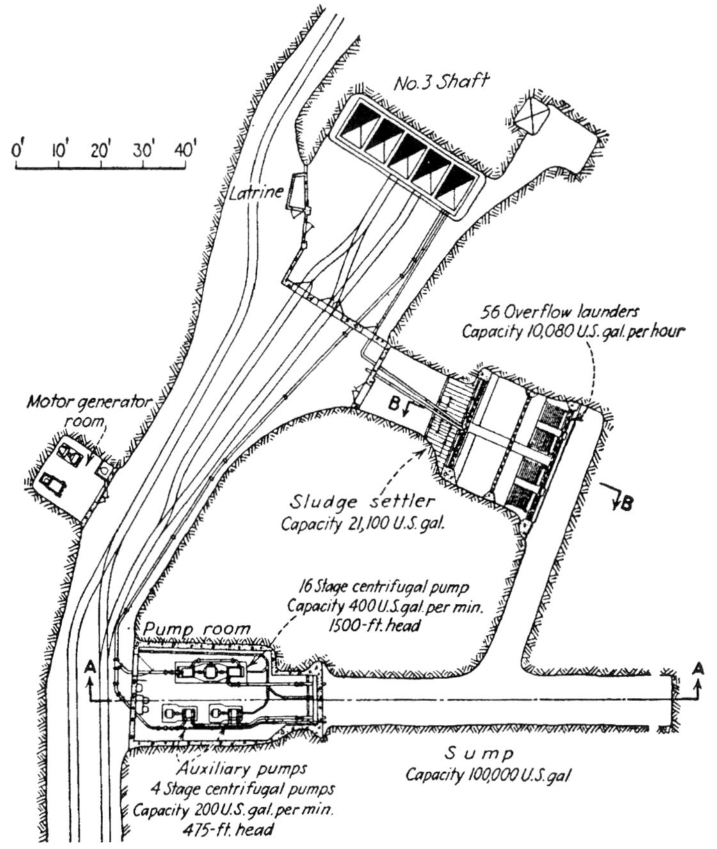

Figure 7.9. Shaft station for direct loading of ore cars including sump and pump rooms (Tillson 1938).

7.6 LEVEL INTERVALS

Except for low angle deposits of moderate thickness (<100-200') multiple levels are necessary to develop a deposit. In Figure 7.1, a 50 m level interval is used to develop the stopes for a high angle vein. In the case of a high angle vein, the height of the stope is defined by the level interval. The drifts on the lower level are used to initiate mining in the stope which will proceed in an overhand fashion with production cuts progressing from just above the lower level toward the upper level. During production, the lower level will be used for drawing off the ore, intake air and haulage. The upper level will be used for exhaust air and access to the working area of the stope via one or more raises that were driven between the levels during initial stope development. In mechanized operations, a ramp is used to provide access to successive cuts or to sublevels.

Level location is often a function of orebody geometry. In highly faulted ground, the stopes, and therefore levels, will be located to avoid mining through displaced ore zones in order to limit dilution. Changes in dip and thickness of the deposit may also determine level location, as will a deposit which consists of scattered high value ore zones. When mining conditions don't dictate the location of levels, then a wide range of stope heights is possible.

Since the levels must be completed prior to production in the stopes, a considerable capital expense must be incurred for drifting, chutes or drawpoints and raises. In addition to the development cost, there is the ongoing maintenance cost. In order to maximize the project's NPV, the number of levels must be kept to a minimum, but the savings that can be realized by increasing the level interval must be balanced against the increased cost of intermediate stope development (raises, orepasses, ramps and sublevels) and the added cost of mining at increasing distances away from a level. For instance, as the level interval increases in height, the life of the stope will also increase, requiring the construction of heavier orepasses and chutes in the raise. In the case of block and sublevel caving operations, heavier support will be required in the underlying mine openings and there will be increased problems with draw control and dilution. In addition, there is the practical limitation of the time spent in getting the miners, equipment and supplies to the production face. A cost comparison over a practical range of level intervals is required to select the interval which compares the cost of development against operating costs. Level intervals in small mines run from 100 to 300 ft and from 300 to 800 ft in large operations (Nilsson 1982).

A generally accepted practical limit for high angle veins is 200 ft, but there are many considerations, such as the mining method, ground conditions, depth and the distance from the shaft. If mining in the stope is mechanized, then LHD and jumbo access to the production face will be by ramp, but if stopers and jacklegs are used, then the time and difficulty involved in using a manway with ladders limits the practical level distance to 200 ft. In some stoping methods, such as longwall, open stoping and stull stoping, the unsupported stope life must be considered when planning the level interval. For these methods, the stopes should be completed and abandoned within the time they can be kept open without excessive maintenance cost for stulls, cribbing or pillars. As a consequence of this, mining on the retreat from the property boundaries should be considered, since this makes it possible to use a greater interval since the drifts do not have to be maintained and the stopes can be allowed to fail.

7.7 RAISES

Raises are used for access, supply, ore passes, draw points, ventilation, backfill placement and cut initiation. Most of these applications will be within the stope and their use will depend upon the mining method. If production takes place on a higher level than haulage, then ore passes will be used to transfer ore and waste from the stopes to the haulage level and from the haulage level to the ore pocket. Since raises are used to connect levels, their inclination depends on the dip of the deposit over the level interval. Raises driven at 45° are less expensive than shallower or steeper angles. For conventionally driven raises, 60° is reported to be the most economical inclination (Peele 1941) in that gravity flow can be used. Since most raises are now bored, high angles are no longer a limitation.

Raise dimensions and inclination depends on the character of the ore. A small raise will tend to hang up if the ore is too coarse and angular. Conventional raises are rectangular with enough space for a manway and an orepass. There is a practical limit to the height of a conventional raise, 200 ft being the limit in terms of ladder access. Beyond 200 ft, an Alimak raise climber can be used. Bored raises are by far the most economical alternative for raises up to 800 ft long 3 to 6 ft in diameter. Chute and orepass inclination should be between 55 to 75° and as low as 40° if the ore is very free flowing and not left standing (Eaton 1934). When the orepass follows the dip, enlarge the dimensions when there is a change in inclination. Raises should not directly intersect drifts for safety, but should be started to the side of the drift.

Raises used for stope development are contained within the limits of the stope. Thus, these openings are not of concern in terms of mine layout and volumetric calculations, although they do represent a major component of stope development cost. Chutes or drawpoints are exterior to the stope and connect the stope to the materials handling system. As such, their spacing needs to be considered. Chute spacing is dependent on the mining system used. For an unmechanized overhand cut and fill stope, chutes are connected to orepasses that are maintained in the fill. In this case, chute spacing is a function of the practical working distance for a scraper, which is around 100 ft, resulting in a 200 ft chute spacing. For shinkage stoping or block caving, the spacing of drawpoints depends on the angle of draw. If the spacing is too great, then excessive material will remain perched above the midpoint between adjacent drawpoints, resulting in lost ore and increased stress on the haulage or scram drifts. Spacing and raise dimensions also depend on the required storage capacity. Thus, the cycle time required for the LHD or train to load, haul, dump and return to the loading point under the stope must be compared against the production rate in the stope and the storage capacity of the raise. The most cost-effective means of avoiding a bottleneck between the production face and the ore pocket may be to decrease the chute interval and/or increase individual chute capacity.

Orepass spacing is a function of the production rate and the cycle time required to haul ore from the stope to the closest orepass. As the average distance from the drawpoints to the orepass increases, so will the haulage time and the time spent by the LHD waiting in queue to dump. The same situation can exist on the main haulage level. As a result, more LHDs will be needed to maintain production on the production level and more or longer trains will be needed on the main haulage level. If there are too few orepasses, then there will be a bottleneck in production. Each orepass location will probably consist of several raises, one for waste and one for each category of ore requiring separate treatment. Additionally, each orepass group will have loading facilities on the haulage level. In order to determine an appropriate spacing, the capital cost of the orepasses must be compared against the capital and operating cost of increasing the number of haulage units (Nilsson 1982).

7.8 RAMPS

Ramps are an essential component of mechanized mining of high angle deposits. In cut and fill stopes, ramps are used to provide equipment access to to the working face. In sublevel caving or stoping mines, the ramp provides access to the sublevels. Multilevel mines will commonly include both shafts and ramps. If large production equipment is used, the ramp provides access to all levels in the mine. In sublevel mining, development and production takes place on several levels simultaneously: on the lowest levels, jumbos drive new sublevels; on successively higher levels there are longhole or ring drill jumbos, bolters, bulk explosive loaders, and LHDs hauling broken ore to the orepass. As production continues to deeper levels, ramps are used by this equipment to shuttle between the different sublevels to continue this cycle of operations.

Ramp cost is similar to the cost of inclines. Ramp inclination is limited to the maximum allowable slope for LHDs over a short haul. Note that when used for ore haulage, the LHD is full on the decline and empty on the incline so that steeper grades can be maintained than in open pit haulage. Otherwise, ramp design parallels switchback haul road design in that turns are 180°, leaving the maximum length for straight incline haulage, curves are flat with decreasing gradients entering and leaving the curve, and inner curve radius is based on the minimum turning radius of the equipment. Except for articulated vehicles, 50 ft should be sufficient.

In addition to interlevel access, ramps can be used in place of orepasses and for access to ore shoots which are not valuable enough to justify either deepening the shaft or driving another level. Over short vertical distances, a ramp may be more economical than an orepass considering the time spent dumping and loading and the first cost of raise boring and loading facilities. Also, when a small secondary deposit or ore shoot is distant from the shaft, a ramp may be more economical than adding several long drifts from the shaft.

7.9 DRIFTS AND CROSSCUTS

Drifts and crosscuts provide low angle access to the deposit. In room and pillar operations, the drifts are driven in the ore as multiple parallel entries which are crosscut, leaving a pattern of chain pillars and rooms. Except for the fact that these chain pillar entries are left unextracted, there is little to distinguish the main access routes from the overall pattern of rooms and pillars. In narrow, steep veins, the drift can be driven directly in the ore in order to develop the stope levels and to offset entry cost by producing ore. For wider deposits, the drifts are best kept in the footwall with crosscuts run through the width of the deposit at an interval which is suitable for the mining method. The determination of a suitable crosscut interval follows on the discussion of the method of drawing ore and drawpoint or chute spacing in the previous section. For instance, for shrinkage stoping, two parallel drifts can be run on each level, one placed in the ore and used as the initial cut in the stope, and the other placed in the footwall. Crosscuts are then driven to connect the two drifts. As mining progresses in the stope, the drift placed in the ore becomes a trench while the crosscuts are draw points for the LHDs. Crosscut spacing is based on the angle of draw. The same method can be used for sublevel caving and sublevel stoping by using the drift in the ore as an undercut and using ring drilling to bell the bottom of the stope into the trench.

The cross sectional area of a drift is a function of the haulage equipment. For tracked haulage, a clearance of 18 inches on the sides of cars and locomotives is suggested (Eaton 1934). In trackless mines, main openings should be 50% larger than the minimum size needed to operate equipment to leave room for ventilation and increased equipment size as the mine life increases and haulage distances lengthen, requiring larger bucket capacities to maintain production (Peele 1941).

A positive grade away from the shaft should be maintained in the drifts. At the very least, a .5 to 1.5% grade is kept for drainage back to the shaft. Tracked haulage grades should be kept below 3% and less than 8% for rubber-tired haulage except for very short distances in which grades can be as high as 15 to 18%.

As per ramps, the radius of curvature for wheeled haulage depends on the turning radius of the vehicle. For rail haulage, an 80 ft radius of curvature can be used for stope development, but a radius of 200 to 500 ft is required on mainlines and 125 ft for limited service mainlines (Bullock 1982).

7.10 USING LYNX FOR UNDERGROUND EXCAVATION DESIGN

As mentioned in the beginning of this chapter, the primary advantage of using mining software for underground design is the ability to integrate geologic and block models with volume models of underground design. The underground mine design can then be used as the basis for all consequent design issues such as ventilation, development rates and production scheduling through intersection of the mine model's volume components with the geologic and block models.

Underground mine design using Lynx is based on volume modelling of two categories of excavations. Development openings, such as drifts, raises and ramps, are usually constant in cross section over large distances. For instance, a shaft will have a fixed inclination and diameter from the collar to the sump. Because of the continuity of many development openings, specialized design tools are available that automate the generation of volume components and speed up the design process. Other openings, such as shops and stopes, are more irregular in shape requiring individual design. For these, the definition of engineering components follows the same process as for geologic components in that their limits are defined on a midplane of suitable orientation and assigned a foreplane and backplane thickness. The foreplanes and backplanes can then be modified to match adjacent components or the design limits of the opening.

The ability to design openings from any viewplane greatly enhances the design process. Since volumetric calculations and reporting are based on the individual volume components, then these components should be oriented in the plane in which they will be extracted. In a cut and fill stope, production occurs in horizontal slices. Thus, the stope should be defined in plan view with a component thickness that corresponds to the expected unsupported opening height between backfilling cycles. In sublevel stoping, with the exception of vertical crater retreat, production occurs in the vertical plane normal to strike. Here, the component thickness should correspond to some multiple of the blast ring spacing. For room and pillar mining, the rooms or pillars will only be one layer thick and will be defined in the plane of the deposit. The midplane orientation for their components depends on convenience.

When defining geologic models in Chapter 4, some attention was paid to the

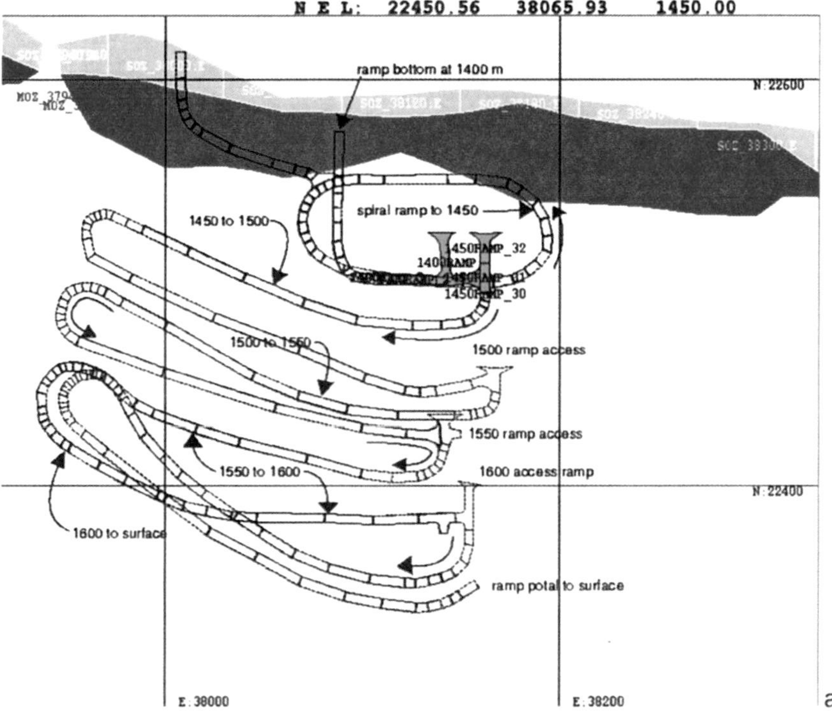

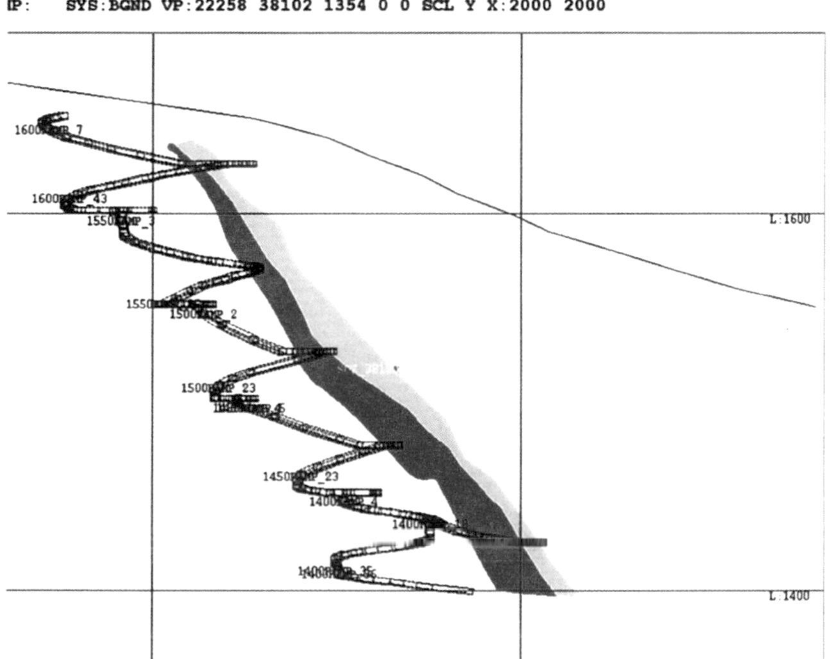

Figure 7.24. Complete ramp haulage and level access system in: a) Plan and b) Side view.

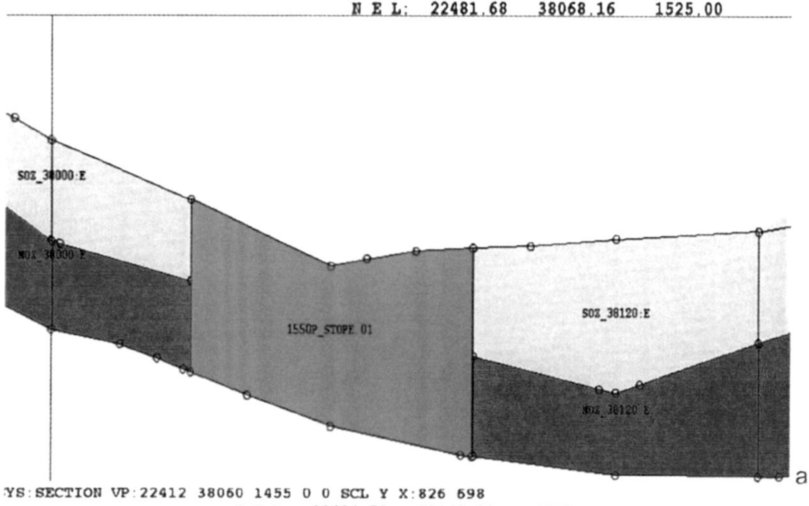

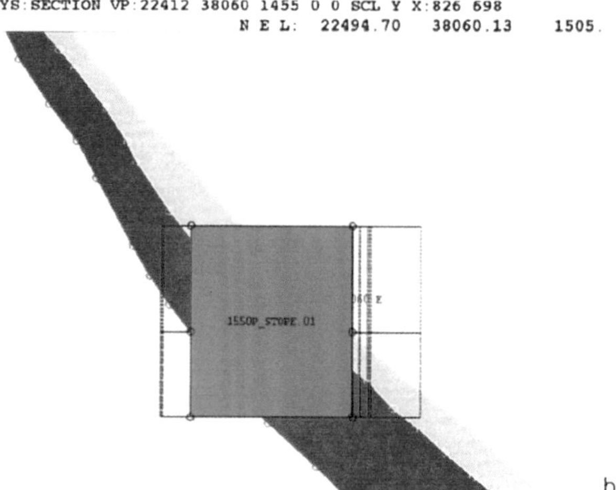

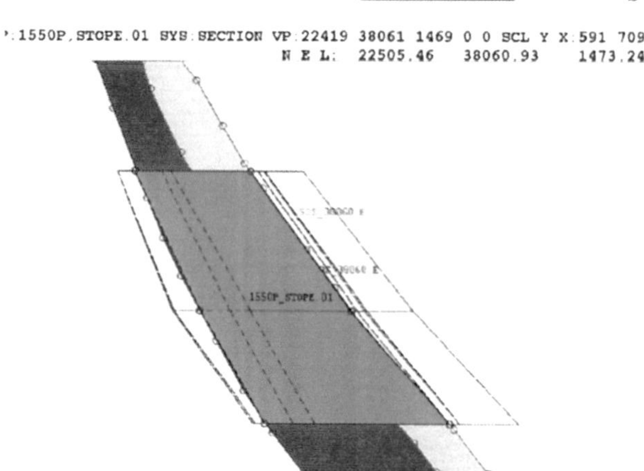

Figure 7.27. Initial stope midplane interpretation in plan (a), in section (b) and final stope component in section (c).

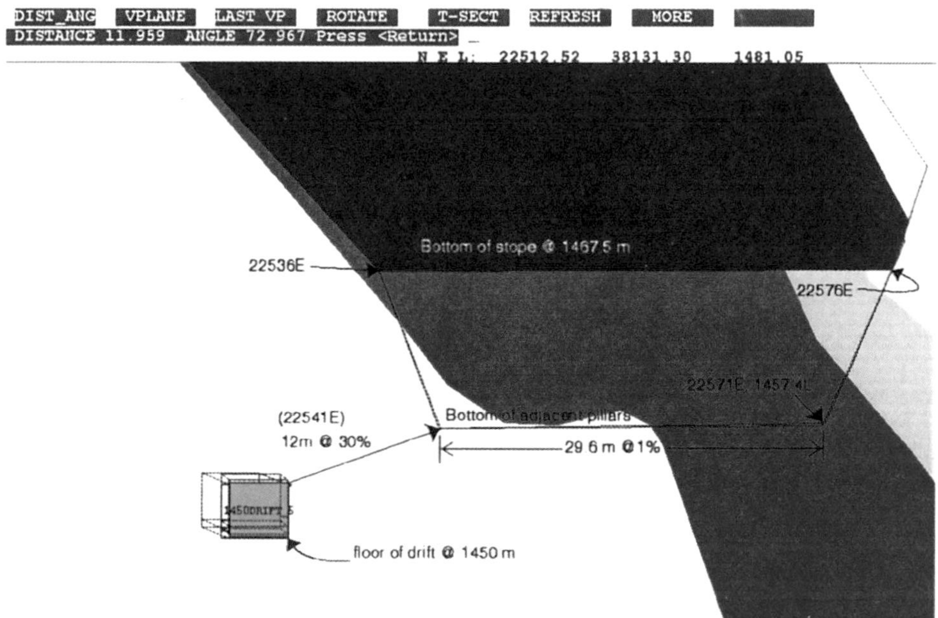

Figure 7.29. Section through drift and stope at 38131E showing measurements required for design.

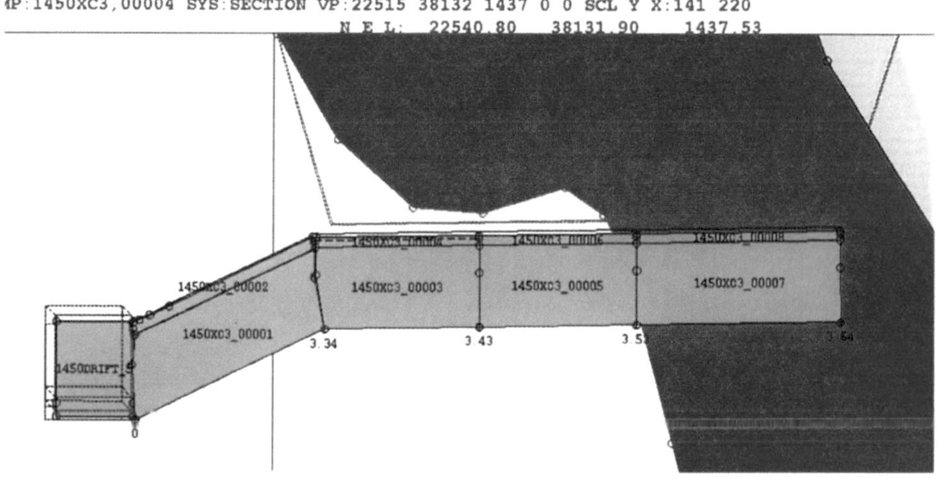

Figure 7.30. Modified x-line showing stacking of components for five breakpoints.

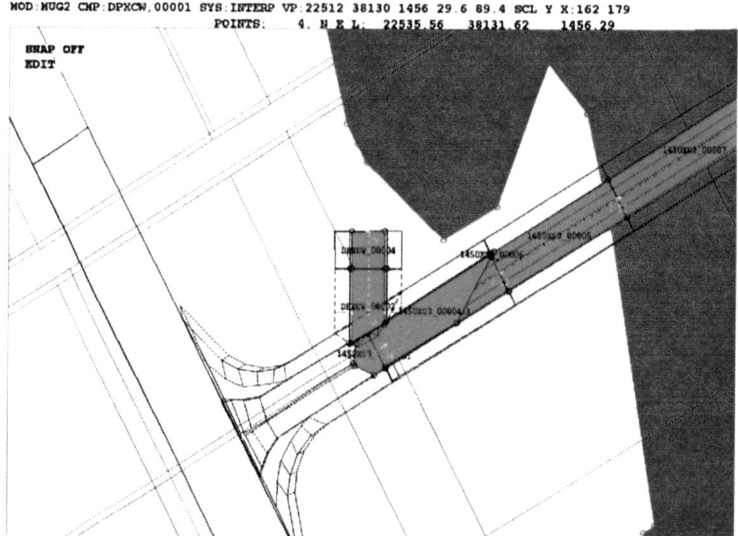

Figure 7.31. Interpretation of the midplane of the DPXCW_00002 drawpoint cross-cut showing rotation of viewplane to match component.

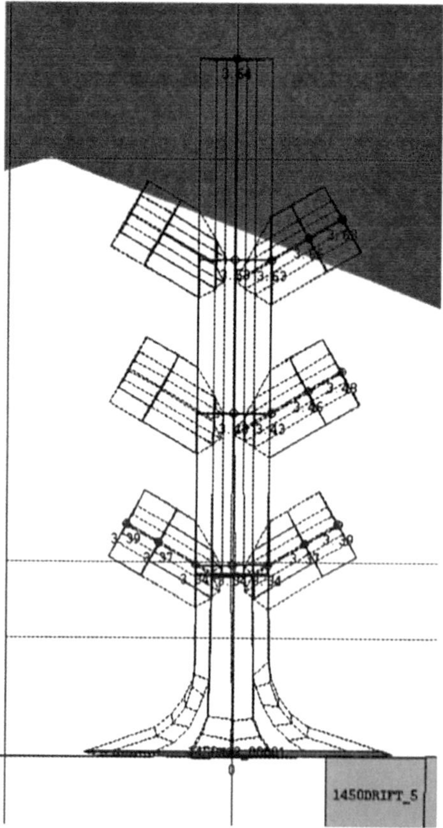

Figure 7.33. Final set of x-line drawpoint cross-cuts.

naming conventions for the volume components. For instance, 'moz,3800E,2' referred to a geologic component in the main ore zone with a vertical midplane at 3800E and a red color code. Close attention must also be given to the naming of underground design components due to the variety of types of openings and the stages in which they are developed. Components comprising the access crosscut for the 1400 level might be named 'mainxcut,1400,3' where a yellow color is used to represent pre-mining development openings. Consistency in applying naming conventions simplifies reporting volumes and displaying related openings. For example, if the color code convention of the previous example was adhered to, requesting that all components '*,*,3' be displayed in background would bring up all pre-mining development. Other colors could be used to discriminate between production and development openings on an annual basis.

The following series of examples demonstrate Lynx's specialized facilities for underground design using the TUTORIAL project. The first examples will focus on development openings followed by irregular stopes and drawpoints and, finally, reserve reporting.

Example 7.1: Model definition and defining shafts and crosscuts using Lynx
A new volume model will be created in the /apps/lynx/projects/TUTORIAL/3D directory to hold volume components for stopes and development openings. From the main menu, select **File ⇒ Volume Data ⇒ Volume Create** and enter the mine model name which will correspond to the file name in the 3D directory <M.UG2> and a description of the contents of the file (has in previous examples, this model has already been generated). Before the underground design facilities can be accessed, a map that can be used to hold center line data must be made active. Use **File ⇒ Map Data ⇒ Map Select** <shaft>, or alternatively create a new map file accepting default project coordinates. The map need only be plan view (Azimuth 90, Inclination 90).

The underground design facilities don't include controls for background display beyond the ability to toggle the various display elements on/off. Therefore, setup a background display that will be suitable for siting a shaft, namely a map of topography with a TIN <TOP> and the volume model of geology <G.OR> and mine workings <M.UG2>. If using the existing mine workings model, you will only want to specify those units that correspond to the components being used in the design to avoid screen clutter.

From the main Lynx menu select **U/G Design ⇒ Centerline Design** bringing up the Underground Centerline Design entry form. The map shaft should be indicated as being the active map in the window title. Next provide a plan **Viewplane** orientation that is centered over the deposit. The shaft and crosscut are best defined in plan view (N, E, L, Azi, Inc: <22300,37850,1500,90,90>). Provide a scale <2500,2500> and global grid spacing <100,100,100>. Finally, display the orebody model (G,OR) and topography (TOP) by selecting **Viewplane ⇒ Background** and toggling on the display for map and volume data structures. The Centerline window and the background display will appear similar to Figure 7.10.

The Centerline Design window contains facilities for the design of centerline layouts for underground excavations using interactive graphics, digitizer or keyboard data entry. As indicated above, centerlines are part of the map data structure and as

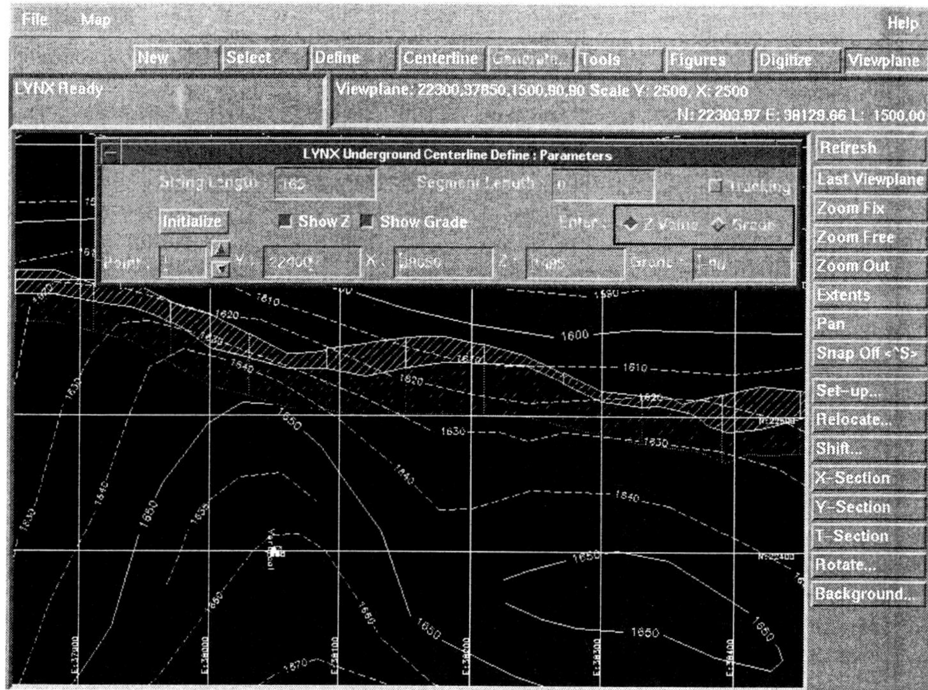

Figure 7.10. Centerline Design window showing shafty location and a background suitable for shaft design.

such are created and manipulated using the same map data facilities that have been used in previous examples. Using these facilities centerlines for various types of underground openings can be laid out with a wide variety of orientations that correspond to shafts, ramps, inclines, raises, drifts and crosscuts. Additional facilities are provided that will take a centerline as input to generate volume components following the path of a centerline. In this example, and several to follow, this methodology will be demonstrated for the major classes of underground development openings.

A shaft location must be found that will be centrally located to the ore body and sufficiently far from it so that there will be no risk of shaft damage. For this example, the shaft will be centered on 22400:N, 38050:E placing it 100 m back in middle of the footwall of the deposit. Since the dip is steep but not too vertical, most subsidence should occur on the hangingwall side.

Before shaft volume components can be generated by accessing the **Shaft** function, a two point centerline must be included in the active map. To generate this centerline click on **Define** ⇒ **Enter** making the parameters menu accessible for keyboard data entry. Centerline entry can also be from a digitizer or mouse, but in most cases accuracy demands keyboard input. Enter the starting point, i.e the collar location of the shaft <22400,38050,1660> for *y:x:z* and the coordinate of the shaft bottom <22400,38050,1495> for the second point. Click on **Save** to save the centerline data. An alternative approach is to click on **Centerline** ⇒ **Straight** bringing up the Centerline Straight entry form as shown in Figure 7.11.

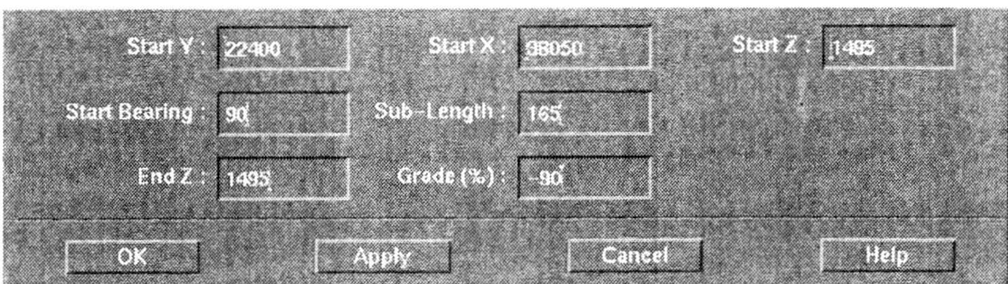

Figure 7.11. Centerline Straight entry form including parameter values for generating a vertical shaft.

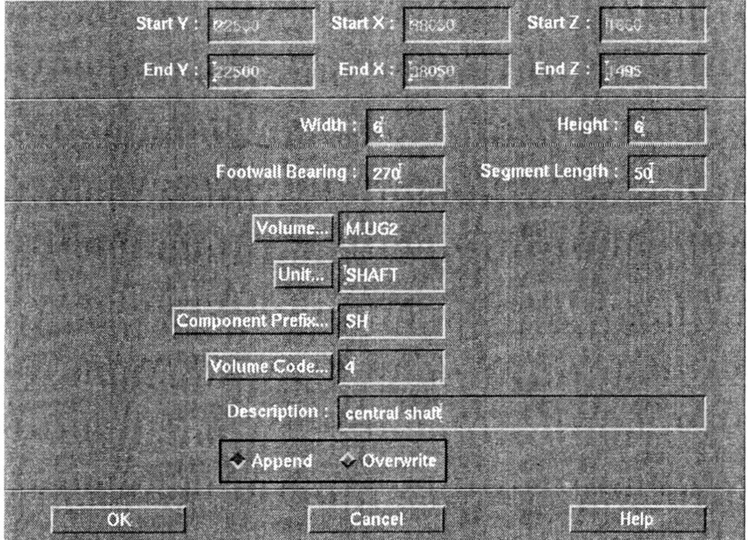

Figure 7.12. Centerline Shaft entry form with value setting for generating volume components for a rectangluar production shaft.

Define a vertical, rectangular shaft by selecting **Centerline** ⇒ **Shaft** bringing up the Centerline Shaft entry form as shown in Figure 7.12. The start and end points are not accessible and are based on the active two point centerline. and entering the unit identity <prodshaft>. Enter a **Footwall Bearing** <270>, a component **Segment Length** <50>, a shaft **Width** and **Height** <6,6>, a unit color code <4> and a description. This will automatically create vertical shaft volume components 6 × 6 m in section and 50 m long running from a collar elevation of 1660 m down to 1495 m. All development openings will use a code of 4 from this point on, and will be named 'prodshaft,00001,4' through 'prodshaft,00003,4' in order of increasing depth from the 'unit start' at 1660 m. Run a **X-sect** through the shaft and deposit to check on the shaft location relative to the deposit footwall. Note that crosscuts of increasing length with depth will have to be used for access and that the shaft bottom is far above the bottom of the deposit. As a justification for this, consider that the combined development cost of additional lower crosscuts, haulage drifts and increased shaft depth may easily exceed the cost of accessing the lowest ore using a ramp. Additionally, below 1450 m the deposit starts to pinch out and becomes more erratic in its thickness.

Next a crosscut will be driven from the 1500 level of the shaft through the deposit leaving 5 m in the shaft bottom for the sump and skip loading facilities. There are two approaches that can be taken for generating related designs. One is to keep all related centerlines in the same map. For instance, all centerlines used for drifts and crosscuts on the same level. The advantage of keeping all related centerlines in the same map is that they're always available to use as elements of design for intersecting openings. The disadvantage of this approach is that the individual identities of the cline features are lost when kept in one map. With separate maps for each centerline, the map name and description identifies its relation to the volume components, and thereby the mine openings, it was used to generate. In order to the identity of individual centerlines, a more involved methodology can be used:

1. When finished with the cline facilities to generate an opening **Save & Exit** back to the main menu and select or create a new map giving it the name of the next opening and an appropriate description.
2. Again go into cline with a new active map.
3. **Fetch** the previous map whose cline is required in order to design the next set of components back into memory as a figure.
4. **Move** the figure into its original position based on the origin and xyz position of some vertex (see Map Report).
5. Use **Pick Figure** to make the vertices of the figure accessible and then use **Enter** with **Snap** on to bind the centerline of the new mine opening to a point of intersection on the figure.

The shaft map will be used to align a shaft crosscut with the shaft's centerline following this methodology. Starting from the main menu, (**Save & Exit** from cline to save the shaft map) use **File** ⇒ **Map Data** ⇒ **Map Creat**e to make a new map <1500xc> active that will hold the centerline of a crosscut starting at the centerline of the shaft at a depth of 1500 m. At this time you may also want to update the background display for volumes so that it includes the shaft components. Select **U/G Design** ⇒ **Centerline Design** to return to centerline design with this new map active. Click on **Figures** ⇒ **Fetch Figures** and select <shaft>. The shaft centerline will appear in cyan at mid screen. The shaft's axis was at 22400:N, 38050:E. This can be confirmed by using **Map Report**. To move the shaft figure back to its correct position use **Pick Figure** ⇒ **Move** bringing up the Figures: Move entry form. Toggle off **Tracking** to allow key board entry of the move point, enter the new coordinates and select **OK** to reposition the shaft. Make sure that the viewplane is in plan view at an elevation near 1500 m since this will be by default the starting elevation of the new centerline. Center-line is used to model openings which are of constant height, only varying in plan view. This works well for drifts and low angle inclines and declines. Center-line components have to be digitized in plan view one component at a time rather than being autogenerated.

Use **Enter** with **Snap** on to digitize the cross cut starting with the first point at the shaft. Make sure that **Tracking** is on and **Grade** is set to 2% in the parameters form. After snapping the first point to the shaft, turn **Snap** off and add segments up to and through the ore body normal to strike. Since there isn't a grid to snap to, there will be some difficulty in maintaining a straight center-line. This can be remedied by creating a map consisting of point data in a square grid, possibly imported as an ASCII file, and then bringing it into background as a figure and setting snap on. Otherwise, digitize an initial attempt at a cross cut and use **Insert** and **Edit** to straighten out the drift. The number and length of components can vary; using too few will make it difficult to intersect the cross cut with other openings. For instance, a drift will be included 10 m behind the footwall. In order for the drift center-line to intersect the center-line of the cross cut, the cross cut center-line will have to be stored with a point 10 m behind the hangingwall and brought into background, then a point on the drift center-line can be snapped to the center-line of the cross cut at this traverse point. Therefore, the end-points of development opening components sometimes have to be chosen with consideration of other intersecting components which have yet to be defined. Too many components will simply occupy too much memory. Except for intersections, there is no reason to generate components which are shorter than the rate of advance that can be achieved during a scheduling period unless a large number of components is needed to capture a complex geometry, as in the case of a curve.

When finished with the crosscut centerline, click on **Save & Clear**. Click on **Generate** and use the mouse to select the centerline. The selected string is highlighted and the Generate entry for is opened as shown in Figure 7.13. Select a rectangular **Profile** Enter a descriptive unit identifier <1500xc>,description, starting elevation for the center line <1500>, grade <2>, cross-cut width, fore-thickness and grad height <4.5,3.5,0>, and unit code <4>. Individual components will be generated with names starting at '1500xc,00001,4'. The starting elevation of the floor of the first

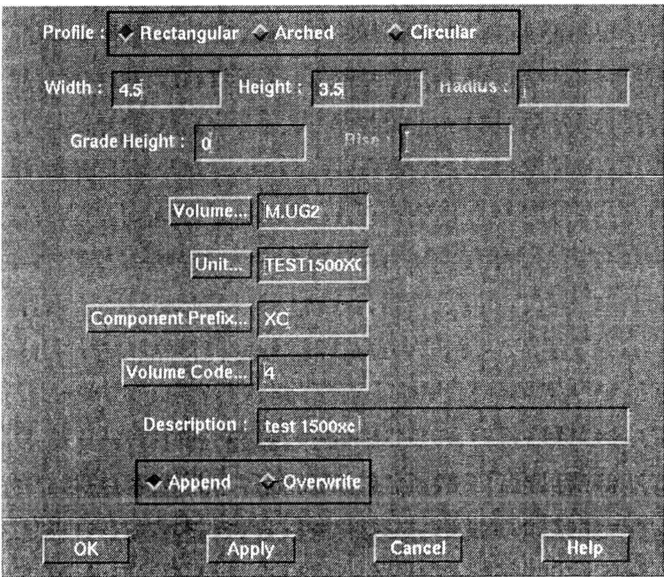

Figure 7.13. Generate entry form setup to generate volume components based on a centerline feature.

component will be at 1500 m and will be centered on the shaft. Each component will be as long as the separation distance of adjacent digitized points and will rise at a grade of 2%, i.e. positive out of the screen towards a viewer who is observing in plan view so that mine water will drain back to the shaft. The cross sectional dimensions will be 4.5 m high by 3.5 m wide. With a back-thickness of zero, the centerline will be located on the floor of the crosscut.

Figures 7.14.a and b show the shaft, 1500 m cross cut and transverse line in plan view and in **Y-sect**. The off-set values displayed on the traverse are relative to the position of the viewplane. Note that the sectional view shows the cross cut passing through the hangingwall. Due to the positive gradient of the cross cut, there is a 3.35 m elevation gain over its length. Since the deposit dips to the north, the hangingwall is moving to the south as elevation increases. This could have been avoided by setting the off-section display thickness to around 3.35 m when the geologic model was brought into background, but would have required some forethought. Instead, the 1500XC,00004 component can be modified using the **Interp** function to interactively edit the components fore, mid and backplanes back behind the hangingwall. This procedure is demonstrated in the following example.

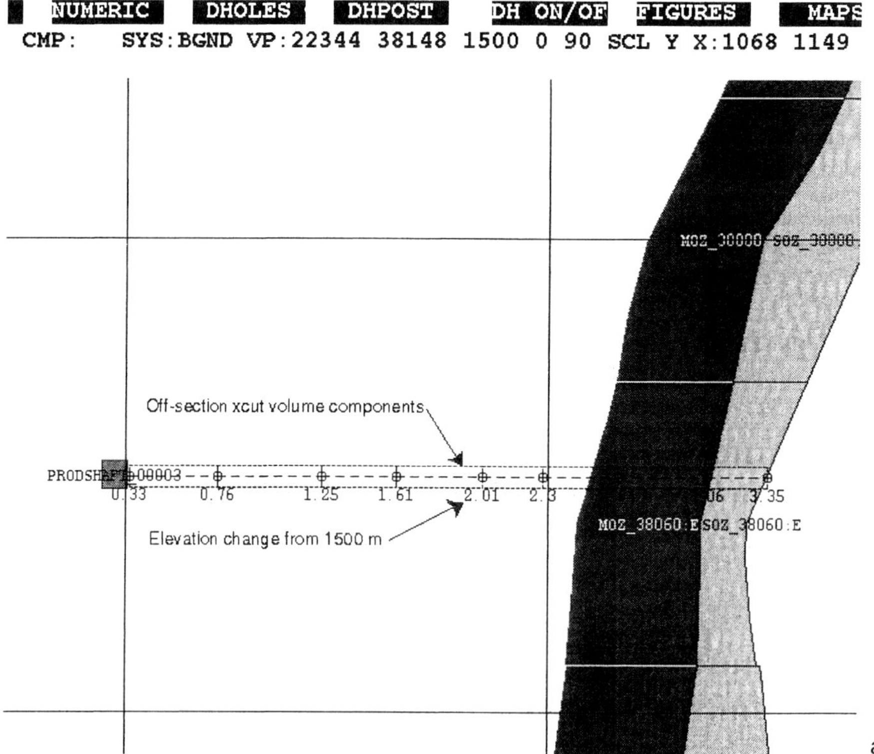

Figure 7.14a. Shaft, off-section 1500xc volume components and 1500 m cross cut center line shown as a traverse in: Plan view.

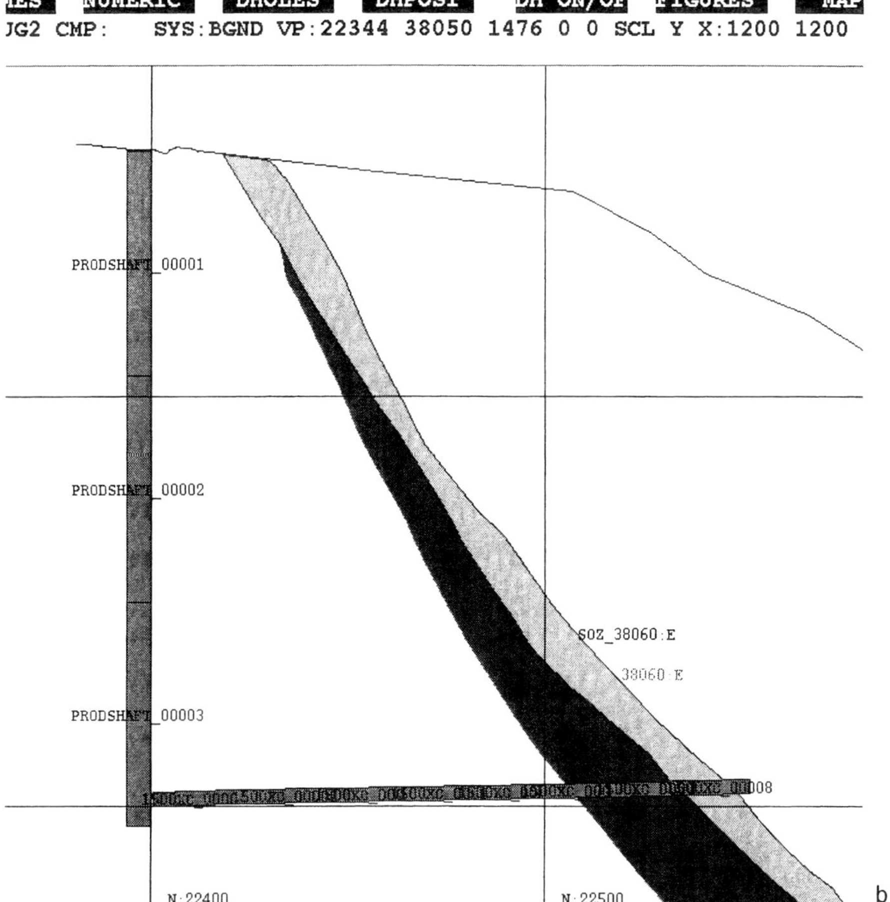

Figure 7.14b. Shaft, off-section 1500xc volume components and 1500 m cross cut center line shown as a traverse in: Y-sect without traverse.

Summary of procedure: Example 7.1

Initial data requirements: information on shaft collar coordinates, depth and dimensions; main crosscut starting and ending coordinates and dimension as well as a triangulated topography and geologic model on which to base these figures.

Procedure:

1. Create a new volume model to hold the underground development volume components (M.UG2 is already available in TUTORIAL as an example)
2. Select or create a new map <shaft> to hold the shaft cline feature.
3. Setup an appropriate background display for shaft design including map <TOP> and geologic model <G.OR>. Place the viewplane over the proposed shaft site (VIEWPLANE N, E, L, AZI, INC: <22300, 37850,1500, 90, 90>, VIEWPLANE SCALES SX, SY: <2500, 2500>, GLOBAL GRID SPACINGS DE, DN, DL: <100,100,100>)
4. Define the shaft **U/G Design** ⇒ **Centerline Design** (Underground Centerline Design entry form)

4.1 Define a two-point centerline **Define Enter** (use the parameters form for keyboard entry of <22400,38050,1660> and <22400,38050,1495> ⇒ **Save**.

4.2 Generate shaft volume components **Centerline** ⇒ **Shaft** (Centerline Shaft entry form)

Width: <6> **Height**: <6>

Footwall Bearing: <270> **Segment Length**: <50>

(Volume, Unit, Prefix and Code based on model used, see Fig. 7.12)

⇒ **OK**

⇒ **Save & Exit** (return to main menu)

5. Make a new map active to hold the 1500 m crosscut centerline.
6. Define the crosscut (Viewplane remains at ELEV:1500) **U/G Design** ⇒ **Centerline Design** (Underground Centerline Design entry form)

6.1 Define the crosscut center line **Figures** ⇒ **Fetch Figures** <shaft>

(Use **Pick Figure** and **Move** with keyboard entry to reposition the shaft at <22400:E, 38050:N and then use **Enter** with **Snap** on to start the crosscut. Toggle **Tracking** on, **Grade** to 2% and **Snap** off. Digitize the crosscut where it exits the shaft and continue to enter component endpoints until the crosscut extends through the ore zone. Locate component endpoints where other they will be needed later for intersections with other mine openings, particularly the footwall drifts. Note that the geology volume model is displayed with zero thickness at 1500 m elevation, but that the crosscut increases in elevation 2 m per 100 m which will result in the opening punching through the hangingwall a short distance. Use **Zoom**, **Insert** and **Edit** to straighten the centerline of the crosscut. When completed use **Save & Clear**).

6.2 Generate crosscut volume components **Generate** (select the centerline with the mouse) (Generate entry form)

Profile: <rectangular>

Width: <4.5> **Height**: <3.5>

Grade Height: <0>

(Volume component naming based on model used, see Fig. 7.13)

⇒ **OK**

Example 7.2: Modifying development openings using Lynx

The next step in development of the 1500 level, following driving the cross-cut, is to place drifts in the footwall for access and development of the stopes, haulage and ventilation. Two drifts will be placed in the footwall with one just inside the ore zone that will be used for ring drilling, and the other 40 m back for access and ventilation. Cross-cuts will connect the two drifts at roughly 60 m intervals.

LHDs will be used for haulage, so the drift intersections must be curved. In this example, the 1500XC,00004 and 1500XC,00005 volume components created with **Cline** in Example 7.1 will be replaced with two new components of identical position and cross section. These new components will be created with the center-line facilities so that the point where one ends and the next begins can be used to exactly specify the beginning of two intersecting drifts. This exercise could be more easily carried out with the existing components of 1500XC, but the creation of a matching replacement will serve to demonstrate the creation of engineering openings that are in alignment.

From the **Centerline Design** window, set up the same viewplane orientation and scale as in Example 7.1. This elevation can be confirmed using **Map Report**. For this example, the elevation was 1499.67 m. Place all the geology components in background (G.OR) without any off-section display. Also place the 1500 level crosscut and its center-line map (cln2) into background (1500XC,*,*) with sufficient off-section display to view all of the components. In the Background Volume Data Selection entry form set DISPLAY COMPONENT LABELLING and DISPLAY COMPONENT POINTS on so that components will be labelled and so that the vertices that define them will be visible.

Select **U/G Eng ⇒ Centerline Design** and define a center-line component (1500XCINT) exactly as was done for 1500XC (width = 4.5 m, height = 3.5 m, a backplane thickness of zero and gradient of +2%) except that this component will start at an elevation of 1499.67 + 1.25 m; the same elevation as the end of 1500XC,00002. At this point the strategy that is being used to create aligned components should be clear: the center-line is saved as a map to be available later for use as a point of alignment with other intersecting openings. By setting the viewplane to the same position as when 1500XC was created, the 1.25 m off-set point can be snapped to start a new component at exactly the same coordinate. Use **Define ⇒ Snap on ⇒ Pick** to activate the background map centerline <cln2> used in Example 7.1 for crosscut design. Set **Snap On** and use **Enter** to start 1500XCINT at the 1.25 m point of cln2. Set **Snap Off** to digitize a second point at what will become the intersection with the footwall drift, and finish the second component of 1500XCINT by setting **Snap On** and entering the last point on the center-line at the 2.01 m off-set. Use **Save & Clear** to store the 1500XCINT center-line as a map (1500XCIN) and **Generate** to create the volume components.

Next, the two new intersection components will be used as the starting point for two intersecting drifts which will slope at a grade of 2% away from 1500XCINT to the east and west and having the same dimensions. Use the vertices where 1500XCINT,00001 and 1500XCINT,00002 and their center-line map (1500XCIN) intersect to define location and elevation. The resulting components are shown in plane view in Figure 7.15a and in side view in Figure 7.15b which illustrates the

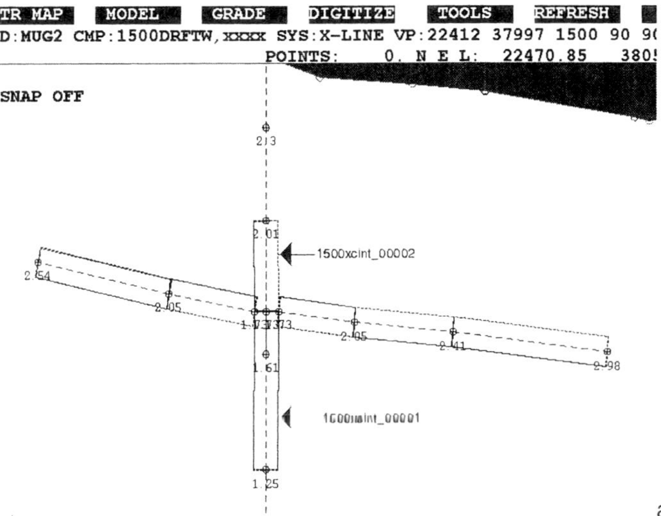

Figure 7.15a. Intersection of shaft cross-cut and footwall drifts based on center-line alignment shown in: Plan view. (Figure in colour, see opposite page 247.)

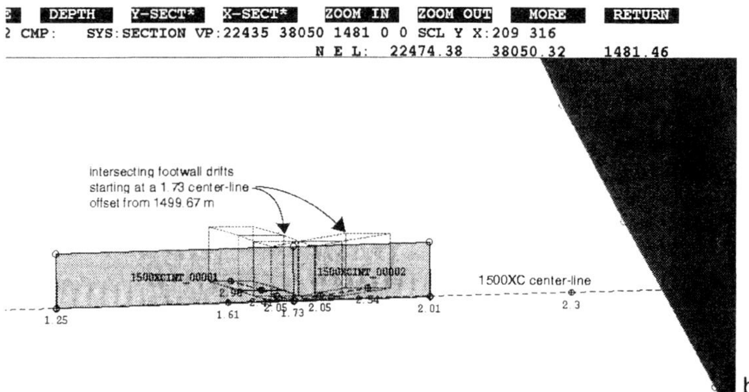

Figure 7.15b. Intersection of shaft cross-cut and footwall drifts based on center-line alignment shown in: Side view. (Figure in colour, see opposite page 247.)

close alignment that can be achieved by making good use of the center-line maps.

Now the two cross-cut drifts will be modified so that the intersection with the footwall drifts will include curved corners. Modification of volume components created with Centerline facilities is carried out in much the same manner as is the generation of irregular volume components as demonstrated in Chapter 4, but in plan view. **Exit & Save** from the Centerline menu, and from the Lynx main menu select **U/G Eng ⇒ Volume Design Tools** opening the Volume engineering Design Tools window. Note that the active volume <M.UG2>. In the upper left corner of the window select **Volume ⇒ Define Mode ⇒ Define**. Under **Viewplane** use **Set-up**, **Shift**, and the zoom facilities to position the viewplane over the intersecting 1500XCINT crosscut components as shown in Figure 7.16. and specify whether the component fore, mid or backplane is to be modified. By starting with the foreplane, the off-set of the modified foreplane from the midplane will be visible and can be used as a guide during subsequent modification of the midplane and backplane. Remember that the midplane is essentially on the floor of the cross-cut, since the back-thickness was defined as being 0 m in Example 7.1, while the foreplane is the roof of the cross-cut (3.5 m high). A good alternative, is to toggle on the **Change all Boundaries** checkbox in the parameters menu. This will cause modifications in any one of the three planes to be copied to the remaining two and makes sense for cline objects that are rectangular in vertical section. Since the intersecting components are going to be modifies to have curved walls at their intersections, using a circular figure and **Snap On** as a template is a good idea. As in Example 7.1, use **Figures ⇒ Fetch Figures** <crc> with **Size** and **Move** to place a circle of the required turning radius into position next to the components. Use **Insert** and **Edit** to modify the foreplane, as shown in Figures 7.15a and 7.15b, (see Fig. 7.16) and then **Save & Clear** to overwrite the component with the modified version. Select **Interp** twice more to modify the midplane and backplane so that they match the foreplane. Once the cross-cut intersection components have been modified, the drift components can also be modified so that they match 1500XCINT,00001 and 1500XCINT,00002 without any overlap. The final intersection is shown in Figure 7.17.

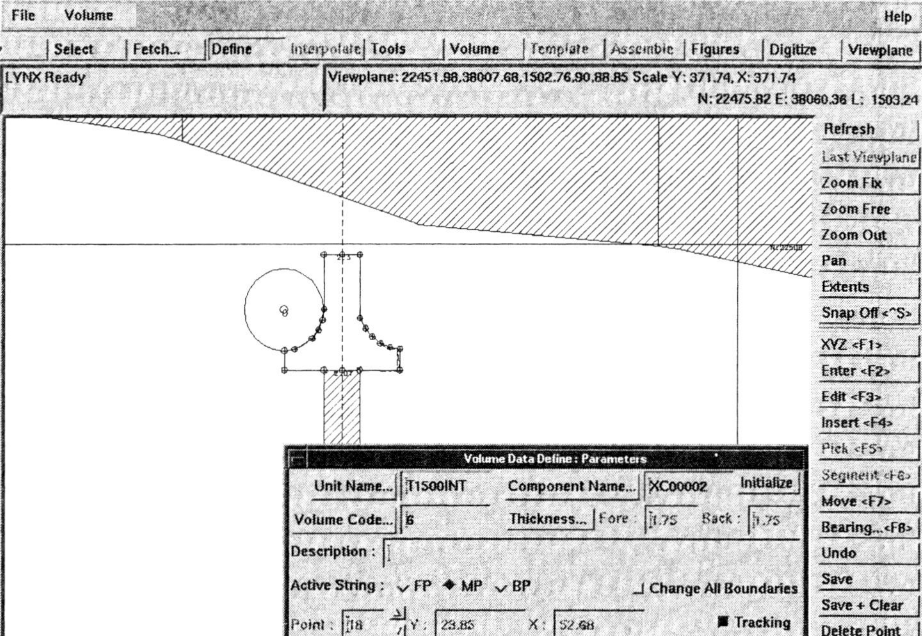

Figure 7.16. Volume Engineering Design Tools in Define mode being used to modify cline objects using a circle figure.

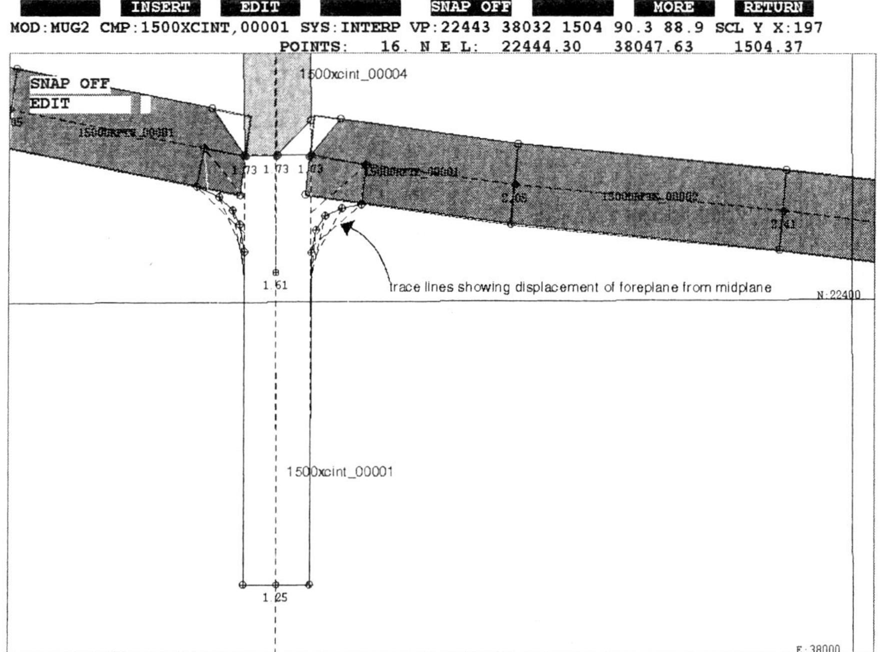

Figure 7.17. Modification of a center-line component's foreplane using interactive interpretation.

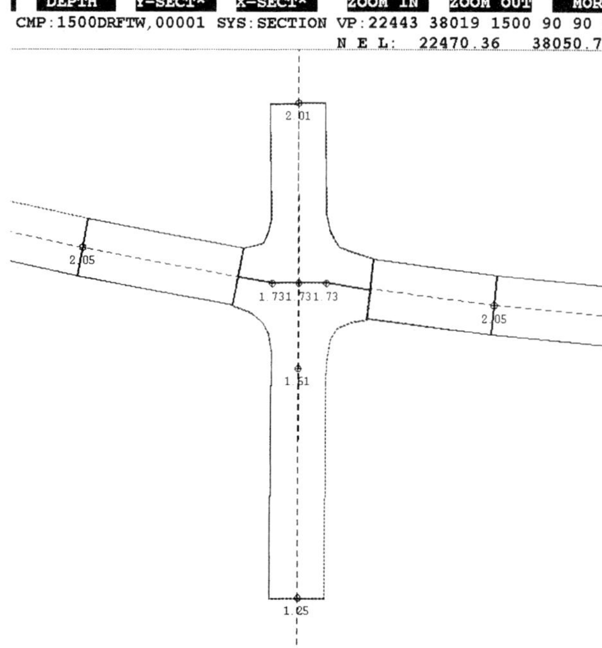

Figure 7.18. Completed intersection of cross-cut and hangingwall drifts.

The **Cline** facility can be used to complete all the development drifts serving the 1500 level and the Volume Engineering Design Tools can be used to modify them as needed. Alternatively, **Define** can be used to generate complex drift components such as intersections and orepass dumps, but one of the advantages of using the center-line facility is that exact dimensions, gradient and center-line path can be specified to ensure a close match between adjacent volume components regardless of their complexity.

The final set of development drifts for the 1500 m sublevel are shown in Figure 7.18. All drifts slope upwards from the entry cross-cut. The filled components are those approximately at an elevation of 1502.5 m.

Summary of procedure: Example 7.2

Initial data requirements: crosscut volume components with which to create an intersection and their defining centerline map and a corresponding centerline map.

Procedure:

1. Set up the viewplane orientation and background display as per steps 2.1 and 2.2, Example 7.1, including map <cln2> and volume model <M,UG2> 1500 level crosscut components <1500xc,*,4>.
2. Create a new map to hold the new components. Be sure to setup a map origin that corresponds to the start location of the new centerlines (see cln2's elevation at the point of intersection and its orientation).
3. Create two matching crosscut components for the intersections per Example 7.1, Step 6 **U/G Eng ⇒ Centerline Design**.

3.1 Activate the 1500 crosscut component's centerline (map cln2) **Enter** ⇒ **Snap On** ⇒ **Pick** (select the defining centerline from map cln2's cline feature)
3.2 Define duplicate 1500 crosscut components
C-LINE UNIT ID: <1500XCINT>
C-LINE START LEVEL: <1500.92> (map elevation for cln plus starting elevation for 1500XC,00003)
CLINE GRADE (% NORMAL TO VIEWPLANE, POSITIVE OUT): <2>
CLINE UNIT WIDTH, FORE THICKNESS, BACKTHICKNESS: <4.5,3.5,0>
CLINE UNIT CODE: <4>
⇒ **Snap On** ⇒ **Enter** (digitize the start of 1500XCINT,00001 at the 1.25 offset)
⇒ **Snap off** ⇒ **Enter** (digitize the start of 1500XCINT,00002 at an offset of 1.73, or thereabouts)
⇒ **Snap On** ⇒ **Enter** (digitize the end of 1500XCINT,00002 at the 2.01 offset)
⇒ **Save** & **Clear** ⇒ **Generate** ⇒ **Save & Exit** (Centerline window)

4. Modify the crosscut components to accommodate a curved intersection. From the main Lynx menu **U/G Eng** ⇒ **Volume Design Tools** (Volume Engineering Design Tools window) ⇒ **Volume** ⇒ **Define mode** ⇒ **Define** (toggle on).
4.1 Position viewplane over crosscut intersection components and activate a circle figure to use as a template **Snap On** ⇒ **Figure** ⇒ **Fetch Figures** <crc> (use **Move** and **Size** to place in position).
4.2 Retrieve a component into active memory. **Select** <1500XCIN_00001> by clicking the mouse on one of its vertices.
4.3 Modify the component using **Define** ⇒ **Snap On** ⇒ **Insert** using the crc figure while setting **Tracking** and **Change All Boundaries** on in the parameters entry form. **Save & Clear** when finished.
3.2 Add the intersecting footwall drifts **U/G Eng** ⇒ **Centerline Design**. The relevant information is as follows.
C-LINE UNIT ID: <1500DRFTE>
C-LINE START LEVEL: <1501.4> (map elevation for cln2 plus starting elevation for a 1.73 m offset)
CLINE GRADE (% NORMAL TO VIEWPLANE, POSITIVE OUT): <2>
CLINE UNIT WIDTH, FORE THICKNESS, BACKTHICKNESS: <4.5,3.5,0>
CLINE UNIT CODE: <4>
⇒ **Snap On** ⇒ **Enter** (digitize the footwall drift as per Example 7.1)

Example 7.3: Creating ramps using Lynx
A ramp is needed for equipment access between sublevels. In this example, the drift of Example 7.2 will be extended to the east remaining in the footwall as shown in Figure 7.19. At about 22460E,38169N ramp system access will be opened on the south side of the drift. From here a ramp will rise from the 1450 level and continue on to the 1550 level.

There are two approaches available for ramp design: the ramp can either spiral continuously like a corkscrew or can be developed as straight inclines connected by level 180° switchbacks. A continuous spiral ramp provides access between levels with the least development and in the most confined space, but is not suited to haulage due to the increased wear and tear on equipment from moving through steep curves and must be negotiated at low speed due to low visibility. For increased haulage speed and safety, long straight ramps connected by flat curves should be used.

Lynx provides facilities suitable for creating ramps of either design. For genera-

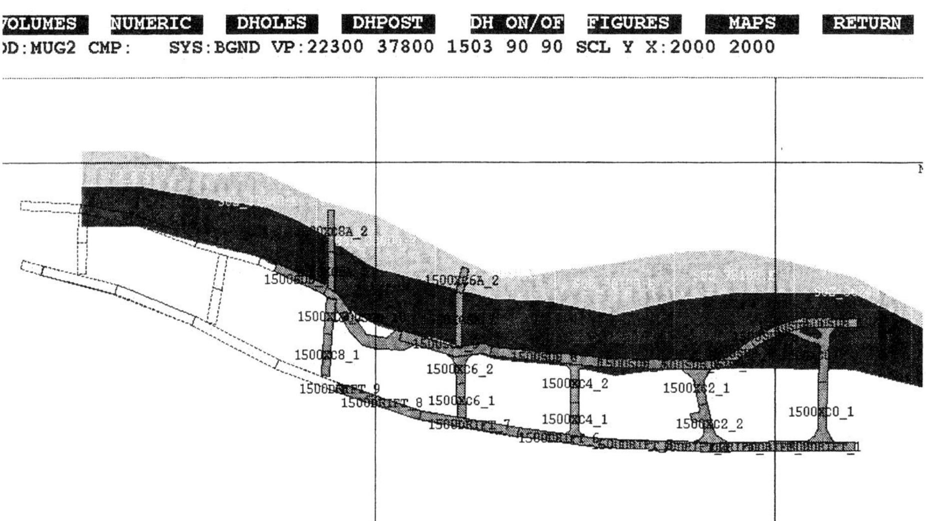

Figure 7.19. An example of development drifts centered on the 1500 m sublevel. (Figure in colour, see opposite page 278.)

tion of curved ramps, **Curve** automates a continuous graded curve or spiral of connected volume components based on either a fixed grade of inclination or a fixed angle of curvature. This provides a handy means of generating a spiral ramp. To produce inclines connected by switchbacks, the center-line facility can be used, as in the previous example, to create drifts of varying inclination. As demonstrated in Examples 6.13 and 6.14, a figure of a circle with a radius equal to the required turning radius can be used as a template with the center-line facility to produce 180° switchbacks of any inclination. Both approaches will be demonstrated in this example, but only the center-line based ramps will be retained in the final design. First, a spiral ramp will be generated connecting the 1500 and 1550 levels.

Before continuing, place the viewplane over the area of development for the 1500 east drift and ramp (N, E, L: 22340, 37925, 1499.67, 90, 90; scale: 1200:1) and put the geologic model (G.OR), drifts (M.UG2; components: 1500xcint and 1500drft), and the drift's center-line map (1500drft) into background display. Using the same methodology as for Example 7.2, use the map definition's elevation (1499.67) and the offset at the end of the center-line (2.98) to extend the drift past the intersection coordinate with the ramp system (22460E,38169N) using the Volume engineering Design Tools (**U/G Eng ⇒ Centerline Design**). The drift's dimensions and inclination should remain unchanged (4.5 m high by 3.5 m wide, backplane thickness of zero, grade of 2% or less). Be sure to end one of the drift's components at 22460E,38169N to provide a point to snap to while creating an intersection with the ramp. Use **Save & Clear** to save the drift center-line. By using the same map name, the new center-line extension can be added to the previous map. **Generate** the drift. The completed extension of the eastern footwall drift is shown in Figure 7.20.

To create the spiral ramp, select **Centerline ⇒ Curve** bringing up the Centerline Curve entry form (see Fig. 7.21). Two types of basic curve definitions can be used.

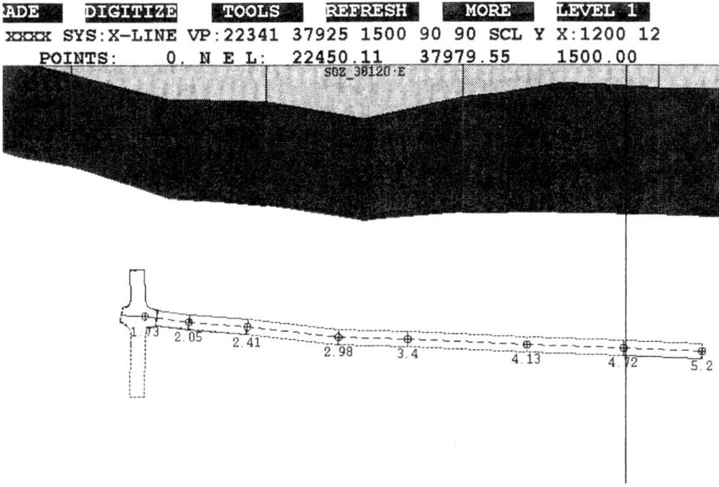

Figure 7.20. Completed extension of 1500 drft E. (Figure in colour, see opposite page 278.)

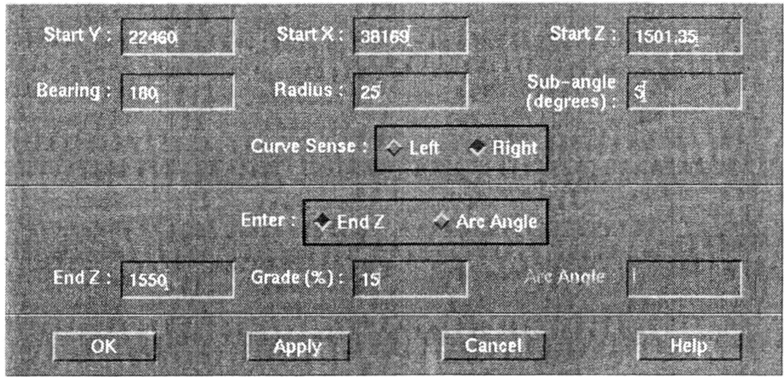

Figure 7.21. Centerline Curve entry form setup to generate a spiraling cline map feature based on an ending elevation.

An **End Z** curve runs at a fixed grade over a vertical interval, while a **Arc Angle** curve runs over a specific horizontal angle. Thus, the *xy* location at which an **End Z** curve will exit is more difficult to control being a function of the starting point, the curve radius, the grade and the elevation change. In contrast, an **Arc Angle** curve is best used for specific change in ramp direction, such as a switchback. Choose an **End Z** curve and enter the starting coordinates of the ramp <22460,38169,1501.36> (the exact value can be obtained from cursor tracking in the parameter menu and the offset on the center-line map), the **End Z** or elevation of the end of the ramp <1550>, the **Bearing** of the center-line of the initial component <180>, the **Curve Sense** for the spiral <R>, the **Curve Radius** <25>, the **Grade** <15>, and the component **Sub-angle** <5> which is the number of degrees subtended by each component. For instance, if a 10° sub-angle was used, then 36 components will be created over a 360° curve. Use **Save** or **Save & Clear** under the **Define** menu. Finally, select **Generate** and pick the ramp centerline to bring up the Centerline Generate entry form. For the cross-sectional Profile, toggle **Rectangular** on. Enter the ramp's **Width** and **Height** <4.5,3.5> and **Grade Height** <0> and provide the **Volume** <M.UG2>, **Unit** <1500RAMP>, **Prefix** <RP> and **Volume Code** <3>. The ramp will be generated as shown in Figure 7.22a and 7.22b. Note that in this example the ramp ends on the 1550 level at 22448E,38167N, a convenient location for intersecting the 1550 foot-

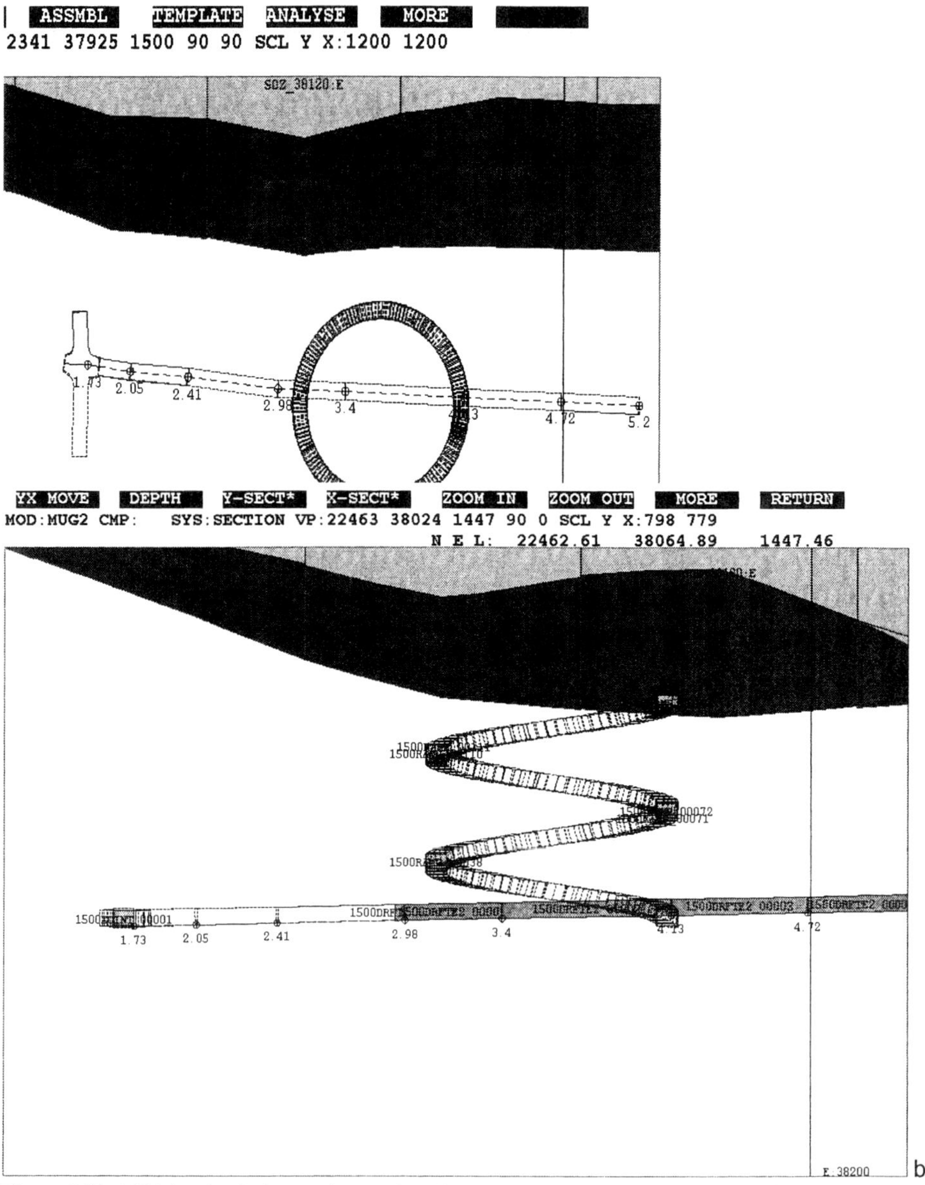

Figure 7.22. 1500 level spiral ramp in: a) plan view, and b) side view. (Figure in colour, see opposite page 279.)

wall access drift, but the ramp is spiralling too closely under the overlying 1550 drift. A spiral ramp is probably not a good choice for this location. Instead, an incline with switchbacks will be used.

To ramp up 50 m in elevation at 15% will require a horizontal incline distance of 333 m, but the switchback cannot be flat due to visibility and drainage requirements. If an 8% incline is used on the curve, then this will account for additional vertical gain. A useful equation for calculating vertical gain and other curve parameters is $d = (\theta/360)(2\pi r)(g/100)$, where d = vertical gain, θ = horizontal angle subtended by the curve, r = curve radius and g = % grade. Thus, a 180° curve, with a 10 m radius at 8% will rise 5.03 m. Therefore, the ramp length outside of the curve will be 328 m, i.e. approximately 164 m to the west rising to 1522.5 m, followed by a 180° switchback to the left (south) ending at 1527.5 m and back 164 m to the east ending at a position 50 m above and 40 m to the south of the starting point of the 1500 ramp to account for the change in position of the footwall due to dip.

Use **Define ⇒ Enter** to digitize an access intersection between the 1500 ramp and access drift, then change the **Grade** in the parameters entry form to 15% and ramp to the west until the elevation of the ramp has increased to 1522.5 m. Select **Figures** and bring the 100 m radius circle <crc> up as a figure. Use **Scale** to reduce the radius to 10 m and **Move** the top of the circle to the start of the curve. Set **Snap On** and **Grade** to 8% and digitize a 180° curve until the elevation reaches 1524.67 m. Set **Snap Off** and **Grade** to 15% and continue digitizing to the east until reaching 1550 m. As always, finish by using **Save & Clear**. Select **Generate** and provide the same dimensions as before, the component name <1500ramp>, the center-line starting level (1499.67 + 4.13 = 1503.8), the initial grade <0.5> for the intersection with the ramp will be kept low, the height, width and back thickness <4.5,3.5,0>, and a color <4> that will be used for ramps. The basic 1500 ramp is shown in plan view in Figure 7.23a and in side view in Figure 7.23b.

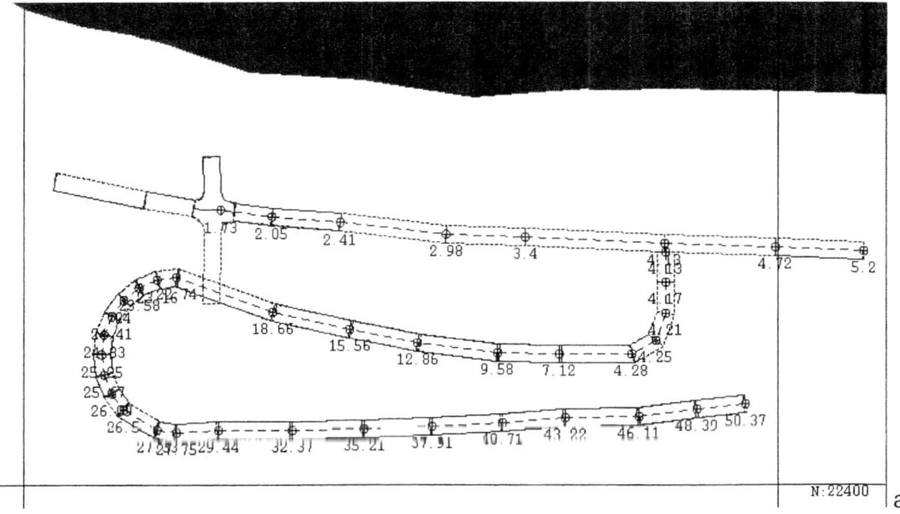

Figure 7.23a. Switchback ramp construction between 1500 and 1550 levels using center-line facilities shown in: Plan view.

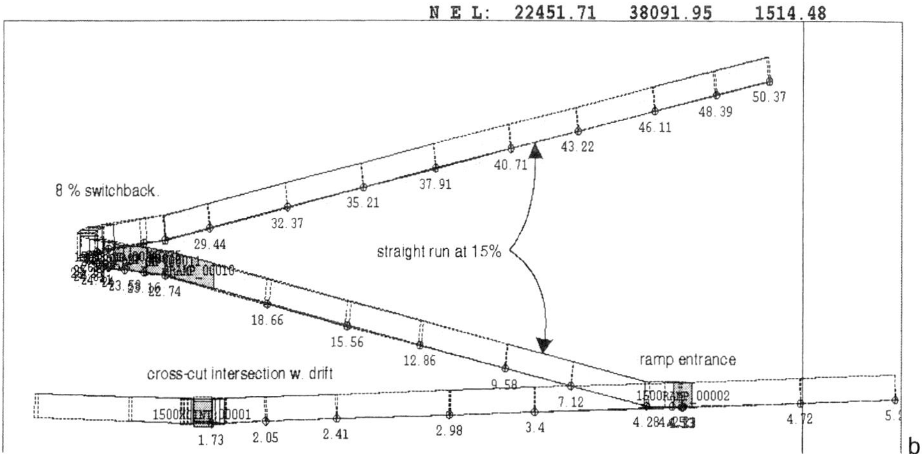

Figure 7.23b. Switchback ramp construction between 1500 and 1550 levels using center-line facilities shown in: Side views.

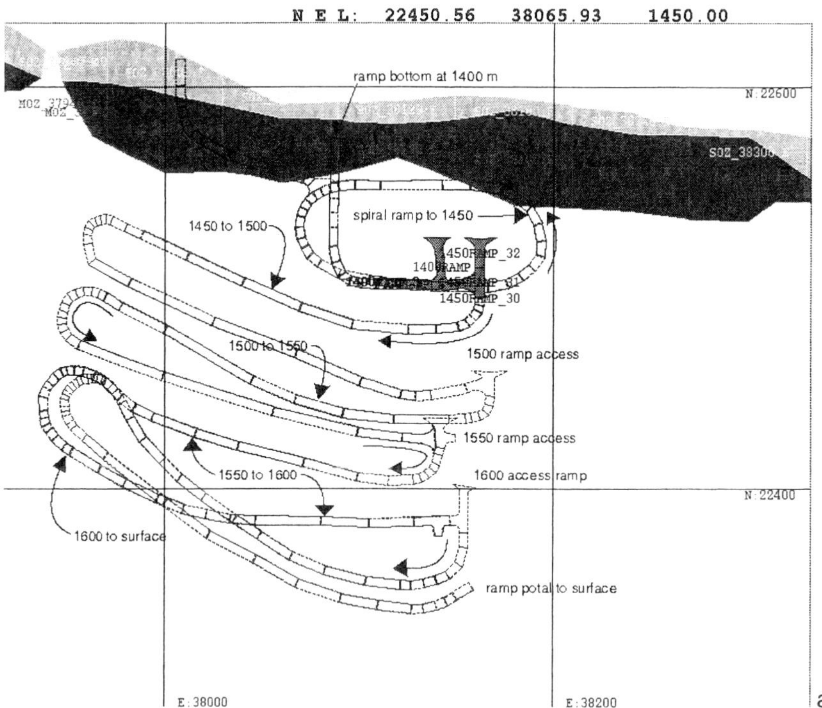

Figure 7.24a. Complete ramp haulage and level access system in: Plan view. (Figure in colour, see opposite page 310.)

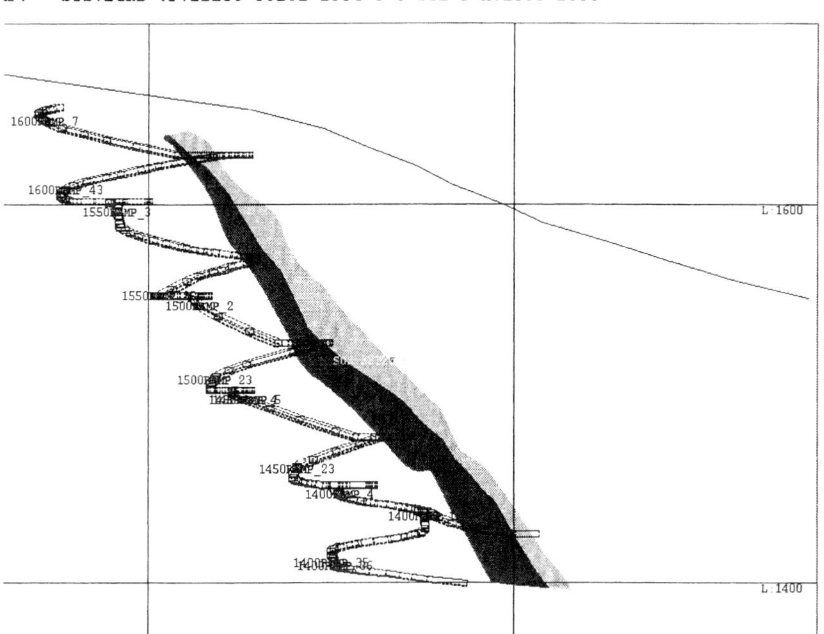

Figure 7.24b. Complete ramp haulage and level access system in: Side view. (Figure in colour, see opposite page 310.)

Once the basic ramp components are in place, the Volume Engineering Design Tools can be used as in Example 7.2 to provide curved intersections between the ramp and the access drifts on the 1500 and 1550 levels.

A complete ramp system from the bottom of the deposit at 1400 m to a surface portal above the 1600 level is shown in plan view at 1450 m in Figure 7.24a and side view through 38100E in Figure 7.24b.

Summary of procedure: Example 7.3

Initial data requirements: a geologic volume model, triangulated topographic surface and centerline design maps used for any other horizontal development openings accessed by the ramp.

Procedure:

1. From the Centerline Design Tools menu create the eastern extension of the footwall drift (provide for this example as 1500DRFT and 1500DRFT2) as per Example 7.2 using **Cline** and saving the centerline as a map <1500drft>.
2. Generate a spiral ramp that exits to the south of the 4.13 offset. Select **Centerline** ⇒ **Curve** (Centerline Curve entry form)
   GLOBAL N, E, Z UNIT START: <22460.04,38169.89,1503.8> (Use **Maps** ⇒ **Map Report** of the centerline map <1500drft> to find the exact coordinates of the 4.13 offset. The map's origin is <22411.57,37997.41,1499.67>. The 4.13 offset is at <50.72,172.48>. The drift is 4.5 m wide. The starting point is 2.25 m to the south of the centerline, i.e. in the wall of the drift at the same floor elevation.)
   GLOBAL ELEVATION OF UNIT END: <1550>

GLOBAL AZIMUTH OF UNIT START: <180>
CURVE SENSE: <R>
UNIT CURVE RADIUS: <25>
UNIT GRADE (%): <15>
COMPONENT SUB-ANGLE (DEG): <5>

3. Generate the spiral's components **Save & Clear** the cline, select **Generate** and **Pick** the centerline (Centerline Generate entry form)
   PROFILE: <rectangular>
   UNIT WIDTH, HEIGHT: <4.5,3.5>
   GRADE HEIGHT: <0>
   VOLUME: <M.UG2>
   UNIT: <1500RAMP>
   COMPONENT PREFIX: <RP>
   UNIT CODE: <3> (The ramp components are generated. Use **Section** ⇒ **X-sect*** to check the components. Note that due to the steep grade, the initial ramp component will lean into the drift. This can be corrected by either using Volume Engineering design Tools on the ramp component's foreplane or by using a much lower grade on the starting components.)
4. Generate an inclined ramp starting from the same location using the following design parameters.
   C-LINE UNIT ID: <1500ramp> (Overwrite the spiral ramp replacing it with an inclined ramp.)
   C-LINE START LEVEL: <1503.8> (map elevation for cln2 plus starting elevation for a 4.13 m offset)
   CLINE GRADE: <0.5> (starting grade only)
   CLINE UNIT WIDTH, FORE THICKNESS, BACKTHICKNESS: <4.5,3.5,0>
   CLINE UNIT CODE: <4> (sys: Xline menu)
   ⇒ **Enter** (Use cursor tracking to digitize an access between the footwall drift and ramp starting approximately at the same location as per the spiral ramp of Step 3, i.e <22460,38170,1503.8>.)
   ⇒ **Grade** (parameters entry form)
      C-LINE GRADE: <15>
   ⇒ **Enter** (continue ramp to west to an elevation of 1522.5 m)
   ⇒ **Figures** (use a circle <crc> as a template for generating a switchback)
      ⇒ **Fetch*** (place the cursor on a vertex of crc and hit F1)
      ⇒ **Size**
         ENTER SIZE SCALE FACTOR: <.1>
         PICK SIZE POINT (pick center of circle)
      ⇒ **Move** (Use F2 to select crc vertex at 0° and then pick last cline point.)
   ⇒ **Snap On** ⇒ **Grade**
      C-LINE GRADE: <8> (use a flatter grade on the curve)
   ⇒ **Enter** (digitize the cline on the circle from 0° to 180° CCW or until the elevation is about 1525 m)
   ⇒ **Grade**
      C-LINE GRADE: <15> (return to a steeper inclination for incline)
   ⇒ **Enter** (continue to 1550 m)
   ⇒ **Save & Clear** ⇒ **Generate** (use the same volume component parameters as in Step 3)

Example 7.4: 3D visualization of underground development openings using Lynx
It is difficult to visualize complex three dimensional openings in plan and sectional view, especially for mine openings: drifts that are essentially horizontal, but extend in layers associated with levels; raises and shafts that are vertical; and ramps that

twist and turn through moderate inclines. Isometric visualization is necessary in order to more easily comprehend an underground design and be able to explore the interactions between mine openings in space. The 3D visualization facilities introduced in Example 4.11 will once again be used to more fully explore the various underground mine development openings that were partially described in Examples 7.1, 7.2 and 7.3.

From the main menu, select **Analysis ⇒ V3 Volumetrics & Visualization ⇒ 3D Visualization ⇒ Volume Model Perspectives**. Enter a null response to the ASCII EXPORT FILE NAME prompt and enter the volume models and model components to be displayed as would be done for a background display of a volume model. First, the ramp system of Example 7.3 will be examined, so for the DISPLAY SOURCE MODEL enter <M,P1>, which is the TUTORIAL development design model, and the ramp component list <*ramp,*,*>. Entering a wild card '*' by itself results in a listing of all the components of M.P1. This can be used to explore the naming conventions used for different elements of the design. By using the wild card in conjunction with the keyword 'ramp' all ramp components will be displayed. The DISPLAY REPRESENTATION sets the type of volume component representation that is to be used for display. The default for mine workings is to use a faceted surface <F>, which presents a solid surface based on the vertices that define the component. A sectional view is best for volumes having a more complex topography, such as veins. Enter an OBSERVER ELEVATION: <1650> which is slightly above the mine workings to provide an oblique viewpoint, and an OBSERVER LOCATION (NORTH, EAST): <22500,37850> that is neither too close or far. From the SYS:3D menu, **Vpoint** can be used to change the position of the observer as can **Azi** and **Inc** to rotate the viewpoint azimuth and inclination. **Hilite** fills each facet with a color defined by the component's code. Figure 7.25 shows the ramps system as observed from near the surface topography to the northwest of the mine workings (Azi, Inc, Vision: 96.4,29.8,45).

The same approach can be used to view all the other mine development opening

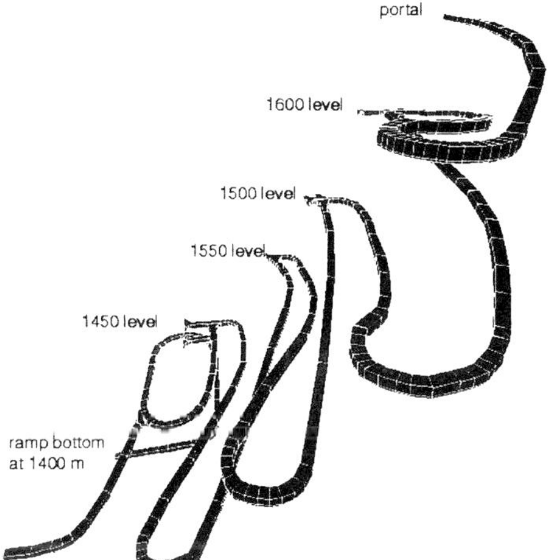

Figure 7.25. Isometric view of ramp system as observed from N, E, L: 22500, 37850, 1650.

MOD:MP1 CMP: SYS:3D VIEW VP:22300 37750 1650 42.6 23.3 VISION: 38.6

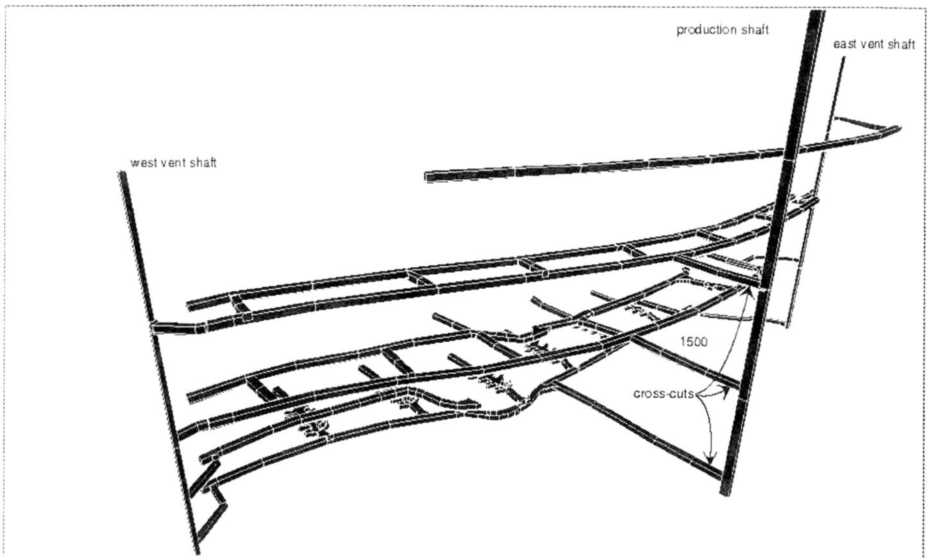

Figure 7.26. Drifts, cross-cuts, ventilation and production shaft mine workings of M.P1.

classes, such as the shafts, drifts and drawpoints. The remainder of the mine development openings for M.P1, less the ramp system, is presented in Figure 7.26.

> **Summary of procedure: Example 7.4**
> Initial data requirements: volume components
>
> Procedure:
> From the main menu **Analysis** ⇒ **V3 Volumetrics & 3D Visualization** ⇒ **3D Visualization** (Version 3 Visualization menu)
> SOURCE MODEL TYPE, MODEL ID: <M,P1>
> UNIT ID, COMPONENT ID, CODE: <*ramp,*,*>
> DISPLAY REPRESENTATION (F: SURFACE, X: SECTIONAL): <F>
> SOURCE MODEL TYPE, MODEL ID: <>
> OBSERVER ELEVATION: <1650>
> OBSERVER LOCATION: <22500,37850> (sys: 3D menu)
> ⇒ **Hilite** (hidden line removal)
> ⇒ **Vpoint** (change viewpoint)

7.11 MINE EXCAVATION OPENINGS

Excavation openings are the areas in an underground mine where the bulk of ore production occurs. While there are specific classes of development openings, such as drifts and ramps, which are essentially the same for mines regardless of the deposit type, the design and layout of excavation openings are entirely a function of the characteristics of the deposit. For instance, the design of a shaft in terms of depth and location is a function of the geometry of the deposit, but from one deposit to another, shafts are largely the same and design is an issue of diameter, depth and inclination. This is not true for excavation openings which may vary in geometry from the face in front of a continuous miner to a well defined open stope.

It is difficult to provide a succinct classification of excavation openings since there are potentially as many different configurations as there are deposits. The difficulty in doing so lies in the primary objective and constraint in underground mining: to recover as high an ore value as possible while retaining openings which are reasonably secure from failure during active mining. Essentially, this means that underground excavation design is not so much a function of the openings themselves as they are a function of the unmined areas that must be left to support those openings, which are in turn a function of the geometry of the deposit, the strength of the ore and surrounding waste, and the ground stresses. For example, in the case of flat lying deposits which are not too thick or deep, the mine design can be based on either supporting the overlying strata during mining by leaving pillars at regular intervals or the design can be based on allowing the roof to collapse immediately behind the mining face and only leaving pillars to support openings needed for access and ventilation. When a regular pattern of pillars is left behind, the mining method is referred to as Room and Pillar, but when ore hardness, thickness and ground conditions are suitable, Longwall mining can be used to further maximize the value of the extracted ore. In either case, there is no distinct excavation opening. Rather, the openings are merely defined by the remaining pillars which bound the rooms or longwall panels which were extracted by mining. Thus, in contrast to development openings whose design is predicated by factors that are common from one mine to the next, most of the factors that determine excavation design are specific to the deposit. Chief among the characteristics used to classify a deposit are: ore strength, surrounding rock strength, deposit shape, deposit dip, deposit size, ore value, ore uniformity and depth (Hartman 1987).

For the purposes of this text, the main concern is the layout and geometry of the excavation openings. Earlier in this chapter in the discussion of development openings, only the type, layout and design was examined within the limits of mine modelling software. So too, the discussion of excavation opening design must be limited to the capabilities of the commercially available packages such as Techbase and Lynx. Whereas ventilation and hoisting were not brought into the discussion of shaft diameter, rock mechanics will not be covered as the basis for the design of the pillars supporting stopes. Specialized finite element software is available for pillar design which is beyond the scope of this text. Much more comprehensive discussions of mining methods selection (Boshkov & Wright 1973; Morrision & Russell 1973; Harmin 1982) and mining methods are available elsewhere.

7.12 STOPE DESIGN USING LYNX

The creation of irregular shaped mine production openings follows the same procedure as for interactive geologic volume modelling as illustrated in Examples 4.9 and 4.10. In the case of stopes, the geologic volume model of the deposit is placed in the background to serve as a template when defining the midplane of the stope. A foreplane and backplane thickness is then assigned that will result in a volume which extends to the full height and width of the stope. The viewplane's orientation must be carefully considered. In this example, only one component will be used to represent an entire stope. As such, the choice of viewplane orientation is based on convenience; since the development openings around the stope were defined in plan, so to will be the stopes, but there are other more important considerations such as volumetric accuracy and production scheduling.

Mining reserves are found by intersecting the mining and geologic volume components. In order to avoid dilution and ore loss, the volume model of the stope should match the hangingwall and footwall of the geologic model as closely as possible. The most effective means of obtaining close agreement between the geologic and mining volume models would be to define them on the same basis, i.e the same midplane orientations and locations. This may be difficult to do since the spacing between geologic model midplanes is based on drillhole availability and geologic complexity, but the midplane location for stopes has to be based on the stope length normal to the viewplane. In Example 4.9, the geologic model's midplane was oriented vertically and approximately normal to strike and the components had foreplane and backplane thicknesses of 30 m. For the following example, sublevel stoping will be used with a stope span on strike of 30 m. This is half the length of the geologic volume components, but a close approximation can be obtained by locating a stope component's midplane between a geologic model component's midplane and either its foreplane or backplane and matching the fore and back plane of the mining component to the geologic component's midplane and either its fore or backplane. If a longer stope was used, then the stope would span more than one geologic component and multiple components could be used to define a single stope with each mining component's mid, back and fore plane matching the location of the geologic model's components. If the stopes are defined in plan view, then it will be difficult to avoid discrepancies between the two volumes that will result in dilution and lost mining reserves.

When defining the stope limits, one must assume that the geologic model is accurate, but in reality, the lack of structural information and the limitations of linearly interpolated surfaces will inevitably result in a stope model that is substantially different in its details from what will be actually mined. Keeping this fact in mind, there may be other considerations in the choice of viewplane orientation which should take precedence over volumetric accuracy. The most relevant consideration is to create components which can be used for production scheduling. For instance, in the case of Cut and Fill mining, horizontal components can be used to represent the height of the stope that is mined between backfilling cycles, or, in the case of sublevel stoping the sublevel interval. A horizontal viewplane will generally be used to represent production sized units in a stope. Also, since development openings are most conveniently defined in plan view, defining the stope boundaries in plan and on the same elevations as the levels will expedite integration of the development openings with the stopes.

Example 7.5: Defining a stope using Lynx

Continuing with the TUTORIAL project, stopes 60 m wide and 50 m high will be defined between the 1500 and 1550 levels. The entire deposit will be mined along strike by leaving alternate stopes as pillars until the active stopes are mined out and backfilled with enough cement so that they can act as rib pillars during the mining of the intervening pillars. In order to check the difference in volumetric accuracy, the stopes will be defined both from a plan and vertical viewplane and the volumes of the two models will be compared against the corresponding geologic component using the **Simple Volumetrics** facilities. Place the geologic model <G.OR> in background.

Follow the same procedure as in Example 7.1 to get to the Excavation and Development Design menu, orient the viewplane in plan at an elevation of 1525 m and place the geologic model (G.OR) in background. The stope will be defined to match the limits of the main and secondary ore zone components centered on 38060 (moz,38060:E and soz,38060:E). Select **U/G Eng ⇒ Volume Design Tools ⇒ Volume ⇒ Define Mode ⇒ Define** and then from the tools menu select **Define.** Zoom in on the 38060:E geologic components, set **Snap On**, **Pick** the boundary of one of the geologic components so that it can be used as a template for the midplane of the stope, and use **Enter** to delineate a stope midplane that matches the limits of the geologic components. Remember to enclose both the primary and secondary ore zones in the stopes limits by picking the other ore zone component and continuing to digitize along the boundaries of both the hanging wall and foot wall. The length of the stope can be made as long as would be consistent with the stoping method. In the parameters entry form enter a fore-thickness of 25 m and back-thickness of 20 m placing the bottom of the stope at 1505. Provide a **Unit Name** <1500p>, **Volume Code** <4>, **Component Name** <00001> and component **Description**. The stope volume component is created upon selecting **Save & Clear** and can be viewed in section if the component is listed in the volume background display. The resulting component is shown in plan and sectional views Figures 7.27a and 7.27b.

The fore and back planes of the stope have to be modified to match the geology model. To modify the foreplane, toggle on the FP checkbox to make it the Active String and use **Viewplane ⇒ Shift** to shift the viewplane up (positive Z) 25 m so that the limits of the geologic model correspond to the foreplane of the stope. Select **Define**, **Pick** the geologic component to make it active, and use **Move**, with **Snap On** to shift the model's foreplane into position and **Edit** the stope foreplane to match the hangingwall of soz,38060:E and the footwall of moz,38060:E. Repeat this procedure for the backplane, remembering to again shift the viewplane down 45 m and finish by using **Save & Clear**. The resulting stope is shown in section in Figure 7.27c.

To provide a comparison of the difference in the volumetric accuracy of a stope defined in plan view as. one defined in the same midplane as the corresponding geology component, once again define the same stope, but this time redesign the viewplane so that the midplane will be on a vertical section at 38060:E. Remember that the top and bottom of the stope are at 1550 and 1505 and that the stope component's fore and backplane thickness will be 30 m in the horizontal. **Define** the alternative stope's midplane <1550s,stope.01> using **Snap On** and **Enter** so that it matches

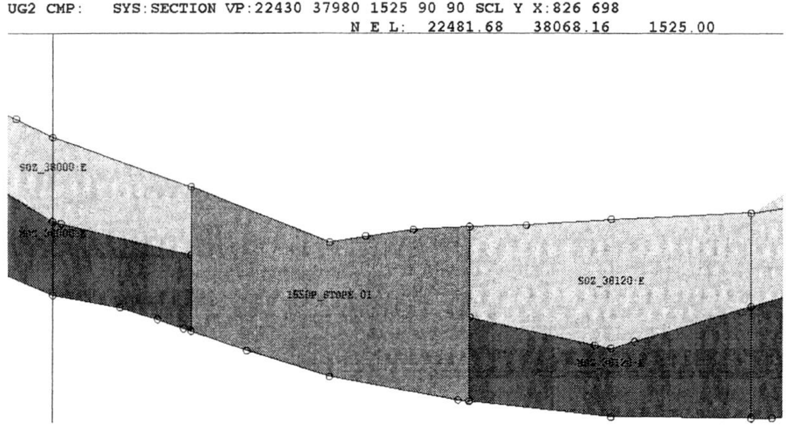

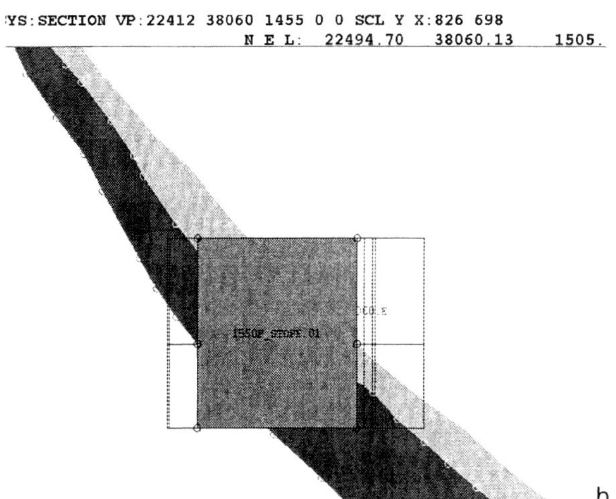

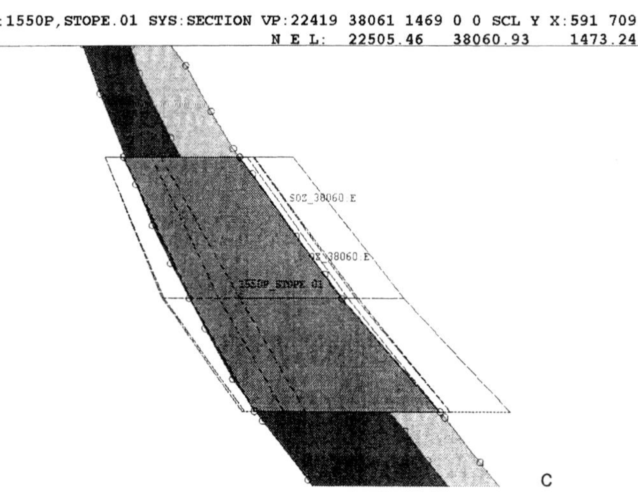

Figure 7.27. Initial stope midplane interpretation in plan (a), in section (b) and final stope component in section (c). (Figure in colour, see between pages 310-311.)

exactly the hangingwall and footwall of the 38060:E geology components and **Save** the preliminary component. Modify the fore and midplanes so that they too match the geology components' fore and midplanes using both **Edit** and **Snap On** to match existing stope vertices to geology vertices and **Insert** to include additional stope vertices so that the stope fore and midplanes are exact matches to the geology and then **Save** the final component. Note that a retrieved component will remain in background display even if the background display is initialized.

The **Simple Volumetrics** can be used to report the volume of a component. The volumes of 1550p,stope.01, 1550s.stope.01 and their intersection with the ore zone will be compared. From the main Lynx window, select **Analysis Volumetrics ⇒ Analysis Simple ⇒ Volumetrics** to bring up the Simple Volumetric entry form. Use the geologic model's components at 38060:E as the **Primary Volume Data** and 1550p.* and 1550s.* as the **Secondary Volume Data**. The results can be output to an ASCII file. The volume of the plan and sectional view stopes are 96312 and 95047 cu. m., respectively. The tonnages of intersection of the plan and sectional view stopes with the deposit are 95110 and 94571, respectively. The discrepancy between the plan view stope's volume and contained ore is 1202 cu. m, while the discrepancy for the sectionally defined stope is 476 cu. m. It would seem that the sectionally defined stope is more accurate, but the exact deposit volume between 1550 and 1505 m is unknown, making a true comparison on the basis of dilution and lost ore impossible without defining another mining component that runs from 1505 to 1550 m and is large enough to entirely contain the two geologic units. By defining another component which encompassed moz,38060:e and soz,38060:e the ore volume between 1505 and 1550 m was found to be 95608 cu. m. Apparently, both stopes lost ore, but the plan view stope lost the least.

The 1450 production level is shown in Figure 7.28 illustrating the location and de-

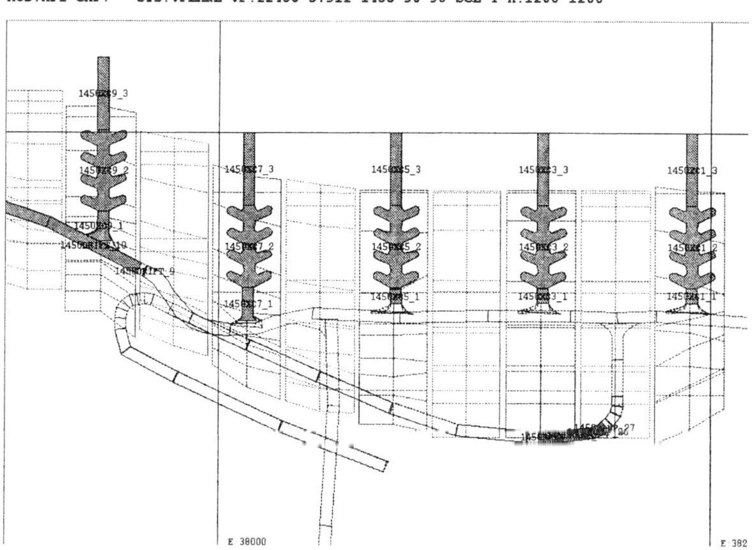

Figure 7.28. Stopes, pillars and draw point drifts in plan view on the 1450 level.

sign of the drawpoint under the stopes, the overlying stopes and their rib pillars as well as the access drift, shaft crosscut and ramp over a vertical interval of 25 m. The drawpoints were created by modifying a center-line drift as demonstrated in Example 7.2.

Summary of procedure: Example 7.5
Initial data requirements: A geologic volume model of the deposit on which the stope design can be based

Procedure:
1. Orient the viewplane to the region of the 38060:E ore zone components (N,E,L,AZI,DIP: <22450,38000,1525,90,90>, SCALE: <500,500>) and put the geologic volume model into background as per Example 7.1.
2. Define the stope **Define** (Volume Design Tools) ⇒ **Snap On** ⇒ **Enter** (digitize a stope mid-plane matching the footwall and hanging boundaries of moz,38060:E and soz,38060:E)
   COMPONENT FORE THICKNESS, BACK THICKNESS: <25,20>
   COMPONENT CODE: <4>
   STORE UNIT ID, COMPONENT ID: <1550p,stope.01>
   ⇒ **Save**
3. Interpret the foreplane and backplane of the stope by shifting the viewplane to the active string, moving the stope boundary into alignment with the geologic components and editing the boundary.
4. Find the stope's volume of intersection with the geologic model using **Analysis** ⇒ **Volumetric Analysis** ⇒ **Simple Volumetrics**

Example 7.6: Replication of mining units and autogeneration of component sequences

Most mine openings are repetitive, having the same dimensions and configuration. For instance, drawpoint spacing is fixed by the flow characteristics of the ore, while the dimensions of the entrance to the drawpoint are a function of the LHD dimensions and turning radius. In a large mine, a great deal of an engineer's time can be spent in replication of equivalent components which will be placed on a regular pattern.

Lynx provides two functions that can be used to reduce the time spent generating similar mining units that will be placed at a fixed interval: the **Assemble** and **Template** mode options in the Volume Design Tools menu. With **Assemble** a component or set of mining units that have already been created can be recalled into memory, shifted into a new position and then saved as new components. **Template** allows a series of components to be auto-generated from a single template. The template will be a previously defined mining unit which has been brought into active memory. The sequence of duplicating components can then be specified by the desired coordinate shift and the number of times that the component is to be replicated.

The drawpoints of Figure 7.28 will be used to illustrate the use of these two functions. A single drawpoint will be defined and stored, then used as a figure and rotated to provide a template drawpoints on the opposite side of the drawpoint cross-cut. **Template** could then be used to complete an initial set of drawpoints on the cross-cut and this will then be used to generate a sequence of drawpoint cross-cuts beneath the stopes.

Place the orebody (model G.OR; components moz_38120:e and soz_38120:e), overlying stope and flanking rib pillars into background (model M.P2; components 1450st_3, 1450pillar_2 and 1450pillar_4) along with the access drift (model M.P1; components 1450drift_*) to provide a basis for designing the drawpoint level. Use **Offsection Linetype** display option in conjunction with a suitable **Display thickness** and place the geologic components first on the list so that they won't obscure the engineering components. **Zoom** in on the components (Viewplane: 22490, 38020, 1450, 90, 90; Scale: 800, 800).

A map with an origin suitable for designing the drawpoint crosscut will be needed. Create a new map with the orientation (0,0,1450,90,90). The drawpoint crosscut will intersect the hangingwall drift (1450drift_*) at 1450 m elevation.

For the purpose of this exercise, it will be assumed that the ramp at the 1500 level has been extended as a decline down to the 1450 production level and that drifts have been driven in the footwall with cross-cuts running through the ore as is illustrated for the 1500 sublevel in Figure 7.19. A cross-cut will be driven under the location of the 1450st_3 stope with drawpoints configured similar to those shown in Figure 7.28. Since these openings will be directly under the stope they will experience much heavier stress. For this reason, the cross-cut will use an arched roof requiring the use of the **Centerline** facility.

From the main Lynx window, select **U/G Eng ⇒ Centerline Design** (Underground Centerline Design window). To include a cline feature of the drawpoint crosscut in the active map click on **Define⇒ Snap On ⇒ Pick** and pick a cline feature of an existing map in background that can be used as the point of intersection of the drawpoint crosscut with the 1450 drift. If none exist, the *Z* value can be entered in the parameters entry form. Lynx's sectional facilities can be used to determine exact coordinates, distances and angles. As shown in Figure 7.29, a vertical section can be run through the center-line of the stope and access drift at 38131.3E and the Tools menu and cursor tracking can be used to determine all the coordinates, distances and grades required for design. Select **Tools ⇒ Vector** while in X-sectional view of the drift and stope and with snap on digitize an initial point on the corner of the top of the footwall drift. Digitize the second point of the vector on the closest edge of the stope. The vector length and angle will be displayed from which the grade and ending xy coordinates of the first drawpoint crosscut component can be estimated. Following the information gleaned from Figure 7.29, use **Enter** to run the cross-cut at 30% to 22542E, i.e. 12 m of incline, and then change the **Grade** to 1 % ending components at 9 m intervals to provide center-line offset data for the starting points of subsequent drawpoint cross-cuts. Since the access drift has enough inclination for drainage the elevation of the starting point will vary depending on where the cross-cut exits the drift. End the cross-cut at 22571E and **Save & Clear**.

To generate the corresponding volume components select **Generate** and pick the cline bringing up the Centerline Generate entry form. Fill this out as in previous example, but select an arched profile and enter **Rise** <1.5>. This is the distance above the floor at which point the arch starts. Enter a volume and unit name <U.G2,

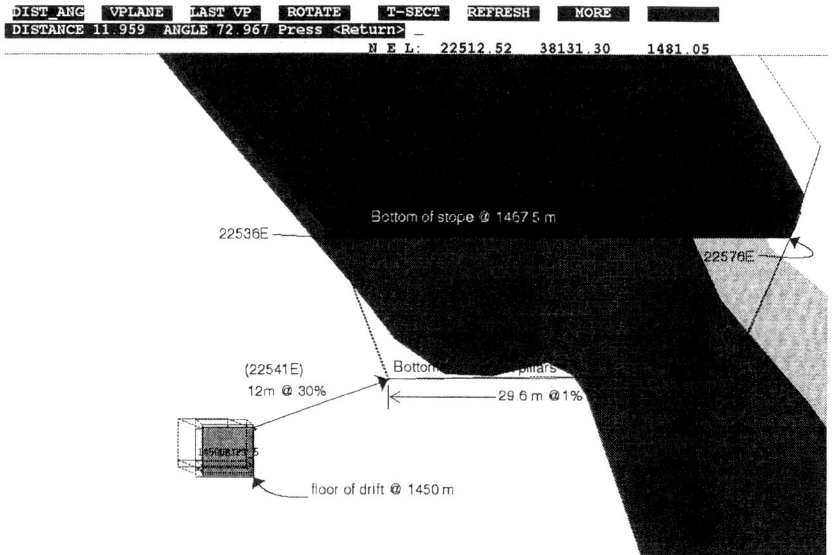

Figure 7.29. Section through drift and stope at 38131E showing measurements required for design. (Figure in colour, see between pages 310-311.)

1450,XC3>, the center-line height above the floor of the cross-cut <0> and the color code of the unit.

Generate will create arched components that actually consist of a set of breakpoints defining the change in cross sectional width by height (e.g., 0,4.5; 2,4.5; 3,2.5; 3.5,0.1). The breakpoints are chosen to provide an approximation of an arched roof and to have the same volume as a true half cylinder. In fact, Lynx lacks the ability to create true circular openings. In the case of an arched cline drift, the mid, front and back planes are defined in plan view as rectangles. At the base (designated CL in the component's naming convention), the width is 4.5 m (0,4.5). From 2 to 3 m height the width tapers from 4.5 m at the midplane to 2.5 m at the foreplane. The definition of breakpoints continues, adding a final width of 0.1 m at 3.5 m. This requires the creation of a second component (with a CU name component) stacked on top with a backplane at 3 m and foreplane at 3.5 m. Since the number of breakpoints assigned was not divisible by three, the midplane of the upper component will be arbitrarily assigned to be midway between midplane and foreplane at a height of 3.25 m. The organization of x-line components must be kept in mind when interpolation is used to modify them or **Assemble** is used to generate duplicate components.

The resulting components still have to be modified to provide a curve at the entrance to the incline and to eliminate the gap at the top of the first two sets of components that resulted from the sharp change in grade. **Exit & Save** from the c-line facilities, and use **U/G Eng ⇒ Volume Design Tools** to modify the cross-cut entrance and the boundary with 1450XC3_00003. Note that the viewplane during interpolation will be oriented to the component's grade and that due to the high angle of 1450XC3_00001 and 1450XC3_00002 the adjacent drift components will be seen from and angle. The resulting cross-cut is shown in Figure 7.30. There are difficul-

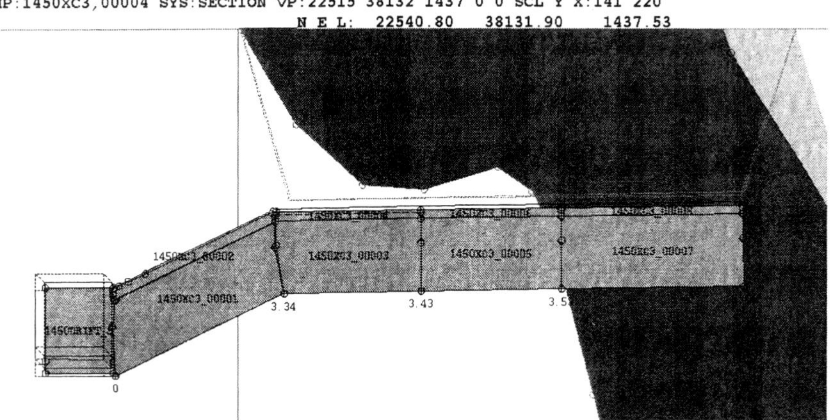

Figure 7.30. Modified x-line showing stacking of components for five breakpoints. (Figure in colour, see between pages 310-311.)

ties that will be encountered at the intersection volume components that are defined using different viewplane angles. In this example, the first cline component has a grade of 30% while subsequent components have grades of 1%. This results in a wedge shaped gap between the first and second components that can be clearly seen in x-section. **Define** with **Snap On** can be used to match the upper and lower components as seen in the viewplane that was used to define the original components, but there will still be some mismatch in the vertical due to the projection of back, fore and midplanes having different angles. This can only be avoided by modifying the defining planes of both components to lines of intersection where their edges will meet in projection. **Tools** and **Vector** can be used to find the angles of projection in order to calculate the points of intersection. Keyboard entry of the points of intersection in the parameters form ensures an exact match.

Now a drawpoint cross-cut can be included. Use **U/G Eng ⇒ Centerline Design** to **Define** the crosscut <DPXCW> starting at an elevation that will align it with 1450XC3_00003. From the center-line map, (Fig. 7.30) this would be 1453.34. Use the same breakpoints and gradient as for the cross-cut, and drive the drawpoint at an angle to about the mid-quarter of the stope. Use **Volume Design Tools** to clean up the intersection of the cross-cut and drawpoint drift as was done for 1450XC*. Don't forget to **Save & Clear** the modified component before retrieving the next component. Note that when a component is displayed in **Interp** that the center-line will be placed in the vertical position. Write down the viewplane orientation at this time, since this is the orientation that will have to be used later to create a duplicate parallel drawpoint using **Assemble**. This viewplane orientation for DPXCW_0001 is shown in Figure 7.31.

Rather that repeat this exercise for each of the drawpoint cross-cuts, **Assemble** can be used to create exact duplicates of a component which can be shifted to a new position. Positioning of a duplicate component is based on shifting the along the midplane. Thus, the new component's midplane will take the azimuth, inclination

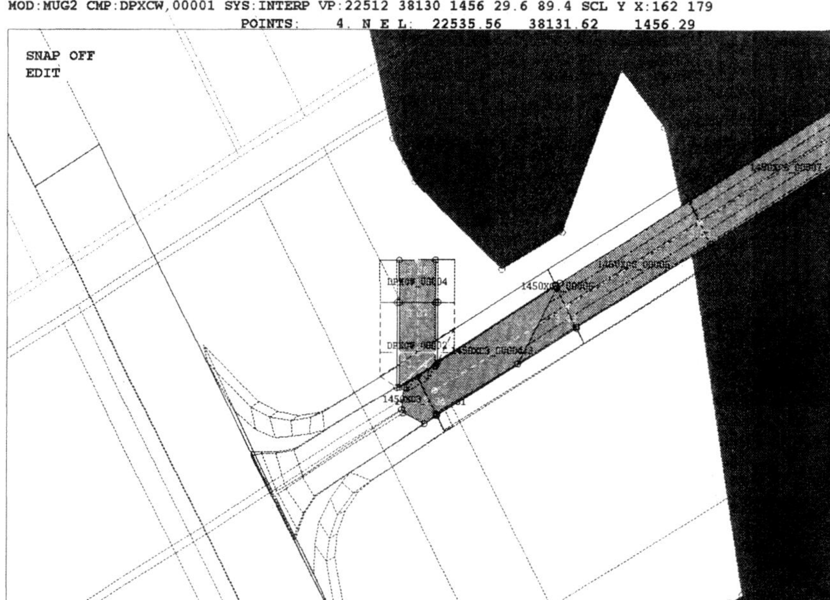

Figure 7.31. Interpretation of the midplane of the DPXCW_00002 drawpoint cross-cut showing rotation of viewplane to match component. (Figure in colour, see opposite page 311.)

and elevation of the current viewplane. To create a duplicate and parallel drawpoint to DPXCW_00001, set the azimuth and inclination to that of the drawpoint's center-line (Az = 29.6, Inc = 89.4). These values can be determined by retrieving the component into memory in Interpolation mode at which point the viewplane will be reset to the component's midplane. This is done in the **Volume Design Tools** by selecting **Volume ⇒ Define Mode ⇒ Interpolate ⇒ Fetch** and providing the name of the components to be interpolated. entering one of the drawpoint crosscut components into the Volume Data Interpolate: Fetch Both parameters form will reorient the viewplane. In this example, the starting elevation of the next drawpoint as a center-line offset of 3.43 m from 1450 m, and the midplane was defined as being 2 m above the center-line. The resulting viewplane elevation is 1455.43 m. Select **Assemble ⇒ Fetch** to bring up the Assemble: Fetch entry form (see Fig. 7.32) and enter the source model <M,UG2> and the volume component to be copied <DPXCW,00001,5>. Multiple source components can be used, but they will all have the same midplane elevation and will act as one unit. One of the component's being duplicated can be used as a **Reference Component**. The movement and rotation of the assembled components is always relative to the **Reference Component**. Click on **OK** to display the selected components. These will be highlighted in cyan and displayed relative to the reference component. The midplane of the duplicate reference component will be positioned in the plane of the current viewplane. Select **Move** with the cursor picking the component midplane vertex that is to be shifted, place the cursor on the mid-point's new position and select the new vertex position. When the new drawpoint is in position, select **Generate** and enter new **Unit** and **Component Prefix** conventions

CONEE,00002 SYS:IDLE VP:22520 38099 1457 90 90 SCL Y X:

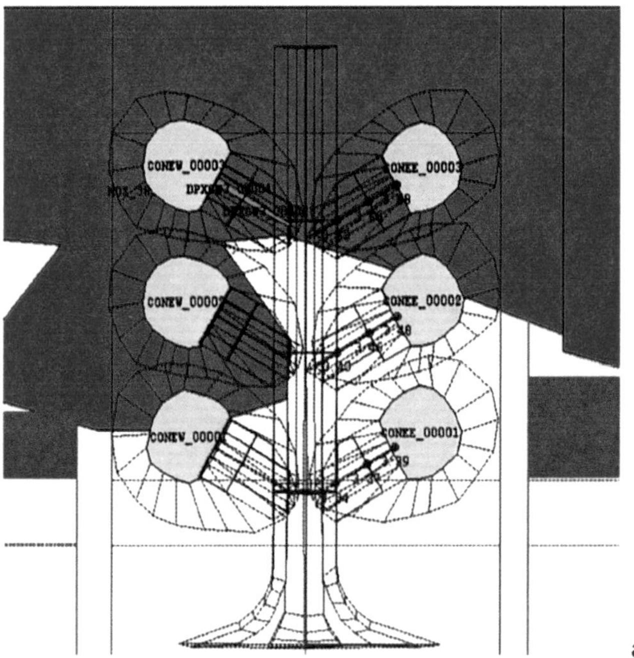

MP:CONEE,00002 SYS:SECTION VP:22517 38141 1441 0 0 SCL Y X:184 218
N E L: 22551.97 38140.50 1458.98

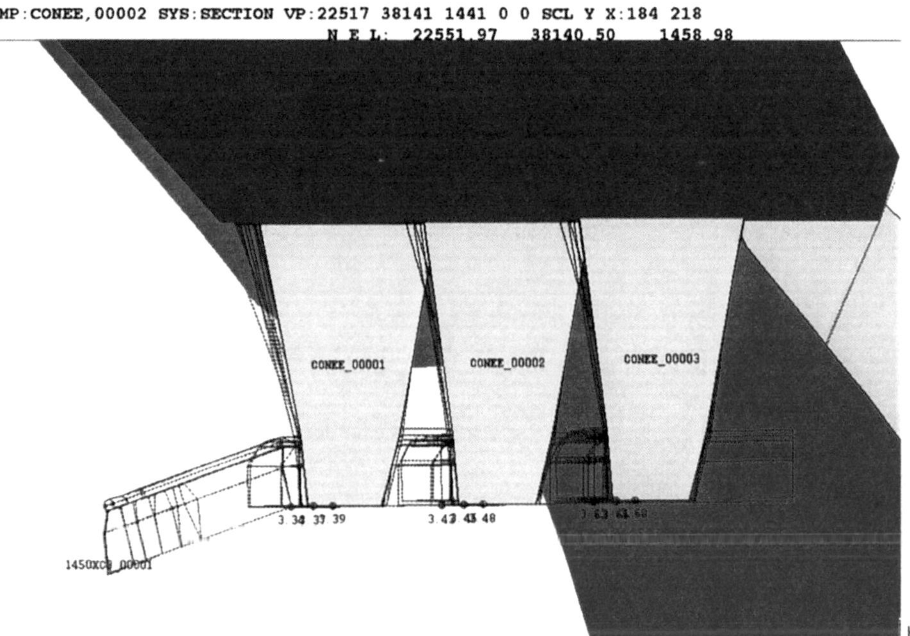

Figure 7.34. Drawpoint cones shown in: a) Plan view and b) Sectional view.

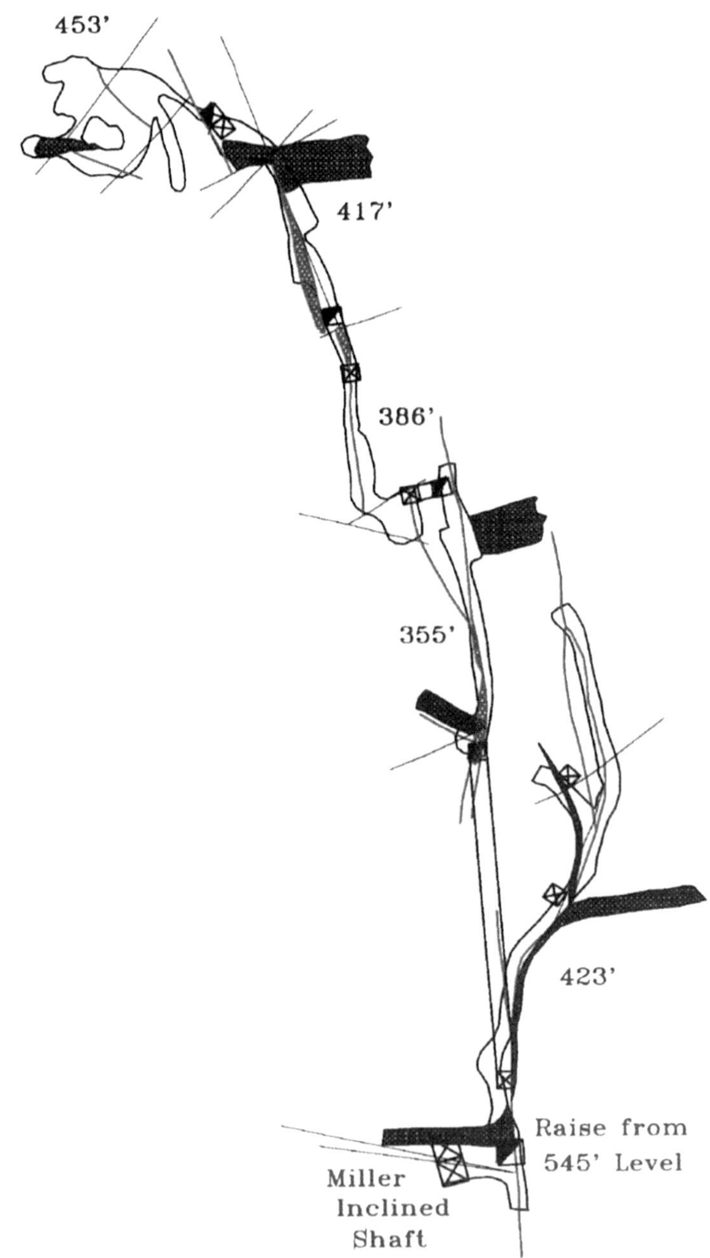

Figure 7.35. The 423 level of the May Day Mine showing veins (red), dikes (blue), faults and various mine openings (courtesy Mark Shutty).

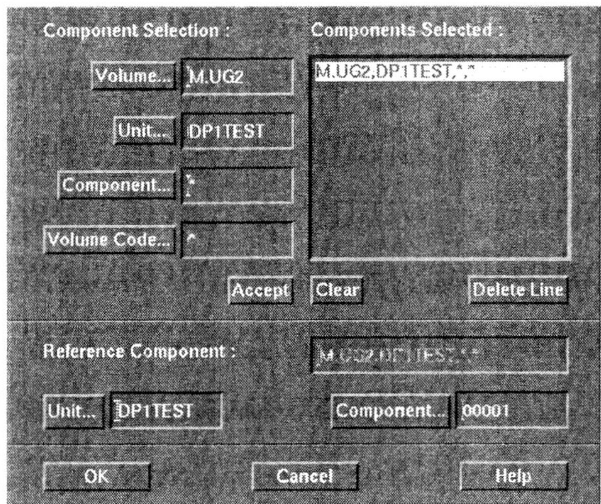

Figure 7.32. Assemble: Fetch entry form set-up for duplicating a drawpoint cross-cut.

for the drawpoint in the Assemble Generate entry form. Toggle on **Append** so that the new set of components will be appended on to the list of existing 1450 drawpoint crosscuts and select **OK** to confirm the generation of these new volume components. Repeat the process for the other three drawpoint components remembering to shift the viewplane up 1.25 m to correspond to the midplane elevations of components 00002 and 00004. The same process can be repeated for the drawpoints on the eastern side of the cross-cut. To ensure that the same angle and distances are being used, determine the correct measurements from **Tools ⇒ Vector**. The completed set of cross-cuts is shown in Figure 7.33. Note that this entire procedure would have been much easier if the base component and all subsequent duplicates were on the same midplane.

Finally, the drawcones themselves must be defined. In practice, a raise would be driven from the drawpoint cross-cut into an undercut slot in the base of the stope and then belled with successive angled shots. Since the stope configuration and the drawpoint intervals are centered under the stope at fixed intervals, the cones will also be of similar dimension. An initial cone will be generated using **Volume ⇒ Define Mode ⇒ Define** and used as the basis for the other two cones on the same side of the cross-cut using **Assemble**. In order to create duplicate cones on the other side of the cross-cut, the initial cone will be saved as a map, and then used as a figure that can be rotated and shifted into position to act as a template.

The placement of the fore, mid and back planes of the cone must also be considered. Since the drawpoint cross-cuts have an arched roof, the backplane cannot be located at the top of the arched drift. Instead, the backplane must start at the floor, or center-line, in front of the drawpoint cross-cut. The midplane can then be located at the roof of the drawpoint cross-cut and the foreplane at the base of the stope. If greater complexity was required, such as intersecting cones, then more than one component could be used to model the cone, or a flat topped center-line component could be used at the end of the drawpoint cross-cut. The cone's backplane could then start on the foreplane of the center-line component and the midplane could be located

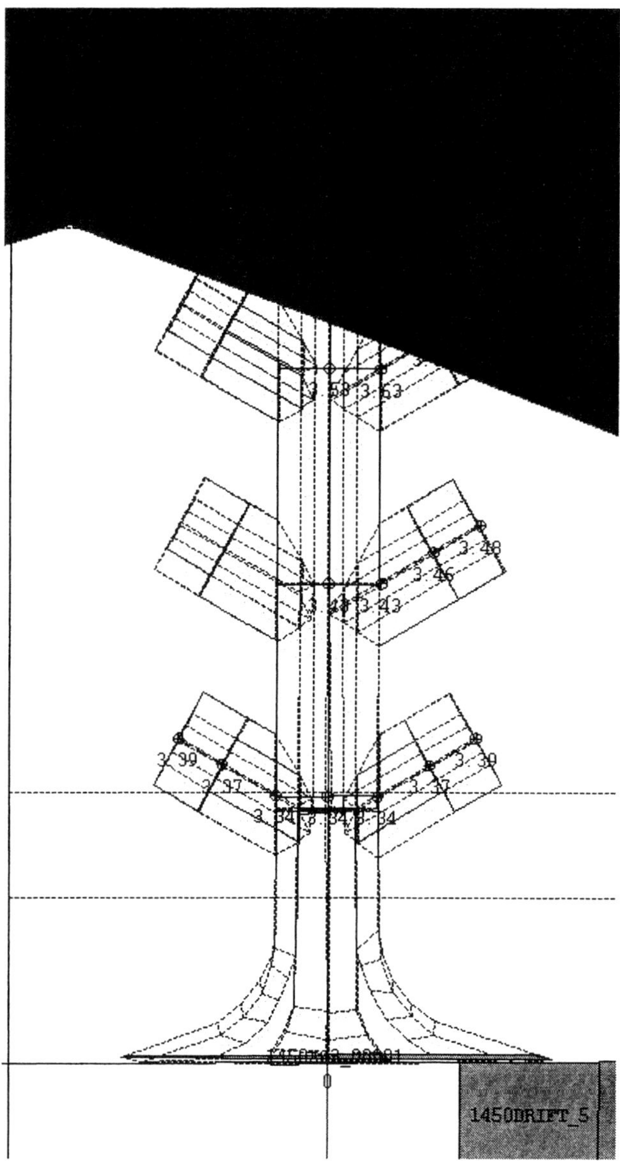

Figure 7.33. Final set of x-line drawpoint cross-cuts. (Figure in colour, see opposite page 311.)

to control the degree of cone intersection. To create simple non-intersecting drawpoints, the easiest procedure would be to define the limits of the cone's midplane at the base of the stope using a foreplane thickness of zero and interpreting the backplane so that it matches the drawpoint.

To position the midplane of the cone at the top of the drawpoint cross-cut (1450XCW) orient the viewplane at 1456.89 m (1450 + 3.39 + 3.5, see Fig. 7.34a). Select **Define** and use **Enter** to define the midplane of the cone. **Save & Clear** the new component (CONEW_00001) with backplane thickness of 3.5 m and a foreplane thickness of 10.6 m to create a vertical cylinder reaching from the floor of the drawpoint cross-cut to the base of the stope. After defining the initial component, **Retrv** it and use **Interpolate** mode to adjust the backplane inwards and to bell the foreplane out to form the cone. **Assemble** can now be used to complete the set of drawpoints to the west of the cross-cut, remembering to adjust the viewplane elevation to account for the 1% grade of the openings as indicated by the displayed center-line maps so that the midplane of the new cone will be at the elevation of the top of its drawpoint cross-cut. Some adjustment of the cones' foreplanes will be required to get the final fit.

When creating the drawcones on the opposite side of the cross-cut, a map of the initial cone can be used as a template. In **Background**, initialize the **Volumes** display and in **Viewplane** turn off the grid so that only CONEW_00001 is in view. Shift the viewplane elevation so that the component is not displayed in solid fill and only shows as vertices and construction lines. From the main Lynx window, select **Output ⇒ Maps** to bring up the Output: Map entry form and enter a map name <CONEW> , description and origin. In the Volume Engineering Design Tools window, select **Figures ⇒ Fetch Figures** and enter the map containing the cone template <CONEW>. Use **Move** and **Rotate** to bring it into position and with **Snap On** use the figure as a template to **Define** the next draw cone. **Save & Clear** the new component <CONEE_00001>. Note that unless the cone is symmetric around the drawpoint cross-cut's center-line axis, that the resulting component will not be identical to the original. A completed drawpoint drift with cones is shown in Figures 7.34b and 7.34c.

A time saving alternative to using **Assemble** or using map figures as templates is

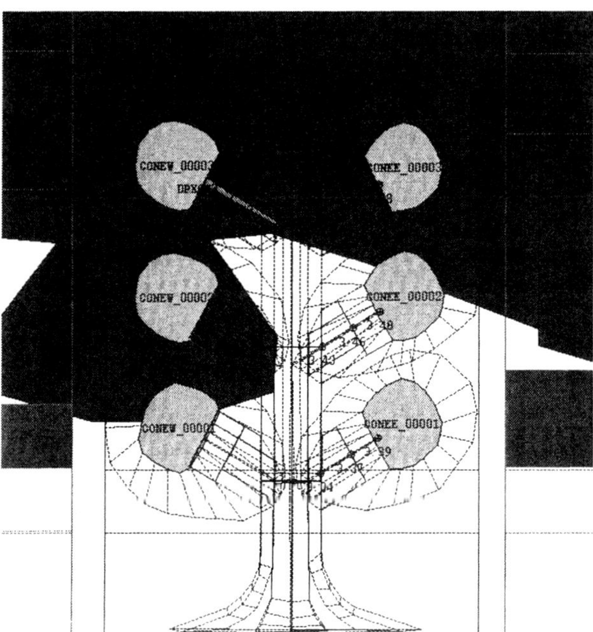

Figure 7.34a. Drawpoint cones shown in: Plan view. (Figure in colour, see opposite page 342.)

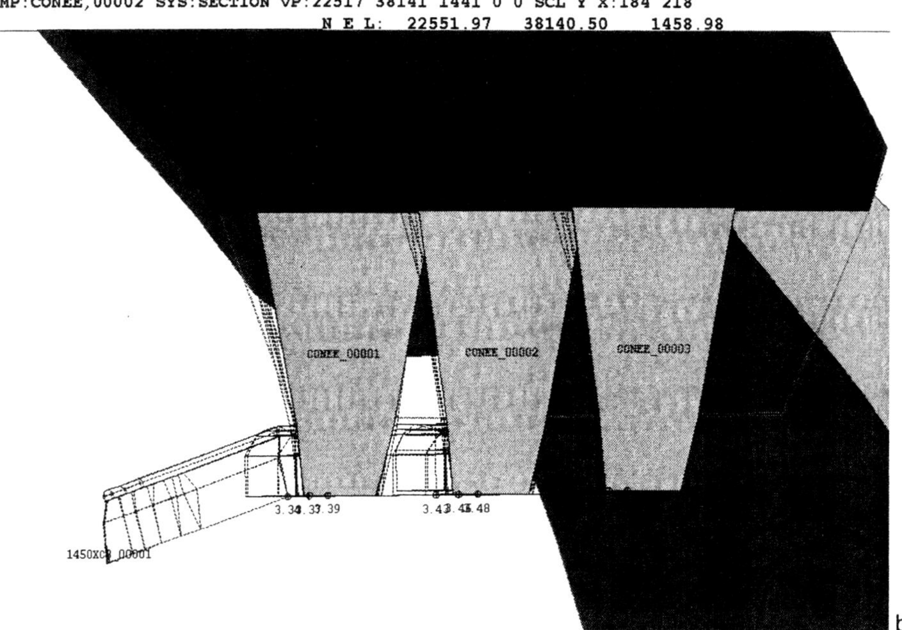

Figure 7.34b. Drawpoint cones shown in: Sectional view. (Figure in colour, see opposite page 342.)

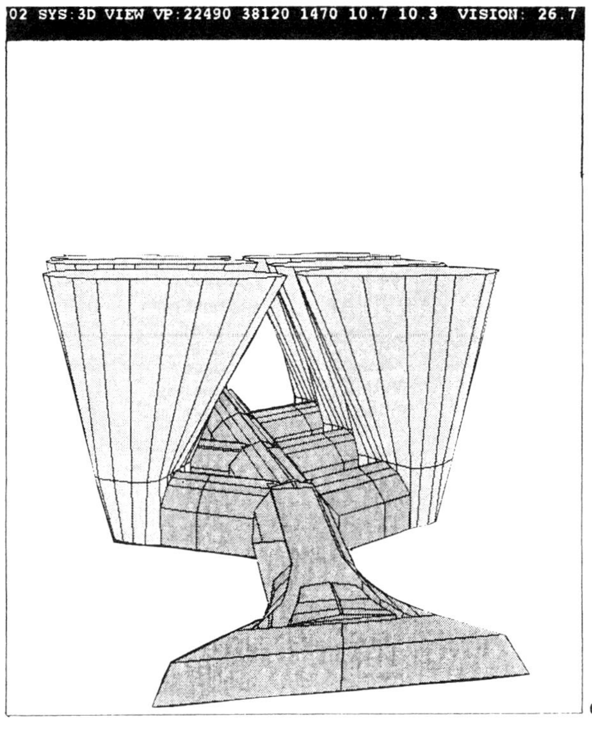

Figure 7.34c. Drawpoint cones shown in: 3D view.

the **Template** option (also found in the Volume Design Tools). the **Template** option can be used to fetch an existing volume component, **Move** and **Rotate** it relative to the current viewplane, and **Generate** a series of replicated components at regular *xyz* intervals. This can be extremely useful when there is a need to produce numerous identical openings on a regular grid, such as drawpoints or sublevels.

Summary of procedure: Example 7.6
Initial data requirements: volume models and cline maps for footwall access drifts and stopes.

Procedure:
1. Orient the viewplane to the region of the 38060:E ore zone components (N,E,L,AZI,DIP: <22450,38000,1525,90,90>, SCALE: <500,500>)
2. Set up the background display **Background** ⇒ **Volumes**
   MODEL ID: <G.OR>
   UNIT, COMPONENT ID, CODE: <*,38120:E,*>
   VOLUME DISPLAY THICKNESS: <0,0>
   ⇒ **Accept**
   MODEL ID: <M,P2>
   UNIT, COMPONENT ID, CODE: <1450st,3,*>
   ⇒ **Accept**
   MODEL ID: <M.P2>
   UNIT, COMPONENT ID, CODE: <1450pillar,2,*>
   ⇒ **Accept**
   MODEL ID: <M.P2>
   UNIT, COMPONENT ID, CODE: <1450pillar,4,*>
   ⇒ **Accept**
   MODEL ID: <M.P1>
   UNIT, COMPONENT ID, CODE: <1450drift,*,*>
   ⇒ **OK**
   Background ⇒ **Maps** (Select the centerline map for the 1450 footwall drift. This will be used as a starting point with Snap On when digitizing the drawpoint crosscut.)
3. Create an arched drawpoint crosscut **U/G Eng** ⇒ **Centerline Design** (Underground Centerline Design window)
   3.1 Define the main entry for the drawpoint crosscut **Define** ⇒ **Snap On** ⇒ **Pick** (select a vertex from the 1450 drift as a starting point)
      ⇒ **Enter** (digitize first component from the drift wall to 22542E) and change the grade in the parameters form to 30%)
      ⇒ **Enter** (change the grade to 1% prior to the entrance of the first set of drawpoint cones' bays)
      ⇒ **Snap Off** ⇒ **Enter** (continue entering breakpoints at three 9 m intervals for the drawpoint bays)
      ⇒ **Save & Clear** ⇒ **Generate** (pick the cline feature bringing up the Centerline Generate entry form and generate 1450XC3)
      ⇒ **Save & Exit** (from Centerline Design Tools)
   3.2 Modify the main drawpoint entry **U/G Eng** ⇒ **Volume Design Tools** ⇒ **Select** <1450XC3,00001> (picking the x-line component with a mouse may be difficult since 1450XC3,00002 is stacked over 1450XC3,00001)
      ⇒ **Define** ⇒ **Snap On** ⇒ **Edit** (Due to the high initial angle, the foreplane and midplane will be tipped into the drift)
      ⇒ **Save & Clear** (overwrite previous interpretation)
   (Repeat Step 3.2 for <1450xc3,00002>. Create a curved entry using **Figures** ⇒ **Fetch Figures** <crc> as a template.)

3.3 Define a arched drawpoint crosscut (bay) using Volume Design Tools
⇒ **Enter** (digitize a two point center-line originating at the drift wall opposite to the 3.34 offset to generate four Cline components)
⇒ **Save & Clear** ⇒ **Generate** <dpxcw> with the following specifications:
Cline Unit Id: <DPXCW>
Cline Start Level (Global Z Value): <1453.34>
Cline Grade: <1>
Cline Height Above Floor: <0>
Cline Unit Code: <5>

3.4 Modify the drawpoint bay <DPXCW> as per Step 3.2 so that the foreplane of DPXCW,0001 and all three planes of DPXCW,00002 are flush with the sides of 1450XC3_00001, 1450XC3_00002, 1450XC3_00003 and 1450XC3_00004. Modify the components as per Step 3.2 **Select** (pick the lower bay) ⇒ **Edit** (shift the vertices back to match the drawpoint drift) ⇒ **Save & Clear** (overwrite for each component)
(Repeat Step 3.4 for all the two components planes.)

4. Use the **Assemble** facility to duplicate the drawpoint bay.

4.1 Reset the viewplane orientation to match that of the original component as noted down in Step 3.4 **Viewplane** ⇒ **Set-up**
N,E,L,Azi,Inc: <22152,38130,1456,29.6,89.4> (The orientation will vary.)
Scale: <162,162> (The scale will vary.)

4.2 Duplicate a component at a new position **Assemble** ⇒ **Fetch** (Assemble: Fetch entry form)
Assemble Source Model Type, Model Id: <M,UG2>
Select Unit Id, Component Id, Code: <DPXCW,*,5>
⇒ **Move** (pick a vertex on the component's midplane and pick the new vertex location)
⇒ **Generate** (enter new component name <DPXCW,00001> and design dimensions matching the drawpoint crosscut)
(Repeat Step 4.2 for each component comprising the bay. Repeat for each bay.)

5. Define the drawcones **U/G Eng** ⇒ **Volume design Tools**

5.1 Orient the viewplane so that the backplane of the cone will be in alignment with the floor of the bay ⇒ **Viewplane** ⇒ **Set-up** (change the elevation to 1456.89 or to whatever the final offset indicates for map <dpxcw>)

5.2 Define the cone's midplane **Define**
⇒ **Enter** (digitize the midplane) ⇒ **Save & Clear** ⇒ **Generate** (pick midplane)
Component Fore Thickness, Back Thickness: <3.5,10.6>
Component Color: <3>
Unit Id, Component Id: <CONEW,00001>
⇒ **OK**

5.3 Modify the foreplane of the cone **Edit** (Bell out the cone) ⇒ **Save & Clear** (overwrite component)

5.4 Use the **Assemble** facility to replicate the cones at each drawpoint on the same side of the drawpoint crosscut as per Step 3.2.

5.5 Create a map of the cone in the main Lynx window to use as a figure **Background** ⇒ **Volumes**
Model Id: <M,UG2>
Unit, Component Id, Code: <CONEW,00001,3>
Volume Display Thickness: <0,0>
Output ⇒ **Maps**
Map Id: <CONEW>

5.6 In Volume Design Tools, use the map as a figure for defining the cones on the eastern side of the drawpoint drift **Figures** ⇒ **Fetch Figures** <CONEW>
⇒ **Move** (into position at the end of the bay) ⇒ **Rotate** ⇒ **Snap On** ⇒ **Enter** (digitize the cone's midplane to match the template cone)

5.7 Use the **Assemble** facility to replicate the cones at each drawpoint on the same side of the drawpoint crosscut as per Step 3.2.

Example 7.7: Underground mining reserves using Lynx

The procedure used to determine underground mining reserves is exactly the same as for surface mining reserves as was demonstrated in Example 6.13. The only difference is that the engineering source model will now be the underground model and that selection of components can be used to report reserves by the class of mine opening. It was noted earlier in the discussion of Lynx component naming conventions, that the unit ID and color code could be used to classify openings in terms of development versus production, ramps, time periods, etc. These naming conventions can be put to good use during the reporting of reserves.

From the Lynx main window, select **Analysis ⇒ Volumetric Analysis ⇒ Simple Volumetrics** and enter the same parameters as in Example 6.13, i.e:

Primary Numeric Variable: <cueq>
Primary Cutoff Value(s): <.7>

Enter as many cutoffs as are to be included in the report.

Other Numeric Attributes To Be Reported: <cu,zn,ag>
Ore Density Value: <2.7>
Undefined Material Density Value: <2.4>

Undefined material refers to areas of intersection between the mining components and geologic models that do not have a color code, i.e waste.

Extraction Ratio: <1>

This is the dilution ratio. For instance, in caving methods a considerable amount of material from outside of the block's initial limits will cave into the block and be drawn. When the amount of dilution by this waste exceeds the allowable dilution ratio, production from the drawpoint will cease. Dilution in a caving operation can be as high as 20% (1.2).

Volume Integration Increments: <5>

The volume of intersection between each block in Mymodel and the mining components will be based on numeric integration on 5 m intervals.

Volume/Tons Factor, Numeric Variable Factor: <1000,1>

The report will be in thousands of tons. The format for the report will only display eight figures.

7.13 USING TECHBASE FOR UNDERGROUND EXCAVATION DESIGN

As of Version 2.52, Techbase has limited facilities for underground mining applications. Techbase is oriented towards two-dimensional modelling of surfaces as grids that are defined in the horizontal plane. The difficulties involved in using a planview, gridded representation of complex, high angle and non-bedded geologic structures also limit the representation of inherently three-dimensional underground openings. Thus, Techbase is awkward to use for mine openings such as ramps, shafts, raises and high angle stopes. Still, these limitations do not apply to the underground mining methods associated with bedded deposits: room and pillar, open stoping and longwall mining which account for the bulk of mineral production, especially for coal, limestone, other industrial minerals and a large percentage of mechanized metal mining. For these types of operations, Techbase's supurb database and analysis facilities can be used to great effect for the layout of mine openings, scheduling and the calculation of mining reserves.

There are no specialized engineering facilities for underground design as there are for strip and open pit mining (**sEam** and **Open-pit** facilities, respectively), no automated generation of ramps or shafts as in Lynx. But, for underground mine workings in a bedded deposit, these facilities, while useful, are far from necessary. In fact, the common and preferred software for mine design in bedded deposits is to use a CAD package. Advanced CAD programs can be used for all aspects of design from the digitizing of 2D data to the design and layout of power transmission with a host of specialized packages available for engineering design. The one essential element of underground mining missing in today's CAD software is the geologic database. Therefore, a good CAD program in combination with Techbase provides a system which is perfectly acceptable for bedded deposits. Engineering design is carried out in CAD, the mine openings are exported to Techbase and incorporated into the database and then can be used in combination with geologic models for volumetric calculations. The reverse procedure can also be used, with maps of geologic data exported to the CAD package from Techbase.

Just as cell and block tables are used to contain geologic models, polygon tables can be used to hold and manipulate engineering designs. Up to this point in this text, the application of polygon tables has been avoided as being an unnecessary complication. A good preliminary example of the use of polygons to represent mine design was given in Example 6.14 in which mid-bench polygons were intersected with a block model in order to determine mining reserves. In that example, only one polygon was needed for each bench and these were stored in 13 ASCII files. In the case of underground mine design, a vastly larger number of polygons are likely to be used to represent all the mine openings and geologic features. In this application, the number of polygons and the variety of characteristics associated with each polygon are too great to manage without a specialized data structure included in the geologic database. Polygon tables serve as the means of integrating polygon-based data, such as property claims and mine openings, into the geologic database.

7.14 POLYGON TABLES

Property boundaries, bench limits, stope walls, drifts, cutoff grade-based diglines, dikes and edges of roads are common examples of the types of data that can be represented as polygons. Herein, a polygon can be either open, as in the case of the hangingwall of a drift, or closed as in the case of a bench limit. Polygon tables can be used to hold these types of data in Techbase. Once a polygon is in a polygon table, it can be displayed with characteristic line types, fill, labeling and color and intersected with either cell or block tables for volumetric calculation. While either two or three dimensional polygons can be stored in polygon tables, 3D polygons are not fully implemented in those facilities that use polygons.

A polygon data structure actually consists of three tables: the polygon table and two supporting tables that contain the vertices and edges of the polygon. The polygon table has one record per polygon and requires only one user defined field, an identifier for the polygon. For instance, a polygon table could be used to hold property boundaries with a text field whose records list the lease boundaries. The table's default fields are: _rec and _nul (as per other table types), _npt (the number of vertices defining the polygon. If the polygon is closed, then the first vertex is included in the first and last edge so that _npt will be one greater than the number of vertices), _edg (the initial record number for the polygon in the edge table), _xc and _yc (the coordinates of the center of the polygon, a 3D polygon table will include a _zc field), _are (the area of the polygon), _per (the length of the polygon's perimeter), _xmn and _ymn (the minimum coordinates of the polygon, a 3D polygon will include a _zmn field), _xmx and _ymx (the maximum coordinates of the polygon, a 3D polygon will include a _zmx field).

The vertex table will be named using the polygon table's name as a prefix and _v as the suffix. There will be one record per vertex with the default fields _v_rec and _v_nul, and _v_xc and _v_yc which are the coordinates of the vertex (a 3D polygon will include _v_zc). The records will be ordered as per the polygon table. For example, if the _npt field in record one of the polygon table is equal to 4, then records 1 to 4 in the vertex table will be the coordinates of the vertices of that polygon. Note that a polygon consisting of four vertices might have _npt = 5, indicating a closed polygon. This can be confirmed by checking the _edg field in the following record of the polygon table. If _edg = 5 in the second record, then the first edge of the second polygon is defined in record 5 of the edge table. Thus, there are four edges (requiring four vertices) in the previous polygon (records 1-4).

The edge table will be named using the polygon table's name as a prefix and _e as the suffix. There will be one record per edge with the default fields: _e_rec and _e_nul as per other table type, _nxt (the next vertex involved in the polygon), _xc1 and _yc1 (the coordinates of the first vertex defining the edge), and _xc1 and _yc2 (the coordinates of the second vertex defining the edge). A 3D polygon will include _zc1 and _zc2. Note that if the polygon isn't closed that the number of edges will be one less than the number of vertices. As discussed above, the _edg field is used to relate the polygon table's record to the edge table records.

Example 7.8: Using polygons to represent underground openings in Techbase

This example is taken from the May Day mine in the San Juan National Forest near May Day, Co. The abandoned mine consists of four major levels and a number of side levels, adits and vertical openings (shafts, raises and winzes). Old maps were used to digitize the mine workings, vein locations and other geologic structures (dikes, contacts and faults). Figure 7.35 shows the 423 level which is based on the

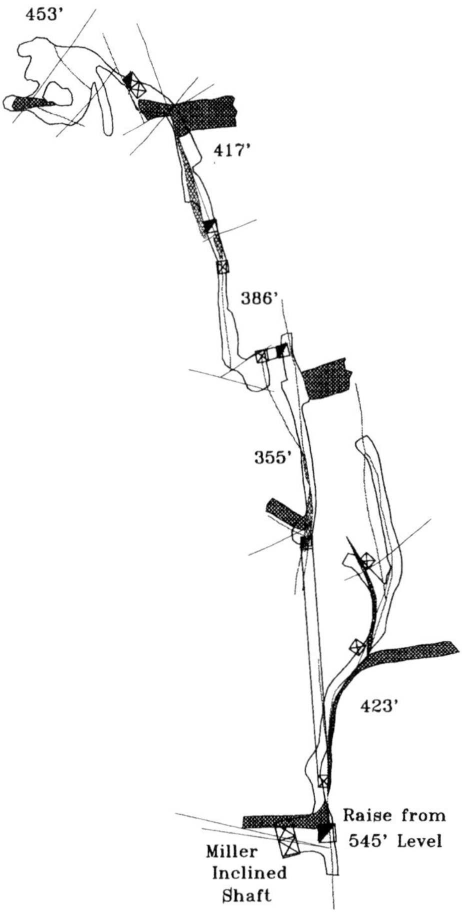

Figure 7.35. The 423 level of the May Day Mine showing veins **(red),** dikes **(blue),** faults and various mine openings (courtesy Mark Shutty). (Figure in colour, see opposite page 343.)

polygon table <level423> which consists of polygons for the mine workings, faults, veins and dikes. The following examples will demonstrate the definition of a polygon table, loading and creating polygon data files, display of polygon table data and using polygon tables for basic volumetric calculations.

Defining and loading a polygon table

The file level2.pol contains polygon data for level2 including coordinates and identities for left and right ribs, faults, the drift and dikes. To define a polygon table to hold this information, select **Techbase** ⇒ **Define** ⇒ **Tables** ⇒ **Create** entering the table name and type <polygon>, followed by the name of the associated edge and vertex tables. For the vertex table, the minimum and maximum vertex coordinates must be supplied along with a coordinate tolerance distance. If two or more vertices are within this tolerance distance, the record will not be read. Add only one user defined field to the table <all2> to hold the polygon identities.

Exit from **Define** and select **Load** ⇒ **Setup** ⇒ **Data** file entering the name of the polygon data file <level2.pol>. Select **Fields** and enter the list of fields as <null null all2 (poly)>. Since the polygon vertices' coordinates are in the vertex table and the identity field is in the polygon table, entering the _xc and _yc field along with <all2> is not acceptable to Techbase. Instead this obscure format for listing fields must be used. Entering 'null null' causes the first and second field in the record to be skipped, the third field <all2> is read and the option '(poly)' results in the first two fields being read into the vertex table. Select **Load** to load in the data.

Posting polygon tables

Once the polygon table has been loaded, the information can be displayed in any of the **Graphics** facilities that accept polygon data, specifically **Poster**, **Section**, **perspecT** and **polyEdit**. The handiest facility for displaying plan view data is Poster. Four default fields are available in the polygon table that can be used to determine the extent of the data, and therefore the scaling. Use **Statistics** ⇒ **Summary** ⇒ **Fields** entering the _xmn, _ymn, _xmx, _ymx fields to determine the range of data in the polygon table. From the main menu select **Graphics** ⇒ **Poster** and set up the scaling. To display the full contents of the polygon table select **Cell/poly** ⇒ **Fields** and enter the polygon ID field <all2>, modify the **Style** as needed and **Draw**. Database filters can be used to control which polygons are posted by placing a filter on the polygon ID. The fill type and color of the polygon can be controlled under **Style** either by posting the polygons in layers using a filter or by creating fields in the polygon table that control line, fill and value color and type. The area and perimeter of the polygons can also be posted using the _are and _per fields.

Volumetric calculations

There are two approaches to volumetric calculations using polygon tables: reporting of the default _are field and using Locate. The _are field value can be used to get an idea of the total contained volume. This is especially useful in the case of 3D polygons, but in the case of a 2D polygons such as those in the level2 table, a drift height must be assumed or taken from a traverse line surveyed along the length of the drift.

A more acceptable approach is to use the **Locate** facility to determine the volume of intersection of the polygon with either a cell or block table. Although this is not the case for the May Day project, let's say that a block model had been estimated from channel sample and other exploratory data. **Locate** could then be used as per Example 6.16 in which a filter is used to restrict Locate to the level of blocks corresponding to the level of the mine workings. Note that the polygons' actual elevation range would have to be contained in one layer of blocks for this approach to work. A more computationally efficient approach would be to use a layer table rather than a block table. Layer tables are essentially a series of joined cell tables, but there can be a different number of cells in each layer. Thus, the extent of each layer could be defined to match the extent of the mine workings and there need only be as many layers as there are levels with a field being used to define the cell height in each layer. For those cells in the vicinity of the mine workings, survey data could be used to estimate cell height.

As mentioned in the beginning of this section, 3D polygons are not fully implemented in Techbase so that using polygons in **Locate** is still limited to 2D. This makes it awkward to do volumetrics with stopes, since series of polygons corresponding to the level of each level in a block or layer table must be used along with a filter to limit location to the appropriate level. **Locate** then must be repeated for each level of the stope changing the value of the filter and the polygon ID at each iteration.

Summary of procedure: Example 7.8
Initial data requirements: digitized polygon data <level2.pol>

Procedure:
1. Define a polygon table **Techbase ⇒ Define ⇒ Tables ⇒ Create**
   TABLE NAME: <level2> TYPE: <polygon>
   EDGE TABLE NAME: <level2_e> VERTEX TABLE NAME: <level2_v>
   MINIMUM X-COORD: <12700> MAXIMUM: <15400> TOLERANCE: <1>
   MINIMUM Y-COORD: <10500> MAXIMUM: <11700> TOLERANCE: <1>
   ⇒ eXit ⇒ Fields ⇒ autoTable
   TABLE NAME: <level2>
   ⇒ Create
   FIELD NAME: <all2> TYPE: <text> CLASS: <actual>
2. Load ASCII data into a polygon table **Techbase ⇒ Load ⇒ Setup ⇒ Data file**
   FILE NAME: <level2.pol>
   ⇒ **Fields**: <null null all2 (poly)>
   ⇒ Load
3. Post the polygon **Graphics ⇒ Poster**
   3.1 Setup the scaling factors ⇒ **Scaling ⇒ Scale**
      X-SCALE: <300> Y-SCALE: <300>
      ⇒ Range
      LEFT X: <12700> RIGHT X: <15400>
      BOTTOM Y: <10500> TOP Y: <11900>
      ⇒ Sheet
      SHEETSIZE: <A> ORIENTATION: <landscape>
      ⇒ Border ⇒ eXit
   3.2 Post all the polygons in the table **Cell/poly ⇒ Fields**
      VALUE 1: <all2>
      ⇒ Draw ⇒ Graphics ⇒ Review
4. Using **Locate** to estimate volumetrics: create either a block or layer table and proceed as per Example 6.16.

CHAPTER 8

Production scheduling

Scheduling of mine production occurs on both long and short-term scales. Long-term schedules are produced for cashflow analysis and as a guide to more detailed mine design and development. Short-term scheduling starts where the long-term, typically annual, plans left off and focuses on the implementation of the mine design over time by planning waste stripping and removal and grade control.

A cashflow analysis requires a schedule of expenses and incomes. In the case of a mine, this will include initial capital investment and annual production operating costs and revenues. Herein, the focus will be on determining the rates of production of ore and waste. For a surface operation, annual pit limits are established within the boundaries of the projected ultimate pit, a main haulage road is included in the design, and ore and waste tonnages for the year are calculated. This process is repeated for each year of the projected mine life until the final and supposedly optimal pit limit is reached. In the case of an underground mine, the schedule includes both ongoing development work and bringing stopes into production. The development of underground openings is carried on in advance of stope production to ensure that no bottlenecks in production will occur. This is roughly analogous to overburden stripping for a pit. Determining development work revolves around scheduling the production areas whether the production face is in a stope or a panel. The level of detail for the long-term schedule is limited mainly to allocating reserves and the associated waste removal and development expenditures by year. Beyond this, the main goal is to maximize the cashflow within the limits of sound engineering practice and full recovery of the deposit.

Short-term production scheduling starts where the long-term schedule left off. This is where the detailed design of the mine takes place and the annual plan is implemented on a level of detail suitable for guiding operations on a time frame which meets the needs of production planning in the mine. In the case of a surface operation, a monthly plan of stripping, ore production and haul road construction will be determined. At this level of design, the goals are to properly utilize production equipment by avoiding idle time and excessive moves to different working levels, ensure that stripping proceeds in advance of ore production, that working slope angles are maintained, that haul road access is provided to all working benches and that ore is blended to avoid excessive quality fluctuations at the mill. Of these, grade control is the most difficult to achieve. Often reserves are classified into several categories: low grade ore may be sent to a heap leach pad or set aside as inventory for periods of increased demand, while high grade ore is sent to the mill. Commonly, all ore is sent to the same destination, a mill, and the efficiency of metallurgical recovery will depend on the quality characteristics of the ore. During comminution, the hardness, friability and size distribution of the fragmented ore will determine the time and energy spent on reducing the particles to mineral liberation size. Likewise, flotation recovery will be dependent on the chemistry of the ore. Finally, the concentrate's composition must be held within acceptable limits for further processing or as a final product. In order for the mill to operate at its best efficiency, an ore feed should be provided from the mine with as little fluctuation in its quality characteristics as possible. It is uncommon for a deposit to be so homogenous in its grade that blending of production from the mine is not a concern. As the number of quality characteristics grows, the difficulty of scheduling production to maintain an acceptable mill feed becomes increasingly complex. Ore grade control is a source of frustration that often leads to contention between the mine and the mill.

In practice, production scheduling is carried out as follows: 1) benches accessible for mining are displayed in plan view, on a bench by bench basis, showing current toe and crest lines; 2) an area of the exposed bench is delineated as a cut which corresponds to a reasonable production volume, for example an area that can be drilled, blasted and loaded within the production period; 3) the grade and tons of ore and waste within the delineated area is determined and 4) additional production cuts are added in order to meet tonnage and grade requirements. This procedure repeats itself with cuts being added or deleted, and is increased or decreased in extent until a reasonable production plan is found for that period. More often than not, the main criterion for selecting a cut is to maintain the progression of the production equipment along the bench face.

Scheduling software greatly simplifies this procedure by making the process of determining the grade and tonnage within a cut much more rapid and accurate. Those blocks which are within the limits of the annual pit and are exposed on a bench are displayed in plan view showing the ore grade within the block. One or more polygons are digitized on the display to delineate the blocks to be mined in the production cuts. The total tonnage selected and average grades are then calculated and displayed. If the grade and tonnage goals are met, then the blocks are treated as having been mined and the block model is adjusted accordingly, with the mined blocks being removed from further consideration and the cut polygons being stored with their associated ore tonnages and grades for future reference. The following example demonstrates how short-term scheduling can be carried out without the use of specialized programs to illustrate the basic procedure.

8.1 LONG-TERM PRODUCTION SCHEDULING

Example 8.1: Moving cone-based long-term production scheduling using Techbase
Before continuing with long-term scheduling of the Boland Banya deposit, the following will be needed: 1) a block model containing fields for ore and waste tonnage or volume, ounces of gold and silver, and the net value of the block; 2) gridded premining and post-mining topography; and 3) a mining sequence file for the ultimate pit. This example will commence with a review of ultimate pit generation using Techbase's moving cone routine. For details see Example 6.6.

From the main menu select **Open-pit** ⇒ **opTopo** ⇒ **design File** ⇒ **Initialize** and define the extent of the sequence file (pit.seq) so that it matches the definition of the block model (pit). Exit the design file menu and under the topography menu (**Topography** ⇒ **Topography**) enter the field containing the estimated cell table topography which is coincident with the block model. The same can be done for the pit's bottom surface (**Bottom**). Exit the **opTopo** routine, and select **opSlope** ⇒ **design File** ⇒ **Open file** or use **Show** to ensure that the previously defined sequence file is open. From the main **opSlope** menu select **Initialize** ⇒ **Initialize** to create a new template file (pit.tem) to hold the allowable slopes in different areas of the block model. Under the **Slope** menu use **Zone**, **Add sector** and **Calculate** to define the pit slopes. Exit from **opSlope** and select **opCone** and repeat the procedure of opening the template and design files. Next select **Apex** ⇒ **apex Cutoffs** and list the field relationships which must be satisfied before a block will be considered as a cone apex. Under **Values** ⇒ **Fields** enter the fields in the block model for the net block value, ore tonnage, etc. as in Example 6.6. Under **Ore cutoffs** enter cutoff definitions to discriminate between different ore and waste categories. **Goals** need not be set for generating an ultimate pit. The default parameters' values can also be accepted unless different values for RESTART and SEARCH will generate the final pit more efficiently. For instance, in the case of Boland Banya the deposit is under very little overburden along its northeastern edge, so a ewnstb (East to West, North to South, Top to Bottom) entry for SEARCH can be expected to result in a more efficient search. Select **Cone** to generate the ultimate pit.

Return to **opTopo** ⇒ **Topography** ⇒ **Save** and enter under TOPO FIELD a cell field having an estimate of the pre-mining topography. This will modify the field to include the post-mining topography in the grid. Given a block model of sufficiently great extent to allow for the expansion of the cone miner with the slopes of the template file, the ultimate pit should appear as in Figure 8.1. Details for grid contouring are given in Example 6.6. Gross mining reserves are as reported in opcone.out. A more detailed report can be generated using **opRept** as demonstrated in Example 6.16.

Once the ultimate pit limit, mining life and mining reserves have been determined, **opCone** can be used in scheduling mode to generate annual pits. Scheduling mode allows the ultimate pit's highwall slope angles to be reduced to working slopes while still honoring the final pit wall that was established while **opCone** was in ultimate mode. For instance, in scheduling mode the working slope for the annual pit limit might be reduced from 50 to 20°. Within the ultimate pit limit, the 20° slope will be used except where it intersects the highwall at which point the slope will revert to 50°.

Select **opTopo** ⇒ **Topography** ⇒ **Reset** and enter <scheduling> under MINING PERIOD. The bottom topography will be reset to the ultimate pit and the surface topography will be reset to its unmined state. The cone mining procedure can now be repeated with a new template file that defines working slopes. Exit **opTopo** and select **opSlope** ⇒ **Template file** ⇒ **Initialize** to create a new *.tem file <working.tem> as before. Under **Slope** define a new set of pit slopes. Exit **opSlope** and select **opCone** ⇒ **design File** ⇒ **Open file** <pit.seq> ⇒ **eXit** ⇒ **Template file** ⇒ **Open file** <working.tem> ⇒ **eXit** ⇒ **Apex** ⇒ **apex Cutoffs** and enter the mining

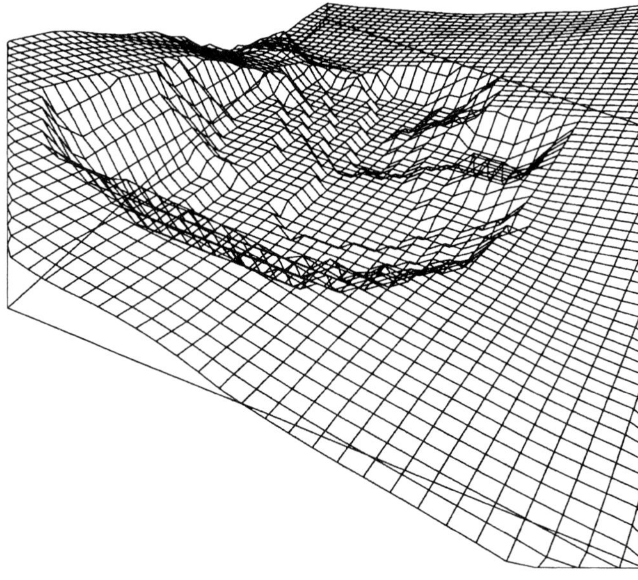

Figure 8.1. Ultimate pit for the Boland Banya deposit based on the Moving Cone Algorithm. Perspective mesh view of pit observed from Azi., Dip, distance of 55°, 35° and 2500', respectively.

criteria that will be used in the first mining period. Note that these criteria may be different from the ultimate pit criteria and the subsequent mining period's criteria. This is especially true in the initial mining period for a deposit that requires substantial waste stripping prior to mining ore. In this case, the criterion that the block's net value be positive may make it difficult for the cone miner to find a solution. If the majority of production will be in waste in the initial periods, it may make more sense to use a negative net value criteria. Another option would be to define a calculated field for the stripping ratio, contour the stripping ratio and use this map as a basis for a criterion based on initiating the pit in an area of low cover and high ore availability. Unfortunately, calculating a true stripping ratio is difficult within the limitations of Techbase since the cone of overlying blocks would have to be used to calculate an average waste thickness. Still, for relatively shallow deposits similar to Boland Banya, a simple cell table calculation based on the ore and waste thickness in each cell should be sufficient for the display of the contours of the stripping ratio, but to place a stripping ratio field in the block model would be much more complex; this would require running the cone miner OUTPUT set at <extra> (see **Parameters**), generating cones for all blocks containing ore (see **Apex ⇒ apex Cutoffs**) and including fields for ore and waste tonnage in the **Fields** list. The resulting output file (define.out) will report ore and waste tonnages for every apex. The output file can then be edited into a format suitable for loading back into the block table and used for calculating a stripping ratio field, a somewhat involved process.

Enter an apex cutoff <netval_pit > 0>, but note that if the production goal for the initial period is based entirely on waste stripping, then the cone miner might not find a positively valued apex. Enter the fields to be included in the output report under **Fields** (NET VALUE = <netval_pit>, NET TONS = <pit_waste>, ORE VALUE = <pitauoz>, ORE TONS = <pittons>), and the cutoffs to be used in reporting under

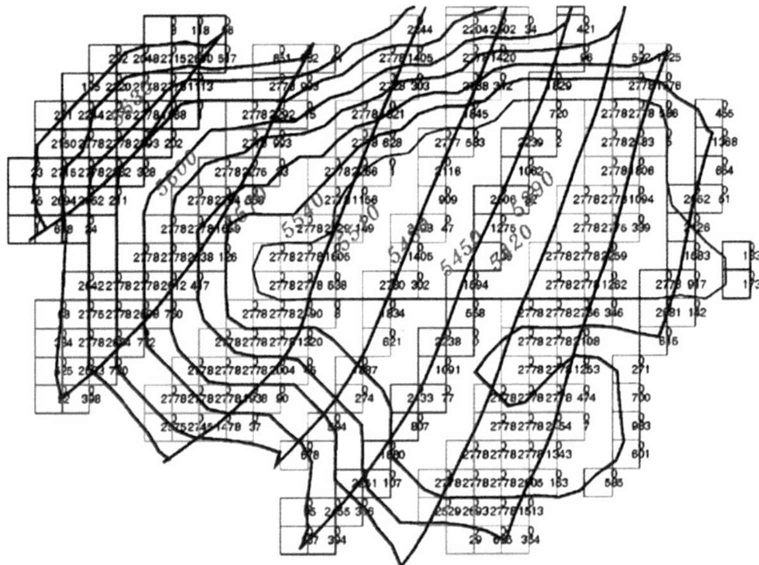

Figure 8.7. Posters of blocks available for mining (dflag = 1) and their confining bench limits at the end of period 1.

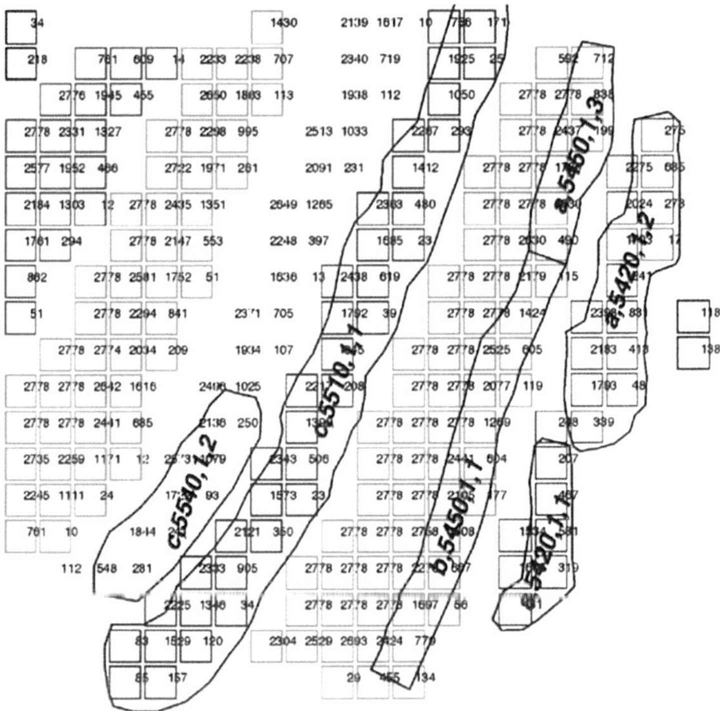

Figure 8.8. Polygons used for selecting blocks to be flagged as being included in the first week of stripping using three shovels.

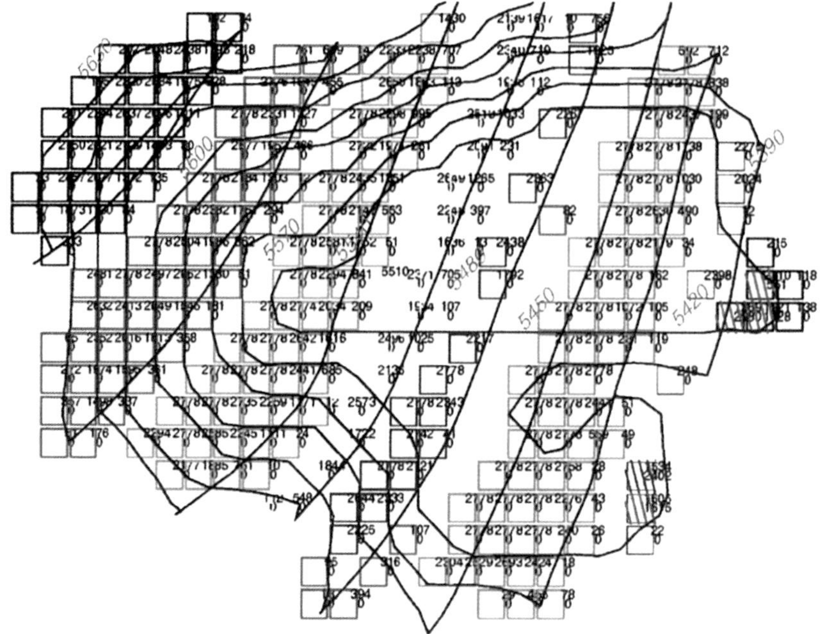

Figure 8.9. Poster of blocks available for mining (dflag = 1) following one week of stripping and their confining bench limits.

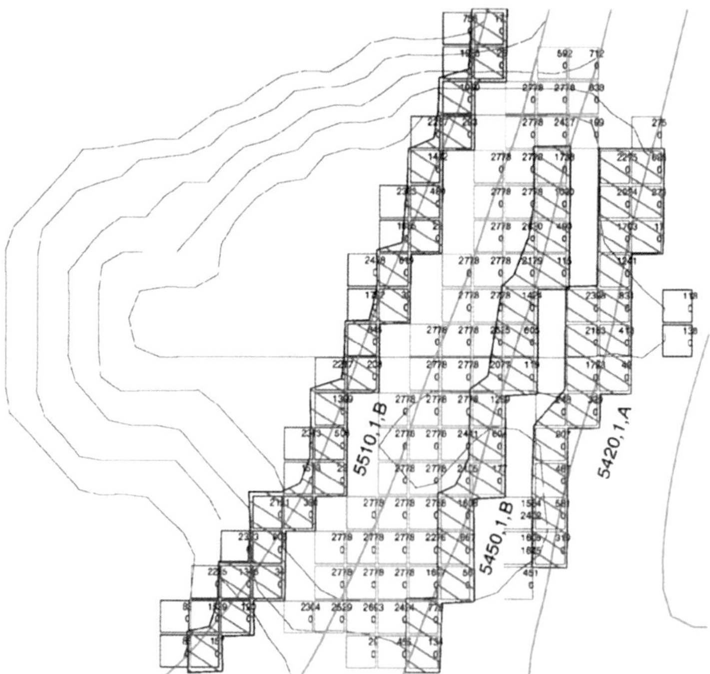

Figure 8.12. Week 1 production schedule cuts (red) showing available blocks with contained waste and ore, and polygons of premine topography (green) and bench limits (black).

ore Cutoffs as before, but using the fields and cutoffs that are relevant to scheduling as opposed to generating an ultimate pit. Under **Goals** enter the production criterion that are to be met during the mining period. Up to three goals can be entered: NET TONS, ORE TONS and WASTE TONS. Mining for the period will cease when any one of these criteria are satisfied. Care must be exercised in entering these goals. If the net tons mined is the criterion, the tonnage of ore may fluctuate widely from period to period causing trouble at the mill and decreasing the value of the project, but production will be kept even over time. If production of ore is the criterion, then the net tonnage might fluctuate, especially in the initial periods when large quantities of waste have to be stripped. As an alternative to actually entering the ore tons field for ORE TONS, another field can be entered such as ounces of gold. It may be that the driving criterion for a mine is to meet a mandated production of metal. In the initial period, waste tonnage might be used.

The mining reserves resulting from running **opCone** in ultimate mode were:

– Ore Value <pitauoz> = 18,449,059 oz Ore tons <pittons> = 5,477,085 st,

– Net value <netval_pit> = $364,556,438 Net tons <pit_waste> = 16,563,600 st.

By way of example, this could be mined in four periods. In each period it would be desirable to maintain a total tonnage of ore and waste of around 5.5 million tons, while maximizing either the ore tons or ounces mined during the earlier periods. Unfortunately, maximizing ounces cannot be used as a goal within the constraint of mining only 5.5 million net tons per period. The best that can be done is to use both net tons and ounces as goals. Due to the overburden, net tons will be satisfied first with more ounces being recovered later. Therefore, enter production goals of: NET TONS = <5,500,000> and ORE TONS = <5,000,000>. In this way the net mining tonnage will certainly be satisfied before mining virtually the total mining reserves of gold, but once the ore has been uncovered the cone miner will seek out ore blocks. This can be further enhanced by gold ounce content as a criterion under ore **Cutoffs** and **apex Cutoffs**.

Select **Parameters** and set MINING PERIOD to <1> through <1>. One period will be mined, the production results noted and the pit limit saved to a cell field. Remember that if the output file name is not specified, then the default naming convention will be used (opcone.out) and the previous period's results will be overwritten. Note that more than one period can be mined at a time, but that the goals for all these periods will be the same in that the per period pit limits will not be available for saving in a cell field. Exit from **opCone** and select **opTopo** ⇒ **Topography** ⇒ **Save** and enter the TOPO FIELD = <ult1>, INCREMENTAL = <yes> and MINING PERIOD = 1. The TOPO FIELD is a field in the cell table that has already been set to the pre-mining topography as per Example 6.6. With INCREMENTAL set to yes, only those cells coincident with a block that has been mined will be reset to the pit elevation. A BLOCK FIELD can be used to save the period mined into the block model. The contoured pit for period one is shown in Figure 8.2. Note the steeper contours on the north side of the pit where the final highwall angle (45°) is being honored. Elsewhere, a working slope of 30° is being used. This procedure is repeated for each mining period, each time saving the pit topography and production totals. The two subsequent mining periods are shown in Figures 8.3 and 8.4. Figure 8.5 shows a section through the four pits. Table 8.1 summarizes production over the three periods.

Table 8.1. Production schedule for Boland Banya.

	Period 1	Period 2	Period 3
Tons Ore	282,152	314,818	1,252,089
Gold (oz)	638,145	901,755	3,937,185
Net value	41,094,476	58,749,956	264,712,006
Tons waste	5,569,200	5,548,800	5,445,600

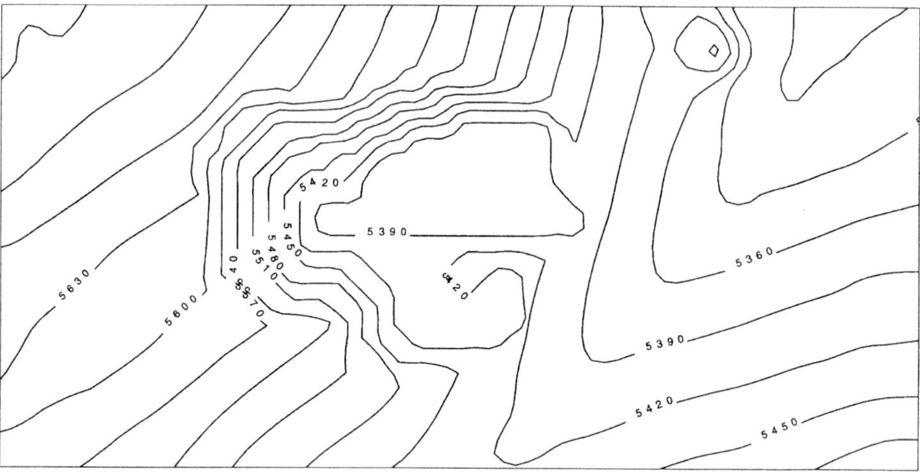

Figure 8.2. Boland Banya pit in period one following the production of 5,569,200 net st.

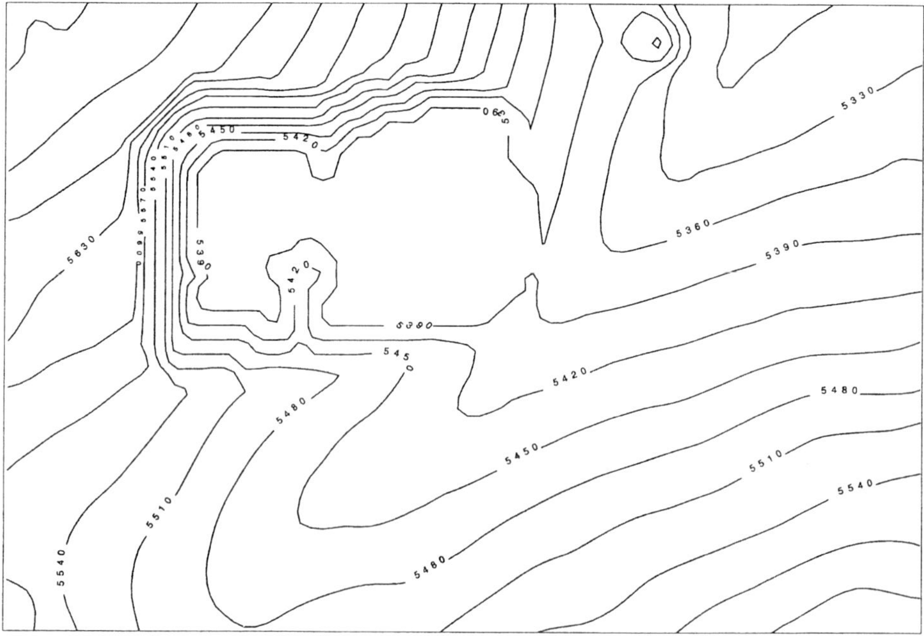

Figure 8.3. Boland Banya pit in period two following the production of 11,118,000 net st.

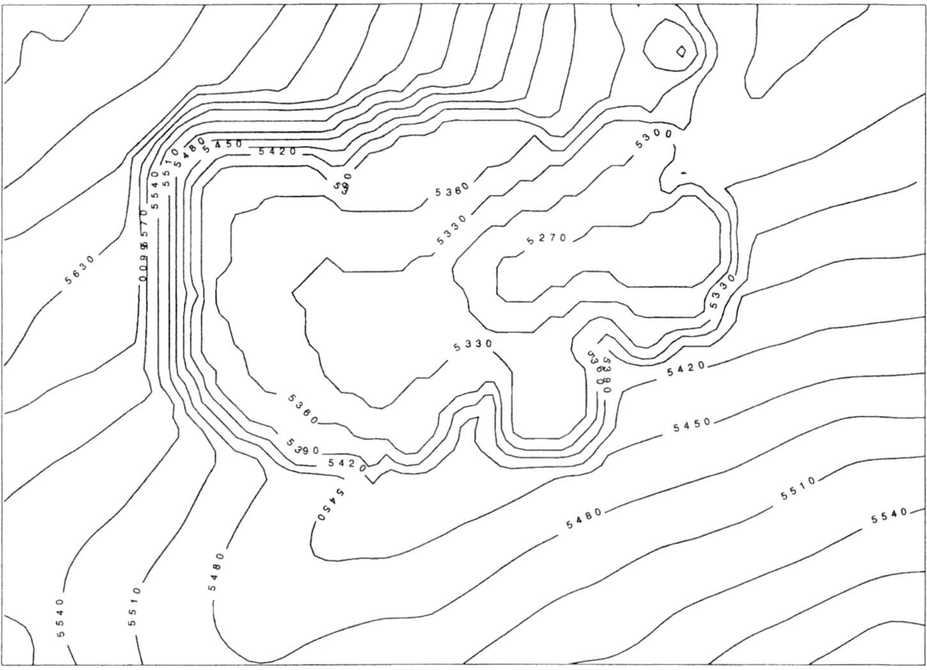

Figure 8.4. Boland Banya pit in period three following the production of 16,563,600 net st.

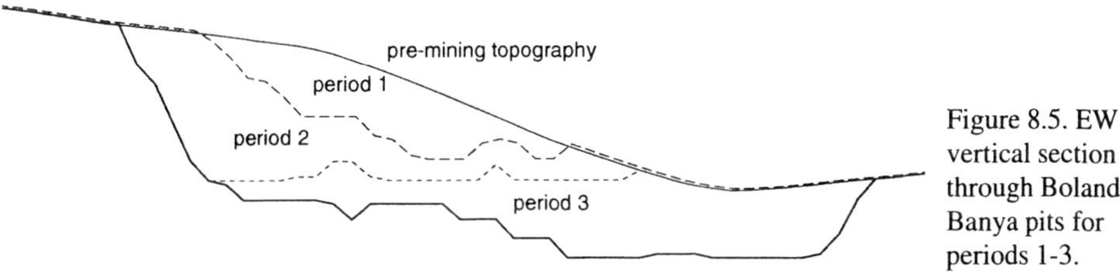

Figure 8.5. EW vertical section through Boland Banya pits for periods 1-3.

As with short-term production scheduling, many iterations of the moving cone miner would be necessary to find a satisfactory solution. One method would be to conduct a sensitivity analysis on the net value field generating a range of likely values by varying metal process and mining and processing costs over their likely range. A series of production schedules can be run over the range of net values.

Summary of procedure: Example 8.1

Initial data requirements: database <boland> with block table <pit> and cell table <surfaces> including the following fields:

1. Pit – auoablk (gold oz/t), agoablk (silver oz/t), oafrac (% ore), pitblkvol (cuft), pitauoz (gold oz per block), pitagoz (silver oz per block), pittons (tons ore), pit_waste (tons waste), netval_pit ($ per block), period (mining period).
2. Surfaces – topo_c (surface topo), oab_c (bottom of OA), ultcone (ultimate), ult1 (period 1), ult2 (period 2), ult3 (period 3).

Procedure:

1. Define a sequence file for the ultimate pit:
   1.1 From the main techbase menu **Open-pit** ⇒ **opTopo** ⇒ **design File** ⇒ **Initialize**
   FILE NAME: <pit.seq>
   (see **Define** ⇒ **Table** ⇒ **Show** <pit.seq> for example parameter values)
   ⇒ **eXit**
   1.2 Define the pre-mining surface **Topography** ⇒ **Topography**
   FIELD/VALUE: <topo_c>
   ⇒ **eXit**
   1.3 Define the base of ore **Bottom**
   FIELD I VALUE: <oab_c>
   ⇒ **eXit** ⇒ **eXit**
2. Define a template file for the maximum pit slopes:
   2.1 From the Open-pit menu **opSlope** ⇒ **design File** ⇒ **Open file**
   FILE NAME: <pit.seq> (may already be open)
   ⇒ **eXit**
   2.2 Initialize the template file **Template file** ⇒ **Initialize**
   FILE NAME: <pit.tem>
   ⇒ **eXit**
   2.3 Define the slopes **Slope** ⇒ **Zone**
   ZONE: <1>
   ⇒ **Add sector**
   ENDING AZIMUTH: <360>
   SLOPE: <45> (default is for all benches)
   Repeat Step 2.3 for as many zones and sectors as needed until done, then select **Calculate**
3. Generate the ultimate pit with the cone miner:
   3.1 From the Open-pit menu **opCone** ⇒ **design File** ⇒ **Open file**
   FILE NAME: <pit.seq> (may already be open)
   ⇒ **eXit**
   3.2 Open the template file **Template file** ⇒ **Open file**
   FILE NAME: <pit.tem>
   ⇒ **eXit**
   3.3 Define a filter for allowable cone apex blocks **Apex** ⇒ **apex Cutoffs**
   <netval_pit > 0>
   ⇒ **eXit**
   3.4 Specify parameters for moving cone
   3.4.1 Define fields to be used in report **Values** ⇒ **Fields**
   NET VALUE: <netval pit> NET TONS: <pit_waste>
   ORE VALUE: <pitauoz> ORE TONS: <pittons>
   (MISSING VALUE and MISSING TONS only need to be entered if the blocks are not completely specified.)
   3.4.2 Define filter for separating ore from waste **ore Cutoff**
   <netval_pit > 0>
   3.5 Define cone miner search **Parameters**
   RESTART: <cone> SEARCH: <ewnstb>
   3.6 Start cone miner **Cone**
4. Save the ultimate pit
   4.1 Modify the pre-mining topography. From the Open-pit menu **opTopo** ⇒ **design File** ⇒ **Open file**
   FILE NAME: <pit.seq> (may already be open)
   ⇒ **eXit**
   4.2 Save grid **Topography** ⇒ **Save**

TOPO FIELD: <ultcone> BLOCK FIELD: <period>
INCREMENTAL: <yes>
MINING PERIOD: <ultimate>

5. To start scheduling select **Reset** (same menu level as Step 4.2)
   MINING PERIOD: <scheduling>
   (can also use to reset to unmined to delete results of cone miner in pit.seq and repeat)
6. Repeat Step 2 to define a new template file <working.tem> with 30° slopes.
7. Repeat Step 3 up to Step 3.4.2 to generate a scheduled pit, but replacing <pit.tem> in Step 3.2 with <working.tem>.
   7.1 Define up to three production goals **Goals**
   NET TONS: <5500000>
   ORE TONS: <5000000>
   ⇒ **eXit**
   7.2 Define cone miner search **Parameters**
   RESTART: <cone> MINING PERIOD: <1>
   SEARCH: <ewnstb>
   7.3 Start cone miner **Cone**
8. Repeat Step 4 up to Step 4.2
   TOPO FIELD: <ult1> BLOCK FIELD: <period>
   INCREMENTAL: <yes>
   MINING PERIOD: <1>
9. Repeat Steps 6-8 for each period using the relevant parameter values and fields.

8.2 SHORT-TERM PRODUCTION SCHEDULING

Chapter 6 presented several computational approaches to generating an ultimate pit: polygon expansion (Example 6.5), moving cone (Example 6.6) and graph theory (Example 6.7). Superficially, it would seem a simple task to extend these methods to multiple production periods rather than generating a final pit limit. Unfortunately, this is not the case and the software available for scheduling, while powerful, is far from optimal or even convenient.

Polygonal expansion can be used to yield a perfectly acceptable pit when the limits of mining are obvious as would be the case when there is a simple deposit in combination with a well defined maximum stripping ratio or cutoff grade. When the location of the initial cut and the subsequent direction of development is also tightly constrained by topography and deposit geometry, then production planning can also be accomplished by using polygonal expansion to generate a series of pushbacks that correspond to annual production requirements. A common example of this would be planning production for a deposit with little compositional variation which is exposed on a hillside. If grade control is not important either due to compositional homogeneity or flexibility in the quality of the final product, then mining can commence and proceed in the most cost effective direction, which in this case would be into the face of the hillside. Polygons can be drawn to represent annual mining limits on each exposed bench. As in Example 6.5, the polygons can be intersected with the block model to determine the tons of ore and waste mined or alternatively the expanded polygon and can be represented as a grid as in Example 6.15.

When there are multiple zones, important spatial variability in grades and numerous points from which a deposit can be opened, production scheduling becomes a much more difficult proposition. As demonstrated in Example 8.1, the moving cone algorithm can be used for production scheduling, but is limited to using one variable to define the demand for production in each period and doesn't ensure an optimal, or even near optimal, solution. A goal can be set to accumulate a block field such as ounces, revenues or tons of ore or waste, but it is difficult to meet other criteria. For instance, a common goal would be to produce a given annual tonnage of ore, but this doesn't account for the variable tonnage of waste that has to be removed. Conversely, if the goal is total tons mined, then the production of ore may vary widely from year to year.

The canned graph theory (Lerches-Grossman) based routines that are included with most mine design packages cannot be applied to scheduling. These routines seek to form a maximum value tree based on a single value, typically the net block value. As a result, only a single maximum value pit can be determined. There are alternative network and binary programming formulations which can be used both for finding an optimal pit and production schedule, but these require a certain degree of specialized knowledge and have their own computational difficulties. Still, the potential for applying mathematical programming formulations to mine design is great if premature.

Examples 8.2 and 8.3 demonstrate the overall process and computational difficulty involved in short-term production planning. The material and procedures of Example 8.2 represent much of the preliminary work that must be done in Techbase prior to using the more automated procedure offered by the **opSched** facility demonstrated in Example 8.3. The nuts and bolts of short-term production scheduling revolve around: 1) displaying the blocks available for mining, 2) digitizing cuts that represent mine production on the benches, 3) storing the cuts in an organized fashion as polygons, 4) intersecting these polygons on the block model to determine progress towards production goals, and 5) modifying the production cuts to improve the production schedule.

Display of the block model data and mining limits must be kept simple or the immense scale and complexity of production scheduling will become overwhelming. Only those blocks that are available for mining should be displayed. Numeric posting of a few critical production related parameters, such as grade and tonnage, should be displayed in the block. Cell colors and fill patterns, even contour lines, can be used to reduce display complexity. Only one bench should be displayed at a time showing the current limit of mining and the final mining limit as defined by the long-term mine plan. The current mine limits on the overlying bench may also be displayed to avoid pushing the current bench beneath the overlying bench.

Starting at the current mine limit on an exposed bench, or at the pre-mining topography, a set of initial cuts are digitized, representing an estimate of the production of the primary production units. For instance, an average cut, as represented by an area that will be drilled, shot and loaded as one unit, may require five days (one production week) to be mucked by a production shovel. If three such shovels are required to maintain mine production, then three corresponding polygons will be digitized on the faces of the exposed benches: one in ore, another in waste, and the third in maintenance, waste or ore as needed. The values displayed from the block model are used to guide cut locations and dimensions with the requirement of digitizing a cut that will match one week of the shovel's production while progressing towards the production goals for ore, waste and grade(s).

The number of polygons that are produced by different scheduling plans can quickly become overwhelming. Following the previous example, if there are only three polygons a week, then there will be 156 polygons needed for a year's scheduling. If alternative production plans are considered, then the number of polygons can rapidly get out of hand. Along with each polygon's identity and vertices must be stored other production information such as area, contained tonnages of ore and waste, average grades and contaminant levels, and recovered mineral content. Techbase's polygon tables provide a format in which scheduling cuts can be conveniently stored as polygons and readily related to other data sets, particularly block models, grids, other polygon tables (e.g. royalty payments) and flat tables containing blasthole data.

When a polygon is digitized on a bench and then stored in a polygon table, nothing is initially known about its contents. The polygon must be intersected with the block model on that bench in order to report totals and averages for the intersection. Trial-and-error is used to find a set of cuts that will meet production goals by modifying the polygons to obtain the desired cut volume and by tying different configurations of polygons in the attempt to find a combination that will produce an acceptable grade and production of mineral. The process of using the facilities **polyEdit**, **Locate** and **Report** for production scheduling, as demonstrated in Example 8.2, is very inefficient since once the polygon is defined in **polyEdit**, the polygon editing program must be exited in order to use **Locate** to intersect the polygon with the blocks on that bench, **Report** the contents of the polygons, and use **tbCalc** to recalculate the remaining block tonnages. **opSched** streamlines this process by calculating and reporting polygon, bench and period production figures while remaining within a polygon editor (see Example 8.3).

Example 8.2: Controlling the background display for production scheduling using Techbase

Consider the problem of short-term scheduling production for the Boland Banya deposit following the assignment of annual pit as in Example 8.1. In order to be able to effectively carry out scheduling, a background display must be generated that shows the availability of mining blocks for excavation and the current level of ore and waste in the block. Following the definition of mining cuts, the display must be updated to reflect the removal of material from the block model so that newly uncovered blocks will be made visible, entirely removed blocks are no longer visible and the amounts of ore and waste in the available blocks are adjusted to reflect any partial mining of the blocks. The procedure used to flag blocks available for mining and updating their contents to reflect the progress of mining should be as automated as possible. This example demonstrates how calculated fields can be used to create a background display for production scheduling that can be updated as needed during the scheduling process.

Before proceeding with this example the following elements of design will be needed:

1. Gridded surfaces of topography and the top and bottom structures of the ore body.
2. A block model with fields containing estimated grades, the fraction of the block which is ore and ore tonnage. The model should cover at least the extent of that portion of the ore body which will be mined, and that has as many levels as mining benches with a block height equal to the bench height.
3. Gridded surfaces of the ultimate and annual pit limits.

The cell table containing the gridded surfaces must match the block model both in its *xy* cell dimensions and location, i.e. the cells must be aligned with the blocks.

Several additional fields will have to be added to the block table (pit) to keep track of the content of the blocks as mining progresses and to indicate the availability of the block for mining:

Actual fields
- *per1bf, per2bf* and *per3bf* the fraction of the block located within the limits of the midbench polygons of the three pits of Figures 8.2-8.4,
- *topof, ult1f,* and *ult2f* the fraction of the block above pre-mining topography for the pit limits of the long-range mine plan,
- *minef*, the fraction of the block mined as part of a production cut,
- *oafrac,* the fraction of the block located within the ore zone,
- *year1, year2,* and *year3* (optional) the year in which the block is mined (assigned as in Example 8.1),
- *bench,* the bench associated with the mining cut,
- *dflag*, a flag indicating the minable status of the block (discussed later).

Calculated fields
- *aucolor* is a color field based on the magnitude of the block's gold grade <auo-ablk>,
- 1 auoablk .1 < 31 skip,
- 4 auoablk .15 < 25 skip,
- 3 auoablk .2 < 19 skip,
- 7 auoablk .25 < 13 skip,
- 6 auoablk .3 < 7 skip,
- 5 auoablk .35 < 1 skip,
- 2.

The tons of ore in the block, *smore = oafrac 6525 **

where there are 6525 tons in a block (50 × 50 × 30 cu. ft * 174 lb/cu. ft / 2000 lb/st).

The block tonnage could also be a calculated field if either the block dimensions were variable, a density field had been estimated into the model or if different densities were used for ore and waste.

per1f, per2f, and per3f the fraction of the block that is both within the mine period limits and beneath the premining topography, i.e.
- *per1f = per1bf 1 topof –& **,
- *per2f = per1bf 1 ult1f –& **, and
- *per3f = per1bf 1 ult2f –& **.

In some cases this may be a negative value and should be set to zero.

Tons of ore remaining in the block during the production schedule,

*or = pitlf pit2f +& pit3f +& smore * 1 pit_lev bench == minef * –& **

where *pitlf pit2f +& pit3f +& smore ** allows only ore tonnages within the ultimate pit limit to be displayed, *pit_lev bench == minef ** returns a one (in dicating true) times the fraction of the block mined only if the current level matches the bench being mined. *minef* is then subtracted from one (and treated as zero if it's a null value by using –&) to determine the fraction of the block remaining unmined which is multiplied by the block's original available ore tonnage.

BCY of waste remaining in the block during the production schedule,

*wr = 1 oafrac – 2778 * 1 pit_lev bench == minef * –& * perlf per2f +& per3f +& **

where *1 oafrac – 2778 * perlf per2f +& per3f +& ** calculates the bcy of waste in the block (2778 bcy per full block) after accounting for ore and *1 smblk_lev bench == minef * –& ** serves the same purpose as for *or*.

Since *wr* is always adjusted downwards by the fractional amount *minef* during scheduling, another calculated field is needed to check the contained content of the cut: *bcy = 1 oafrac –& pitblkvol 27 / * minef * perlf ** where *pitblkvol = pit_rsz pit_csz * pit_lsz **, i.e. the cu. ft in a full block. Note that fields such as *bcy* have limited value for tracking cut production parameters since *minef* cannot be tied to any one cut polygon: since cuts may include partial blocks and will have shared boundaries the *minef* fraction is accumulated from all polygons that intersect a block. Only when all the blocks in a mining cut exclusively belong to that polygon can the volume of the cut be determined. The only solution to this problem would be to have a separate fractional field for every cut so that the portions of the block associated with each polygon could be tracked. While possible, this is an unacceptably clumsy approach to production scheduling.

dflag is an ACTUAL field but is assigned a value in **tbCalc** as was the block netvalue field in Example 6.4. In general, it is preferable to use calculated fields rather than **tbCalc** since calculated fields are automatically updated whenever a field that is referenced in their equation changes value. The limitations of calculated fields are that there can only be 32 steps in the equation (the (f,filename) reference can also be used for calculated fields) and that only field values in the same record can be referenced in the equation. *dflag* requires more than 32 steps and references other records in the block table and therefore must be an actual field whose values are assigned using **tbCalc**. A file should used to assign values in the block model since the calculation will have to be repeated for each set of mining cuts.

*wr[–1696] or[–1696] +& wr[–1748] or[–1748] +& +& wr[–1749] or[–1749] +& +& wr[–1750] or[–1750] +& +& wr[–1802] or[–1802] +& +& 20 <= wr or +& 10 >= * = dflag*

Here, the five overlying mining blocks are summed for their total content of remaining ore and waste: *wr[–1696] or[–1696] +& ... 20 <=*. The summation is compared against 20 so that if the overlying blocks are nearly mined out or are above the surface topography a value of one is placed in the stack. Finally, the content of the block itself is compared against zero, *wr or +& 10 >=*, i.e if the minable content of the block is significantly greater than zero a one is returned and placed in the stack. The results of the two comparisons are multiplied together and assigned to *dflag* which only takes a value of one if the block contains both ore or waste and is uncovered. Note that the value in brackets, such as *wr[–1749]*, is referring to a field value displaced by –1749 records in the block table. In a block model each record represents a block with record 1 being located in the SW corner of the first bench. Block records increment by row, column and then level. Since the table smblk consists of 33 rows, 53 columns and 13 benches, moving back –1749 records places us at the same block in the next level up. The procedure for inter-record references in a block model is illustrated in Figure 8.6. In this example, consider a block model consisting of three rows, three columns and two levels. If the current record is 14 and field values in the five overlying blocks are to be referenced, then the record number of the block directly above record 14 will be 14 – (3*3) = 5 or [–9] relative to record 14. Likewise, the overlying blocks one row in advance and one row behind [–9] will be at [–9+3] = [–6] and [–9–3] = [–12], records 8 and 2, respectively. In general, for any dimension of block model, the overlying block will be at [–rows*cols] relative to the current record.

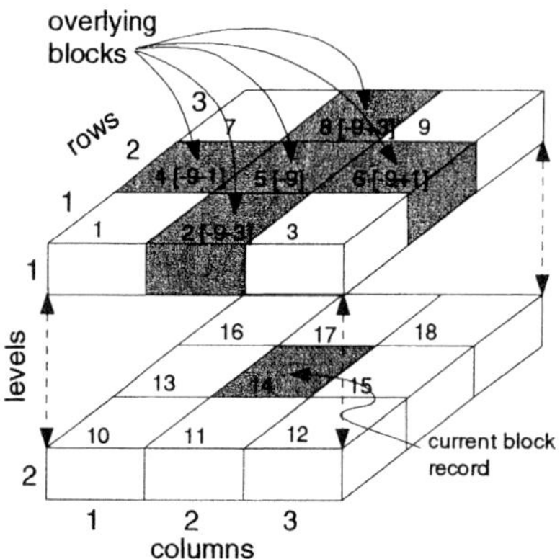

Figure 8.6. Referencing other records from the current record in a block model.

Once *or*, *wr*, and *dflag* have been defined, the blocks available for mining can be displayed in plan view and saved as a metafile. Select **Graphics** ⇒ **Poster** ⇒ **Database** ⇒ **Add filter** and set *dflag* and *year1* both equal to one so that only those blocks available on the surface and within the mining limits of the first year will be drawn. After assigning a metafile name and setting the scaling, select **Cell/poly** ⇒ **Fields** and enter the fields to be displayed as an aid to production scheduling. In this initial cut, the blocks will only contain waste, so only *wr* will be displayed, but once the cut is into ore, then *or*, *wr*, and the assays will have to be displayed. The mid-bench polygons for period 1 have been generated using the **Write polygons** facility in **Gridcont** and then loaded into a polygon table <sched>. These can be displayed along with the flagged cells by entering the polygon identity field <ult1cont> as

VALUE 1 under **Cell/poly ⇒ Fields**. The resulting posting of available blocks is given in Figure 8.7. Cell color corresponds to the bench level (e.g. red {pen 2} for bench 2). A calculated color field <aucolor> for the block grade was used to control cell fill. Note that none of the ore zone is available for mining at the beginning of the first mining period.

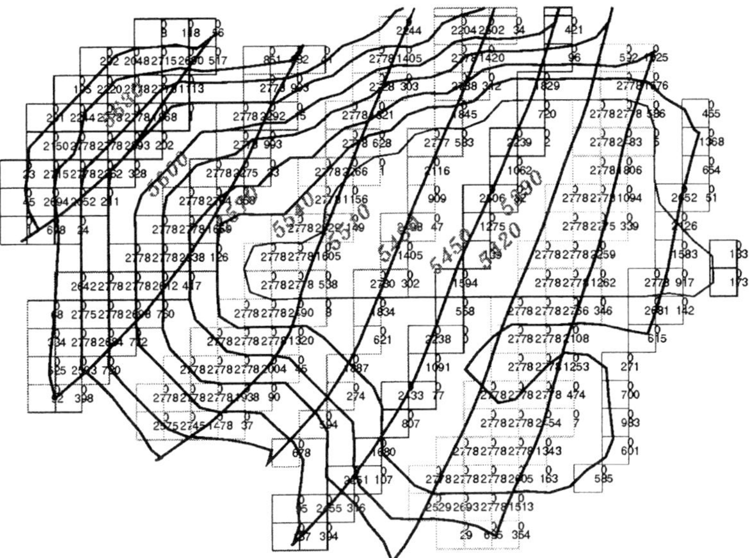

Figure 8.7. Poster of blocks available for mining (dflag = 1) and their confining bench limits at the end of period 1. (Figure in colour, see opposite page 358.)

Now the initial set of mining cuts opening up the deposit can be defined. From the main menu select **Graphics ⇒ polyEdit ⇒ Load** and load the **background Metafile** <sched.met>. Select **Edit ⇒ Line ⇒ Create** and enter the polygon identity and layer. The polygon naming convention is critical to maintaining an organized production schedule. In this simple example the shovel, midbench elevation, the week and the sequence of cut number within that week will be used to track production. For example, two cuts will be needed on bench 5420 due to the break in topography into northern and southern sections of the bench that will have to be taken as two separate cuts. Shovel A will be assigned to the initial stripping on this bench and the cuts will be identified as 'a, 5420, 1, 1' and 'a, 5420, 1, 2' to indicate the shovel, bench, week and cut. Later, this identity can be treated as a text field and filters used to report production by shovel, bench and week. An additional cut will be needed on 5450 to complete the week's production for shovel A. Shovel B will complete the initial cut across the face of 5450, while shovel C will work cuts in benches 5510 and 5540. Note that the shovels must work across a large number of blocks in this initial period since these surface exposed blocks are mostly in the air. As the shovels dig deeper into the benches, few cuts over a smaller areal extent will be needed to maintain production.

Use **Add**, **Insert**, **Delete**, **Move point** and **Close line** as needed to digitize the cuts on 5420. Referring to Table 8.1, the weekly production goal for both ore and waste is 51,893 bcy or 17,297 yards per shovel. There are 11,825 bcy of waste (no ore) on 5420 so the exposed bench can be finished. Complete the week's production for A on 5450. Likewise, digitize the production cuts for the other shovels on 5450 and 5510 to complete both production and the necessary pushback of stripping for the week. The resulting cuts are shown in Figure 8.8. Note that the polygons in Figure 8.8 extend beyound the limits of the blocks in the display. Normally, this will result in a polygon that intersects more than the displayed blocks and a resulting cut tonnage which is much higher than planned. Later, it will be shown how filters can be used to limit the selection of mining blocks to the blocks displayed. This example is aimed only at demonstrating how to use **tbCalc** and calculated fields to create a dynamic display for scheduling using **opSched**, not for scheduling itself. Digitizing oversized polygons helps to ensure that the selected blocks will be completely mined so that the tonnage, as determined from examination of the posted volumes, will be close to the week's production requirement. In Example 8.3, care will be taken to digitize cuts that are within the bench limits and do not overlap.

When finished digitizing the cuts, **eXit** from the **Edit** facilities, select **Save** ⇒ **polygon Table** and enter the polygon ID field <per1sch> and six new records will be appended to the polygon table. Note that subsequent updates of these polygons in the

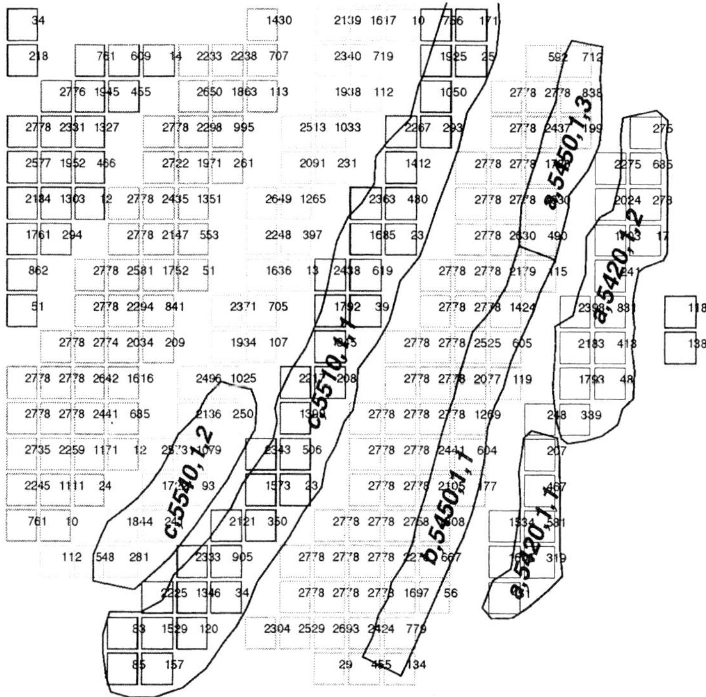

Figure 8.8. Polygons used for selecting blocks to be flagged as being included in the first week of stripping using three shovels. (Figure in colour, see opposite page 358.)

polygon table while saving polygons will require a filter <per1sch != 'null'> where 'null' is a null, i.e blank entry for FIELD/VALUE. This filter will limit the save to those polygon table records which already have an entry for the cut ID so that these records will be overwritten. Otherwise, a new set of cut polygons having the same identities will be appended to the table. Having near duplicate polygons with the same name will result in multiple intersections of the blocks in that cut and mined fractions <minef> exceeding 100%.

The cut polygons will now be intersected with the block model. From the main menu select **Modelling** ⇒ **Locate** ⇒ **Database** ⇒ **Filter** and set up two filter groups: one to restrict intersection to only those blocks that are uncovered <dflag = 1> and on 5420 <pit_zc = 5420>, and the second to base intersection on the two 5420 polygons <per1sch () 5420>. Select **Setup** ⇒ **Fields** and enter the block tables *x* <pit_xc> and *y* <pit_yc> coordinate fields. No surfaces will be used for intersection to determine the vertical fraction, negating the use of the *z* coordinate field. Select Polygon and enter both the polygon table and identity field name as @tablename.fieldname <@sched.per1sch> (this is the only facility requiring entry of polygon table fields in this format). The resulting message should show that two of the size cut polygons have been loaded. Under **Assignment** set <bench = 8> and <minef> = POLYGON/VERTICAL FRACTION. **Locate** will intersect 'a,5420,1,1' and 'a,5420,1,2' with the uncovered blocks on 5420. Repeat this procedure for the remaining four polygons on the three other benches by changing the two filters, resetting **Polygon** and the bench field and selecting **Locate**. This will have to be done once for each bench to avoid intersecting the wrong combination of bench and polygon. Even after changing the polygon filter, **Polygon** must be used to re-initialize the polygons that are in memory.

The volume of material contained in the cuts can be confirmed using the Report facility. From the main menu select **Techbase** ⇒ **Report** and add a filter (deleting of modifying the filter used during **Locate**) so that only intersected blocks will be displayed <minef > 0>. The filter can be extended to displaying the cut volumes on each bench, but without using a separate fractional field value for every polygon there is no way to track the results for each polygon. Under **Setup**, supply a **Report file** name, setup a **Page layout** that includes display of the field values (optional), and enter a **Field list** that will total the volume of waste <bcy (t)>. Selecting **Report** generates an ASCII file that can be viewed with **View file**. The result will hopefully be in the neighborhood of one week's production. If not, **polyEdit** can be used to modify, delete or add new polygons and the locate/report procedure can be repeated. Note that subsequent runs of **Locate** for <minef> and <bench> will overwrite all previous instances of these fields unless these records are protected from location by a filter. Thus, in practice, a block can only belong to one polygon unless there is no overlay between adjacent polygons and location is carried out simultaneously for all polygons and blocks on a single bench. As mentioned earlier, using these facilities for scheduling is very problematic.

Once <minef> and <bench> have been assigned for the production period, the block availability flag <dflag> can be re-initialized to reflect changes in the available waste volumes <wr> and ore tonnage <or>. To do this select **Techbase** ⇒ **tbCalc** (delete all filters) ⇒ **Setup** ⇒ **Equations** and enter the equation file holding the calculation steps <(f,dflag3.calc)>. Selecting **Calculate** will reset <dflag> for all records in the block model. The background metafile showing updated block availability can now be generated to use in the scheduling of the second week of mine production as was done in the beginning of this example. The resulting poster is shown in Figure 8.9.

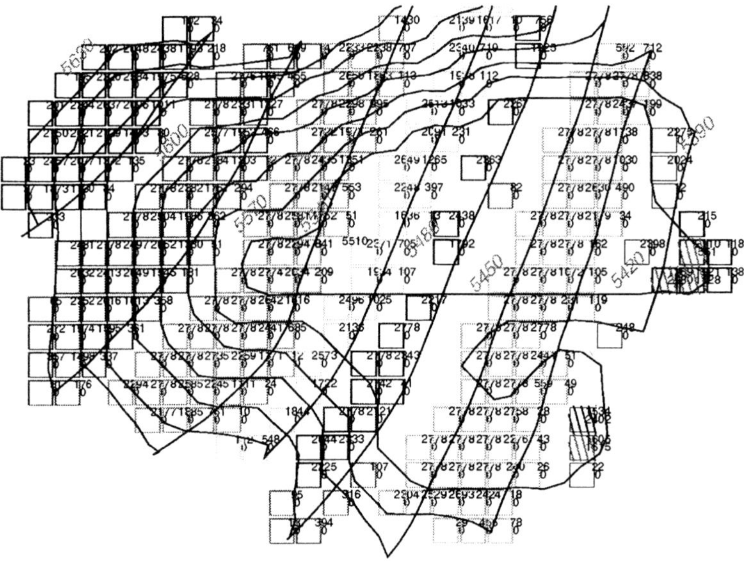

Figure 8.9. Poster of blocks available for mining (dflag = 1) following one week of stripping and their confining bench limits. (Figure in colour, see opposite page 359.)

Summary of procedure: Example 8.2

Initial data requirements:

1. A cell table <surfaces> containing pre-mining topography <topo_c> and post-mining topography <ult1, ult2, ul3> over a series of long-range mine plans
2. A block table <pit> containing fractional fields indicating the portion of each block contained within the limits of the long-range schedule's pit polygons <per1bf, per2bf, per3bf>, the fraction of the block model lying above the pre-mining topography <topof> and annual pits <ult1f, ult2f>, a field to record the fraction of the block contained within the mining cuts <minef>, the fraction of a block contained within the ore zone <oafrac>, the bench that corresponds to a cut <bench>, and a flag indicating the minable status of a block <dflag>, and a field that will give the waste contents of a cut <bcy> as long as only one polygon intersects the blocks contained in the cut. Additionally, calculated fields for the relation between gold grade and cell color <aucolor>, the fraction of a block contained within the limits of the scheduled pit and pre-mining topography <per1f, per2f, per3f>, and the ore and waste remaining in the block during production scheduling <or, wr>.
3. A polygon table <sched> containing the bench polygons for the long-range pit limits <premine, ult1cont, ult2cont, ult3cont>.
4. ASCII files containing the RPN calculation steps for assigning pen numbers to <aucolor> (aucolor.calc) and determining the minable status of a block <dflag> (dflag3.calc).

Procedure:

1. Assign values for the cell color and flag **Techbase** ⇒ **tbCalc** ⇒ **Setup** ⇒ **Equations**
   Calculation Steps: (f,aucolor.calc)
   ⇒ **eXit**
   ⇒ **Calculate**
   (Repeat Step 1 using the file dflag3.calc)
2. Create a poster of minable blocks to use as a background metafile **Graphics** ⇒ **Poster** (name the metafile <dflag.met> and setup scaling if necessary)
   2.1 Post the midbench polygons for the first period's pit **Cell/poly** ⇒ **Fields**
   Value 1: <ult1cont>
   ⇒ **Style** (accept defaults) ⇒ **Draw** ⇒ **eXit**
   2.2 Restrict posting to uncovered blocks **Database** ⇒ **Add filter**
   Field: <dflag> Relation: <=> Field/Value: <1>
   2.3 Post the flagged blocks **Cell/poly** ⇒ **Fields**
   Value 1: <or> Value 2: <wr> (note that there will only be waste blocks available)
   ⇒ **Style**
   Cell Color: <pit_lev> (color of cell outline will correspond to level)
   ⇒ **Draw**
3. Generate a set of mining cuts.
   3.1 Set up the background **Graphics** ⇒ **polyEdit** ⇒ **Load** ⇒ **background Metafile**
   File Name: <sched.met>
   ⇒ **eXit**
   3.2 Digitize the initial set of mining cuts ⇒ **Edit** (zoom into the area to be mined) ⇒ **Line** ⇒ **Create**
   New Polygon Id: <a,5420,1,1> (ID polygons by bench and number)
   Layer: <1> (polygon visibility can be controlled by using different layers)
   ⇒ **Add** (Digitize the initial sinking cut along the 5420 bench's SE boundary where it meets the pre-mining topography staying within the limits of the available mining blocks. Note that mining much beyond the limits of these blocks will result in the mining of buried ground. Use the posted waste volumes to determine the size of the cut. One week's production is 17,297 bcy per shovel. Attempt to establish a polygon whose perimeter encloses a week's shovel production on a single bench.)
   (Repeat Step 3.2 for polygons 'a,5420,1,2', 'a,5450,1,3', 'b,5450,1,1', 'c,5510,1,1' and 'c,5540,1,2' or as needed.)
   3.3 Save the polygons in a polygon table **eXit** from Edit and **Save** ⇒ **polygon Table**
   Polygon Id: <per1sch> (Note that any modifications of these polygons and subsequent saves will require a filter so that the polygon records are updated rather than appended as a second set of polygons. Using a filter <per1sch = > will ensure record update instead of append.)
4. Intersect the cut polygons with the mining blocks **Modelling** ⇒ **Locate**
   4.1 Setup two filters ⇒ **Database** ⇒ **Add filter** (first group)
   Field: <per1sch> Relation: <()> Field/Value: <5420>
   ⇒ **Add filter** (second group)
   Field: <dflag> Relation: <=> Field/Value: <1>
   Field: <pit_zc> Relation: <=> Field/Value: <5420>
   4.2 Establish the location search parameters and fields and locate blocks ⇒ **Setup** ⇒ **Fields**
   X: <pit_xc> Y: <pit_yc> (no z coordinate)
   ⇒ **Polygon**
   File Name: <@sched.per1sch>
   ⇒ **Assignment**
   Field: <bench> = Field/Value: <8>
   Field: <minef> = Polygon/Vertical Fraction
   ⇒ **eXit** ⇒ **Locate**

(Repeat Steps 4.1 and 4.2 changing the filter values, assignment of the bench value and initializing **Polygon** to locate the blocks contained within the polygons on the other benches.)

5. Check the resulting production volumes **Techbase** ⇒ **Report**
   5.1 Use a filter to include only the located blocks **Database** ⇒ **Add filter**
   FIELD: <minef> RELATION: <<>> FIELD/VALUE: <0>
   5.2 Set up the report's format and contents **Setup** ⇒ **Report file**
   FILE NAME: <week1.rpt>
   ⇒ **Page layout**
   FIELD NAMES DISPLAYED: <y> FIELD VALUES DISPLAYED: <y>
   ⇒ **Fields**
   FIELD AND FORMAT LIST: <bcy (t)>
   ⇒ **eXit**
   5.3 Generate the report and view results **Report** ⇒ **View file**
6. Reset the availability flag **Techbase** ⇒ **tbCalc** (remove any filters) ⇒ **Setup** ⇒ **Equations**
   CALCULATION STEPS: <(f,dflag3.calc)>
   ⇒ **eXit** ⇒ **Calculate**
7. Repeat Step 2 to produce a background metafile for the scheduling of the second week's production.

Example 8.3: Short-term production scheduling using Techbase's opSched
Example 8.2 illustrates the complexity of production scheduling based only on block model intersection, database manipulation and reporting intersection results. While this methodology can be used for scheduling, the need for complex calculated fields and working between several facilities makes the process cumbersome. The greatest drawback of the procedure is that cut definition and reporting of production totals are not combined into a single facility, requiring that the planning engineer work between several programs. **opSched** combines polygon editing, cell posting and production reporting into a single facility, thereby greatly enhancing productivity and simplifying the process. Still, there are some weaknesses.

Only one bench of mining blocks can be displayed at one time. This can be overcome by either displaying the polygon of mining limits on the adjacent benches or by making use of an availability flag <dflag> as in Example 8.2. As mining progresses, it is up to the user to keep track of the extent of production and bench limit polygons on adjacent benches to ensure that production on one bench doesn't violate slope limits or undercut the overlying bench. Judicious use of the display facilities, filters and flagging of available blocks simplifies scheduling, but these facilities are not interactive within **opSched**. For instance, the user must still exit the program to update block availability flagging or the values that are posted in the blocks. The greatest limitation of **opSched** is shared by all trial-and-error based scheduling programs: typically, there are far too many alternative solutions to the production scheduling problem when more than one or two variables are considered. Simplification of the problem from an operational point of view will greatly reduce the number of alternative solutions. For instance, in this example production will commence with three shovels working in stripping overburden. Since the deposit is closest to the surface directly under the eastern side of the 5420 and 5390 benches for the period 1 pit (see Fig. 8.2) where they meet the premining topography, then this is the logical place to

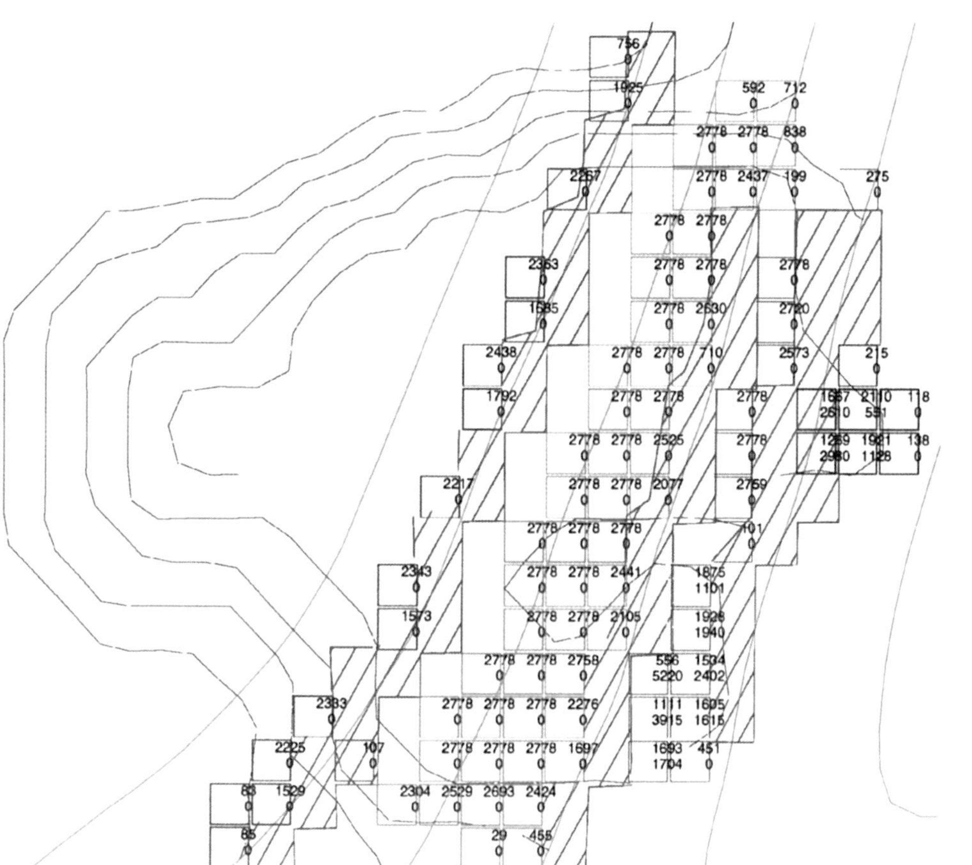

Figure 8.13. Week 2 production schedule cuts (yellow) showing available blocks with contained waste and ore, and polygons of week 1 production (red), premine topography (green) and bench limits (black).

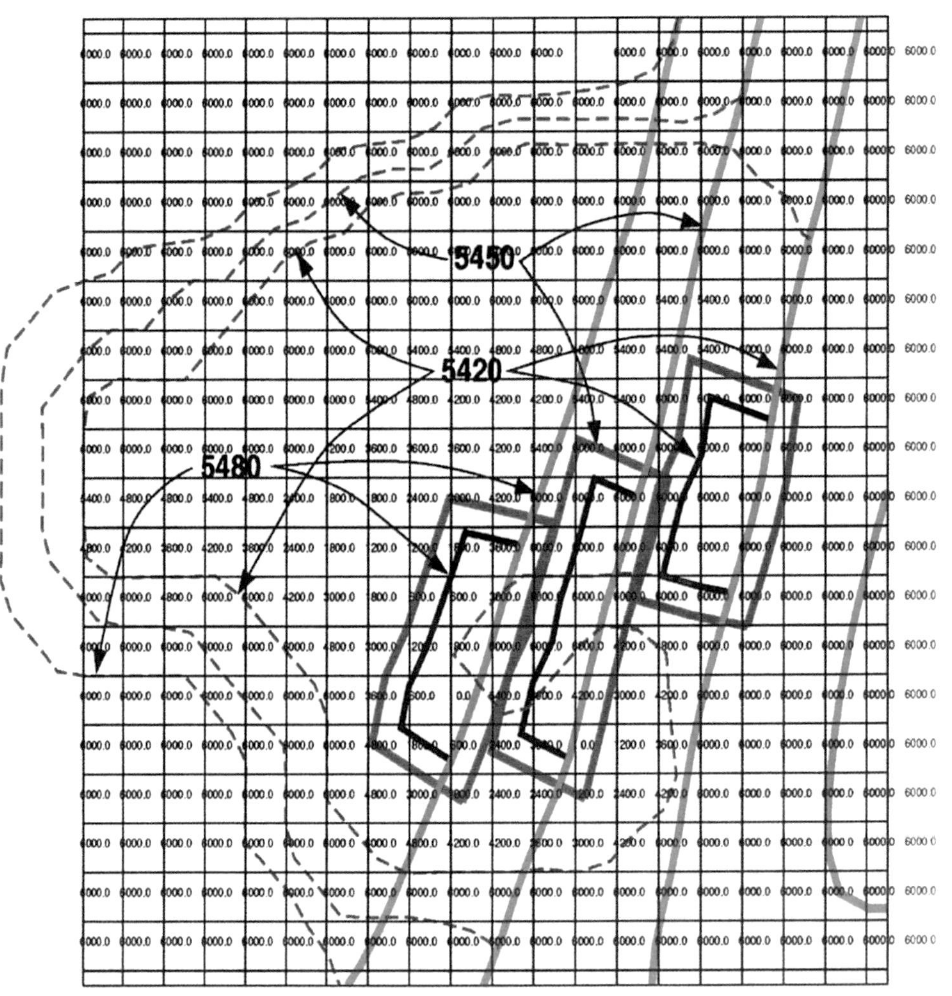

Figure 8.14. Production scheduling based on cut expansion (red) and pit limits (pit brown, topography = green).

commence stripping so as to expose ore blocks as early as possible. Once production has started in this area, it must continue as a series of pushbacks to the western period 1 pit limit. After a few weeks of stripping there will be sufficient ore uncovered for one shovel to remain in ore while the other two remain in waste. In this case, the need to uncover ore in the lower benches will drive the decision on where to locate the two other shovels. Later in the life of this project when there is more ore available than waste, the scheduling problem will be more complex, since at this time there will be two or three active production faces on different benches in the ore. At this time maintaining a stable mill feed along with meeting gold and silver production targets will greatly complicate the production schedule. Obviously, as the mine's size and geologic complexity increases, finding a satisfactory solution to the production scheduling problem will become much more problematic.

Figure 8.10 illustrates a problem that arises during the scheduling of production

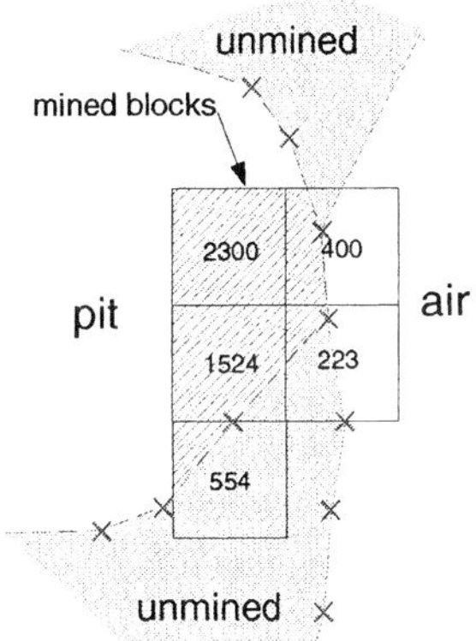

Figure 8.10. The convergence of mined, unmined and air in blocks located during scheduling.

from blocks that intersect the pre-mining topography. These blocks sit above the topography and can consist mostly of air as in the case of the two blocks containing 400 and 223 tons of material. Logically, scheduling would proceed by digitizing the bench limit or following the topographic contour where the bench meets the topographic contour, but even though it is understood that the 400 tons of material lies in the portion of the block that is interior to the pit, computationally it doesn't. Scheduling is based on locating the fraction of a block located within a 2D polygon, shown in Figure 8.10 in hatching. Assume that 60% of the 400 ton block is air, 10% is left unmined and the remainder is included in the cut polygon. Following mining of the cut, the fraction assigned to the block is .3, i.e. 0.3*400 tons from the block will have been mined. The majority of the block's material will remain hanging in midair! The failure to clean up these blocks will result in production cuts with lower tonnage reported than will be encountered during actual production and a thin skin of blocks that are never fully mined. These mostly air blocks can complicate the scheduling process by seeming to continue to cover underlying blocks in the following production periods. The only solution to this problem is to completely encompass these

these partial air blocks by extending the cut polygon beyond the topographical limit of a bench.

Two approaches to production scheduling using **opSched** will be demonstrated in this Example. In the first case (Case A), the methodology of Example 8.2 will be used to control the display of blocks available for mining and the tonnage contained in the blocks following mining. This methodology simplifies the scheduling process but requires the use of complicated field calculations which can only be updated by exiting **opSched**. The second case (Case B) demonstrates the more basic application of **opSched** in which all blocks contained within the bench limits are displayed and the values posted in the blocks are unaffected by mining. In this case it is the scheduler's responsibility to avoid undercutting overlying blocks or mining blocks covered by unmined material.

Case A

For this example, scheduling will be limited to the relatively simple case of stripping waste as demonstrated in Example 8.2. From the main menu select **Open-pit ⇒ opSched ⇒ Database** and add filters to control the display of mining blocks and polygons. In one filter group set <dflag = 1> so that only the uncovered mining blocks are displayed as per Example 8.2. Note that this filter can be removed when the cut has been established so that those blocks that are very near the surface and have very little tonnage can be displayed and cleaned up by adjusting the cut polygon. A second filter group is needed to control the display of scheduling polygons if a polygon table is used as in Example 8.2. In that example, the polygon table <sched> contains both the period 1 bench limits <ult1cont> and the production cuts as defined in Example 8.2 <per1sch>. Add a filter that eliminates any polygons that aren't relevant to <per1sch> (<per1sch ! = 'null'>) and the previous set of production polygons (e.g. <per1sch !) a>).

The results of **opSched** can be recorded as a metafile, so use **Graphics ⇒ Metafile name** and **Scaling** to define the name and size of the metafile. **Highlights** can be used to alter the default settings for displaying polygons.

Under **Load ⇒ Topography** enter the polygon file name for the premining contours <premine.pol> and under **Ultimate** enter the file name for the bench limits at the end of the scheduling period <postmine.pol>. Of course, these two polygon files can have any origin and need not be the original topography and the ultimate pit limits. A common application would be that the two bounding polygon sets for two consecutive long-term production schedules would be loaded. Both sets of polygons should be based on the same contouring intervals corresponding to midbench elevations in the block model <pit_zc>. If a polygon file is also being used for storing the production cuts, enter its name under **Schedule**, but in this example a **polygon Table** is being used for the production cuts. Enter the name of the field in the polygon table that is going to be used to store the production cuts. Since this example starts with the beginning of mining, no scheduling polygons will be loaded, but in later periods the display of the previous period's production cuts will be important. The polygon table filter should be set up so that existing polygons will neither be displayed nor overwritten.

Select **Values** to enter the values to be displayed in the production summary during scheduling. The production summary will list the parameters entered for NET VALUE: <netval_pit> and NET TONS: <pit_waste> under **Fields**, where either a field or constant value can optionally be entered for missing block values and tons that are within the digitized cuts but outside of either the filter or block model. Select **Add class** to define production classes that will be displayed in the production summary. For instance, grade cutoffs or ore types might have to be included in the production summary as part of scheduling. The EXCLUSIVE parameter determines if a block can be included in more than one class. In the case of scheduling based on grade cutoffs in a single ore type the appropriate response would be <yes> since there is only one grade level estimated for a block, but if multiple ore types and grades were estimated into the block model, then the response might be negative. Note that in the initial periods, most of the blocks will have a negative net value. If a class definition was <netval_pit > 0>, then the reported production would be zero for all negatively valued blocks in the polygon. Therefore, in these initial periods at least two classes should be used, one for waste <netval_pit <= 0> and another for ore <netval_pit > 0>. **Delete class** must be used to eliminate classes from the production summary.

Display Setup controls the display of blocks during production scheduling and is the same as for setting up the **Cell/poly** block model display in **Poster** in Example 8.2. The cell color can be set to a calculated color field based on ore grade <aucolor> but in these initial periods when nearly all the blocks are overburden this won't be very helpful. Instead, the cell color can be set to the bench level <pit_lev>. Likewise, the cell fill style can be set to a calculated field representing the ore type.

Under **Fields** enter the block model fields that will be displayed in the block. Initially, there will only be waste to strip and only the calculated waste volume field <wr> from Example 8.2 need be displayed. Select **Cell style** to enter the cell color <pit_lev> to post blocks by the pen number corresponding to their bench, character size <.12> and fill style <inset>. These will have to be modified later when ore is available in the uncovered blocks. **Line style** controls which of the bench and production cuts will be displayed and their boundary line color and style. CURRENT refers to the scheduling polygons on the current bench being displayed. UNMINED and ULTIMATE are the polygons entered under **Load** ⇒ **Topography** and **Ultimate**. PREV BENCH and NEXT BENCH refers to the current scheduling polygons on the benches above and below the current bench being displayed. PREV PERIOD and NEXT PERIOD refers to the display of scheduling polygons on the current bench for the scheduling periods immediately before and after the current period. To avoid confusion, a limited number of these options should be accepted and different line styles and colors should be used for those that will be displayed. Since this example starts with the initial stripping of overburden in scheduling period 1, only the CURRENT, UNMINED and ULTIMATE polygons will be of any use.

Select **Parameters** to enter the current mining period <1>, the bench labelling <middle>, initial bench to be displayed <8> (the lowest bench in this example) and the RECALCULATE and PARTIAL parameter values. Entering 'yes' to RECALCULATE will cause the parameters included in the production summary to be recalculated whenever the production cuts are modified. Otherwise, recalculation must be explicitly selected. PARTIAL controls whether or not fractional blocks are used. If set to 'yes', then when a block is only partially within a cut, its contribution towards the cut will be based on its fractional content. If set to 'no', then any block whose centroid is contained within the polygon will be included fully and exclusively to that cut. Setting PARTIAL to 'no' has advantages during the initial stages of stripping since it allows for the complete removal of the many small volume blocks that intersect the topographic grid to be cleaned up, but the cut boundaries will not correspond exactly to actual production. In this example, start by setting PARTIAL to 'yes' since the back of each cut will be running diagonally across the edges of the mining blocks. With PARTIAL set to 'yes' the contents of the cut will reflect the actual contained tonnage. Note that in terms of scheduling having the axes of the block model oriented roughly parallel to the benches will simplify scheduling since the production cuts will include fewer partial blocks. Eventually, in this example, the benches will be largely parallel to the rows and columns of the block model, but in the initial stages of overburden stripping the cuts must follow the topography which runs diagonally to the blocks.

Report fields controls the manner in which the fields included in the production summary will be reported. In general, totals for the calculated waste volume and ore tonnage remaining in the blocks and the recovery of gold and silver will be of interest, i.e <wr (t) or (t) pitauoz (t) pitagoz (t)>. The production report will have one line for each class.

Select **sChedule** to start the scheduling process by digitizing the initial cuts as per Example 8.2. The display will be similar to that shown in Figure 8.11 in which the period 1 mining limit polygon is displayed in brown, the corresponding premining polyline is green and the available mining blocks are posted with their contained waste volumes. Zoom in on the bench and create the initial cut by selecting **Line** ⇒ **Create** ⇒ **Add** and digitizing the polygon as in Example 8.2 using **Bind** and **Follow**

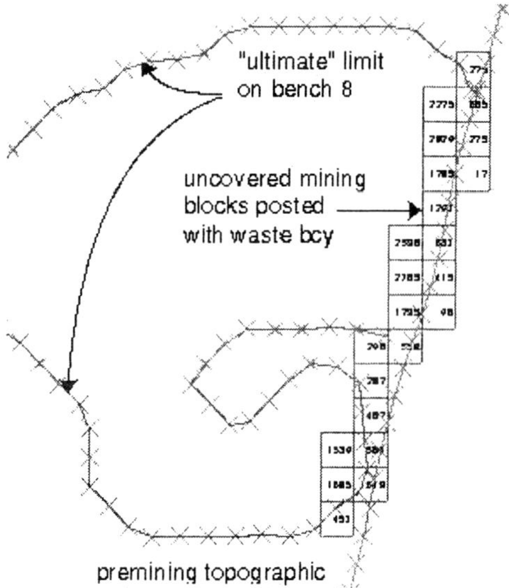

Figure 8.11. Scheduling display of bench 8 limit, midbench topographic polygon and uncovered mining blocks (dflag = 1).

to match the cut boundary to the topography and bench limit. Note that opSched's polygon editing facilities do not prompt for a polygon ID when a new polygon is created. Instead, a default naming convention is used in which the period is concatenated with the bench ID resulting in this case in the polygon ID being '5420,1'. All polygons digitized on the same bench and in the same period will have the same designation. Use **new Id** in the level 2 Line menu to change polygon names for subsequent cuts, i.e '5420,1,A', '5450,1,B' and '5510,1,C' to denote the bench period and shovel. Only one cut will be needed per shovel for this example. If there were more than one polygon on a bench during a period, then individual cut production totals could not be obtained.

The three production cuts for week 1 are shown in Figure 8.12 with corresponding production totals in Table 8.2. Note that the cuts extend beyond the topographic limit and were carefully digitized to contain all of each block. This had to be done to clean up the numerous blocks of small volume which intersect the topography. If the cuts had been bound to the topographic limit, those blocks that extend beyond the topography would have still contained material following the recalculation of <wr> and <or>. Since Techbase has no volume modelling facilities, there is no way to spatially

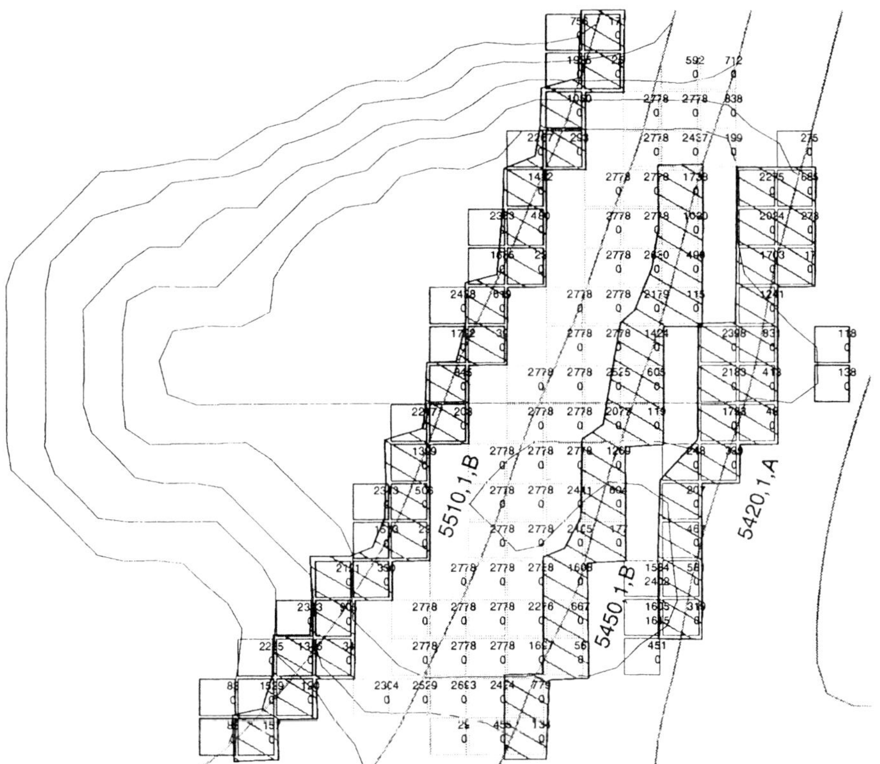

Figure 8.12. Week 1 production schedule cuts (red) showing available blocks with contained waste and ore, and polygons of premine topography (green) and bench limits (black). (Figure in colour, see opposite page 359.)

Table 8.2. Weeks 1 and 2 production totals (no NET_TONS field).

POLYGONS	NET_VALUE	NET_TONS	wr	or	Pitauoz	Pitagoz
Week 1						
5420,1,A	-145503	0.00	18202	317	5797	93985
5450,1,B	-169088	0.00	18001	0	1620	24000
5510,1,C	-223602	0.00	17714	0	0	0
Total	-538193	0.00	53918	317	7417	117985
Week 2						
5420,2,A	743138	0.00	8143	9318	8371	131296
5450,2,B	- 68613	0.00	17795	0	0	0
5510,2,C,1	- 40414	0.00	5221	0	0	0
5510,2,C,2	- 12717	0.00	2868	0	0	0
5510,2,C,3	- 6280	0.00	1497	0	0	0
5510,2,C,4	- 16214	0.00	3741	0	0	0
Total	598700	0.00	39255	9318	8371	131296

discriminate between ore, waste and air within a block. **opSched** uses the **Locate** facilities which determine the fraction of the block contained within the polygon. Even if a block is 90% air and 10% waste, the waste volume essentially is distributed evenly across the block. Therefore, to completely mine these surface blocks the polygon must be digitized completely around the block even if this seems to violate the topographic limit. For blocks that are within the topographic boundary, cuts can be digitized without regard to block limits.

The waste and ore quantities displayed in the cells can be used to approximate the cut tonnage as was done in Example 8.2, but more exact totals can be obtained by returning to the level 1 **Line** menu which includes function keys 4-9 (F4, ..., F9). F4 and F5 moves the display up or down one bench from the currently displayed bench. F6 and F7 change the display and production totals to the previous and following period. F8 toggles the report totals between *Bench*, the total of all polygons on the bench, *Period*, the total of all polygons in the period, *Bench/period*, the total for all polygons in the current bench and period, *Total*, the total for all benches and periods and *Polygon*, the contents of the selected polygon (there may be more than one polygon with the same identity). F9 recalculates the report values and must be used to update the totals if the RECALCULATE parameter is set to 'no'. To report the contents of 5420,1,A use F9 to recalculate and F8 to toggle to *Polygon*. The report will consist of one line with a listing of the field values defined in **Report** under the **display sEtup** menu (as opposed to Report in the level 1 **sChedule** menu). The list is repeated in the line for each class.

The production requirement for one shovel during the week should be obtained with as little displacement of the production equipment as possible. Generally, this means that the shovel should be kept in the same cut on the same bench until maintenance of the pit slope, stripping of waste or ore blending necessitates a move. To adhere to this principle, the 5420 cut shown in Figure 8.12 is continuous and extends through the neck of waste left unmined in Examples 8.1 and 8.2. Similarly, digitize cuts on 5450 and 5510 using F8 and F9 recalculate the contained waste until the week's production is approximately met. Select **Report** to output the production summary to the output file (opsched.out) as shown in Table 8.2. **eXit** the scheduling menu and select **Save** ⇒ **polygon File** to save the polygons into an ASCII polygon file or select **polygon Table**.

At this point **opSched** must be exited in order to update the scheduling polygon table with the values recorded in opsched.out and the fields <wr>, <or> and <dflag>. As in Example 8.2, the calculated fields <wr> and <or> are based on <minef>, the fraction mined. The fraction mined is assigned using **Locate**. The availability flag <dflag> is based on <wr> and <or> and is assigned values using **tbCalc** and the calculation file field dflag3.calc.

To update the polygon table with the production from week 1, create fields in the table to hold values for the cut netvalue <cutvalue> and contained waste <cutwr>, ore <cutor>, gold <cutau> and silver <cutag>. Modify the scheduling report (opsched.out) reducing it to the three lines of polygon statistics and then **Load** this into the polygon table setting UPDATE TYPE in the **Parameters** menu to <overlay> and using three database filters (e.g. <per1sch = 5420,1,A>) to restrict loading to the three week 1 cuts.

Next, update the cut fraction <minef> and bench mined <bench> fields. From the main Techbase menu select **Model** ⇒ **Locate** ⇒ **Database** ⇒ **Add filter** and set a filter on the block model so that <minef> will only be assigned for blocks on bench 8 (pit_lev = 8) and calculated for the polygon on that bench (per1sch = 5420,1,A). Select **Setup** ⇒ **Fields** entering the block *X* <pit_xc> and *Y* <pit_yc> coordinate fields. The *Z* coordinate is only entered when elevations or grids are used for locating the block. Under **Polygon** provide the polygon table and identity field <@sched.per1sch>. Select **Assignment** to set the bench associated with the cut (bench = 8) and the POLYGON/VERTICAL FRACTION <minef>. **Locate** determines the fraction of the block model mined on bench 8. Repeat this procedure for the two other cuts by modifying the two filter groups, initializing the polygon table by again selecting **Polygon** and changing the bench field value under **Assignment**. As in Example 8.2, this one bench at a time methodology must be used to limit cell location to the correct bench. The cut polygons could be converted to 3D polygons, but in the current version of Techbase (Version 2.52) 3D polygons have not been implemented in **Locate**.

With the calculated fields <wr> and <or> updated the availability flag <dflag> can be updated. Select **Techbase** ⇒ **tbCalc** (delete all block model filters) ⇒ **Setup** ⇒ **Equations** and enter the file containing the RPN calculation steps <dflag3.calc>. Selecting **Calculate** updates the availability flag.

Production scheduling can now continue using **opSched**. A new set of cells will be available and the ore and waste values in the remaining blocks will reflect the mining that took place in the previous week. The first week's cuts can be loaded from the polygon table and used as the starting points for a new set of polygons. If a sufficient number of blocks are exposed, production can continue for more than one period. Remember to set the MINING PERIOD parameter to 2 for the second set of cuts. Figure 8.13 shows a set of polygons that could be used for the second week of production. Table 8.2 gives the corresponding production.

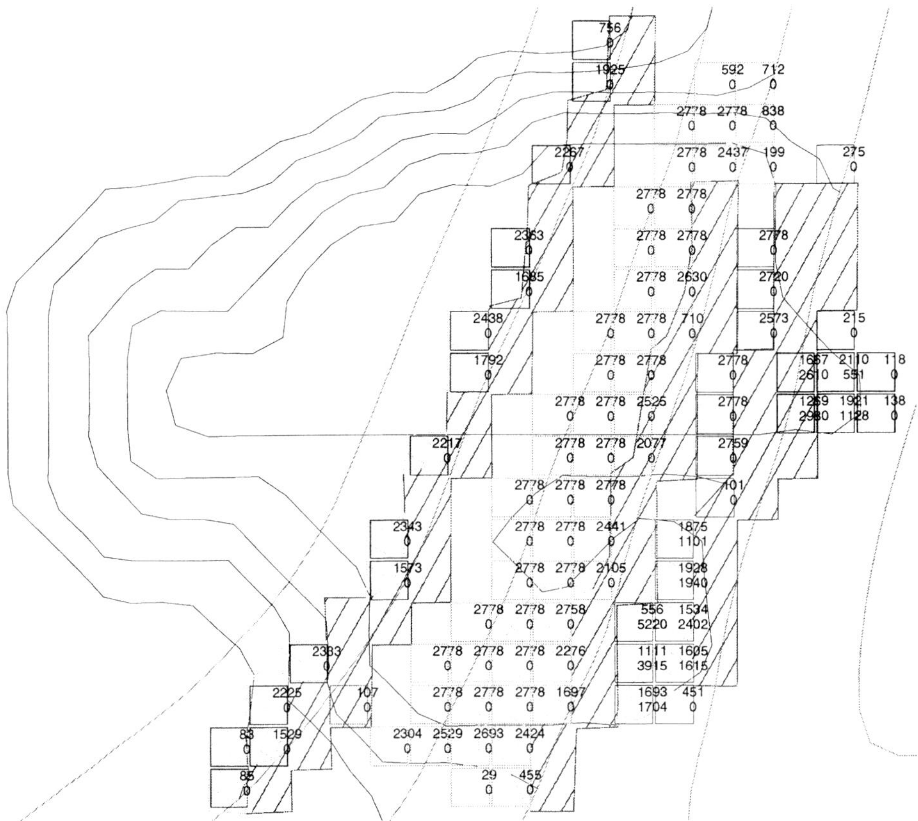

Figure 8.13. Week 2 production schedule cuts (yellow) showing available blocks with contained waste and ore, and polygons of week 1 production (red), premine topography (green) and bench limits (black). (Figure in colour, see opposite page 374.)

Case B

Again execute **opSched** as in Case A, but without the database filter. **Load** the **Topography** <premine> and **Ultimate** <postmine> polygon files. Under **Values** ⇒ **Fields** enter the netvalue field <netval_pit> and use **Add class** to display all cells containing waste <pit_waste > 0>. Select **display Setup** and enter the **Field** to be displayed in the cell <pit_waste>, the same **Cell style**, **Line style** and **Parameters** as in Case A. For **Report** use non-calculated fields <pittons (t) pit_waste (t) pitauoz (t) pitagoz (t)>. Select **Schedule** to digitize the first cut on bench eight as in Case A. The production goal is now based on waste tonnage <pit_waste> rather than bcy <wr> as in Case A so the weekly production goal for one shovel will be 37,362 bcy.

Note that without the availability flag all blocks on the bench are displayed and the tonnage of waste is the same in all blocks. In this situation, the scheduler must rely on the pit limit polygon in laying out the cuts. Digitize the first polygon on the bench as per Case A, but this time remain within the topographic contour: the mining of partial air blocks is not a concern since <pit_waste> is not intersected with either the pit limits or surface grid as were <wr> and <or>.

Once the cut has been digitized on bench 8 it must be expanded to the overlying benches. The expanded polygon can be used to identify blocks that must be mined in order to maintain stripping requirements. From the level 1 opsched menu select **Line ⇒ Advance up** and enter a polygon expansion distance <30>. This is the horizontal toe to crest distance that will be covered over the height of one bench given the final slope angle (45°). The display is now changed to bench 7 and shows the new polygon (5450,1) which encompasses the area that must be free of overburden for the underlying cut (5420,1) to be mined. Checking this against the location of the pit limit on bench 7 shows that 5420,1 is overlaid only by air blocks and that 5450, 1 can be deleted. Repeating this exercise for 5450 and 5480 shows that there will be no problem with maintaining the working pit slope in the initial period due to the low angle of the pre-mining surface. The three production cuts and their expanded polygons are shown in Figure 8.14. At later stages of production, as the pit starts to take shape, cut expansion will have to be used to guide the location of cuts.

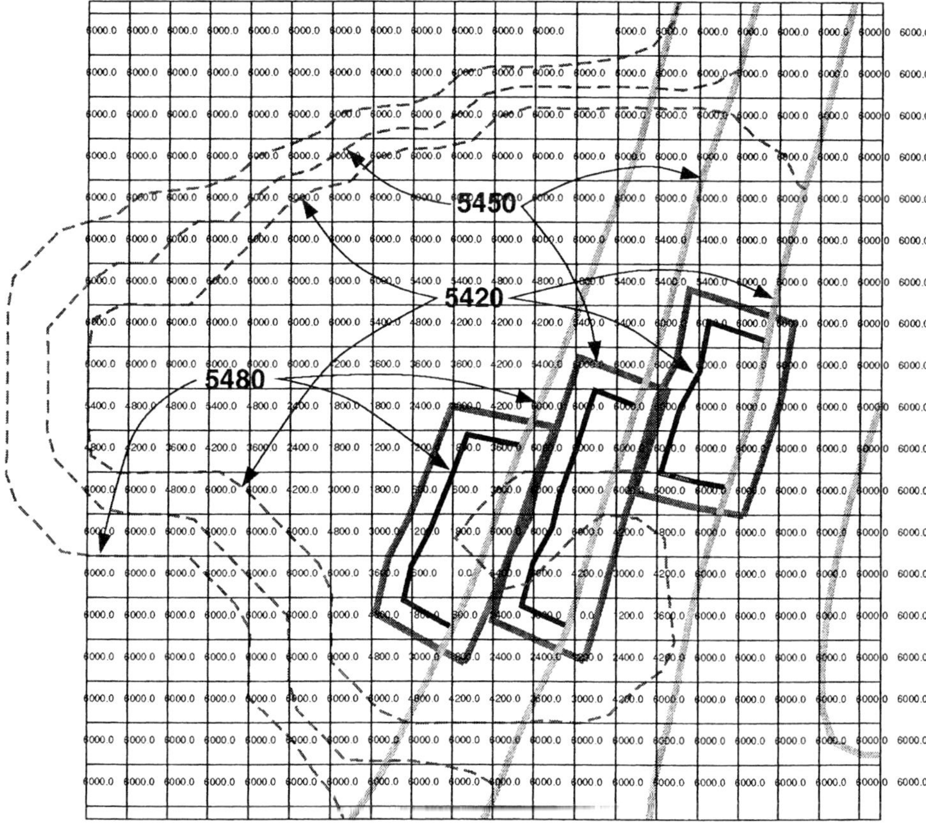

Figure 8.14. Production scheduling based on cut expansion (red) and pit limits (pit = brown, topography = green). (Figure in colour, see opposite page 375.)

Summary of procedure: Example 8.3

In this example, two approaches will be given for production scheduling. In the first, Case A, Steps 1-9, the methodology of Example 8.2 will be combined with the facilities of **opSched**. This will limit the cell display and volumetric calculations to those blocks which are available for mining in the period and uses fields which can be updated to reflect the contents of the blocks following mining activity in a previous period. Case B presents a simplified methodology that dispenses with block availability flagging, mining of fractional blocks and the use of complex calculated fields.

Initial data requirements: polygon files of pre-mining <premine.pol> and post-mining <post-mine.pol> contours; a polygon table <sched> with a field for storing cut polygons <per1sch>; a block model <pit> with fields for netvalue <netval_pit>, net block volume or tonnage <bcy>, waste volume <wr>, ore tonnage <or> and grades <pitauoz>, and, optionally, a flag to indicate block availability for mining <dflag> as in Example 8.2.

Procedure: Case A

1. Set filters to limit the polygons and blocks displayed **Open-pit ⇒ opscheD ⇒ Database ⇒ Add filter**
   FIELD: <dflag> RELATION: <=> FIELD/VALUE: <1>
   ⇒ **Add filter** (second filter group for the polygon table, optional if tables not being used to store scheduled cuts)
   FIELD: <per1sch> RELATION: <!=> FIELD/VALUE: <> (eliminates polygons having other field identities whose values under <per1sch> will be null)
   FIELD: <per1sch> RELATION: <!)> FIELD/VALUE: <a>
   FIELD: <per1sch> RELATION: <!)> FIELD/VALUE: <b>
   FIELD: <per1sch> RELATION: <!)> FIELD/VALUE: <c> (avoids display of cut polygons for shovel a, b and c generated in Example 8.2)
2. Provide a metafile name and scaling factors so that metafiles of production limits can be generated during scheduling.
3. Load polygons **Load ⇒ Topography**
   FILE NAME: <premine> (can be either topographic contours or bench limits of previous mining period)
   ⇒ **Ultimate**
   FILE NAME: <postmine>
   ⇒ **polygon Table** (use **Schedule** if cut polygons are stored in an ASCII file)
   POLYGON ID: <per1sch> (if an existing set of cuts are being displayed provide distinctive display parameters)
4. Set the database fields to be used for block values in the scheduler and define classifications to be used in reporting those fields during scheduling ⇒ **Values**.
   4.1 Fields to be used for production report ⇒ **Fields**
   NET VALUE: <netval_pit> NET TONS: <bcy> (assign a constant for MISSING VALUE and MISSING TONS for blocks which might be included in the cut limits but have null field values due to the filter or because of being external to the ultimate pit during field value assignment)
   4.2 Define reporting classifications ⇒ **Add class**
   TITLE: <waste blocks> EXCLUSIVE: <n> (negatively valued block total – blocks can be in multiple report classes)
   FIELD: <netval_pit> RELATION: <<=> FIELD/VALUE: <0>
   ⇒ **Add class** (a second class to report ore tons in positively valued blocks)
   TITLE: <ore blocks> EXCLUSIVE: <n>
   FIELD: <netval_pit> RELATION: <>> FIELD/VALUE: <0>
   (Additional lines can be added to make the classes more restrictive as in using database filters. Use **Delete class** to drop report classes.)

⇒ **eXit**

5. Define the block model display parameters ⇒ **display sEtup**
   5.1 List block fields to display in cells ⇒ **Fields**
   VALUE 1: <wr> VALUE 2: <or> (if value plots to right of cell limit max. value in field definition)
   5.2 Define cell display ⇒ **Cell style**
   CELL COLOR: <pit_lev> (pen equals bench)
   CHAR. SIZE: <.12> (sufficiently small to stack values inside a cell)
   FILE STYLE: <inset> (can use a calculated field for ore type, e.g. oxide versus sulfide)
   5.3 Set polygons to be displayed and their style ⇒ **Line style**
   CURRENT COLOR: <black> LINE STYLE: <solid>
   UNMINED: <y> COLOR: <green> LINE STYLE: <solid>
   (polygon from **Topography** polygon file whose ID matches bench elevation)
   ULTIMATE: <y> COLOR: <brown> LINE STYLE: <solid>
   (polygon from **Ultimate** polygon file whose ID matches bench elevation)
   5.4 Set parameters for initial display and calculation of polygon totals ⇒ **Parameters**
   MINING PERIOD: <1> BENCH: <8>
   BENCH LABEL TYPE: <number>
   RECALCULATE: <y> PARTIAL: <y> (automatic report updating and partial block values)
   5.5 List fields and formats for interactive production reporting ⇒ **Report fields**
   FIELD AND FORMAT LIST: <wr (t) or (t) bcy (t) pitauoz (t)> (Note that the totals given in the production report will depend on the report class definitions given in Step 4.2 and therefore might not be in agreement with the NET VALUE and NET TONS fields.)
   ⇒ **eXit**
6. Use polygon editing facilities to define production cuts ⇒ **Schedule** (zoom in on the exposed blocks)
   6.1 Create the initial cut interior to the pre-mining contour and the post-mining contour ⇒ **Line** ⇒ **Create** ⇒ **Add** (Don't mine beyond the brown period 1 mining limit. Close the polygon when finished.) ⇒ **Close**.
   ⇒ **eXit** ⇒ **eXit** (to the level one menu which includes F8 to F9 options) ⇒ **F8** (toggle to polygon totals) ⇒ **F9** (explicitly recalculate the production report)
   6.2 Modify the polygon until approximately 17900 bcy of waste had been mined **Line** ⇒ **Modify** (use **Move** and **Insert** and F9 to check total).
   Repeat Steps 6.1 and 6.2 for all the cuts required for one week's production.
   6.3 Report production totals to the output file from the level 1 menu ⇒ **Report**
   6.4 Save the polygons **Save** ⇒ **polygon File** or **polygon Table**. In the case of using a polygon table
   POLYGON ID: <per1sch>
7. Load the output file containing the production summary into the polygon table.
   7.1 Edit the ASCII output file opsched.out so that it only contains the polygon data (as shown in Table 8.2) and store as a new file, schweek1.rpt, to avoid overwriting later on.
   7.2 Define fields to hold values for waste production **Techbase** ⇒ **Define** ⇒ **Fields** ⇒ **auto-Table**
   TABLE NAME: <sched>
   ⇒ **Create**
   FIELD NAME: <cutvalue> TYPE: <real> CLASS: <actual>
   (Repeat Step 7.2 for the contained waste <cutwr>, ore <cutor>, gold <cutau> and silver <cutag>.)
   7.3 Load the edited output file into the polygon table **Techbase** ⇒ **Load** ⇒ **Database** ⇒ **Add filter**
   FIELD: <per1sch> RELATION: <=> FIELD/VALUE: <5420,1,A>
   (Repeat Step 7.3 adding filters for 5450,1,B and 5510,1,C – 3 filter groups.)

⇒ **Setup** ⇒ **Data file**
FILE NAME: <schweek1.rpt> SKIP: <3>
⇒ **Fields**
FIELD AND FORMAT LIST: <perlsch cutvalue (10,10) cutwr (35,10) cutor (45,10) cutau (55,10) cutag (65,10)
⇒ **Parameters**
UPDATE TYPE: <overlay>
⇒ **Input filter**
FIELD: <perlsch> RELATION: <!=> FIELD/VALUE: <TOTAL>
⇒ **eXit** ⇒ **Load**

8. Update fractional value mined and related calculated fields.
   8.1 Set filters for restricting polygons and benches **Modelling** ⇒ **Locate** ⇒ **Database** ⇒ **Add filter**
   FIELD: <perlsch> RELATION: <=> FIELD/VALUE: <5420,1,A>
   ⇒ **Add filter** (second filter group)
   FIELD: <pit_lev> RELATION: <=> FIELD/VALUE: <8>
   8.2 Define location parameters **Setup** ⇒ **Fields**
   COORDINATE FIELDDSX: <pit_xc> Y: <pit_yc> (only enter x and y for a 2D polygon)
   ⇒ **Polygon**
   FILE NAME: <@sched.perlsch>
   ⇒ **Assignment**
   FIELD: <bench>= FIELD/VALUE: <8>
   FIELD: <minef> = POLYGON/VERTICAL FRACTION
   ⇒ **eXit** ⇒ **Locate**
   (Repeat Steps 8.1 and 8.2 changing the filters for the appropriate bench polygon combination.)
   8.3 Recalculate the availability flag <dflag3> as per Example 8.2 Step 6.
9. Repeat Steps 1-8 for the second week's production, but now using the first week's cuts as the basis of the second week's using **Bind** and **Follow**.

Procedure: Case B

1. Repeat Step 3, Case A except for loading the polygon table.
2. Repeat Step 4, Case A.
3. Repeat Step 5, Case A as follows:
   3.1 List block fields to display in cells ⇒ **display sEtup** ⇒ **Fields**
   VALUE 1: <pittons> VALUE 2: <pit_waste>
   3.2 Define cell display ⇒ **Cell style** (as per Step 5.2, Case A)
   3.3 Set polygons to be displayed and their style ⇒ **Line style** (as per Step 5.3, Case A)
   3.4 Set parameters for initial display and calculation of polygon totals ⇒ **Parameters** (as per Step 5.3, Case A)
   3.5 List fields and formats for interactive production reporting ⇒ **Report fields**
   FIELD AND FORMAT LIST: <pit_waste (t) pittons (t) pitauoz (t)>
4. Repeat Step 6, Case A.
5. Check the limits of the 5420,1 cut expanded on the 5440 bench.
   5.1 From the level 1 **Schedule** menu **Line** ⇒ **Advance up**
   EXPANSION DISTANCE: <30> (5450,1 is displayed on bench 7)
   (**Delete** 5450,1 if in air or **Modify** to match topography if stripping is required.)
   5.2 Repeat Steps 6.3 and 6.4, Case A.
6. Repeat Step 7, Case A.

APPENDIX 1

Project databases

Five project databases are used to illustrate the subject matter for this text: the Boland Banya, Pasir, and May Day projects for the Techbase examples and the Tutorial and Smoot projects for the Lynx examples. Of these, the Boland and Tutorial projects are used for the bulk of the examples and will be covered in the greatest detail in this appendix.

A1.1 USING THE PROJECT DATABASES

In most cases, the project databases exactly duplicate the structure and names used in the examples, most notably in the Summary of procedure sections which are design to provide a step-by-step template for duplication of the examples and application to other data sets. As such, the databases are provided only for reference sake: the ASCII files provided for the Boland and Tutorial projects are all that is needed to complete all the relevant examples in this text. Do not use the original databases provided with this text to repeat the examples as this will most likely corrupt the databases with the addition of numerous near duplicate tables and fields. Instead, make copies of the projects and use these for experimentation and practice.

Some of the examples, such as the use of the PNL well logs of Example 4.9, do not involve significant modification of the original database. This is also the case for the use of the Pasir database in Example 6.6 and the May Day database in Example 7.9. In these cases, the original files can be used to run through the examples, but use of the same naming conventions will not work due to duplication. For instance, in the Summary of procedure for Example 7.9 the polygons tables level2, level2_e and level2_v are created with the field all2. These fields already exist in the May Day database. Techbase will not allow the creation of duplicate table and field names and will report an error if the original database is used along with the same names. Thus, either the original names must be changed, different names used or the database must be built from the ground up.

For minor projects that are used only once or twice, the user can use alternative names when creating data structures. For the two main projects, this should only be done on copies of the original databases so that the originals will retain their value as reference sources. The following guidelines should be used copying databases and renaming data structures in Techbase and Lynx. Also, see the material in Chapter 2, especially Examples 2.1, 2.2, and 2.4.

A1.1.1 *Suggestions for using Techbase databases*

Prior to following the examples in the text, the project databases should be transferred into a duplicate database so that the original is preserved. Database transfer (**Techbase** ⇒ **Transfer** ⇒ **Put**) allows the definitions and records of a database to be replicated into a new database. **Transfer** also provides a means of breaking up large database files into more easily handled files of smaller size. Under **Output file** the new transfer format files naming convention and maximum size is specified. **Tables** is used to specify which tables will be transferred. The default is to transfer all tables, fields and records. Otherwise, only those tables listed will have their records transferred, while tables not included in the list will only have their definitions transferred. To transfer only definitions and records of tables included in the list include the (–) option anywhere in the table list.

A good practice is to create database versions that only contain the information required for the exercise. For instance, to carry out Example 4.1, length compositing, only three flat tables are needed from the Boland database: collars, surveys and assays. These three flat tables can be transferred with the (–) option and then rebuilt (**Techbase ⇒ Transfer ⇒ Get**) into a new database. This new database can then be used to exactly replicate the Summary of Procedure by creating a flat table for the composite data using the same table and field names.

A1.1.2 *Suggestions for using Lynx databases*

The same strategy for using Techbase databases should be followed for the Lynx examples in that copies of the database should be generated that only have the data necessary for carrying out the exercises.

The procedure for copying a Lynx database requires that the new duplicate database be first created (**File ⇒ Project create**). Example 2.4 details the procedure used for creating and copying databases. To copy files into the new database from the source project select **File ⇒ Project copy** to bring up the Project Copy entry form. From here the source project along with the desired data can copied into the new project.

In Techbase, all the data is imbedded into the database and cannot be separated from the database except by using the Transfer or Report facilities. Note that data elements in Lynx exist as distinct directories and files (see description of Tutorial project in Section 4.2.4) that can be copied or moved from Unix. Thus, an accomplished Unix user may find it easier to generate example databases without using the Project Copy facility.

A1.2 PROJECT DESCRIPTIONS

The following sections provide some background to the projects used as examples in this text and an overview of their contents.

A1.2.1 *Boland Banya*

The Boland database is used for the majority of the Techbase examples. Boland is a Nevada gold deposit with a volcanic setting. There are two ore zones, designated OA and OB, with OA being the main deposit. OA lies close to the surface, nearly outcropping on the western side of the properties' northeast trending valley. Both deposits are flat laying and lensoidal in shape. Gold and silver values are very high and the resulting pit removes nearly all the OA deposit.

ASCII data files for the project include separate topographic map data and assay, collar and survey drillholes files (topo.dat, assays.dat, collar.dat and survey.dat). The Techbase database, Boland, includes flat tables corresponding to each of these data files. Other flat tables are used to hold the value composited top and bottom surface locations for the ore zones (OAtop, OAbotm, OBtop and OBbotm) and length composites of assays (OAcomp and OBcomp). Estimated surfaces for the topography, deposit and mine limits are held in a single cell table (surfaces). Various block models are used to hold grade estimates, volumes of intersection and block netvalues (pit). Polygon tables are used to hold bench and cut boundaries for scheduling (sched_*).

A1.2.2 *Smoot*

The Smoot database is derived from a PNNL data set of well logs. Four lines of wells were arranged on a star pattern, logged for sediment type and measured for moisture content using a neutron probe. The original study sought to correlate moisture level with lithologic zones. Lynx was used for geologic modelling of the sediment layers and kriging of moisture levels as a function of sediment type. The data set was also used for a demonstration of generating experimental variograms in Example 3.5.

A1.2.3 *Pasir*

The Pasir data set is derived from an Indonesian surface coal mine. Only two of the numerous steep and highly folded coal seams are included in the database. While the database includes a set of drillholes, these were not used to define any geologic structures. Instead, a set of digitized plan view sections showing the seams were used to locate those blocks that had coal. A variable block model was used to help capture in better detail the steeply dipping coal seams on both wings of a major anticline. This database provides an example of how Techbase can be used to handle volumetric calculations for very steep and complex coal deposits.

A1.2.4 *Tutorial*

This forms the basis for the majority of the Lynx examples. The deposit is a steeply dipping polymetallic vein from six up to one hundred meters thick, but with an average thickness between 20 to 30 m. There are two ore zones, zone7 and zone8, referred to as the primary and secondary ore zones. The deposit strikes nearly NWW and dips at around 45° with little noticeable plunge and outcrops for most of its length continuing to a depth of about 1400 m. The surface elevation slopes downwards to the north and east from an elevation of 1680 m down to 1540 m over a distance of 600 m.

Project organization in Lynx is defined by the directory structure, which for all projects is

```
3D/       designs/      geostats/   models/    sessions/
GTAB.Z       dh_prj.INDX.Z images/     overlays/   tmp/
PRJCT.Z      dh_prj.Z     maps/      plots/     wave/
avs/      dholes/     misc/       reports/    wavefront/
```

in the appropriate project subdirectory (e.g. default hpux installation /apps/lynx/projects/TUTORIAL).

/3D holds 3D or volume models of geologic structures and mine openings. The key models include: G.ZZ*, G.OR* and G.GEOL*, which are versions of zones 7 and 8; Underground mine development openings, M.P1*; stopes, M.P2* and open pit volume models, M.PIT*. G.ORST* contains intersection volume components between the stopes and geologic model.

/designs contains design files for various open pit configurations.

/geostats holds data file derived from either a drillhole subset or residuals from trend surface fitting. The file piezo.DAT is used to demonstrate variography and trend surfaces in Example 3.6.

/misc holds miscellaneous files, mainly ASCII data files for map and drillholes data import/export, other data sets, AWK data processing scripts, and the like.

/models contains 3D block models, 2D map conversions into grids and pos files resulting from pit optimization.

/sessions holds various session files that are used for loading commonly used viewplane orientations and background displays.

/overlays contains the map data files.

/dholes contains the drillhole data files

APPENDIX 2

Techbase database automatic fields

The following is a list of the AUTOMATIC field suffixes and their functions:

_are = area of the POLYGON
_col = column number for CELL, BLOCK or LAYER
_csz = column size for CELL, BLOCK or LAYER
_edg = points to the first edge of the POLYGON
_lay = layer number for LAYER
_lev = level number for BLOCK
_lsz = level size for BLOCK
_nam = layer name (text) for LAYER
_npt = number of points in the POLYGON
_nul = null (missing) value
_nxt = next point of the EDGE
_per = perimeter of the POLYGON
_rec = record number
_row = row number for CELL, BLOCK or LAYER
_rsz = row size for CELL, BLOCK or LAYER
_v1 = first vertex number of the EDGE
_v2 = adjoining VERTEX number of the EDGE
_xc = X coordinate of the center of the current CELL, BLOCK, LAYER, POLYGON, or VERTEX
_xc1 = first *X* coordinate of the line segment of the EDGE
_xc2 = second *X* coordinate of the line segment of the EDGE
_xmn = minimum *X* coordinate of the center of the POLYGON
_xmx = maximum *X* coordinate of the center of the POLYGON
_yc = *Y* coordinate of the center of the current CELL, BLOCK, LAYER or POLYGON
_yc1 = first *Y* coordinate of the line segment of the EDGE
_yc2 = second *Y* coordinate of the line segment of the EDGE
_ymn = minimum *Y* coordinate of the center of the POLYGON
_ymx = maximum *Y* coordinate of the center of the POLYGON
_zc = *Z* coordinate of the center of the current BLOCK
_zc1 = first *Z* coordinate of the line segment of the EDGE
_zc2 = second *Z* coordinate of the line segment of the EDGE

APPENDIX 3

Techbase value settings

VALUE LISTS

Throughout TECHBASE, various menu *fields* request that a value be entered from a list of legal values, for example the color for a plot title. Rather than list all the values for each menu *field*, the manual notation '⇒**Listname**' has been adopted. When this appears, refer to this page for the full list of legal values.

⇒**Colors** – Colors are entered by a name or by a number. The name may be abbreviated, usually to a single letter. Exceptions are BLUe and BLAck, which require three letters each, while BRown requires two. In some TECHBASE programs, the color may be specified by number using an INTEGER database field. Note that the number next to the name is the pen number used in pen plotters.

BACKGR	0 (Background color)
BLACK	1
RED	2
GREEN	3
BLUE	4
PURPLE	5
ORANGE	6
YELLOW	7
BROWN	8
nnn	Specify color by number (0-4000+)

⇒**Fillstyles** – Fillstyles are entered by name or by a number. The name may be abbreviated, usually to a single character. The numbers from 0 to 3 are predefined, and numbers from 4 to 255 may be defined using the *MAKEPATT* program in TAP-Graph-II. Pattern number 0 is actually a request to not fill. In some TECHBASE programs, the fillstyle may be specified by number using an INTEGER database field.

BLANK	0 (no fill)
HOLLOW	1 (outline)
INSET	2 (outline slightly inside original polygon)
SOLID	3 (filled)
nnn	0-255 (fill with pattern number 0-255)

⇒**Fonts** – Fonts are entered by number. Font number 1 is the default and may vary in appearance by device. Fonts 3-15 are available with TAP-GraphII, (see Fig. A3.1).

1 Default	6 Italic - duplex small	11 Gothic German
2 Roman - simplex	7 Italic - triplex	12 Gothic - Italian
3 Roman - duplex - small	8 Script - simplex	13 Greek - simplex
4 Roman - duplex	9 Script - duplex	14 Greek - duplex
5 Roman - triplex	10 Gothic - English	15 Cyrillic - Russian

⇒**Gridstyles** – Gridstyles are entered by name. The name may be abbreviated to a single character.

LINES
CROSSES
TICKS

⇒**Linestyles** – Linestyles are entered by name or by a number (see Fig. A3.2). The name may be abbreviated, usually to a single character. Note that a linestyle of BLANK is actually a request not to draw the line, and may not be appropriate in all cases.

BLANK	0 (no line)
SOLID	1
DOTTED	2
MIXED	3 (dot-dashed)
DASHED	4
WIDE	5
nnn	0-255

⇒**Locations** – Locations for text (relative to a point) are entered by name or by a number. The name may be abbreviated to the first two characters. The origin for the text will be at a distance of one half the text size plus one half the marker size, in the direction indicated, (see Fig. A3.3).

TOP	1
LEFT	2
RIGHT	3
BOTTOM	4
RCENTER	5 Relative Center
RTOP	6 Relative Top
RLEFT	7 Relative Left
RRIGHT	8 Relative Right
RBOTTOM	9 Relative Bottom

⇒**Markers** – Markers are entered by number (see Fig. A3.4). The numbers from 0 to 10 are predefined, and numbers from 11 to 255 may be defined using the *MAKEMARK* program. Marker number 0 is actually a request not to plot a marker. In some TECHBASE programs, the marker may be specified by number using an INTEGER database field. The predefined markers are:

0 No marker	6 Octagon with Plus
1 Square	7 Octagon with X
2 Plus	8 Square with Plus
3 Asterisk	9 Square with X
4 Octagon	10 Triangle
5 X	nn 0-255

⇒**Relations** – Relations between values are entered by symbol.

<	Less Than
>	Greater Than
=	Equal To
<=	Less Than or Equal To
>=	Greater Than or Equal To
!=	Not Equal To
()	Contains Substring
!)	Does NOT Contain Substring

⇒**Axistypes** – Axistypes are entered by name. The name may be abbreviated, usually to a single letter. Exceptions are LInear and LOgarithmic, which require two letters.

Logarithmic
Linear
Date
Time
None
Blank

Note: NONE prints a line with no annotation, and BLANK prints no line and no annotation.
⇒**Logtypes** – Logtypes are entered by name. The name may be abbreviated, usually to a single letter, (see Fig. A3.5).

None	
TIcks	
Halfticks	
Bar	RBbar
TRace	RTrace
Markers	RMarkers

1 ABCDEFGHIJKLMNOPQRSTUVWXYZ1234567890
2 ABCDEFGHIJKLMNOPQRSTUVWXYZ1234567890
3 ABCDEFGHIJKLMNOPQRSTUVWXYZ1234567890
4 ABCDEFGHIJKLMNOPQRSTUVWXYZ1234567890
5 ABCDEFGHIJKLMNOPQRSTUVWXYZ1234567890
6 *ABCDEFGHIJKLMNOPQRSTUVWXYZ1234567890*
7 *ABCDEFGHIJKLMNOPQRSTUVWXYZ1234567890*
8 *ABCDEFGHIJKLMNOPQRSTUVWXYZ1234567890*
9 *ABCDEFGHIJKLMNOPQRSTUVWXYZ1234567890*
10 ABCDEFGHIJKLMNOPQRSTUVWXYZ1234567890
11 ABCDEFGHIJKLMNOPQRSTUVWXYZ1234567890
12 ABCDEFGHIJKLMNOPQRSTUVWXYZ1234567890
13 ΑΒΓΔΕΖΗΘΙΚΛΜΝΞΟΠΡΣΤΥΦΧΨΩ 1234567890
14 ΑΒΓΔΕΖΗΘΙΚΛΜΝΞΟΠΡΣΤΥΦΧΨΩ 1234567890
15 АБВГДЕЖЗИЙКЛМНОПРСТУФХЦЧШЩ1234567890

Figure A3.1. Fonts.

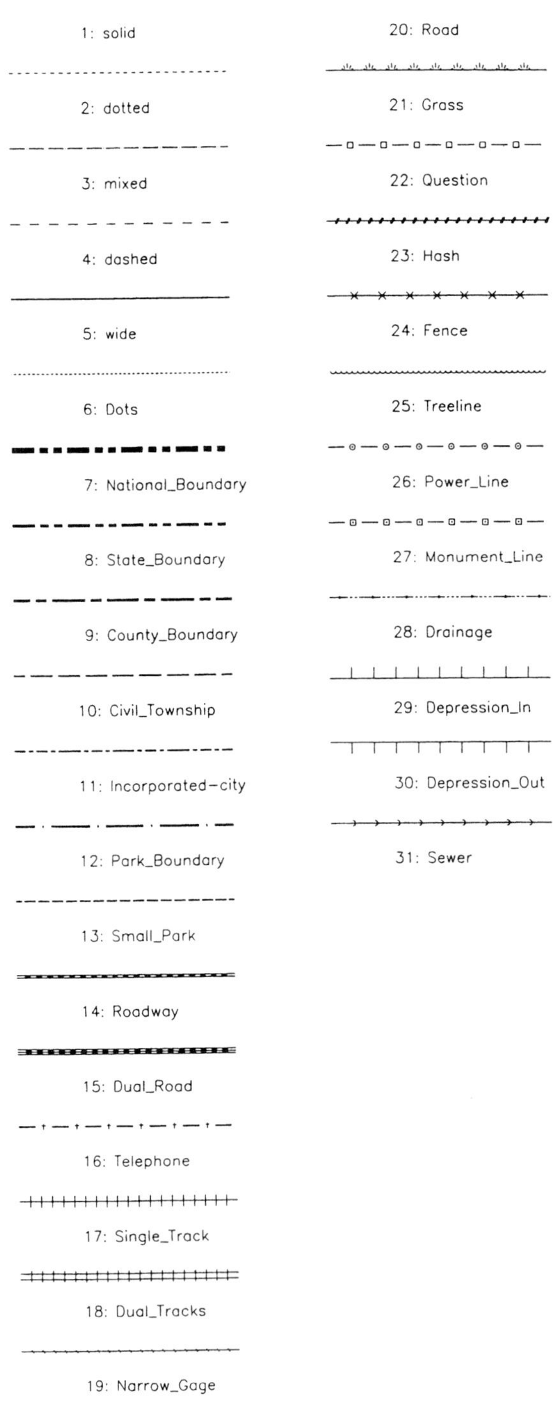

Figure A3.2. Linestyles.

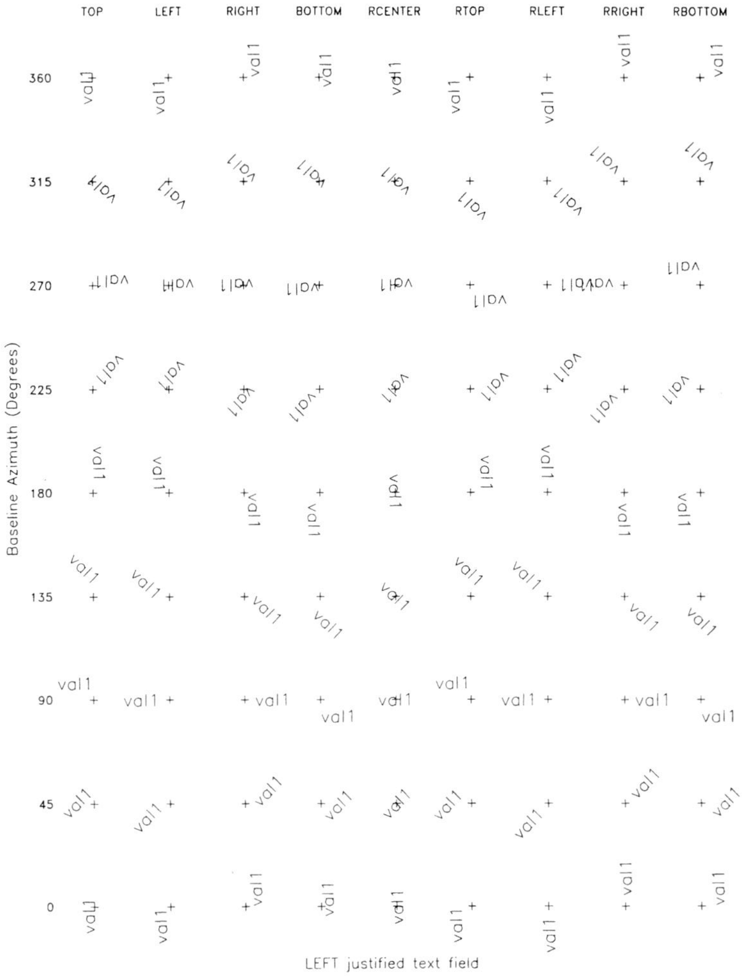

TOP LEFT RIGHT BOTTOM RCENTER RTOP RLEFT RRIGHT RBOTTOM

Baseline Azimuth (Degrees)

360
315
270
225
180
135
90
45
0

val1

LEFT justified text field

Figure A3.3. Locations

2: plus	22: oil_prod	42: quarter_br	62: left_triangle	82: heart	102: vertical	
3: asterisk	23: gas_prod	43: quarter_bl	63: down_triangle	83: diamond	103: horizontal	123: ar
4: octagon	24: og_prod	44: quarter_tl	64: right_triangle	84: club	104: plunge_anti	124: ar
5: X	25: drilling	45: top_half	65: up_triangle		105: plunge_syn	125: v_
6: octagon_plus	26: dr_&_ab	46: bottom_half	66: left_triangle	86: highway	106: plunge_ahinge	126: re
7: octagon_X	27: oil_show	47: left_half	67: down_triangle	87: interstate	107: plunge_shinge	127: x_
8: square_plus	28: gas_show	48: right_half	68: right_triangle	88: dollar_sign	108: incl_joint	128: h
9: square_X	29: service	49: tmp_abd_loc	69: star_david		109: vert_joint	
10: triangle	30: h2o_injection	50: oil_gas_show	70: cross		110: horiz_joint	
11: diamond	31: shutin_unkn	51: dot	71: palm		111: fault	
12: star	32: shutin_oil	52: circle_dot	72: pine		112: strike_slip	
13: star	33: abnd_oil	53: square_dot	73: oak		113: foliation	
	34: abnd_gas	54: octagon_dot	74: maple		114: vert_foliation	
15: exclam	35: pit	55: triangle_dot	75: grass		115: dual_strike	
16: question	36: headframe	56: diamond_dot				
	37: shaft					
	38: filled_square					
		60: water_lev				

Figure A3.4. Markers.

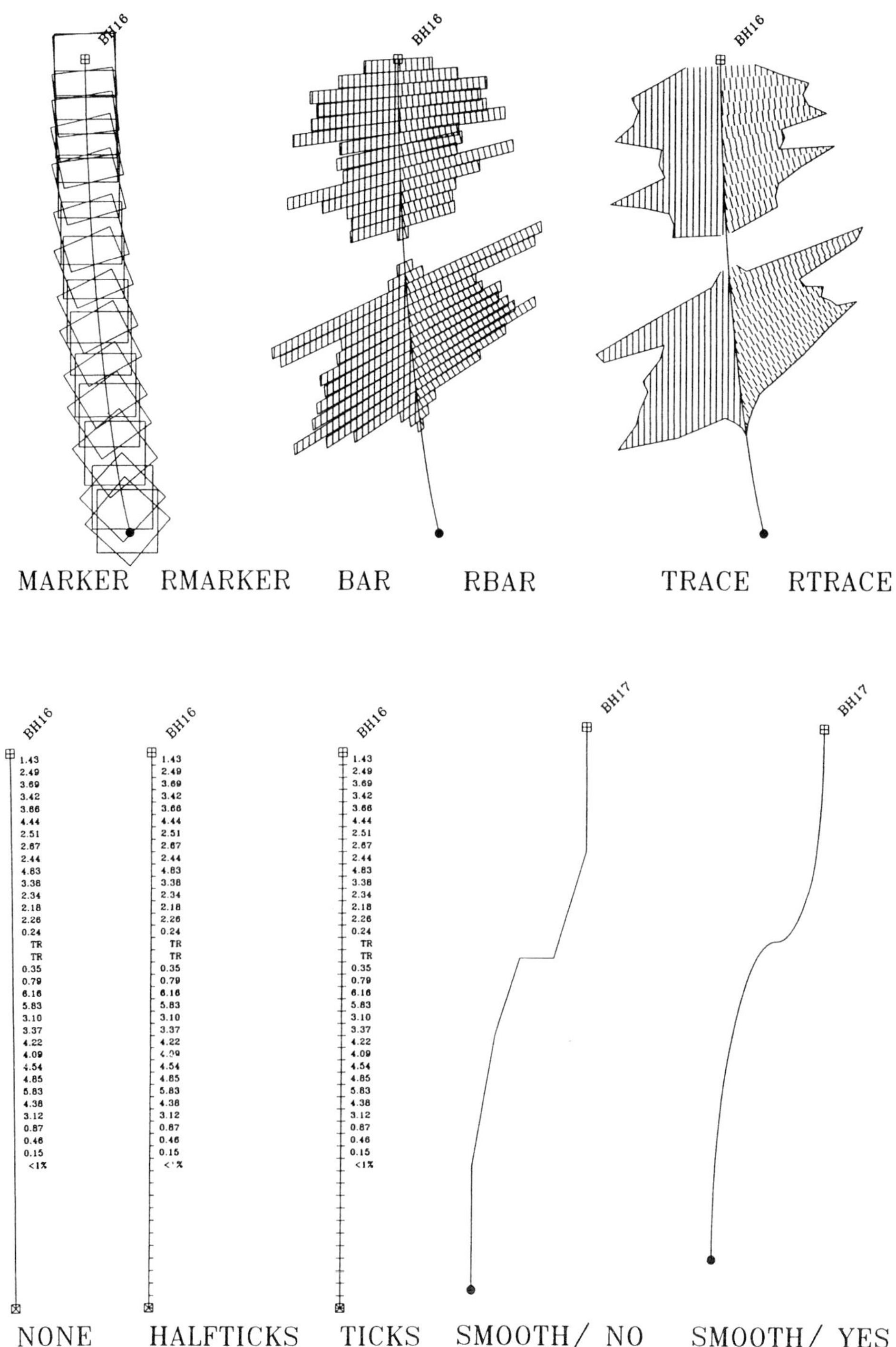

Figure A3.5. Logtypes

References

Aho, A.V., Kernigban, B.W. & P.J. Weinberger 1988. *The AWK Programming Language*, Addison-Wesley Service in Computer Science.

Boshkov, S.H. & F.D. Wright 1973. Basic and parametric citeria in the selection, design and development of underground mining systems. In Cummins & Givens (eds*), SME mining engineering handbook (first edition)*: 12-2. AIME/SME, New York.

Briggs, I.C. 1974. Spline contouring using minimum curvature. *Geophysics*, Vol. 39(1): 39-48.

Bullock, R.L. 1982. General mine planning. In W.A. Hustralid (ed.), *Underground mining methods handbook*: 113-137. SME. Littleton, Co.

CANMET 1986. *Underground metal mining*. SP86-11E.

Dravo Corporation 1974. Analysis of large-scale non-coal underground mining methods. *US Bureau of Mines,* Contract No. S0122059.

Eaton, L. 1934. *Practical mine development*. McGraw-Hill, New York.

Englund, E. & A. Sparks 1991. Geo-EAS 1.2.1 user's guide, EPA Report 600/8-91/1008, EPA-EMSL, Las Vegas.

Harmin, H. 1982. Choosing an underground mining method. In W.A. Hustralid (ed.), *Underground mining methods handbook*: 88-112. SME, Littleton, Co.

Hartman, H. 1987. *Introductory mining engineering*. Wiley Interscience.

Holding, S.W. 1994. *3D geoscience modeling*. Springer-Verlag, New York.

Hustralid, W.A. (ed.) 1982. *Underground mining methods handbook*. SME, Littleton, Co.

Hustralid, W.A. & M. Kuchta 1995. *Open pit mine planning and design*, Vol. 1: Fundamentals. A.A. Balkema, Rotterdam.

Isaaks, E.H. & M.R. Srivastava 1989. *An introduction to applied geostatistics*. Oxford University Press.

Jackson, C.F. & J.H. Hedges 1939. Metal mining practice. *US Bureau of Mines, Bull.* p. 419.

Matheron, G. 1971. *The theory of regionalized variables and its applications,* cahiers au centre de Morphologie mathématique de Fontainebleau.

McArthur, G. 1982. Development of open stoping at the Carr Fork mine. In W.A. Hustralid (ed.), *Underground mining methods handbook*: p. 375. SME, Littleton, Co.

Morrison, R.G.K. & P.L. Russell 1973. Selecting a mining method – rock mechanics, other factors. In Cummins & Givens (eds*), SME mining engineering handbook (first edition)*: 9-2. AIME/SME, New York.

Nilsson, D. 1982. Planning economics of sublevel caving. In H. Hartman (ed.), *SME mining engineering handbook (second edition)*, Vol. (2): 953.

Parks, R.D. 1949. *Examination and valuation of mineral property* (third edition). Addison-Wesley, Cambridge, MA.

Peele, R. 1941. *Mining engineers handbook (third edition)*. J. Wiley & Sons.

Sichel, H.S. 1966. The estimation of means and associated confidence limits for small samples from lognormal populations. *Symposium on mathematical statistics and computer applications in ore valuation*, 106-122, S. Afr. Inst. Mining & Metall, Johannesburg.

Smith, M.L. 1992. Spatial characterization and sampling of mine wastes, *23rd Conf. on Applications of Computers and Operations Research in the Minerale Industry*, 219-224. Phoenix, Az, SME, Littleton, Co.

Stewart, D.R. (ed.) 1981. *Design and operation of caving and sublevel stoping mines*. SME, Littleton, CO.
Tillson, B.J. 1938. *Mine plant*. Soc. Mining Engineers, AIME/SME, New York.
Unrug, K. 1992. Construction of development openings. In H. Hartman (ed.), *SME mining engineering handbook (second edition)*, V2: 1580. SME, Littleton, Co.
Venables, W.N. & B.D. Ripley 1994. *Modern applied statistics with S-Plus*. Springer-Verlag, New York.

Index

T - #0049 - 101024 - C20 - 254/178/24 [26] - CB - 9789054106913 - Gloss Lamination